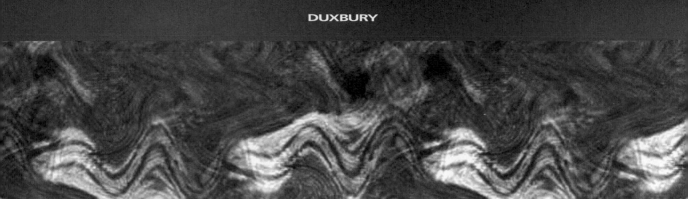

DUXBURY

Applied Regression Analysis

A Second Course in Business and Economic Statistics

Fourth Edition

Terry E. Dielman

M. J. Neeley School of Business
Texas Christian University

THOMSON

™

BROOKS/COLE

Australia • Canada • Mexico • Singapore • Spain
United Kingdom • United States

THOMSON

BROOKS/COLE

Publisher: *Curt Hinrichs*
Assistant Editor: *Ann Day*
Editorial Assistant: *Katherine Brayton*
Technology Project Manager: *Burke Taft*
Marketing Manager: *Tom Ziolkowski*
Advertising Project Manager:
 Nathaniel Bergson-Michelson
Project Manager, Editorial Production: *Kelsey McGee*
Print/Media Buyer: *Barbara Britton*

Permissions Editor: *Kiely Sexton*
Production Service: *The Book Company*
Illustrator: *Lori Heckelman*
Compositor: *Laserwords*
Cover Designer: *Roy Neuhaus*
Cover Image: *Sandra Baker/Getty Images*
Cover Printing: *Phoenix Color Corp.*
Printing and Binding: *Quebecor World/Kingsport*

For more information about our products,
contact us at:
Thomson Learning Academic Resource Center
1-800-423-0563

For permission to use material from this text,
contact us by:
Web: http://www.thomsonrights.com

Brooks/Cole Thomson Learning
10 Davis Drive
Belmont, CA 94002
USA

Asia
Thomson Learning
5 Shenton Way #01-01
UIC Building
Singapore 068808

Australia/New Zealand
Thomson Learning
102 Dodds Street
Southbank, Victoria 3006
Australia

Canada
Nelson
1120 Birchmount Road
Toronto, Ontario M1K 5G4
Canada

Europe/Middle East/Africa
Thomson Learning
High Holborn House
50/51 Bedford Row
London WC1R 4LR
United Kingdom

Library of Congress Control Number: 2004101384

ISBN 0-534-46548-X

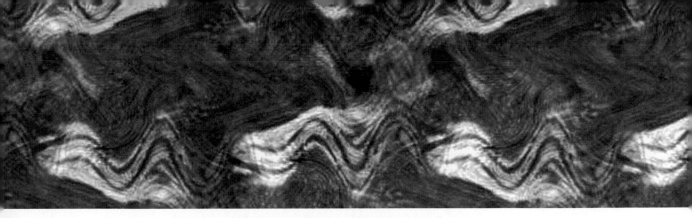

About the Author

Terry E. Dielman, professor in the M.J. Neeley School of Business at Texas Christian University (TCU), focuses his research on regression analysis, time-series forecasting, and statistical applications to business and economic problems. Dr. Dielman, winner of the TCU Deans' Teaching Award in 2003, teaches statistics, operations research, forecasting, statistical quality control, and regression at TCU to MBAs and undergraduate students.

Dr. Dielman has published more than 30 articles in refereed journals such as *American Statistical Association Journal, Communications in Statistics, Computational Statistics and Data Analysis, Decision Sciences, Journal of Business Research, Journal of Financial and Quantitative Analysis, International Journal of Forecasting, Journal of Forecasting,* and *Journal of Statistical Computation and Simulation.* He is a member of the Decision Sciences Institute, the American Statistical Association, and the International Institute of Forecasting.

Dr. Dielman earned his Ph.D. degree in statistics and management science from the University of Michigan Graduate School of Business in 1979, his B.A. in mathematics from Emporia State University, and his M.S. in mathematics from the University of Cincinnati.

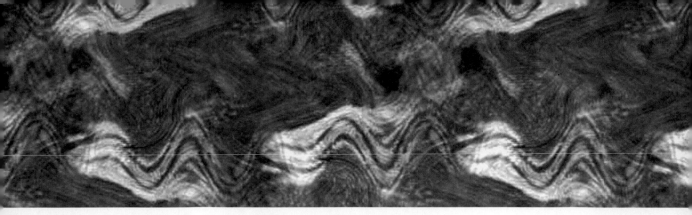

Brief Contents

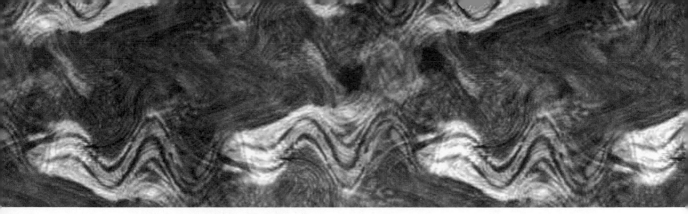

Contents

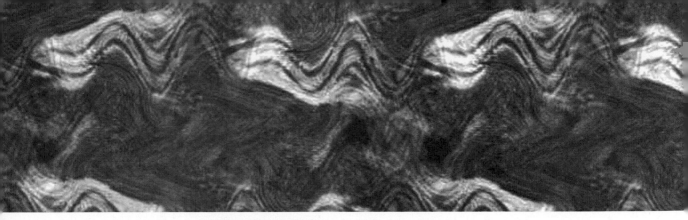

Preface

Applied Regression Analysis: A Second Course in Business and Economic Statistics presents the fundamental concepts of regression, time-series forecasting, and analysis of variance through examples and with an emphasis on applications, real data, and the use of statistical software and spreadsheets. Because of the practical value of these statistical techniques, the text takes great care to explain the proper application of the methods and the importance of correctly interpreting computer output. Emphasized are understanding the assumptions of the regression model, knowing how to validate these assumptions for a selected model, knowing when and how regression might be useful in a business setting, and understanding and interpreting output from statistical packages and spreadsheets. The goal of the text is to present statistical concepts and techniques in a way that helps students see the value and application while avoiding unnecessary mathematical rigor. The prerequisite is a one-semester course in elementary statistics, which is quickly reviewed in Chapter 2 of this text.

RATIONALE AND IMPORTANT FEATURES

I wrote this text out of necessity for my own courses. At the time I began writing there was no shortage of regression analysis textbooks. In fact, I used most of them. However, in teaching business and economics students, I found the existing texts to be less than ideal for my students. As a result, I decided to write a text that addressed their important needs and the type of course I thought would be most valuable to them.

Approach: It is important to illustrate regression concepts in the context of real examples and to show students how the concepts are correctly applied and interpreted. I attempted to present the entire process of data analysis with regression. At the time I found that most textbooks emphasized the mathematics of the techniques rather than the concepts and correct application to real problems.

Scope and Coverage: At Texas Christian University, the regression course serves as a second course in statistics for business and economics students. As such, there is a need to expose students to other important and related topics such as time-series forecasting and analysis of variance. I have included just enough of these related topics to help students recognize the types of problems these methods solve and how to apply them intelligently. I also wanted a book that I could reasonably cover in one semester. While each edition has added topics and now features more than most could cover in a semester, the coverage has been carefully divided into manageable and concise chunks, allowing maximum flexibility.

Real Data: Also, I found in my consulting experience that data sets were often large and in some cases difficult to deal with due to missing data or the like. As a result, I started collecting larger real data sets that I thought my students would be interested in. While they are intentionally tractable in size for instruction, these data sets provide a more realistic view of the data sets students may encounter in their future studies or careers.

Accessibility: In my own classes, I found that students have a broad range of quantitative skills. Some are quite good at mathematics, modeling, and analysis but others are much weaker. I have tried to maintain a focus on applications while providing just enough of the mathematics to understand the key concepts. Matrix algebra is not required although an appendix at the end of the book provides a brief introduction to this material.

AUDIENCE

Applied Regression Analysis can be used for a one-semester course in regression analysis for business and economics undergraduates or MBAs. Alternatively, it can be used as a text for a second-semester course in statistics at either the undergraduate or MBA level. Although the coverage is heavily weighted toward regression concepts, time-series forecasting and analysis of variance are also covered to provide a broader survey of important statistical methods.

ROLE OF THE COMPUTER

To effectively teach the application of regression analysis, one must use the computer. This text assumes that software will be used for most of the calculations and presents output and instructions for Microsoft® Excel, MINITAB® (Version 14), and SAS® (Version 8.2). However, the text can be used with any major statistical software containing regression procedures. Using the Computer sections at the end of each chapter present the corresponding Excel, MINITAB, and SAS procedures used to perform the analyses in the chapter, and Appendix C provides a brief general introduction to these programs. The accompanying CD stores the data used in the text in a variety of formats, including Minitab, SAS, JMP, SPSS, eViews, Stata, Excel, and ASCII.

To facilitate the use of the computer, I have described Excel, MINITAB, and SAS computer output when introducing a technique. This allows students to see and understand actual output from these software packages. However, I have used generic output in subsequent examples and exercises. The content of the generic output is very similar to that found in most statistical software. I have made this change to the fourth edition to avoid the distraction caused by seeing repetitive outputs from specific software.

The text concentrates on using the computer to do calculations, but the student is responsible for knowing what to do with the resulting computer output. Although the text could be used without computer access, I believe that actually analyzing data is an important component of the learning process.

LEVEL AND PREREQUISITES

The mathematical level of this text assumes only basic facility with the algebra that might be covered in a finite mathematics or college algebra course. No knowledge of linear algebra is assumed. For the interested reader, Appendix D does provide a summary of matrices and matrix operations and a brief introduction to the use of matrices in presenting the least-squares method. An introductory (or first-semester) course in statistics is assumed as a prerequisite. Chapter 2 does, however, contain a brief review of most of the concepts covered in an introductory statistics course.

REAL DATA

The examples and exercises throughout the book use actual data drawn from various sources. When data are simulated, I have made an attempt to provide realistic data and situations in which these data might occur. In this way, the relevance of the techniques being presented is highlighted for students.

Data sets for the exercises in this text are available on the CD. Available file formats include ASCII, JMP, MINITAB, SAS, and SPSS files and also Excel spreadsheets. Filenames needed to access the data are shown with each exercise and example.

ORGANIZATION AND COVERAGE

Chapter 2 provides a quick review of most concepts covered in a first-semester statistics course. Chapters 3 through 8 provide the material on linear regression. Chapter 3 introduces simple linear regression, including Excel, MINITAB, and SAS regression output. Chapter 4 provides the extension to multiple linear regression. Chapter 5 discusses the fitting of curves with regression. Chapter 6 discusses the implications of violations of assumptions of the regression model, presents ways to recognize possible violations, and suggests corrections for violations. Chapter 7 describes the use of indicator and interaction variables. Chapter 8 discusses several techniques used to aid in selecting explanatory variables for the regression.

Chapters 9, 10, and 11 can be viewed as optional in a course on regression, but may be used in a second course in statistics. Chapter 9 presents an introduction to analysis of variance. One-way analysis of variance and its relationship to regression with indicator variables are discussed. The chapter concludes with an examination of randomized block designs and two-way complete factorial designs. Chapter 10 introduces two procedures that can be used when qualitative dependent variables are encountered: discriminant analysis and logistic regression. The chapter concentrates on the two-group case. Chapter 11 introduces several extrapolative time-series forecasting techniques that supplement various extrapolative regression models discussed throughout the text. The extrapolative techniques in Chapter 11 include moving averages and single, double, and Winters' exponential smoothing and decomposition.

CHANGES IN THE FOURTH EDITION

These are the major changes in the fourth edition:

—Many of the data sets in the text have been updated and additional problems and examples involving real data have been included. Some of these data sets come from actual business settings; others are taken from journals and popular publications.

—The total number of exercises in the text has been increased by about 20 percent.

—A new chapter, Chapter 11, on time-series forecasting methods has been added, including moving averages and single, double, and Winters' exponential smoothing and decomposition.

—SAS output is included and discussed in addition to Excel and MINITAB output.

—Excel, MINITAB, and SAS computer output is described when a technique is first encountered. However, generic output very similar to that found in most statistical software is used in subsequent examples and exercises. The use of generic output in all but the examples that initially introduce the technique has reduced the quantity of output throughout the text.

—An add-in called SMARTReg is available on the CD that supplements the regression options available in Excel.

—Additional material on multiple comparisons has been added to the discussion of analysis of variance (Chapter 9).

—A section on multicollinearity has been added to the multiple regression chapter (Chapter 4).

ACKNOWLEDGMENTS

I would like to thank those who reviewed the manuscript at various stages of the revision:

Bob Getty, University of North Texas; Melissa Groves, Towson University; Constantine Loucopoulos, Northern Illinois University; Jeffrey Michael, Towson University; Galit Shmueli, University of Maryland; and Patrick Thompson, University of Florida.

I would also like to express my appreciation to the staff and associates of Duxbury, especially Curt Hinrichs, Ann Day, Kelsey McGee, Tom Ziolkowski, Dusty Friedman, Lori Heckelman, Bob Kauser, Pam Rockwell, Kiely Sexton, and Burke Taft.

Finally, thanks to my wife, Karen, for her support, patience, and input throughout this process and to the MBA students at TCU who have taken my regression class and helped me to refine various aspects of the book through the experience of teaching from it.

Terry E. Dielman
Fort Worth

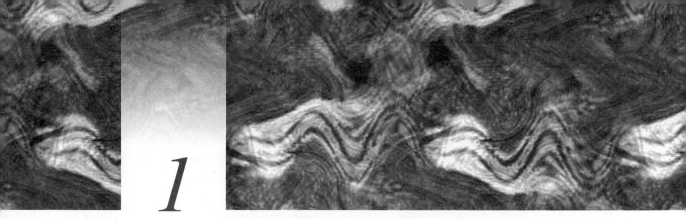

1

An Introduction to Regression Analysis

Advances in technology including computers, scanners, and telecommunications equipment have buried present-day managers under a mountain of data. Although the purpose of these data is to assist managers in the decision-making process, corporate executives who face the task of juggling data on many variables may find themselves at a loss when attempting to make sense of such information. The decision-making process is further complicated by the dynamic elements in the business environment and the complex interrelationships among these elements.

This text has been prepared to give managers (and future managers) tools for examining possible relationships between two or more variables. For example, sales and advertising are two variables commonly thought to be related. When a soft drink company increases advertising expenditures by paying entertainers or professional athletes millions of dollars to do its advertisements, it expects this outlay to increase sales. In general, it is comforting to have some evidence that past increases in advertising expenditures indeed led to increased sales.

Another example is the relationship between the selling price of a house and its square footage. When a new house is listed for sale, how should the price be determined? Is a 4000-square-foot house worth twice as much as a 2000-square-foot house? What other factors might be involved in the pricing of houses, and how should these factors be included in determining the price?

In a study of absenteeism at a large manufacturing plant, management may feel that several variables have an impact. These variables might include job complexity, base pay, the number of years a worker has been with the plant, and the age of that worker. If absenteeism can cost the company tens of thousands of dollars, then the importance of identifying its associated factors becomes clear.

Perhaps the most important analytic tool for examining the relationships between two or more variables is regression analysis. *Regression analysis* is a statistical technique for developing an equation that describes the relationship between two or more variables. One variable is specified to be the *dependent variable,* or the variable to be explained. The other one or more variables are called the *independent* or *explanatory variables*. Using the previous examples, the soft drink firm would identify sales as the dependent variable and advertising expenditures as the explanatory variable. The real estate firm would choose selling price as the dependent variable and size of the house as the explanatory variable.

There are several reasons business researchers might want to know how certain variables are related. The soft drink firm may want to know how much advertising is necessary to achieve a certain level of sales. An equation expressing the relationship between sales and advertising is useful in answering this question. For the real estate firm, the relationship might be used in assigning prices to houses coming onto the market. To try to lower the absenteeism rate, the management of the manufacturing firm wants to know which variables are most highly related to absenteeism. Reasons for wanting to develop an equation relating two or more variables can be classified as follows: (a) to describe the relationship, (b) for control purposes (what value of the explanatory variable is needed to produce a certain level of the dependent variable), or (c) for prediction.

Much statistical analysis is a multistage process of trial and error. A good deal of exploratory work must be done to select appropriate variables for study and to determine the relationships between or among them. A variety of statistical tests and other procedures must be performed and sound judgments made in order to arrive at satisfactory choices of dependent and explanatory variables. The emphasis in this text is on this multistage process rather than on computations or an in-depth study of the theory behind the techniques presented. In this sense, the text is directed at the applied researcher or the consumer of statistics.

Except for a few preparatory examples, it is assumed that the reader has a computer available to perform the actual computations. The use of statistical software frees the user to concentrate on the multistage "model-building" process.

Three software packages, Microsoft® Excel 2000, MINITAB™ (Version 14), and SAS® (Version 8.2), are discussed in this text. MINITAB and SAS are included because they are widely used as teaching tools in universities and are also used in industry. Excel is included because it is the prevalent spreadsheet in businesses throughout the world. In a business environment, not all managers have access to a statistical package, but nearly all have access to a spreadsheet package, and that package is usually Excel. Knowing how to perform statistical routines with Excel enables the manager who lacks a statistical package to still conduct some statistical analyses. The output from Excel, MINITAB, and SAS is fairly standard and is described in the text and illustrated in at least one example. Most examples and exercises use generic output, but the content of the generic output is very similar to that found in most statistical software. Many of the exercises are intended to be done with the aid of a computer, and most statistical software packages or spreadsheets could be used for this purpose. Certain options available in the software discussed in this text may not be present in other packages, but this should not create a problem in completing the exercises.

One note of caution at this point: Excel is a spreadsheet, not created specifically for statistical analysis. It can be useful for many analyses, but it cannot take the place of a true statistical package such as MINITAB or SAS. If the reader is involved in a substantial amount of data analysis, I recommend using a statistical package rather than a spreadsheet. This text contains many examples of regression analysis where Excel, MINITAB and SAS are used on the same data set. Unless specifically noted in the example, Excel produces the same answers as MINITAB and SAS. There are situations where Excel may fail to produce correct answers, however. The data sets where this occurs are likely to be those that impose a severe computational burden on Excel. The reader is cautioned to make sure the results of any analysis make sense. Results that are contrary to intuition should be questioned and verified with a statistical package. Also some procedures that are useful in more advanced analyses are not available in Excel. These require the use of a package specifically designed for statistical analysis.

Data sets for all exercises are available on the CD that accompanies this text. Data sets come in a variety of formats, including ASCII, EViews®, JMP®, MINITAB, SAS, SPSS and Stata® files, and also as Excel spreadsheets. In each problem where data sets are provided, the file names required to read the data are given.

A section called Using the Computer is included at the end of each chapter. It presents the procedures used in Excel, MINITAB, and SAS to produce the statistical analyses discussed in the chapter. Appendix C provides a brief, general discussion of the use of Excel, MINITAB, and SAS. However, this book is not intended to provide full information on the use of these software packages. For further information on Excel, MINITAB, and SAS, the interested reader should consult one of the following references:

Berk, K., and Carey, P. *Data Analysis with Microsoft® Excel: Updated for Office XP*. Pacific Grove, CA: Duxbury Press, 2004.

Freund, R., and Littell, R. *SAS® System for Regression* (3rd ed.). Cary, NC: SAS Institute, 2000.

McKenzie, J., and Goldman, R. *The Student Edition of MINITAB™ for Windows 95/NT™, Release 12* (4th ed.). Reading, MA.: Addison-Wesley, 1999.

Middleton, M. *Data Analysis Using Microsoft® Excel: Updated for Microsoft Office XP*. Pacific Grove, CA: Duxbury Press, 2004.

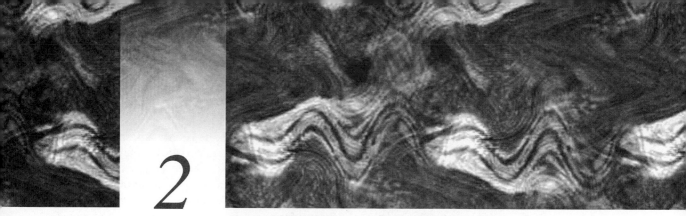

2

Review of Basic Statistical Concepts

2.1 INTRODUCTION

This chapter summarizes and reviews many of the basic statistical concepts taught in an introductory statistics course. For the most part, introductory courses in statistics deal with three main areas of interest: descriptive statistics, probability, and statistical inference.

Typically, statistics is used to study a particular population. A *population*, for purposes of this text, may be defined as the collection of all items of interest to a researcher. The researcher may want to study the sales figures for firms in a particular industry, the rates of return on public utility firms, or the lifetimes of a new brand of automobile tires. But because of time limitations, cost, or the destructive nature of testing, it is not always possible to examine all elements in a population. Instead, a subset of the population, called a *sample*, is chosen, and the characteristic of interest is determined for the items in the sample.

Descriptive statistics is that area of statistics that summarizes the information contained in a sample. This summary may be achieved by condensing the information and presenting it in tabular form. For example, frequency distributions are one way to summarize data in a table. Graphical methods of summarizing data also may be used. The types of graphs discussed in introductory statistics courses include histograms, pie charts, bar charts, and scatterplots.

Data also may be summarized by numerical values. For example, to describe the center of a data set, the mean or median is often used. To describe variability, the variance, standard deviation, or interquartile range might be used. Each of the numerical values is a single number computed from the data that describe a certain characteristic of a sample.

Describing the information contained in a sample is only a first step for most statistical studies. If the study of a population's characteristics is the goal, then the researcher wants to use the information obtained from the sample to make statements about the population. The process of generalizing from characteristics of a sample to those of a population is called *statistical inference.* The bridge leading from descriptive measures computed for a sample to inferences made about population characteristics is the field of probability.

Statistical sampling is an additional topic discussed in introductory statistics. By choosing the elements of a sample in a particular manner, objective evaluations can be made of the quality of the inferences concerning population characteristics. Without proper choice of a sample, there is no way to evaluate these inferences objectively. Thus, the manner in which the sample is chosen is important.

The most common type of sampling procedure discussed in introductory statistics is simple random sampling. Suppose a sample of *n* items is desired. To qualify as a *simple random sample* (SRS), the items in the sample are selected so that each possible sample of size *n* is equally likely to be chosen. In other words, each possible sample has an equal probability of being the one actually chosen for study. This is one of the pieces of the bridge that links descriptive statistics and statistical inference. Another piece of the bridge is a description of the behavior of certain numerical summaries that are computed as descriptive statistics.

Any numerical summary computed from a sample is called a *statistic.* A researcher often computes a single statistic from one sample chosen from the population of interest and uses the numerical value of this statistic to make a statement about the value of some population characteristic. For example, suppose a particular brand of tires is to be studied to determine their average life. If the average life is known, the tire company might use this information to establish a warranty for its tires. A SRS of *n* tires is chosen, and each tire is tested to determine its individual lifetime. Then the sample average lifetime is computed. This sample average can be used as an estimate of the population average lifetime of these tires.

The statistic computed, however, is the sample average lifetime for one particular sample of tires chosen. If a different set of *n* tires had been chosen, a different sample average would have resulted because of individual variation in the tire-lifetimes. Thus, the sample means themselves vary depending on which set of *n* tires is chosen as the sample. If this variation in the sample means was without any pattern, then there is no way to relate the value of the sample mean obtained to the unknown value of the population mean. Fortunately, the behavior of the sample means (and other statistics) from random samples is not without a pattern. The behavior of statistics is described by a concept called a *sampling distribution.* Probability enters the picture because sampling distributions are simply probability distributions. Through knowledge of the sampling distribution of a statistic, procedures can be developed to objectively evaluate the quality of sample statistics used to approximate population characteristics.

In this chapter, many of these concepts are reviewed, including descriptive statistics, random variables and probability distributions, sampling distributions, and statistical inference. Because most or all of these topics are covered in an introductory course in statistics, the coverage here is brief.

For detailed discussions of introductory statistics, the interested reader is referred to such texts as

Albright, S., Winston, W., and Zappe, C. *Data Analysis for Managers with Microsoft® Excel* (2nd ed.). Pacific Grove, CA: Duxbury Press, 2004.

Keller, G., and Warrack, B. *Statistics for Management and Economics* (6th ed.). Pacific Grove, CA: Duxbury Press, 2003.

Mendenhall, W., Beaver, R., and Beaver, B. *A Brief Course in Business Statistics* (2nd ed.). Pacific Grove, CA: Duxbury Press, 2001.

Weiers, R. *Introduction to Business Statistics* (4th ed.). Pacific Grove, CA: Duxbury Press, 2002.

2.2 DESCRIPTIVE STATISTICS

Table 2.1 shows the 5-year returns as of July 1, 2002, for a random sample of 83 mutual funds. Examining the 83 numbers in this list provides little useful information. Just looking at a list of numbers is confusing even when the sample size is only 83. For larger samples, the confusion becomes even greater.

The field of descriptive statistics provides ways to summarize the information in a data set. Summaries can be tabular, graphical, or numerical. One common tabular method of summarizing data is the frequency distribution. A *frequency distribution* is a table that is used to summarize quantitative data. The frequency distribution is set up by defining *bins* or *classes* that contain the data values. An examination of the returns in Table 2.1 shows that the largest 5-year rate of return is 17.0% and the smallest is –7.8%. We want to make sure that we include all the data in our frequency distribution, so the bins of the frequency distribution must begin at or below the smallest value and end at or above the largest. One example of how we might set up the frequency distribution is as follows: Start the first bin at –8.0%, end the last bin at 20.0%, and use a total of seven bins. The resulting frequency distribution is shown in Figure 2.1.

The bins are constructed so that there is no confusion about where a data value should go. Each bin excludes the lower limit, but includes the upper limit. A 5-year return of 4% belongs in the third bin; a 16% return belongs in the sixth bin. Also note that each of the bins has the same width—4%. Two guidelines for constructing an effective frequency distribution are (1) make sure each data value belongs in a

FIGURE 2.1
Frequency Distribution for 5-Year Rates of Return.

Five-Year Rates of Return	Number of Funds
Greater than –8% but less than or equal to –4%	5
Greater than –4% but less than or equal to 0%	6
Greater than 0% but less than or equal to 4%	17
Greater than 4% but less than or equal to 8%	34
Greater than 8% but less than or equal to 12%	12
Greater than 12% but less than or equal to 16%	8
Greater than 16% but less than or equal to 20%	1

TABLE 2.1 Five-Year Rates of Return for Mutual Funds

Name of Fund	Five-Year Return	Name of Fund	Five-Year Return	Name of Fund	Five-Year Return
AAL Capital Growth A	5.7	Fifth Third Quality Growth A	2.7	T. Rowe Price Blue Chip Growth	3.4
Aim Constellation A	1.0	First American Small Cap Value A	6.6	T. Rowe Price Mid-Cap Value	11.6
Aim Weingarten A	−4.0	FPA Paramount	−5.0	Principal MidCap A	5.4
Alpine US Real Estate Equity Y	9.6	Franklin Small Mid Cap Growth A	4.2	Prudential US Emerging Growth A	6.3
American AAdvantage Balanced Plan	5.2	Gabelli Small Cap Growth	9.8	Putnam Classic Equity A	0.9
American Century Income & Growth	5.0	Goldman Sachs Capital Growth A	4.3	Putnam New Value A	5.4
Ariel Appreciation	14.2	J. Hancock Large Cap Equity A	4.0	Rainier Core Equity	5.0
AXA Rosenberg US Small Cap	9.5	Hartford Growth Opportunities L	4.3	RS Diversified Growth A	14.0
AXP Small Companies Index A	7.1	ICAP Equity	5.7	Salomon Brothers Investors Value A	6.3
Berger Growth	−5.0	Invesco Health Sciences	6.4	Scudder Capital Growth AARP	0.7
BlackRock Small Cap Growth A	−1.5	Janus Core Equity	10.6	Security Equity A	0.2
Calamos Convertible A	10.6	Janus Twenty	4.3	Sit Mid Cap Growth	−0.9
CG Capital Markets Large Cap Value	3.9	Liberty Fund A	2.4	Smith Barney Large Cap Value A	1.7
Columbia Balanced	4.5	Lord Abbett Developing Growth A	2.9	State Street Research Aurora A	15.9
Davis Financial A	7.7	Mairs & Power Growth	10.8	Strong Growth	4.1
Deutsche Flag Value Builder A	5.6	Merrill Lynch Balanced Capital D	3.0	Third Avenue Value	8.9
Dreyfus Growth & Value Emerging		MFS Emerging Growth A	−4.4	Turner Small Cap Value	14.4
Leaders	10.1	MFS Value A	10.5	US Global Investors World Precious	
Dreyfus Premier Third Century Z	0.4	Morgan Stanley Global Utilities B	5.9	Minerals	−7.8
Evergreen Fund A	−3.2	Morgan Stanley Total Return B	2.7	Van Kampen Comstock A	10.4
Evergreen Small Cap Value A	9.4	Nations Convertible Securities A	7.9	Vanguard Capital Opportunity	15.6
Federated Equity Income A	0.0	Needham Growth	17.0	Vanguard LifeStrategy Conservative	
Fidelity Blue Chip Growth	2.5	Nuveen Large Cap Value A	4.8	Growth	6.0
Fidelity Equity-Income II	5.1	One Group Equity Income A	3.2	Vanguard Small Cap Index	5.2
Fidelity New Millennium	14.5	Oppenheimer Cap Appreciation A	6.5	Vanguard Utilities Income	4.6
Fidelity Advisor Balanced T	1.9	Oppenheimer Total Return A	4.1	Waddell & Reed Continental	4.1
Fidelity Advisor Technology T	−1.1	PBHG Growth	−3.9	Wasatch Small Cap Growth	15.2
Fidelity Select Air Transportation	14.1	Pimco Capital Appreciation A	6.3	WM Growth & Income A	4.9
Fidelity Select Health Care	7.3	Pioneer Fund A	6.6		

unique bin and (2) if possible, make each bin width the same. Intervals covering the range of the data are constructed, and the number of observations in each interval is then tabulated and recorded.

If the proportion or percentage of items in each class is noted rather than the number, the table is referred to as a *relative frequency distribution*. It is also possible to construct a *cumulative frequency distribution* in which the number of items at or below each class limit is noted.

A *histogram* is a graphical representation of a frequency distribution. The horizontal axis of the graph is marked off into classes or bins over the full range of the data, and the vertical axis represents the number or proportion (relative frequency) of observations in each of the classes. The bin limits for the horizontal axis are the limits established in the frequency distribution. Rectangles (bars) are drawn over the bin limits, with the area of the bar proportional to the frequency in that particular bin. If the bin limits are all the same width, the height of the bars can be equal to the frequency in each bin. If the bin limits differ in width, adjustments must be made. It is recommended that bin widths be made the same whenever possible. The adjustments for unequal bin widths are not discussed here.

From the frequency distribution or the histogram, one can obtain a quick picture of certain characteristics of the data. For example, the center of the data and how much variability is present can be observed. The data have been summarized so that these characteristics are more obvious. When the frequency distribution in Figure 2.1 was constructed, we arbitrarily decided to use seven bins. In general, you do not want to have too few bins because the data will be oversummarized and it will be hard to see patterns in it. Also, too many bins make the frequency distribution confusing and difficult to read. Various rules have been suggested concerning the appropriate number of bins. A good guideline is to use between 5 and 20 bins. As a rough idea of the number of bins to try, start with the square root of the number of observations. We have 83 5-year rates of return. The square root of 83 is between 9 and 10, so we might start with 10 as a possible number of bins. There is no right or wrong number of bins. However, there are better and worse numbers—too few or too many bins are not good—and there are better choices than others (my choices are better than yours, for example, because I wrote this book), but constructing a frequency distribution is in large part a matter of preference. It may take two or three tries to get the table the way you believe is most helpful in representing the data or clearest for presentation purposes.

Most statistical software packages and spreadsheets provide various tabular and graphical methods of summarizing data. Figure 2.2a shows the frequency distribution constructed by Excel for the mutual-fund-return data. This is the distribution without any modifications and would obviously be modified before use in a presentation. Such a modification might appear as in Figure 2.2b. Figures 2.3 and 2.4 show histograms constructed using Excel and MINITAB, respectively. I chose to use ten bins for the Excel frequency distribution and histogram, starting the first bin at –8% and giving each bin a width of 2.5%. The MINITAB histogram was allowed to choose its own limits.[1] Note that the numbers shown in the bin column of the

FIGURE 2.2A Excel Frequency Distribution for 5-Year Rates of Return for Mutual Funds.

bins	Frequency
−5.5	1
−3	6
−0.5	3
2	8
4.5	18
7	22
9.5	7
12	9
14.5	5
17	4

[1]Excel can be allowed to choose its own bin limits as well. However, I find the limits chosen by Excel are often (actually, always) not to my liking. I prefer to set up the bin limits myself to make them easier to work with. This process is discussed in the Using the Computer section at the end of this chapter.

FIGURE 2.2B
Modified Excel
Frequency Distribution
for 5-Year Rates of
Return for Mutual Funds.

Five-Year Returns	Number
Greater than −8% but less than or equal to −5.5%	1
Greater than −5.5% but less than or equal to −3%	6
Greater than −3% but less than or equal to −0.5%	3
Greater than −0.5% but less than or equal to 2%	8
Greater than 2% but less than or equal to 4.5%	18
Greater than 4.5% but less than or equal to 7%	22
Greater than 7% but less than or equal to 9.5%	7
Greater than 9.5% but less than or equal to 12%	9
Greater than 12% but less than or equal to 14.5%	5
Greater than 14.5% but less than or equal to 17%	4

FIGURE 2.3 Excel
Histogram for 5-Year
Rates of Return for
Mutual Funds.

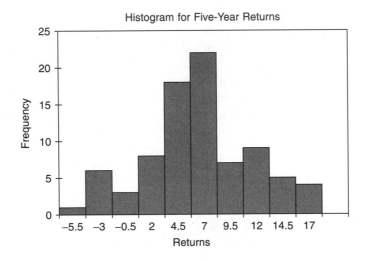

FIGURE 2.4 MINITAB
Histogram for 5-Year
Rates of Return for
Mutual Funds.

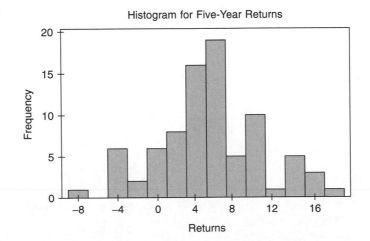

Excel frequency distribution are the inclusive upper bin limits. Excel includes a final bin that is represented by the label "More" (indicating values more than the last bin limit shown). When constructing a frequency distribution or histogram in Excel, I prefer not to use the More class but to use numerical bin limits instead. I have eliminated the More class in the frequency distribution shown. On the Excel histogram, the numbers shown on the horizontal axis also represent the upper limits of the bin under which they are printed. Even though they are printed in the middle of the bin, they do not represent midpoints, but upper limits.

On the MINITAB histogram, the numbers shown on the horizontal axis are the midpoints of the bin intervals. For example, the first bin has a midpoint of –8. The bin limits are –9 and –7. The next bin has no fund returns in it, but would have a midpoint of –6 and bin limits of –7 and –5. Each bin limit is halfway between two of the midpoints. MINITAB includes the value of the lower limit in the bin, but excludes the upper limit.

The idea of a graph such as a histogram is to summarize the data so that the viewer can get a quick picture of what is going on but without masking too much of the information. Using too few classes on a histogram oversummarizes the data, whereas using too many does not summarize the data sufficiently. In either case, a histogram that is difficult to read is produced. Again, there is no right or wrong number of bins to use when constructing a histogram.

The histograms show that the returns for these funds vary from a low of about –8.0% (the lower boundary of the first interval is –9.0% on the MINITAB histogram and –8.0% on the Excel histogram) to a high of around 17.0% (17.0% is the upper boundary of the last interval on the Excel histogram; 18.5% is the upper limit of the last interval on the MINITAB histogram). Very high returns and very low returns are rare. Most returns cluster toward the center of the histogram (we might identify the center as roughly around 5%).

Numerical summaries are single numbers computed from a sample to describe some characteristic of the data set. Some common numerical summaries are sample mean, sample median, sample variance, and sample standard deviation. The *sample mean* and *sample median* are measures of the center, or central tendency, of the data. The sample mean is the average of all the observations in the data set:

$$\bar{y} = \frac{\sum_{i=1}^{n} y_i}{n}$$

where y_i represents the ith observation in the data set. (See Appendix A for an explanation of summation notation.) The sample median is the midpoint of the data after they have been ordered. If n, the number of observations, is even, then the median is the average of the two middle observations after the observations have been ordered from smallest to largest. If n is odd, the unique middle value in the ordered data set is the median.

The *sample variance* and *sample standard deviation* are measures of variability. The sample variance is computed as

$$s^2 = \frac{\sum_{i=1}^{n} (y_i - \bar{y})^2}{n - 1}$$

This is the average squared distance of each data point, y_i, from the center of the data, $\bar{y}$. The divisor $n - 1$ is used, rather than n, to provide an unbiased estimator (one that neither consistently overestimates nor underestimates the true parameter) of the population variance. Because s^2 expresses variability in squared units, an intuitively more appealing measure is the sample standard deviation, s, which is simply the square root of s^2. Although many other numerical summaries exist, they are not discussed in this review.

Figure 2.5 shows the results of using MINITAB to compute several descriptive measures for the mutual-fund-return data, including the sample mean, sample median, sample variance, and sample standard deviation. Figure 2.6 shows similar results using Excel.

FIGURE 2.5 MINITAB Numerical Summaries for the 5-Year Rates of Return.[a]

Descriptive Statistics: Five-Year Return

Variable	N	N*	Mean	SE Mean	StDev	Minimum	Q1	Median	Q3	Maximum
5yr ret	83	0	5.371	0.574	5.229	−7.800	2.700	5.100	8.900	17.000

[a]The summaries computed include the number of observations (N), the number of missing observations (N*), mean, standard error of the mean (SE Mean), sample standard deviation (StDev), minimum, first quartile (Q1), median, third quartile (Q3), and maximum.

FIGURE 2.6 Excel Numerical Summaries for the 5-Year Rates of Return.[a]

Five-Year Return	
Mean	5.37
Standard Error	0.57
Median	5.10
Mode	4.30
Standard Deviation	5.23
Sample Variance	27.35
Kurtosis	0.07
Skewness	0.02
Range	24.80
Minimum	−7.80
Maximum	17.00
Sum	445.80
Count	83.00

[a]The summaries computed include the mean, standard error of the mean (Standard Error), median, mode, sample standard deviation (Standard Deviation), sample variance, kurtosis, skewness, range, minimum, maximum, the sum of all the data values (Sum), and the number of observations (Count).

EXERCISES

1. **Highway Mileages.** The highway mileages of 147 cars are in a file named CARS2. These cars are 2003 models. Find the mean, standard deviation, and median for the mileages. Construct a histogram of the data.
(Data from *Road & Track: The New Cars.* October 2002. Copyright 2002 by Hachette Filipacchi Media, Inc. Used with permission.)

2. **Graduation Rates.** The National Collegiate Athletic Association (NCAA) is concerned with the graduation rate of student athletes. Part of an effort to increase graduation rates for student athletes involved implementing Proposition 48, beginning with the 1986–1987 school year. Proposition 48 mandated that student athletes

obtain a 700 SAT or a 15 ACT test score to be eligible to play.

   Graduation rates are provided in a file named GRADRATE2. The file contains graduation rates (percentages) for several groups of students.

   all students entering freshman classes in
         1983–1984 through 1985–1986 (AS83)
   student athletes entering in 1983–1984
         through 1985–1986 (SA83)
   all students entering freshman classes in
         1986–1987 (AS86)
   student athletes entering in 1986–1987
         (SA86)
   all students entering freshman classes in
         1987–1988 (AS87)
   student athletes entering in 1987–1988
         (SA87)
   all students entering freshman classes in
         1988–1989 (AS88)
   student athletes entering in 1988–1989
         (SA88)
   all students entering freshman classes in
         1989–1990 (AS89)
   student athletes entering in 1989–1990 (SA89)
   all students entering freshman classes in
         1990–1991 (AS90)
   student athletes entering in 1990-1991
         (SA90)
   all students entering freshman classes in
         1991–1992 (AS91)
   student athletes entering in 1991–1992
         (SA91)
   all students entering freshman classes in
         1993–1994 (AS93)

student athletes entering in 1993–1994
   (SA93)
all students entering freshman classes in
   1994–1995 (AS94)
student athletes entering in 1994–1995 (SA94)

   All Division I schools with complete data for all years are represented. [The data were obtained from the Fort Worth Star-Telegram (July 2, 1993; May 20, 1993; July 1, 1994; June 30, 1995; June 28, 1996; June 27, 1997; and November 9, 1998 issues)] and from the NCAA web site for the remaining years. Note that the 1983–1984 through 1985–1986 data provide graduation rates prior to the implementation of Proposition 48. All other years provide graduation rates after its implementation. (Note that rates were not available for the 1992 groups.)
(*Source*: Data used courtesy of the *Fort Worth Star-Telegram* and the NCAA.)

a. Examine the groups by finding the mean and median graduation rate for each. Construct a histogram for each set of graduation rates.

b. U.S. District Judge Ronald Buckwalter invalidated the NCAA's academic eligibility standards for incoming freshmen athletes on March 8, 1999. The NCAA still believes in the standards and in the positive effect of Proposition 48 on graduation rates. To help support its position, the NCAA has asked for your input concerning the effect of Proposition 48 on graduation rates. Using any graphical or numerical summaries to support your position, what would you report to the NCAA based on the data you have been given?

2.3 DISCRETE RANDOM VARIABLES AND PROBABILITY DISTRIBUTIONS

A *random variable* can be defined as a rule that assigns a number to every possible outcome of an experiment. A *discrete random variable* is one with a definite distance between each of its possible values. For example, consider the toss of a coin. The two possible outcomes are head (H) and tail (T). A random variable of interest could be defined as

$$X = \text{number of heads on a single coin toss}$$

Then X assigns the number 1 to the outcome H and the number 0 to outcome T. As another example, suppose two cards are randomly drawn without replacement from a deck of 52 cards. Let

$$Y = \text{number of kings on two draws}$$

Then Y assigns the number 0, 1, or 2 to each possible outcome of the experiment.

In each of these examples, the outcome of the experiment is determined by chance. Probabilities can be assigned to the outcomes of the experiment and thus to the values of the random variables. A table listing the values of a random variable and the probabilities associated with each value is called a *probability distribution* for the random variable.

For the coin toss, the probability distribution of X is

x	$P(x)$
0	1/2
1	1/2

Here, the notation $P(x)$ means "the probability that the random variable X has the value x" or $P(x) = P(X = x)$. The function $P(x)$ is called the *probability mass function* (pmf) of X.

For the card-drawing experiment, the probability distribution of Y is

y	$P(y)$
0	188/221
1	32/221
2	1/221

Note that probabilities must satisfy the following conditions:

1. They must be between 0 and 1. $0 \leq P(x) \leq 1$

2. They must sum to 1. $\sum P(x) = 1$

When we discussed a sample of observations drawn from a population in the previous section, certain characteristics were of interest, primarily center and variability. Numerical summaries were used to measure these characteristics. Describing the center and the variation in a probability distribution also is often useful. The measures most commonly used to do this are the mean and variance (or standard deviation) of the random variable.

As an example, consider two random variables X and Y, representing the profit from two different investments. Suppose the two probability distributions have been set up as follows:

x	$P(x)$	y	$P(y)$
−2000	.05	0	.40
−1000	.10	1000	.20
1000	.10	2000	.20
2000	.25	3000	.10
5000	.50	4000	.10

If only one of the investments can be chosen, some methods to compare the two would be useful. As can be seen, the chances of a loss are greater for investment X than for investment Y, although the chances for a large profit are also greater for investment X.

One way to compare the investments might be to use the expected value, or mean, of the random variables representing the outcomes of the investments. The *expected value* of a discrete random variable X is defined as the sum of each value of X times the probability associated with that value:

$$E(X) = \mu_X = \sum xP(x)$$

The subscript X on μ_X often is dropped if it is clear which random variable is being discussed. The computation of the expected values of X and Y is shown in Figure 2.7. The expected value of X is greater than the expected value of Y. Thus, on the basis of maximizing expected values, investment X would be chosen.

The expected value of a random variable deserves some additional explanation. Consider again the coin-toss experiment with X equal to the "number of heads" and probability distribution:

x	$P(x)$
0	1/2
1	1/2

Computing the expected value of X gives $E(x) = 1/2$.

Obviously, if a coin is tossed once, either the outcome 0 (tail) or 1 (head) will appear. The expected value of X represents the average obtained over a large number of trials. If the coin is tossed a large number of times and 0s were recorded for tails and 1s for heads, then the average of these 0s and 1s will be close to one-half. The same interpretation can be made for the case of the investments. The expected outcomes represent the averages obtained over a large number of trials rather than the outcome of a single trial. Thus, in the long run, investment X will provide a higher average profit than investment Y.

There are, of course, other criteria for choosing between investments than simply maximizing the expected returns. A measure of each investment's risk also might be important. The variability of the outcomes is sometimes used as a measure of risk. One measure of a random variable's variation is the *variance*, defined for a discrete random variable, X, as

$$Var(X) = \sigma_X^2 = \sum (x - \mu)^2 P(x)$$

To compute $Var(X)$, the mean is subtracted from each possible value of X, and the differences are squared and then multiplied times the probability of the associated value of X. The resulting sum is the variance, which represents an average squared distance of each value of X to the center of the probability distribution. Note that no division is used in computing this "average." The division used to compute a sample variance has been replaced by the weighting of each outcome by its

FIGURE 2.7
Computation of $E(X)$ and $E(Y)$.

x	$P(x)$	$xP(x)$	y	$P(y)$	$yP(y)$
−2000	0.05	−100	0	0.40	0
−1000	0.10	−100	1000	0.20	200
1000	0.10	100	2000	0.20	400
2000	0.25	500	3000	0.10	300
5000	0.50	2500	4000	0.10	400
	$E(X) = \sum xP(x) = 2900$			$E(Y) = \sum yP(y) = 1300$	

x	$P(x)$	$xP(x)$	$x - \mu_X$	$(x - \mu_X)^2$	$(x - \mu_X)^2 P(x)$
-2000	0.05	-100	-4900	24,010,000	1,200,500
-1000	0.10	-100	-3900	15,210,000	1,521,000
1000	0.10	100	-1900	3,610,000	361,000
2000	0.25	500	-900	810,000	202,500
5000	0.50	2500	2100	4,410,000	2,205,000
		$\mu_X = 2900$			$\sigma_X^2 = 5,490,000$

y	$P(y)$	$yP(y)$	$y - \mu_Y$	$(y - \mu_Y)^2$	$(y - \mu_Y)^2 P(y)$
0	0.40	0	-1300	1,690,000	676,000
1000	0.20	200	-300	90,000	18,000
2000	0.20	400	700	490,000	98,000
3000	0.10	300	1700	2,890,000	289,000
4000	0.10	400	2700	7,290,000	729,000
		$\mu_Y = 1300$			$\sigma_Y^2 = 1,810,000$

FIGURE 2.8 Computation of and σ_X^2 and σ_Y^2.

probability. An alternative formula for computing the variance of a discrete random variable is

$$\sigma_X^2 = \sum x^2 P(x) - \mu^2$$

This formula is sometimes preferred when doing computations on a calculator. Both formulas provide the same answer. The variances of X and Y are computed in Figure 2.8. The variances are

$$\sigma_X^2 = 5,490,000 \quad \text{and} \quad \sigma_Y^2 = 1,810,000$$

Obviously, investment X is more variable than investment Y. The variances are somewhat difficult to interpret, however, because they measure variability in squared units (squared dollars for the investments). To return to the original units of the problem, the square root of the variance, called the *standard deviation*, may be used:

$$\sigma_X = 2343.07 \quad \text{and} \quad \sigma_Y = 1345.36$$

The standard deviations are expressed in the original units of the problem (dollars for the investments).

All of the random variables discussed so far have been discrete random variables. A *continuous random variable* is one whose values are measured on a continuous scale. It is measured over a range of values with all numbers within that range as possible values (at least in theory). Examples of quantities that might be represented by continuous random variables are temperature, gas mileage, and stock prices. In the next section, a very useful continuous random variable in statistics, the normal random variable, is introduced.

EXERCISES

3. Consider the roll of a single die. Construct the probability distribution of the random variable $X =$ number of dots showing on the die. Find the expected value and standard deviation of X. How would you interpret the number obtained for the expected value?

4. Let X be a random variable defined as

$X = 1$ if an even number of dots appears on the roll of a single die

$X = 0$ if an odd number of dots appears on the roll of a single die

Construct the probability distribution of X. Find the expected value and standard deviation of X. How would you interpret the number obtained for the expected value?

5. Consider the roll of two dice. Let X be a random variable representing the sum of the number of dots appearing on each of the dice. The probabilities of each possible value of X are as follows:

x	$P(x)$
2	1/36
3	2/36
4	3/36
5	4/36
6	5/36
7	6/36
8	5/36
9	4/36
10	3/36
11	2/36
12	1/36

Determine the expected value and standard deviation of X.

6. The game of craps deals with rolling a pair of fair dice. In one version of the game, a field bet is a one-roll bet based on the outcome of the pair of dice. For every $1 bet, you lose $1 if the sum is 5, 6, 7, or 8; you win $1 if the sum is 3, 4, 9, 10, or 11; or you win $2 if the sum is 2 or 12.

a. Using the probability distribution in Exercise 5, construct the probability distribution of the different outcomes available in a field bet.

b. Determine the expected value of this probability distribution. How would you interpret this number?

7. A computer shop builds PCs from shipments of parts it receives from various suppliers. The number of defective hard drives per shipment is to be modeled as a random variable X. The random variable is assumed to have the following distribution:

x	$P(x)$
0	0.55
1	0.15
2	0.10
3	0.10
4	0.05
5	0.05

a. What is the expected number of defective hard drives per shipment?

b. If each defective drive costs the company $100 in rework costs, what is the expected rework cost per shipment?

c. What is the probability that a shipment has more than two defective hard drives?

2.4 THE NORMAL DISTRIBUTION

A continuous random variable is a random variable that can take any value over a given range. An example that is important in statistical inference is the normal random variable. The probability distribution of the normal random variable, called the *normal distribution*, is often depicted as a bell-shaped symmetric curve, as shown in Figure 2.9. The normal distribution is centered at the mean, μ. Variation in the distribution is described by the variance, σ^2, or standard deviation, σ.

Figure 2.10 shows two normal distributions with different means but equal standard deviations, and Figure 2.11 shows two distributions with the same mean but different standard deviations. The location of the distribution is determined by the mean; the spread of the distribution (how compressed or spread out it appears) is determined by the standard deviation.

For a continuous distribution such as the normal distribution, the probability that the random variable takes on a value within a certain range can be determined by computing the area under the curve that defines the probability distribution between the limits of the range. To determine the probability that a normal random

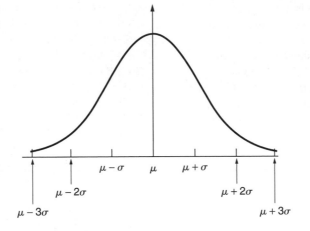

FIGURE 2.9 The Normal Distribution with Mean μ and Standard Deviation σ.

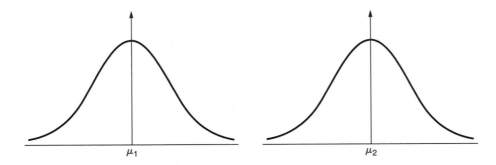

FIGURE 2.10 Normal Distributions with Equal Standard Deviations but Different Means.

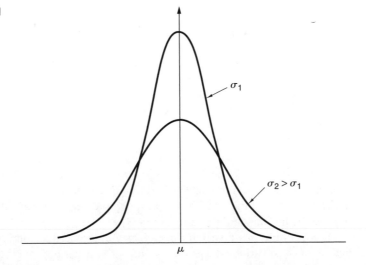

FIGURE 2.11 Normal Distributions with Equal Means but Different Standard Deviations.

variable is between 0 and 2, the area under the normal curve between these values must be computed. This computation is a fairly difficult task if done from scratch. Fortunately, a table of certain areas or probabilities under the normal curve is available to simplify these computations considerably.

Table B.1 in Appendix B lists probabilities between certain values of the standard normal distribution. The standard normal distribution has a mean of $\mu = 0$ and a standard deviation of $\sigma = 1$. Throughout this text, the standard normal random variable is denoted by the letter Z. This table is set up to show the probability that the normal random variable is between 0 and some number z written

$$P(0 \leq Z \leq z)$$

The z numbers are given by the values in the far left-hand column of the table to one decimal place. A second decimal place is provided by using the values in the top row of the table. For example, to find the probability that Z is between 0 and 1, the value 1.0 is located in the left-hand column and the probability is read from the .00 column of the table:

$$P(0 \leq Z \leq 1) = 0.3413$$

This area is illustrated by the shaded region in Figure 2.12.

Similarly, the probability between 0 and 2.3 is

$$P(0 \leq Z \leq 2.3) = 0.4893$$

To compute the probability between 0 and 1.96, first find 1.9 in the left-hand column. The probability is then read from the .06 column of the table as

$$P(0 \leq Z \leq 1.96) = 0.4750$$

Because the standard normal curve has a mean of 0, the numbers to the right of the mean are positive, as illustrated in the examples thus far. The numbers to the left of the mean are negative. How is the table used to find the probability that Z is between, say, -1.0 and 1.0? There are no negative z values in the table. But the fact that the curve is symmetric can be used to determine the probabilities for numbers to the left of the mean.

The probability between 0 and 1.0 has been determined to be 0.3413. Because the curve is symmetric, the half of the curve to the left of the mean is a mirror image of the half to the right. Thus, in an interval between 0 and -1.0, there is exactly the same probability as in the interval between 0 and 1.0 because these regions are mirror images of each other. So,

$$P(-1.0 \leq Z \leq 1.0) = 0.3413 + 0.3413 = 0.6826$$

This probability is illustrated in Figure 2.13.

Now, consider finding the following probability

$$P(Z \geq 1.7)$$

The area between 0 and 1.7 can be found from Table B.1 in Appendix B as

$$P(0 \leq Z \leq 1.7) = 0.4554$$

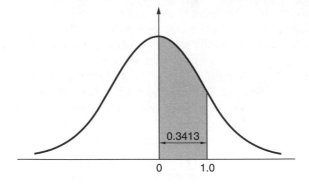

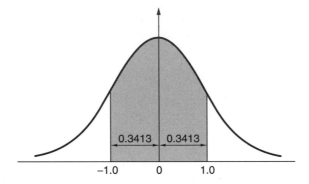

The desired area is to the right of 1.7, however. Here, we use the facts that the total area under the curve must be 1.0 and that the curve is symmetric. The total area under the standard normal curve must be 1.0 because this area represents probability, and probability must sum to 1. Because the curve is symmetric, the area to the right of the mean (0) must be 0.5. In Figure 2.14, if the unshaded area between 0 and 1.7 is subtracted from the total area to the right of 0, the remainder is the area in the shaded region:

$$P(Z \geq 1.7) = 0.5 - 0.4554 = 0.0446$$

The probabilities in the standard normal table also can be used to find probabilities for normal distributions other than the standard normal distribution. For example, suppose X is a normal random variable with mean $\mu = 10$ and standard deviation $\sigma = 2$. Find the probability that X is between 10 and 12: $P(10 \leq X \leq 12)$. The standard normal table cannot be used to find this probability as it is currently stated. But the problem can be solved by translating it into *standardized units*, or units of standard deviation away from the mean. Referring to Figure 2.15, first recognize that 10 is the mean of the normal distribution represented by X. Because $\sigma = 2$, 12 is 1 standard deviation above the mean ($10 + 2 = 12$). Then, in standardized units, the problem becomes

$$P(10 \leq X \leq 12) = P(0 \leq Z \leq 1)$$

FIGURE 2.14 Area or Probability Under the Standard Normal Curve Between 0 and 1.7 and Above 1.7.

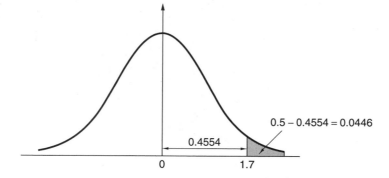

FIGURE 2.15 Finding $P(10 \leq X \leq 12)$ When X Is a Normal Random Variable with $\mu = 10$ and $\sigma = 2$.

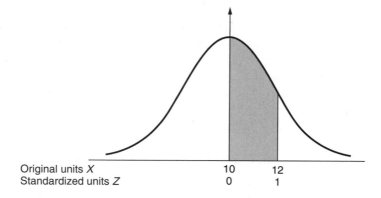

The area between 10 and 12 under the normal curve with $\mu = 10$ and $\sigma = 2$ is the same as the area between 0 and 1 under the standard normal curve. By translating the original units into standardized units, any probability can be determined from the standard normal table. The general transformation is given by the formula

$$Z = \frac{X - \mu_X}{\sigma_X}$$

To translate a number, X, into standardized units, Z, simply subtract the mean and divide by the standard deviation. The following examples should help further illustrate.

EXAMPLE 2.1 Suppose X is a normal random variable with $\mu = 50$ and $\sigma = 10$. What is $P(30 \leq X \leq 60)$?

Answer:

$$P(30 \leq X \leq 60) = P\left(\frac{30 - 50}{10} \leq Z \leq \frac{60 - 50}{10}\right) = P(-2 \leq Z \leq 1)$$

$$= 0.4772 + 0.3413 = 0.8185$$

The solution is illustrated in Figure 2.16.

FIGURE 2.16

Finding $P(30 \leq X \leq 60)$ When X Is a Normal Random Variable with $\mu = 50$ and $\sigma = 10$.

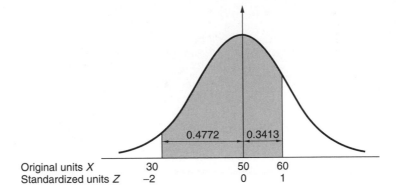

Original units X	30	50	60
Standardized units Z	−2	0	1

EXAMPLE 2.2   A large retail firm has accounts receivable that are assumed to be normally distributed with mean $\mu = \$281$ and standard deviation $\sigma = \$35$.

1. What is the proportion of accounts with balances greater than $316?
Answer:

$$P(X > 316) = P\left(Z > \frac{316 - 281}{35}\right) = P(Z > 1) = 0.5 - 0.3413 = 0.1587$$

Thus, 0.1587 or 15.87% of all accounts have balances greater than $316. The solution is illustrated in Figure 2.17.

2. Above what value do 13.57% of all account balances lie?
Answer: Figure 2.18 illustrates the problem to be solved. Find an account balance, k, such that the probability above k is 0.1357. This means the probability between the mean and k must be $0.5 - 0.1357 = 0.3643$. Looking up 0.3643 in the standard normal table, k has a z value of 1.1, so k is 1.1 standard deviations above the mean:

$$k = \mu + z\sigma = 281 + (1.1)(35) = \$319.50$$

FIGURE 2.17

Finding $P(X > 316)$ When X Is a Normal Random Variable with $\mu = 281$ and $\sigma = 35$.

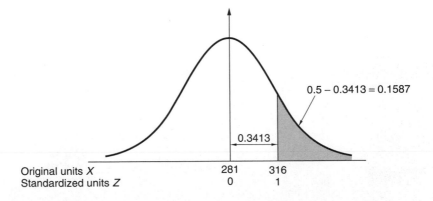

Original units X	281	316
Standardized units Z	0	1

FIGURE 2.18

Above What Value Do 13.57% of All Account Balances Lie When $\mu = 281$ and $\sigma = 35$?

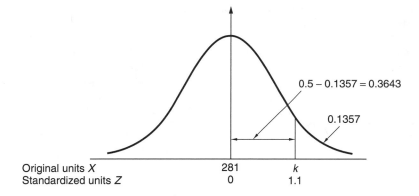

0.5 − 0.1357 = 0.3643

0.1357

Original units X	281	k
Standardized units Z	0	1.1

EXERCISES

8. Calculate the following probabilities using the standard normal distribution:

  a. $P(0.0 \le Z \le 1.2)$

  b. $P(-0.9 \le Z \le 0.0)$

  c. $P(0.0 \le Z \le 1.45)$

  d. $P(0.3 \le Z \le 1.56)$

  e. $P(-2.03 \le Z \le -1.71)$

  f. $P(-0.02 \le Z \le 3.54)$

  g. $P(Z \ge 2.50)$

  h. $P(Z \le 1.66)$

  i. $P(Z \ge 5)$

  j. $P(Z \ge -6)$

9. All applicants at a large university are required to take a special entrance exam before they are admitted. The exam scores are known to be normally distributed with a mean of 800 and a standard deviation of 100. Applicants must score 700 or more on the exam before they are admitted.

  a. What proportion of all applicants taking the exam is granted admission?

  b. What proportion of all applicants will score 1000 or higher on the exam?

  c. For the coming academic year, 2500 applicants have registered to take the exam. How many do we expect to be qualified for admission to the university?

10. A manufacturer produces bearings, but because of variability in the production process, not all of the bearings have the same diameter. The diameters have a normal distribution with a mean of 1.2 centimeters (cm) and a standard deviation of 0.01 cm. The manufacturer has determined that diameters in the range of 1.18 to 1.22 cm are acceptable. What proportion of all bearings falls in the acceptable range?

11. Find the value of Z from the standard normal table such that the probability between Z and –Z is

  a. 0.9544

  b. 0.9010

  c. 0.9802

  d. 0.9902

12. A large manufacturing plant uses lightbulbs with lifetimes that are normally distributed with a mean of 1000 hours and a standard deviation of 50 hours. To minimize the number of bulbs that burn out during operating hours, all bulbs are replaced at once. How often should the bulbs be replaced so that no more than 1% burn out between replacement periods?

13. A company that produces an expensive stereo component is considering offering a warranty on the component. Suppose the population of lifetimes of the components is a normal distribution with a mean of 84 months and a standard deviation of 7 months. If the company wants no more than 2% of the components to wear out before they reach the warranty date, what number of months should be used for the warranty?

14. The average amount of time that students use computers at a university computer center is 36 minutes with a standard deviation of 5 minutes. The times are known to be normally distributed. Around 10,000 uses are recorded each week in the computer center. The computer center administrative committee has decided that if more than 2000 uses of longer than 40 minutes at each sitting are recorded weekly, some new terminals must be purchased to meet usage needs. Should the computer center purchase the new computers?

2.5 POPULATIONS, SAMPLES, AND SAMPLING DISTRIBUTIONS

A *population* is a group to be studied and may consist of people, households, firms, automobile tires, and so on. A *sample* is a subset of a population. In other words, a sample is a group of items chosen from the population. Statistics is concerned with the use of sample information to make generalizations or inferences about a *population.*

Typically, the study of every item in a population is not feasible. It may be too time-consuming or too expensive to examine every item. As an alternative, a few items are chosen from the population. From the information provided by this sample, we hope to make reliable generalizations about characteristics of the population.

In this section, we assume that the items of the sample are randomly chosen from the population. By choosing the sample in this way, it is possible to objectively evaluate the quality of the generalizations made about the population characteristics of interest.

As discussed in Section 2.1, many possible random samples can be chosen from a particular population. In practice, typically only one such sample is chosen and examined. (In some applications, such as statistical quality control, repeated samples may be used.) To understand the processes that govern how inferences should be made, it is necessary to imagine all possible random samples of a given sample size n chosen from a particular population. Suppose the characteristic of interest for this population is the mean, μ. To estimate the population mean, the statistic most often used is the sample mean, $\bar{y}$; each possible random sample has an associated value of $\bar{y}$. Thus, the sample mean acts just like a random variable: It assigns a number (the value of the sample mean for each sample) to each of the possible outcomes of an experiment. The experiment, in this instance, is the process of choosing samples of size n from the population. Because the samples are chosen randomly, each one has an equal probability of being chosen. Thus, each value of $\bar{y}$ has a probability associated with it.

Because the sample mean, $\bar{y}$, can be viewed as a random variable, it has a probability distribution. This probability distribution is called the *sampling distribution of the sample mean.* In this section, some of the properties of the sampling distribution of the sample mean are reviewed. These are discussed in more detail in most introductory statistics courses. The sampling distribution of the sample mean is important because it allows us to make the link between population characteristics and sample values that make it possible to assess the quality of our inferences.

First, suppose the population of interest has a mean μ and variance σ^2. The mean of the sampling distribution of $\bar{y}$, written $\mu_{\bar{y}}$, is equal to the population mean:

$$\mu_{\bar{y}} = \mu$$

The variance of the sampling distribution of $\bar{y}$, written $\sigma_{\bar{y}}^2$, is

$$\sigma_{\bar{y}}^2 = \frac{\sigma^2}{n}$$

The standard deviation of the sampling distribution of $\bar{y}$, $\sigma_{\bar{y}}$, is the square root of the variance:

$$\sigma_{\bar{y}} = \frac{\sigma}{\sqrt{n}}$$

Thus, if all possible sample means for samples of size n could be collected, the average of the sample means would be the same as the average of all the individual population values. The sample mean values, however, would be less spread out than the individual population values because $\sigma_{\bar{y}}$ is always less than σ.

If the original population from which the samples were drawn is a normal distribution, then the sampling distribution of the sample mean is also a normal distribution for any sample size n. Thus, if μ and σ are known and the population to be sampled is normal, probability statements could be made about the sample mean, $\bar{y}$. Consider the following example.

EXAMPLE 2.3   In a certain manufacturing process, the diameter of a part produced is 40 centimeters (cm) on average, although it varies somewhat from part to part. This variation is thought to be well represented by a normal distribution with a standard deviation of 0.2 cm. If a random sample of 16 parts is chosen, what is the probability that the average diameter of the 16 parts is greater than 40.1 cm?

Answer: Because the population is normal, the sampling distribution of sample means also is normal. The mean of the sampling distribution is $\mu_{\bar{y}} = 40$ and the standard deviation is:

$$\sigma_{\bar{y}} = \frac{0.2}{\sqrt{16}} = 0.05$$

Thus,

$$P(\bar{y} > 40) = P\left(Z > \frac{40.1 - 40}{0.05}\right) = 0.5 - 0.4772 = 0.0228$$

Knowledge of the sampling distribution provides information about how sample means from a particular population should behave. But what if the population does not have a normal distribution, or what if the actual distribution of the population is unknown? In this case, there is an important result in statistics called the *central limit theorem* (CLT) that states

As long as the sample size is large, the sampling distribution of the sample mean is approximately normal, regardless of the population distribution.

The CLT states that probabilities still can be computed concerning sample means, even though the population does not have a normal distribution, as long as the sample size is large enough. How large "large enough" is varies somewhat from one population distribution to another, but a generally accepted rule is to treat a sample size of 30 or more as large. The next example illustrates the use of the CLT.

EXAMPLE 2.4   A cereal manufacturer claims that boxes of its cereal weigh 20 ounces (oz) on average with a population standard deviation of 0.5 oz. The manufacturer does not know whether the population distribution is normal. A random sample of 100 boxes is selected. What is the probability that the sample mean is between 19.9 and 20.1 oz?

Answer: Because the sample size is large ($n = 100$), the sampling distribution is approximately normal even though the population distribution may be nonnormal. The mean and standard deviation of the sampling distribution are

$$\mu_{\bar{y}} = 20 \quad \text{and} \quad \sigma_{\bar{y}} = \frac{0.5}{\sqrt{100}} = 0.05$$

Thus,

$$P(19.9 \leq \bar{y} \leq 20.1) = P\left(\frac{19.9 - 20}{0.05} \leq Z \leq \frac{20.1 - 20}{0.05}\right) = P(-2 \leq Z \leq 2)$$

$$= 0.4772 + 0.4772 = 0.9544$$

EXERCISES

15. The daily receipts of a fast-food franchise are normally distributed with a mean of $2200 per day and a standard deviation of $50. A random sample of 25 days' receipts is chosen for an audit.

   a. What is the probability that the sample mean is larger than $2220?

   b. What is the probability that the sample mean differs from the true population mean by more than ±$10?

16. The accounts receivable of a large department store are normally distributed with a mean of $250 and a standard deviation of $80. If a random sample of 225 accounts is chosen, what is the probability that the mean of the sample is between $232 and $268?

17. When a certain manufacturing process is correctly adjusted, the length of a machine part produced is a random variable with a mean of 200 cm and a standard deviation of 0.1 cm. The individual measurements are normally distributed.

   a. What is the probability that an individual part is longer than 200.2 cm?

   b. Suppose a sample of 25 parts is chosen randomly. What is the probability that the mean of the sample is bigger than 200.2 cm?

   c. Is it correct to use the normal distribution to compute the probability in part b? Why or why not?

18. Suppose we have a large population of houses in a community. The average annual heating expense for each house is $400 with a population standard deviation of $25. A random sample of 25 houses had a sample mean of $380 and a sample standard deviation of $35.

   a. What is the mean of the sampling distribution of the sample mean for samples of size 25 chosen from the population of all houses?

   b. What is the standard deviation of the sampling distribution of the sample mean for samples of size 25 chosen from the population of all houses?

   c. Is it safe to assume that the sampling distribution of the sample mean for samples of size 25 is normally distributed? Why or why not?

19. The average time to complete a certain production-line task is assumed to be normally distributed with a standard deviation of 5 minutes (min). A random sample of 16 workers' times is selected to estimate the average time taken to complete the task. What is the probability that the sample mean is within ±1 min of the population's true mean time?

20. The speed of automobiles on I-20 west of Fort Worth, Texas, is being investigated by the Texas Department of Public Safety (DPS). If the average speed of cars on the highway exceeds 80 miles per hour (mph), the DPS plans to add more patrol cars to the area. To decide what to do, it takes a random sample of 150 cars and finds the sample average speed to be 80.5 mph. Assuming that the population standard deviation is 8 mph, should the DPS add patrol cars to the area?

21. Suppose past evidence shows that the lifetimes of hard drives from a certain production line have a population standard deviation of 700 hours. But a modification has been made in the material used to manufacture the hard drives. The manufacturer wants to know if the average lifetime of the modified hard drives is longer than the previous average lifetime. It is believed that the modification will not affect the population standard deviation of the lifetimes, but it may affect the average life. The previous average lifetime was 3250 hours. A random sample of 50 hard drives with the modification is taken and the drives are tested. The sample average lifetime for the drives from the new process is found to be 3575 hours. Should the manufacturer conclude that the new process produces hard drives with longer average lifetimes?

2.6 ESTIMATING A POPULATION MEAN

Two types of estimates can be constructed for any population parameter: point estimates and interval estimates. *Point estimates* are single numbers used as an estimate of a parameter. To estimate the population mean μ, the sample mean $\bar{y}$ typically is used. The sample mean is a point estimate.

An *interval estimate* is a range of values used as an estimate of a population parameter. The width of the interval provides a sense of the accuracy of the point estimate. The interval tells us the likely values of the population mean.

Assuming the population standard deviation, σ, is known, a confidence interval for the population mean, μ, can be constructed as

$$\left(\bar{y} - z_{\alpha/2}\frac{\sigma}{\sqrt{n}}, \bar{y} + z_{\alpha/2}\frac{\sigma}{\sqrt{n}} \right)$$

where $z_{\alpha/2}$ is a standard normal value chosen so that the probability above $z_{\alpha/2}$ is $\alpha/2$. Thus, between $z_{\alpha/2}$ and $-z_{\alpha/2}$, there is a probability of $1 - \alpha$ (see Figure 2.19). For this reason, the confidence interval written in its general form is referred to as a $(1 - \alpha)100\%$ (read "one minus alpha times 100 percent") confidence interval estimate of μ. By replacing $z_{\alpha/2}$ by the appropriate standard normal value, the desired level of confidence can be achieved. For example, to achieve 95% confidence, use $z_{0.025} = 1.96$ because the probability under the standard normal curve between -1.96 and 1.96 is 0.95. Note that lowercase z's are used here to represent the specific values chosen from the standard normal distribution as opposed to uppercase Z's, which represent the standard normal random variable.

The term "95% confidence interval" means that if repeated samples of size n are taken from the same population, and a confidence interval is constructed in the manner just described for each sample, 95% of those intervals contain the

FIGURE 2.19
Choosing z Values for a
(1 − α)100% Confidence
Interval.

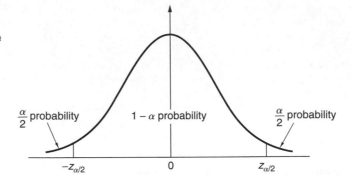

$\frac{\alpha}{2}$ probability   1 − α probability $\frac{\alpha}{2}$ probability

$-z_{\alpha/2}$ 0 $z_{\alpha/2}$

population mean. The use of the interval, assuming that σ is known, is justified for any sample size if the population is normal. When the population is normal, the sampling distribution of sample means is also normal, which is the basis for the construction of the interval. If the distribution of the population is unknown or if it is known to be nonnormal, the interval still can be used as long as the sample size is large (generally $n \geq 30$) because the CLT guarantees that the sampling distribution of sample means is approximately normal.

EXAMPLE 2.5   Managers of Newman-Markups Department Store want a 90% confidence interval estimate of the current average balance of charge customers. With a random sample of 100 accounts, a sample mean of $245, and a population standard deviation of $45, what is the 90% interval estimate of the true average balance?
Answer: The 90% confidence interval estimate of the true average balance is

$$\left(245 - 1.65\frac{45}{\sqrt{100}}, 245 + 1.645\frac{45}{\sqrt{100}}\right)$$

or ($237.58, $252.43)

In Example 2.5, the population standard deviation σ was assumed to be known. In most instances, however, σ is unknown. In this case, σ can be estimated by the sample standard deviation:

$$s = \sqrt{\frac{\sum_{i=1}^{n}(y_i - \bar{y})^2}{n - 1}}$$

Replacing σ by s in the previous interval and $z_{\alpha/2}$ by $t_{\alpha/2,n-1}$ gives

$$\left(\bar{y} - t_{\alpha/2,\, n-1}\frac{s}{\sqrt{n}}, \bar{y} + t_{\alpha/2,\, n-1}\frac{s}{\sqrt{n}}\right)$$

Changing $z_{\alpha/2}$ to $t_{\alpha/2,n-1}$ reflects the fact that s, an estimator of σ, is being used to construct a confidence interval estimate for μ. The value of $t_{\alpha/2,n-1}$ is chosen from the t table (Table B.2) in Appendix B. The t value chosen depends on the number of *degrees of freedom* (df) and on the confidence level desired. The number of degrees of freedom for estimating μ is $n - 1$. These values are listed on the left-hand side of the t table. The confidence levels are reflected through the upper-tail areas at the top of the table. Note that the 0.025 column is used for a 95% level of confidence ($\alpha/2 = 0.025$).

The shape of the t distribution depends on the number of degrees of freedom. The t distribution has fatter tails than the normal distribution and, thus, has greater probability in its tails. The t value for a given level of confidence is therefore larger than the standard normal value, producing wider confidence intervals (less precise estimates) because s rather than σ is used to construct the interval estimate. Also, as the number of degrees of freedom increases, the t distribution begins to look more like the normal distribution. In the last row of the t table, which is the ∞ degree of freedom row, the z values corresponding to the given upper-tail areas are shown. This indicates that, for a very large sample, the t and Z distributions are identical.

When the number of degrees of freedom is 30 or more, the t values are sometimes replaced by z values because there is little difference between the two in these cases. Although using t values is more correct in all cases when σ is unknown, replacing the t values by z values for large degrees of freedom is often adopted simply for convenience. Note that Excel, MINITAB, and SAS continue to use t values in certain applications when σ is unknown, regardless of the sample size. Also, the estimated standard deviation of the sampling distribution ($s/\sqrt{n}$) is often referred to as the *standard error of the mean*.

In small samples ($n < 30$), the population should be normal, or close to a normal distribution, before the interval is used. If the population is nonnormal and small samples are used, nonparametric methods may be appropriate. These methods are discussed in many introductory statistics texts (for example, see Chapter 17 of *Statistics for Management and Economics* by Keller and Warrack[2]). When the sample size is large, the assumption of a normal population is not necessary. The flowchart in Figure 2.20 describes the interval choices when σ is unknown. The σ known case is not shown because it rarely occurs in practice.

EXAMPLE 2.6 A manufacturer wants to estimate the average life of an expensive electrical component. Because the test to be used destroys the component, a small sample is desired. The lifetimes in hours of five randomly selected components are

$$92, 110, 115, 103, 98$$

Find a point estimate and a 95% confidence interval estimate of the population average lifetime of the components. The population of lifetimes is assumed to be normal. Answer: Because σ is unknown, the interval to be used is

$$\left(\bar{y} - t_{0.025,4} \frac{s}{\sqrt{n}}, \bar{y} + t_{0.025,4} \frac{s}{\sqrt{n}} \right)$$

---

[2] See References for complete publication information.

FIGURE 2.20
Estimating μ Assuming
σ **is Unknown.**

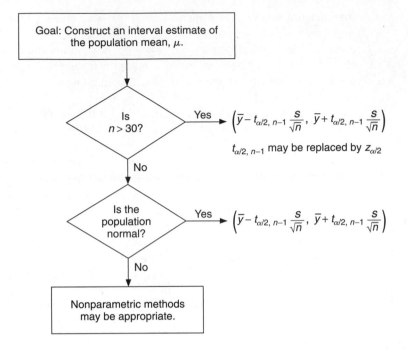

FIGURE 2.21 Using
MINITAB to Construct
Confidence Intervals.[a]

One-Sample T: hours

Variable	N	Mean	StDev	SE Mean	90% CI
hours	5	103.600	9.182	4.106	(94.846, 112.354)

One-Sample T: hours

Variable	N	Mean	StDev	SE Mean	95% CI
hours	5	103.600	9.182	4.106	(92.200, 115.000)

One-Sample T: hours

Variable	N	Mean	StDev	SE Mean	99% CI
hours	5	103.600	9.182	4.106	(84.695, 122.505)

[a]The quantities shown include the sample size (N), the sample mean (Mean), the sample standard deviation (StDev), the standard error of the mean (SE Mean) and the confidence interval for the requested level of confidence.

The quantities needed to construct the interval are

$$\bar{y} = \frac{\sum_{i=1}^{n} y_i}{n} = 103.6$$

$$s = \sqrt{\frac{\sum_{i=1}^{n}(y_i - \bar{y})^2}{n - 1}} = 9.18$$

$$t_{0.025,4} = 2.776$$

The point estimate of μ is $\bar{y} = 103.6$. The interval estimate of μ is

$$\left[103.6 - 2.776\left(\frac{9.18}{\sqrt{5}}\right),\ 103.6 + 2.776\left(\frac{9.18}{\sqrt{5}}\right) \right] \quad \text{or} \quad (92.2, 115.0)$$

Computer packages such as Excel, MINITAB, and SAS also can be used to construct confidence intervals. Figure 2.21 shows the MINITAB output for requests for 90%, 95%, and 99% confidence intervals using the data from Example 2.6. There is more information on such procedures in the Using the Computer section at the end of this chapter.

EXERCISES

22. A local department store wants to determine the average age of the adults in its existing marketing area to help target its advertising. A random sample of 400 adults is selected. The sample mean age is found to be 35 years with a sample standard deviation of 5 years. Construct a 95% confidence interval estimate of the population average age of the adults in the area.

23. The management of a manufacturing plant is studying the number of times employees in a large population of workers are absent. A random sample of 25 employees is chosen, and the average number of annual absences per employee in the sample is found to be six. The sample standard deviation is 0.6. Assuming the population of absences is normally distributed, construct a 99% confidence interval estimate of the population average number of absences.

24. A quality control inspector is concerned with the average amount of weight that can be held by a type of steel beam. A random sample of five beams is tested with the following amounts of

weight added before the beams begin to show stress (in thousands of pounds):

9, 11, 10, 10, 8

Assuming that the population of weights is normally distributed, construct a 95% confidence interval estimate of the population average weight that can be held.

25. Refer to the data in Exercise 1. Highway mileages of 147 cars are in a file named CARS2. Assume these cars represent a random sample of all new cars produced in 2003. Find a 95% confidence interval estimate for the population mean miles per gallon.

26. The July 1, 2002, one-year returns for a random sample of 83 mutual funds are available in a data file named ONERET2. Find a 95% confidence interval estimate for the population mean rate of return.

(*Source*: The data are used by permission from the September issue of *Kiplinger's's Personal Finance*. Copyright © 2002 The Kiplinger Washington Editors, Inc.)

2.7 HYPOTHESIS TESTS ABOUT A POPULATION MEAN

Section 2.6 discussed estimation of a population mean. Estimation was the first of our two main topics of statistical inference. In this section, we discuss the second topic: hypothesis tests. Again, the population mean is used to demonstrate tests of hypotheses.

The following definitions are useful in testing hypotheses:

Null Hypothesis, H_0: The null hypothesis states a hypothesis to be tested.

Alternative Hypothesis, H_a: The alternative hypothesis includes possible values of the population parameter not included in the null hypothesis.

Test Statistic: A number computed from sample information.

Decision Rule: A rule used in conjunction with the test statistic to determine whether or not the null hypothesis should be rejected.

In setting up a hypothesis test, the null hypothesis is initially assumed to be true. Under this assumption, a decision rule is constructed based on the sampling distribution of the test statistic. The decision rule states a range of values for the test statistic that are plausible if H_0 is true and a range of values for the test statistic that seem implausible if H_0 is true. Depending on the test statistic value, the statistical decision made is either "reject H_0" (the test statistic falls in the implausible range) or "do not reject H_0" (the test statistic falls in the plausible range).

Because the decision is based on sample information, it is not possible to be certain that the correct decision has been made. A statistical decision does not prove or disprove the null hypothesis with certainty, although it does present support for one of the two hypotheses. In a business environment, decisions are typically made on the basis of limited information. Hypothesis-testing results provide support for alternative possible courses of action based on such limited (sample) information.

Two types of errors are possible in hypothesis testing; these are illustrated in Figure 2.22. On the left-hand side of the figure are the two possible states of nature: Either H_0 is true or H_0 is false. The statistical decisions "do not reject H_0" and "reject H_0" are listed at the top of the figure. If H_0 is true and the sample information says do not reject H_0, a correct decision has been made. Also, if H_0 is false and the sample information says to reject H_0, the decision is correct. If H_0 is true, however, and the sample information says to reject H_0, the decision is incorrect. Rejecting the null hypothesis when it is true is called a *Type I error*. A *Type II error* occurs if the null hypothesis is actually false but the sample information says do not reject H_0.

FIGURE 2.22
The Risks of Hypothesis Testing.

Decision

State of Nature	Do Not Reject H_0	Reject H_0
H_0 True	Correct Decision	Type I Error
H_0 False	Type II Error	Correct Decision

Note that the decision made is always stated with reference to the null hypothesis: Either reject H_0 or do not reject H_0. Also note that because only two possibilities are considered (either H_0 or H_a), rejecting H_0 implies agreement with H_a.

Note that the phrase "do not reject H_0" (or "fail to reject H_0") rather than "accept H_0" is used. When the data suggest that we should not reject H_0, this is not proof that H_0 is true. This is a statistical decision and may imply simply that we do not have enough evidence to reject H_0. The expression "accept H_0" is viewed as too strong in suggesting that H_0 has been proven true. "Do not reject H_0" emphasizes that our result is not proof that the null hypothesis is true. Our sample may simply not provide enough evidence to reject it.

When testing any hypothesis, it is desirable to keep the chances of an error occurring as small as possible. Typically, when setting up the test, a desired level for the probability of a Type I error is established. By specifying a small probability, control can be exercised over the chances of making such an error. The probability of a Type I error is called the *level of significance* (or *significance level*) of the test and is denoted α.

To illustrate, suppose the following hypotheses are to be tested:

$$H_0: \mu = 10$$
$$H_a: \mu \neq 10$$

Also assume that the population standard deviation is known to be 2. This assumption will be relaxed later. The sample to be drawn consists of 100 items, and the desired level of significance is $\alpha = 0.05$, or a 5% chance of making a Type I error.

The hypotheses have now been set up and a level of significance chosen. The next step is to establish the decision rule for the test. To do this, consider the sampling distribution of sample means as shown in Figure 2.23. If the null hypothesis is true ($\mu = 10$), then a region can be determined in which 95% of all sample means will fall. The upper bound of this region is denoted C_1 and the lower bound C_2. Above and below these bounds, there is a combined probability of only 0.05 of obtaining a sample mean. The decision rule for the test can be set up as

Reject H_0 if $\bar{y} > C_1$ or $\bar{y} < C_2$

Do not reject H_0 if $C_2 \leq \bar{y} \leq C_1$

FIGURE 2.23
Sampling Distribution of Sample Means Assuming $H_0: \mu = 10$ Is True.

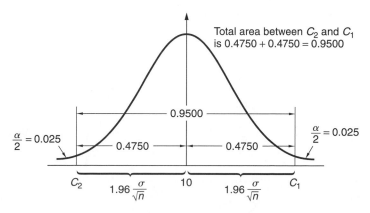

Total area between C_2 and C_1 is $0.4750 + 0.4750 = 0.9500$

$\frac{\alpha}{2} = 0.025$

$\frac{\alpha}{2} = 0.025$

0.9500

0.4750 0.4750

C_2 $1.96 \frac{\sigma}{\sqrt{n}}$ 10 $1.96 \frac{\sigma}{\sqrt{n}}$ C_1

If the null hypothesis is true, 95% of all sample means will fall between C_1 and C_2. So there is only a 5% chance of obtaining a sample mean that falls in the rejection region. In other words, there is a 5% chance of rejecting H_0 if H_0 is true. Thus, the desired level of significance for the test has been achieved.

The critical values C_1 and C_2 in the decision rule can be determined because the sampling distribution is normal (as guaranteed by the CLT because the sample size is large). Between C_1 and C_2, we want a probability of 0.95; the associated z value to produce this probability is 1.96. The upper critical value is

$$C_1 = 10 + 1.96 \frac{2}{\sqrt{100}} = 10.392$$

and the lower critical value is

$$C_1 = 10 - 1.96 \frac{2}{\sqrt{100}} = 9.608$$

The decision rule becomes

Reject H_0 if $\bar{y} > 10.392$ or $\bar{y} < 9.608$

Do not reject H_0 if $9.608 \le \bar{y} \le 10.392$

Now suppose a sample of size 100 is randomly selected, and the sample mean computed is $\bar{y} = 11.2$. Based on this value of the sample mean, the null hypothesis is not supported. A sample mean of 11.2 is too extreme to believe that the individual values which produced it came from a population with mean $\mu = 10$. So the null hypothesis is rejected. Note that the null hypothesis has not been proved false. There is simply contradictory evidence, and so a statistical decision to reject was made. The alternative hypothesis, $\mu \ne 10$, seems more plausible given the evidence obtained.

The previous hypothesis-testing problem was set up in terms of the sampling distribution of the sample mean. An alternative and more typical way of performing hypothesis tests is with a standardized test statistic. The basic philosophy and structure of the test are the same. The only difference is that the test statistic is standardized and compared directly with the z value. For example, the standardized test statistic for testing hypotheses about the population mean is

$$z = \frac{\bar{y} - \mu_0}{\sigma/\sqrt{n}}$$

where μ_0 is the hypothesized value of the population mean. For the previous example, the standardized decision rule is

Reject H_0 if $\dfrac{\bar{y} - \mu_0}{\sigma/\sqrt{n}} > 1.96$ or $\dfrac{\bar{y} - \mu_0}{\sigma/\sqrt{n}} < -1.96$

Do not reject H_0 if $-1.96 \le \dfrac{\bar{y} - \mu_0}{\sigma/\sqrt{n}} \le 1.96$

The standardized test statistic value is

$$\frac{\bar{y} - \mu_0}{\sigma/\sqrt{n}} = \frac{11.2 - 10}{2/\sqrt{100}} = 6$$

Because the test statistic value falls in the rejection region ($6 > 1.96$), the null hypothesis is rejected.

Whether the standardized or unstandardized form of the test is used, the decision made (reject H_0 or do not reject H_0) will always be the same. Throughout this text, the standardized form is used unless otherwise noted.

If σ is unknown, the same adjustment is used as with confidence intervals. The unknown population standard deviation, σ, is replaced by an estimate, s, and the z value is replaced by the t value with $n - 1$ degrees of freedom.

The hypothesis structure previously discussed is called a *two-tailed test* because rejection occurs in both the upper and lower tails of the sampling distribution. Two other hypothesis structures need to be considered: *upper-tailed* and *lower-tailed tests.* Both of these involve rejection in only one tail of the sampling distribution and are therefore referred to as *one-tailed tests.* The hypothesis structures, test statistic, and decision rules for the case when σ is unknown are shown in Figure 2.24. The σ known case is omitted because it is rarely encountered in practice.

As was the case for confidence intervals, the t tests are constructed with the assumption that the sampling distribution is normally distributed. The population should be normal (or nearly so) before tests are used with small samples ($n < 30$), or nonparametric methods may be appropriate. Because the CLT guarantees that the sampling distribution is close to normal when n is large, the assumption of a normal population is unnecessary in cases with large sample sizes.

The following examples help illustrate hypothesis-testing techniques.

Hypotheses	Test Statistic	Decision Rules
$H_0: \mu = \mu_0$	$t = \dfrac{\bar{x} - \mu_0}{s/\sqrt{n}}$	Reject H_0 if $t > t_{\alpha/2,n-1}$ or if $t < -t_{\alpha/2,n-1}$
$H_a: \mu \neq \mu_0$		Do not reject H_0 if $-t_{\alpha/2,n-1} \leq t \leq t_{\alpha/2,n-1}$
$H_0: \mu \leq \mu_0$	$t = \dfrac{\bar{x} - \mu_0}{s/\sqrt{n}}$	Reject H_0 if $t > t_{\alpha,n-1}$
$H_a: \mu > \mu_0$		Do not reject H_0 if $t \leq t_{\alpha,n-1}$
$H_0: \mu \geq \mu_0$	$t = \dfrac{\bar{x} - \mu_0}{s/\sqrt{n}}$	Reject H_0 if $t < -t_{\alpha,n-1}$
$H_a: \mu < \mu_0$		Do not reject H_0 if $t \geq -t_{\alpha,n-1}$

FIGURE 2.24 Hypotheses, Test Statistics, and Decision Rules for Testing Hypotheses About Population Means (σ unknown)

EXAMPLE 2.7   Consider again the manufacturer of electrical components in Example 2.6. Suppose the manufacturer wishes to test whether the population average life of the components is 110 hours or more. If it is less than 110 hours, the components do not meet specifications, and the production process must be adjusted to increase the average life of the components.

As in Example 2.6, five components are randomly selected, and the lifetimes in hours for these components are determined to be

$$92, 110, 115, 103, 98$$

Assume the population of lifetimes is known to be normally distributed. The hypotheses to be tested are

$$H_0\colon \mu \geq 110$$

$$H_a\colon \mu < 110$$

Using a 5% level of significance, the decision rule for the test is

Reject H_0 if $t < -t_{0.05,4} = -2.132$

Do not reject H_0 if $t \geq -t_{0.05,4} = -2.132$

The sample mean and standard deviation are

$$\bar{y} = 103.6, s = 9.18$$

and the standardized test statistic, t, is

$$t = \frac{\bar{y} - \mu_0}{\sigma/\sqrt{n}} = \frac{103.6 - 110}{9.18/\sqrt{5}} = -1.56$$

so we do not reject the null hypothesis. Note that the sample mean of 103.6 hours is below the desired average lifetime of 110 hours. However, it is possible that the population mean could be 110 hours and our sample will produce a sample mean of 103.6 hours. We have not proved the null hypothesis is true, but there is not enough evidence in our small sample to reject it.

EXAMPLE 2.8   A company that manufactures rulers wants to ensure that the average length of its rulers is 12 inches (for obvious reasons). From each production run, a random sample of 25 rulers is selected and their lengths determined by very accurate measuring instruments. On one particular run, the average length of the 25 rulers is determined to be 12.02 inches, with a sample standard deviation of 0.02 inch.

Using a 1% level of significance, is the average length of the rulers produced by this manufacturer equal to 12 inches?

Answer: The hypotheses to be tested are

$$H_0: \mu = 12$$
$$H_a: \mu \neq 12$$

Using a 1% level of significance, the decision rule for the test is

Reject H_0 if $t > t_{0.005,24} = 2.797$ or if $t < -t_{0.005,24} = -2.797$

Do not reject H_0 if $-2.797 \leq t \leq 2.797$

From the sample information, the standardized test statistic is

$$t = \frac{\bar{y} - \mu_0}{\sigma/\sqrt{n}} = \frac{12.02 - 12}{0.02/\sqrt{25}} = 5.0$$

resulting in a decision to reject the null hypothesis. This decision suggests that the average length of rulers is not 12 inches. Some adjustment to the production process is necessary.

Most statistical software packages perform tests of hypotheses. Instead of reporting a decision to reject or fail to reject the null hypothesis, however, the output often includes a number called a p value. By comparing the p value to the level of significance, α, an alternative decision rule can be constructed:

Reject H_0 if p value $< \alpha$

Do not reject H_0 if p value $\geq \alpha$

The p value is the computed area under the sampling distribution at or beyond the value of the standardized test statistic. That is, it is the probability of observing a t value or z value as extreme as, or more extreme than, the sample test statistic. For example, suppose the hypotheses to be tested are

$$H_0: \mu \leq 10$$
$$H_a: \mu > 10$$

A sample size of 100 is to be used and σ is known to be 10 (for illustrative purposes, we assume σ is known; the process is similar when σ is unknown). From the random sample of 100 items, a sample mean of $\bar{y} = 12$ is obtained. The standardized test statistic is

$$z = \frac{12 - 10}{10/\sqrt{100}} = 2.0$$

In Figure 2.25, the standard normal distribution is shown, and the position of the test statistic has been located. The p value for this test is

$$p \text{ value} = P(Z \geq 2.0) = 0.5 - 0.4772 = 0.0228$$

FIGURE 2.25
Computation of *p* Value.

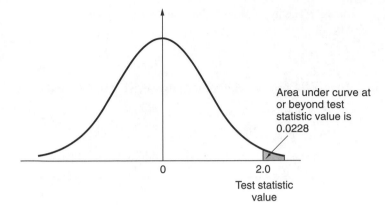

Area under curve at
or beyond test
statistic value is
0.0228

0

2.0

Test statistic
value

If we require a 5% level of significance, our decision is to reject the null hypothesis because the *p* value of 0.0228 is less than 0.05.

Regardless of which approach to hypothesis testing is chosen (unstandardized or standardized test statistic or *p* value), the decision made is identical for a given level of significance.

The way *p* values are computed differs depending on whether the test is an upper-tailed, lower-tailed, or two-tailed test. In all three cases, the first step is to compute the standardized test statistic. For an upper-tailed test (as in the previous illustration), compute the probability to the right of the standardized test statistic. For a lower-tailed test, compute the probability to the left of the standardized test statistic. For a two-tailed test, compute the probability in the tail area closest to the standardized test statistic, and then multiply this area by 2. By computing *p* values in this manner, the following decision rule can always be used:

Reject H_0 if *p* value $< \alpha$

Do not reject H_0 if *p* value $\geq \alpha$

EXAMPLE 2.9 Figure 2.26 shows the MINITAB output for testing the hypothesis discussed in Example 2.7. The output shows $\bar{y} = 103.600$, $s = 9.182$, $s/\sqrt{n} = 4.106$, $t = -1.56$, and the *p* value $= 0.097$. Using the *p* value decision rule, note that the null hypothesis is not rejected for levels of significance of 1% or 5%, but is rejected at

FIGURE 2.26
MINITAB Output for Test of Hypothesis in Example 2.9.

One-Sample *T*: **hours**

Test of mu = 110 vs < 110

Variable	N	Mean	StDev	SE Mean	95% Upper Bound	T	P
hours	5	103.600	9.182	4.106	112.354	−1.56	0.097

10%. Reporting the p value is somewhat more informative in a case such as this than simply reporting a decision to reject or fail to reject the null hypothesis.[3]

EXERCISES

27. If a null hypothesis is rejected at the 5% level of significance, what decision would have been made at the 10% level? Why?

28. To investigate an alleged unfair trade practice, the Federal Trade Commission (FTC) takes a random sample of sixteen "5-ounce" candy bars from a large shipment. The mean of the sample weights is 4.85 ounces and the sample standard deviation is 0.1 ounce. Test the hypotheses

$$H_0: \mu \geq 5$$
$$H_a: \mu < 5$$

at the 5% level of significance. Assume the population of candy bar weights is approximately normally distributed. Based on the results of the test, does the FTC have grounds to proceed against the manufacturer for the unfair practice of short-weight selling? State the decision rule, the test statistic, and your decision.

29. A quality inspector is interested in the time spent replacing defective parts in one of the company's products. The average time spent should be at most 20 minutes (min) per day according to company standards. The following hypotheses are set up to examine whether the standards are being met:

$$H_0: \mu \leq 20$$
$$H_a: \mu > 20$$

where μ represents the population average time spent replacing defective parts. To conduct the test, a random sample of 16 employees is chosen. The average time spent replacing defective parts for the sample was 20.5 min, with a sample standard deviation of 4 min. Perform the test at a 5%

level of significance. Assume the population of service times is approximately normally distributed. State the decision rule, the test statistic, and your decision. Are company standards being met?

30. Refer to the data in Exercise 1. Highway mileages of 147 cars are in a file named CARS2. Assume these cars represent a random sample of all new cars produced in 2003. The corporate average fuel economy (CAFE) standards set by the government require that average fuel economy for cars be more than 27.5 miles per gallon. To examine whether the CAFE standard is being met, the following hypotheses are tested:

$$H_0: \mu \leq 27.5$$
$$H_a: \mu > 27.5$$

where μ is the population average gas mileage for all 2003 cars. Use a 5% level of significance. State the decision rule, the test statistic, and your decision. What is your conclusion regarding the average mileage of 2003 cars?

31. Refer to the data in Exercise 26. The July 1, 2002, one-year returns for a random sample of 83 mutual funds are available in a file named ONERET2. The return for the S&P 500 stock index for the same one-year time period was −18.0%. Test to see if there is evidence that the average one-year return for the population of funds is more than the return for the S&P 500 stock index. Use a 5% level of significance. State the hypotheses to be tested, the decision rule, the test statistic, and your decision. What is your conclusion regarding the average one-year return for mutual funds?

---

[3]MINITAB also reports a 95% upper bound of 112.354 for the population mean. This number is the upper limit of a one-sided confidence interval for the mean. We are 95% sure that the population mean is less than 112.354. One-sided confidence intervals were not discussed in this text but would be consistent with the use of a one-sided test.

2.8 ESTIMATING THE DIFFERENCE BETWEEN TWO POPULATION MEANS

The comparison of two separate populations often is of more concern than the estimation of the parameters of a single population as discussed in Section 2.6. In many cases, the comparison is between the means of the two populations. In this section, point and interval estimates of the difference between two population means are discussed.

Throughout this section, we assume that two populations with parameters μ_1, σ_1 and μ_2, σ_2 are being studied. Random samples are drawn independently from the two populations. Summary statistics from the two samples are:

	Sample 1	*Sample 2*
Sample size	n_1	n_2
Sample mean	$\bar{y}_1$	$\bar{y}_2$
Sample standard deviation	s_1	s_2

A point estimate of the difference between the two population means, $\mu_1 - \mu_2$, is given by the difference between the two sample means, $\bar{y}_1 - \bar{y}_2$.

If σ_1 and σ_2 are known, the standard deviation of the sampling distribution of $\bar{y}_1 - \bar{y}_2$ is written as

$$\sqrt{\frac{\sigma_1^2}{n_1} + \frac{\sigma_2^2}{n_2}}$$

In addition, the sampling distribution of $\bar{y}_1 - \bar{y}_2$ can be shown to be normally distributed if each of the populations is normal. The sampling distribution is approximately normal if both sample sizes are large ($n_1 \geq 30$ and $n_2 \geq 30$) even if the populations are not normal. A $(1 - \alpha)100\%$ confidence interval for $\mu_1 - \mu_2$ would be

$$\bar{y}_1 - \bar{y}_2 \pm z_{\alpha/2} \sqrt{\frac{\sigma_1^2}{n_1} + \frac{\sigma_2^2}{n_2}}$$

As in the previous situations we discussed, it is unlikely that the population variances, σ_1^2 and σ_2^2, are known in practice. They must be estimated by the sample variances, s_1^2 and s_2^2. These estimates are substituted into the formula for the standard deviation of the sampling distribution of $\bar{y}_1 - \bar{y}_2$, which results in

$$\sqrt{\frac{s_1^2}{n_1} + \frac{s_2^2}{n_2}}$$

An approximate $(1 - \alpha)100\%$ confidence interval for $\mu_1 - \mu_2$ is then

$$\bar{y}_1 - \bar{y}_2 \pm t_{\alpha/2,\Delta} \sqrt{\frac{s_1^2}{n_1} + \frac{s_2^2}{n_2}}$$

The interval is approximate because the population variances may differ. When $\sigma_1^2 \neq \sigma_2^2$, an exact interval cannot be constructed. This interval is referred to as the *approximate* interval.

The approximate degrees of freedom for the sampling distribution of $\bar{y}_1 - \bar{y}_2$ are given by

$$\Delta = \frac{(s_1^2/n_1 + s_2^2/n_2)^2}{[(s_1^2/n_1)^2/(n_1 - 1)] + [(s_2^2/n_2)^2/(n_2 - 1)]}$$

If the population variances can be assumed equal, $\sigma_1^2 = \sigma_2^2$, then an exact interval can be constructed. Because $\sigma_1^2 = \sigma_2^2$, it is no longer necessary to provide separate estimates of the two variances. The information in both samples can be combined, or pooled, to estimate the common variance. The pooled estimator of the population variance is

$$s_p^2 = \frac{(n_1 - 1)s_1^2 + (n_2 - 1)s_2^2}{n_1 + n_2 - 2}$$

The $(1 - \alpha)100\%$ confidence interval is given by

$$\bar{y}_1 - \bar{y}_2 \pm t_{\alpha/2, n_1 + n_2 - 2} \sqrt{s_p^2 \left(\frac{1}{n_1} + \frac{1}{n_2} \right)}$$

This interval is referred to as the *exact* interval.

Which of these intervals, exact or approximate, should be used in practice? The answer depends on what we know about the population variances, σ_1^2 and σ_2^2. If σ_1^2 and σ_2^2 are known to be equal, choose the exact interval. If σ_1^2 and σ_2^2 are known to be unequal, choose the approximate interval. But what if we have no information on whether or not the variances are equal? In this case, current research recommends using the approximate interval.[4]

EXAMPLE 2.10   Table 2.2, panel A, lists the 5-year returns for a random sample of 51 load mutual funds. (Load funds require the payment of an up-front sales charge to invest in the fund.) Table 2.2, panel B, shows the returns over the same 5-year period for a random sample of 32 no-load funds (no up-front sales charge is required). Construct a 95% confidence interval estimate of the difference between the population average 5-year returns for no-load and load funds.

---

[4] We could test for equality of the variances and choose an approach based on the test result, but this procedure has been shown to be less powerful than simply using the approximate interval. See, for example, "Homogeneity of Variance in the Two-Sample Means Test" by Moser and Stevens in *The American Statistician* 46(1992): 19–21.

TABLE 2.2 5-Year Rates of Return

Panel A: Load Funds				Panel B: No-Load Funds	
Name of Fund	5-yr Return	Name of Fund	5-yr Return	Name of Fund	5-yr Return
Dreyfus Growth		Aim Weingarten A	−4	American AAdvantage Balanced Plan	5.2
& Value Emerging Leaders	10.1	Deutsche Flag Value Builder A	5.6	American Century Income & Growth	5
Vanguard Capital Opportunity	15.6	Federated Equity Income A	0	Ariel Appreciation	14.2
Fidelity New Millennium	14.5	First American Small Cap Value A	6.6	AXA Rosenberg US Small Cap	9.5
Fidelity Select Air Transportation	14.1	Goldman Sachs Capital Growth A	4.3	Berger Growth	−5
Fidelity Select Health Care	7.3	Pimco Capital Appreciation A	6.3	CG Capital Markets Large Cap Value	3.9
Fidelity Advisor Balanced T	1.9	WM Growth & Income A	4.9	Columbia Balanced	4.5
Fidelity Advisor Technology T	−1.1	AXP Small Companies Index A	7.1	Dreyfus Premier Third Century Z	0.4
AAL Capital Growth A	5.7	Evergreen Fund A	−3.2	Fidelity Blue Chip Growth	2.5
MFS Emerging Growth A	−4.4	Evergreen Small Cap Value A	9.4	Fidelity Equity-Income II	5.1
BlackRock Small Cap Growth A	−1.5	Franklin Small Mid Cap Growth A	4.2	Gabelli Small Cap Growth	9.8
Fifth Third Quality Growth A	2.7	Liberty Fund A	2.4	ICAP Equity	5.7
Calamos Convertible A	10.6	Lord Abbott Developing Growth A	2.9	Invesco Health Sciences	6.4
Davis Financial A	7.7	MFS Value A	10.5	Janus Core Equity	10.6
Hartford Growth Opportunities L	4.3	Nations Convertible Securities A	7.9	Janus Twenty	4.3
Principal MidCap A	5.4	Nuveen Large Cap Value A	4.8	Mairs & Power Growth	10.8
Alpine US Real Estate Equity Y	9.6	Oppenheimer Cap Appreciation A	6.5	Needham Growth	17
J. Hancock Large Cap Equity A	4	Oppenheimer Total Return A	4.1	PBHG Growth	−3.9
Morgan Stanley Global Utilities B	5.9	Pioneer Fund A	6.6	T. Rowe Price Blue Chip Growth	3.4
Morgan Stanley Total Return B	2.7	Putnam Classic Equity A	0.9	T. Rowe Price Mid-Cap Value	11.6
Prudential US Emerging Growth A	6.3	Putnam New Value A	5.4	Rainier Core Equity	5
Smith Barney Large Cap Value A	1.7	Salomon Brothers Investors Value A	6.3	RS Diversified Growth A	14
FPA Paramount	−5	Security Equity A	0.2	Scudder Capital Growth AARP	0.7
Merrill Lynch Balanced Capital D	3	State Street Research Aurora A	15.9	Sit Mid Cap Growth	−0.9
One Group Equity Income A	3.2	Van Kampen Comstock A	10.4	Strong Growth	4.1
Aim Constellation A	1	Waddell & Reed Continental	4.1	Third Avenue Value	8.9
				Turner Small Cap Value	14.4
				US Global Investors World	
				Precious Minerals	−7.8
				Vanguard LifeStrategy	
				Conservative Growth	6
				Vanguard Small Cap Index	5.2
				Vanguard Utilities Income	4.6
				Wasatch Small Cap Growth	15.2

The MINITAB output shown in Figure 2.27 can be used to obtain the confidence interval. As will be discussed in Section 2.9, this output also can be used to test hypotheses about the difference between the two population means. In panel A of Figure 2.27, the results correspond to the approximate interval (assuming $\sigma_1^2 \neq \sigma_2^2$). Panel B shows the results for the exact interval (assuming $\sigma_1^2 = \sigma_2^2$).

Which interval is appropriate in this problem? Since we really have no information on the population variances, the approximate interval is the better choice. In this case, there is little difference between the two intervals, but sizable differences

FIGURE 2.27
MINITAB Output for
Examples 2.10 and 2.11.

Panel A: Assumes Variances Are Unequal

```
Two-sample T for FIVEYRRET

LOAD   N    Mean    StDev   SE Mean
0      32   5.95    5.88     1.0
1      51   5.01    4.80     0.67


Difference = mu (0) - mu (1)
Estimate for difference:  0.942157
95% CI for difference:   (-1.538493, 3.422807)
T-Test of difference = 0 (vs not =): T-Value = 0.76   P-Value = 0.450   DF = 56
```

Panel B: Assumes Variances Are Equal

```
Two-sample T for FIVEYRRET

LOAD    N    Mean    StDev   SE Mean
0       32   5.95    5.88     1.0
1       51   5.01    4.80     0.67


Difference = mu (0) - mu (1)
Estimate for difference:  0.942157
95% CI for difference:   (-1.409585, 3.293899)
T-Test of difference = 0 (vs not =): T-Value = 0.80   P-Value = 0.428   DF = 81
Both use Pooled StDev = 5.2411
```

can occur that can produce misleading conclusions if the exact interval is used in an inappropriate situation. The approximate interval provides a conservative result when the population variances are equal but also provides protection against the case when the variances are not equal.

The interval produced by the approximate method is (−1.54%, 3.42%). This result suggests that we can be 95% confident that the difference in population average 5-year returns for load and no-load funds ($\mu_{NoLoad} - \mu_{Load}$) is between −1.54% and 3.42%. What does this result suggest to you regarding the two types of funds?

EXERCISES

32. A graduate school of business is interested in estimating the difference between mean GMAT scores for applicants with and without work experience. Independent random samples of 50 applicants with and 50 applicants without work experience are chosen. The following results were obtained:

	With Work Experience	Without Work Experience
Sample size	50	50
Sample mean	545	510
Sample standard deviation	104	95

Construct a 95% confidence interval estimate of the difference between the mean GMAT scores for the two groups.

33. Two suppliers are being considered by a manufacturer. Independent random samples of ten parts from shipments from each supplier are selected, and the lifetime in hours for each part is determined for each sample. Use the following information to construct a 98% confidence interval estimate of the difference in the population average lifetimes. Assume that the population variances are equal and the populations are normally distributed.

	Supplier 1	Supplier 2
Sample size	10.0	10.0
Sample mean	15.0	11.0
Sample standard deviation	1.5	1.0

34. To help validate a new employee-rating form, a company administers it to independent random samples of employees in two different divisions. The following information is obtained from the scores on the forms:

	Division 1	Division 2
Sample size	15.0	15.0
Sample mean	82.0	78.0
Sample standard deviation	3.0	2.5

Use the information to construct a 95% confidence interval estimate of the difference in mean scores between the two divisions. Assume that the population variances are equal and the populations are normally distributed.

35. The one-year returns for a random sample of 51 load mutual funds and 32 no-load funds were obtained. The returns are in a file named RE-TURNS2. Construct a 95% confidence interval estimate of the difference between the population mean returns.

These data are arranged in the file in two columns. The first column contains the returns for the funds and the second column indicates to which sample (load = 1, no-load = 0) each value in column 1 belongs (in the Excel spreadsheet, the returns are in two separate columns).

(*Source*: Used with permission from the September issue of *Kiplinger's Personal Finance*. Copyright © 2002. The Kiplinger Washington Editors, Inc.)

36. The file named PRIVATE2 contains the graduation rates for 195 schools and a variable coded 1 for private schools and 0 for public schools. Construct a 99% confidence interval estimate for the difference between population average graduation rates for public and private schools.

These data are arranged in the file in two columns. The first column contains the graduation rates and the second column contains the private school variable (in the Excel spreadsheet, the graduation rates are in two separate columns for private and public schools).

(*Source*: Used by permission from the November and December 2003 issues of *Kiplinger's Personal Finance*. Copyright © 2003 The Kiplinger Washington Editors, Inc. Visit our website at www.kiplingers.com for further information).

2.9 HYPOTHESIS TESTS ABOUT THE DIFFERENCE BETWEEN TWO POPULATION MEANS

We may be interested in testing hypotheses about the difference between two population means rather than estimating that difference. The most common hypotheses tested in comparing two populations are

$$H_0: \mu_1 = \mu_2$$

$$H_a: \mu_1 \neq \mu_2$$

The null hypothesis states that the means of the two populations are equal, whereas the alternate states that the two population means differ. These hypotheses can be restated in terms of the difference between two means as

$$H_0: \mu_1 - \mu_2 = 0$$
$$H_a: \mu_1 - \mu_2 \neq 0$$

The decision rule for the test is

Reject H_0 if $t > t_{\alpha/2,df}$ or $t < -t_{\alpha/2,df}$

Do not reject H_0 if $-t_{\alpha/2,df} \leq t \leq t_{\alpha/2,df}$

The construction of the test statistic, t, depends on whether the population variances can be assumed equal. If $\sigma_1^2 = \sigma_2^2$, then

$$t = \frac{\bar{y}_1 - \bar{y}_2}{\sqrt{s_p^2\left(\dfrac{1}{n_1} + \dfrac{1}{n_2}\right)}}$$

and the critical value, $t_{\alpha/2,df}$, is chosen with df $= n_1 + n_2 - 2$ degrees of freedom. The pooled estimate of the population variance, s_p^2, is used in computing the standard deviation of the sampling distribution.

If $\sigma_1^2 \neq \sigma_2^2$, then

$$t = \frac{\bar{y}_1 - \bar{y}_2}{\sqrt{\left(\dfrac{s_1^2}{n_1} + \dfrac{s_2^2}{n_2}\right)}}$$

and the approximate critical value is chosen with

$$\Delta = \frac{(s_1^2/n_1 + s_2^2/n_2)^2}{[(s_1^2/n_1)^2/(n_1 - 1)] + [(s_2^2/n_2)^2/(n_2 - 1)]}$$

degrees of freedom.

The justification for using two different standard errors in constructing the test statistics is the same as that for constructing the confidence intervals in the previous section.

Figure 2.28 shows the three possible hypothesis structures, the test statistic to be used, and the decision rules for the case when $\sigma_1^2 = \sigma_2^2$. Figure 2.29 presents similar information for $\sigma_1^2 \neq \sigma_2^2$.

EXAMPLE 2.11   Consider again the 5-year return data for the random sample of mutual funds examined in Example 2.10 of Section 2.8 (✏).

Let μ_L represent the population mean 5-year return for load funds and μ_N represent the population mean 5-year return for no-load funds. Then the hypotheses

$$H_0: \mu_N - \mu_L = 0$$
$$H_a: \mu_N - \mu_L \neq 0$$

Hypotheses	Test Statistic	Decision Rules
$H_0: \mu_1 - \mu_2 = 0$	$t = \dfrac{\bar{y}_1 - \bar{y}_2}{\sqrt{s_p^2\left(\dfrac{1}{n_1} + \dfrac{1}{n_2}\right)}}$	Reject H_0 if $t > t_{\alpha/2,\, n_1+n_2-2}$ or if $t < -t_{\alpha/2,\, n_1+n_2-2}$
$H_a: \mu_1 - \mu_2 \neq 0$		Do not reject H_0 if $-t_{\alpha/2,\, n_1+n_2-2} \leq t \leq t_{\alpha/2,\, n_1+n_2-2}$
$H_0: \mu_1 - \mu_2 \leq 0$	$t = \dfrac{\bar{y}_1 - \bar{y}_2}{\sqrt{s_p^2\left(\dfrac{1}{n_1} + \dfrac{1}{n_2}\right)}}$	Reject H_0 if $t > t_{\alpha,\, n_1+n_2-2}$
$H_a: \mu_1 - \mu_2 > 0$		Do not reject H_0 if $t \leq t_{\alpha,\, n_1+n_2-2}$
$H_0: \mu_1 - \mu_2 \geq 0$	$t = \dfrac{\bar{y}_1 - \bar{y}_2}{\sqrt{s_p^2\left(\dfrac{1}{n_1} + \dfrac{1}{n_2}\right)}}$	Reject H_0 if $t < -t_{\alpha,\, n_1+n_2-2}$
$H_a: \mu_1 - \mu_2 < 0$		Do not reject H_0 if $t \geq -t_{\alpha,\, n_1+n_2-2}$

FIGURE 2.28 Hypotheses, Test Statistics, and Decision Rules for Testing Hypotheses About Differences Between Population Means When $\sigma_1^2 = \sigma_2^2$.

Hypotheses	Test Statistic	Decision Rules
$H_0: \mu_1 - \mu_2 = 0$	$t = \dfrac{\bar{y}_1 - \bar{y}_2}{\sqrt{\left(\dfrac{s_1^2}{n_1} + \dfrac{s_2^2}{n_2}\right)}}$	Reject H_0 if $t > t_{\alpha/2,\Delta}$ or if $t < -t_{\alpha/2,\Delta}$
$H_a: \mu_1 - \mu_2 \neq 0$		Do not reject H_0 if $-t_{\alpha/2,\Delta} \leq t \leq t_{\alpha/2,\Delta}$
$H_0: \mu_1 - \mu_2 \leq 0$	$t = \dfrac{\bar{y}_1 - \bar{y}_2}{\sqrt{\left(\dfrac{s_1^2}{n_1} + \dfrac{s_2^2}{n_2}\right)}}$	Reject H_0 if $t > t_{\alpha,\Delta}$
$H_a: \mu_1 - \mu_2 > 0$		Do not reject H_0 if $t \leq t_{\alpha,\Delta}$
$H_0: \mu_1 - \mu_2 \geq 0$	$t = \dfrac{\bar{y}_1 - \bar{y}_2}{\sqrt{\left(\dfrac{s_1^2}{n_1} + \dfrac{s_2^2}{n_2}\right)}}$	Reject H_0 if $t < -t_{\alpha,\Delta}$
$H_a: \mu_1 - \mu_2 < 0$		Do not reject H_0 if $t \geq -t_{\alpha,\Delta}$

FIGURE 2.29 Hypotheses, Test Statistics, and Decision Rules for Testing Hypotheses About Differences Between Population Means When $\sigma_1^2 \neq \sigma_2^2$.

can be tested to determine if there is a difference between the average returns for these two groups.

Figure 2.29 shows the MINITAB output for testing the hypotheses. Figure 2.30 shows the Excel output and Figure 2.31 shows the SAS output. Assuming that the population variances are unequal (or that we have no information about the variances), the panel A output is appropriate in MINITAB and Excel. In the SAS output, the two versions of the test are referred to as the Pooled Method and the Satterthwaite Method. Both results are printed whenever a two-sample comparison is requested.

FIGURE 2.30 Excel Output for Examples 2.10 and 2.11.

Panel A: Assumes Variances Are Unequal

t-Test: Two-Sample Assuming Unequal Variances

	NOLOAD5	LOAD5
Mean	5.95	5.01
Variance	34.61	23.04
Observations	32.00	51.00
Hypothesized Mean Difference	0.00	
df	56.00	
t Stat	0.76	
P(T<=t) one-tail	0.22	
t Critical one-tail	1.67	
P(T<=t) two-tail	0.45	
t Critical two-tail	2.00	

Panel B: Assumes Variances Are Equal

t-Test: Two-Sample Assuming Equal Variances

	NOLOAD5	LOAD5
Mean	5.95	5.01
Variance	34.61	23.04
Observations	32.00	51.00
Pooled Variance	27.47	
Hypothesized Mean Difference	0.00	
df	81.00	
t Stat	0.80	
P(T<=t) one-tail	0.21	
t Critical one-tail	1.66	
P(T<=t) two-tail	0.43	
t Critical two-tail	1.99	

The TTEST Procedure

Statistics

Variable	type	N	Lower CL Mean	Mean	Upper CL Mean	Lower CL Std Dev	Std Dev	Upper CL Std Dev	Std Err
FIVEYR	0	32	3.8289	5.95	8.0711	4.7166	5.8833	7.8217	1.04
FIVEYR	1	51	3.6578	5.0078	6.3579	4.0163	4.8001	5.9669	0.6721
FIVEYR	Diff(1-2)		-1.41	0.9422	3.2939	4.5435	5.2411	6.1938	1.182

T-Tests

Variable	Method	Variances	DF	t Value	Pr>|t|
FIVEYR	Pooled	Equal	81	0.80	0.4277
FIVEYR	Satterthwaite	Unequal	56.2	0.76	0.4499

Equality of Variances

Variable	Method	Num DF	Den DF	F Value	Pr>F
FIVEYR	Folded F	31	50	1.50	0.1963

FIGURE 2.31 SAS Output for Examples 2.10 and 2.11.

Based on the test results, the null hypothesis should not be rejected. The t statistic or the p value can be used to reach this decision. If the t statistic is used, the decision rule is

Reject H_0 if $t > 1.96$ or $t < -1.96$

Do not reject H_0 if $-1.96 \leq t \leq 1.96$

There are 56 degrees of freedom for this test, so the z value of 1.96 was used as the critical value. Note that Excel provides the t value for 56 degrees of freedom for a two-tailed test: $t = 2.00$. This value could be used (and is in fact preferred) rather than the z value if it is available. The test statistic value is $t = 0.76$.

If the p value is used, the decision rule is

Reject H_0 if p value < 0.05

Do not reject H_0 if p value ≥ 0.05

where the p value is 0.45. As always, both procedures lead to the same decision.

The statistical decision is do not reject the null hypothesis, so we conclude that there is no difference between the average 5-year returns for load and no-load mutual funds.

EXERCISES

37. Consider again Exercise 32 in Section 2.8. Two independent random samples of applicants to business schools who had and did not have work experience were chosen. Each sample contained 50 applicants. To determine whether there is a difference in the population average test scores, the following hypotheses should be tested:

$$H_0: \mu_1 - \mu_2 = 0$$
$$H_a: \mu_1 - \mu_2 \neq 0$$

Use a 5% level of significance. State the decision rule, the test statistic, and your decision. What implication do these test results have for admissions officers in MBA programs?

38. Use the information in Exercise 33. Suppose that the manufacturer currently uses supplier 2. A change to supplier 1 will be made only if the average lifetime of parts for supplier 1 is greater than the average for supplier 2. Using a 1% level of significance, conduct the appropriate test. State the hypotheses to be tested, the decision rule, the test statistic, and your decision. Assume that the population variances are equal and the populations are normally distributed. On the

basis of the test result, which supplier will the manufacturer choose?

39. Use the information in Exercise 34. Is there a difference in population mean rating scores for the two divisions? State the hypotheses to be tested, the decision rule, the test statistic, and your decision. Assume that the population variances are equal and the populations are normally distributed. Use a 5% level of significance.

40. The 1-year returns for a random sample of 51 load mutual funds and 32 no-load funds were obtained. Test to see if there is any difference between the population average 1-year returns for load and no-load funds. Use a 5% level of significance. State the decision rule, the test statistic, and your decision. Is there a difference in the population averages? Based on the test results, what conclusions do you draw concerning investment in load versus no-load funds? The returns are in a file named RETURNS2. See Exercise 35 for a further description of the data in the file.

41. Let μ_0 = the population average graduation rate for public schools and μ_1 = the population average graduation rate for private schools. Suppose that a claim is made that private schools have higher graduation rates, on average, than public schools. Examine the claim by testing the following hypotheses:

$$H_0: \mu_0 - \mu_1 \geq 0$$

$$H_a: \mu_0 - \mu_1 < 0$$

Use a 5% level of significance. State the decision rule, the test statistic, and your decision. Is the claim supported? The file named PRIVATE2 contains the graduation rates. See Exercise 36 for a further description of the data in the file.

ADDITIONAL EXERCISES

42. A university wants to examine starting monthly salaries for its finance and marketing graduates. Independent random samples of 12 finance graduates and 12 marketing graduates are selected from the files of last year's graduates. The following starting salaries are obtained from these people:

Finance Marketing

1850	1675
2150	1275
1700	1800
1500	2100
2200	2200
1650	2250
2100	1950
2140	1850
1790	2000
1650	1800
2300	2100
2000	2150

Use these data to determine the following:

a. Find the sample mean starting salaries for finance graduates and marketing graduates (separately).

b. Construct a histogram for starting salaries in both finance and marketing.

c. Construct a 95% confidence interval estimate of the population mean starting salary for finance majors. Do the same for marketing majors. Assume that the populations of starting salaries for both groups are normally distributed.

The data are available in the file named SALARY2. The salary data are in column one. The second column indicates to which sample each column one value belongs: 1 = finance; 0 = marketing (in the Excel spreadsheet, the salary data are in two separate columns.)

43. Consider again the finance and marketing starting salaries in Exercise 42.

a. Conduct a test to determine whether the population mean starting salaries for the two majors are equal. State the hypotheses to be tested, the decision rule, the test statistic, and your decision. Use a 5% level of significance.

b. What conclusion can be drawn from the result in part a?

44. A telemarketing firm is considering two different sales approaches for selling magazine subscriptions over the phone. Two independent random samples of 20 salespeople each are selected to use either approach 1 or 2 for the sample period and for the same number of household contacts. The sales data (number of subscriptions sold per period) are given here.

Approach 1	Approach 2
12	8
15	10
28	24
14	14
18	10
10	20
15	20
20	15
5	0
4	7
12	10
10	16
24	17
16	20
13	12
14	4
18	12
22	18
6	3
12	16

a. Is there a difference in average sales produced by the two approaches? Assume that the populations are normally distributed. State the hypotheses to be tested, the decision rule, the test statistic, and your decision. Use a 10% level of significance.

b. What does the result in part a suggest to a sales division manager?

The data are available in the file named SALES2. The sales data are in the first column; the number in the second column indicates whether the sales value belongs to approach 1 or approach 2 (in the Excel spreadsheet, the sales data are in two separate columns.)

45. The file named HARRIS2 contains 1977 annual starting salary data for 93 employees of Harris Bank of Chicago. The column of data denoted MALE indicates whether the employees were MALE(1) or FEMALE(0) (in the Excel spreadsheet, the salary data are in two separate columns.) Let μ_0 = average starting salary for females and μ_1 = average starting salary for males.

a. Set up and test hypotheses to determine whether there is evidence of wage discrimination for the Harris Bank employees. Use a 5% level of significance. Set up the hypotheses assuming that discrimination is represented by an average wage for females that is less than the average wage for males.

b. What implications do your test results have for Harris Bank?

c. Are there other factors that might need to be considered in this analysis? If so, state them and why you believe they are important.

(*Source:* These data were obtained from D. Schafer, "Measurement-Error Diagnostics and the Sex Discrimination Problem," *Journal of Business and Economic Statistics.* Copyright 1987 by the American Statistical Association. Used with permission. All rights reserved.)

46. Can expert stock analysts pick stocks that perform better than stocks chosen at random? Or better than a stock market index? There are no definitive answers to these questions, but people have lots of fun trying to find out. For example, in Fort Worth, TX, we have Rusty, a 1700-pound steer. At the start of each calendar year, Rusty is pitted against several of the state's top stock analysts. The analysts pick their portfolio of stocks and Rusty picks his. Rusty makes his picks in a special corral in Sundance Square in downtown Fort Worth with rectangles representing local companies. He lets the chips (as they say) fall where they may and his stocks are chosen accordingly. The results for 1997: Rusty's stocks gained 62.87% and the experts' stocks gained 37.09% (the S&P 500 gained 31% in 1997). For 1997–2003, Rusty came out ahead four times to three for the experts.

In another comparison, the *Wall Street Journal* formed a panel of four experts each month and compared the performance of their four stock picks to four stocks chosen by throwing darts at stock tables. A 6-month holding period was used for the comparison. Table 2.3 shows the results of the first five contests along with the return for the Dow Jones Industrial Average for each time period. The complete data set is available in the file named DARTS2. The three columns of data represent returns for the pros, darts, and the Dow Jones Industrial Average, respectively.

a. Assume that the returns for the experts' portfolios and the dartboard portfolios represent two independent random samples of results. Test to see if there is a difference in the average returns for the experts and the average of the randomly chosen stocks.

b. Perform the same test to compare the experts' average performance with the average performance of the Dow Jones Industrial Average.

c. Based on the results of the tests in parts a and b, what do you conclude about the stock-picking ability of the experts?

d. What problems might there be in comparing performance in this manner? What other comparisons might be useful in deciding whether expert stock analysts can pick stocks that perform better than stocks chosen at random or better than a stock market index?

47. The following is a relative frequency distribution that appeared in the *Wall Street Journal* (1/31/96 issue, reprinted courtesy of *The Wall Street Journal).* Use the distribution to help answer the questions that follow.

TABLE 2.3 Returns for Experts (PROS), Dartboard Portfolio (DARTS), and Dow Jones Industrial Index (DJIA) for First Five *Wall Street Journal* Darts Contests

Contest Period	PROS	DARTS	DJIA
1 January–June 1990	12.7	0.0	2.5
2 February–July 1990	26.4	1.8	11.5
3 March–August 1990	2.5	–14.3	–2.3
4 April–September 1990	–20.0	–7.2	–9.2
5 May–October 1990	–37.8	–16.3	–8.5

The Going Rate

Price distribution of homes sold through the Multiple Listing Service in Texas for the year ended October 1995.

Price of Home	% of Homes Sold
$29,999 or less	5.3
$30,000–39,999	5.1
$40,000–49,999	7.8
$50,000–59,999	10.0
$60,000–69,999	10.6
$70,000–79,999	9.9
$80,000–89,999	8.9
$90,000–99,999	6.4
$100,000–119,999	9.5
$120,000–139,999	7.4
$140,000–159,999	4.9
$160,000–179,999	3.5
$180,000–199,999	2.4
$200,000–299,999	5.2
$300,000–399,999	1.6
$400,000–499,999	0.7
$500,000 and more	0.8

a. What percentage of homes sold for less than $50,000?

b. What percentage of homes sold for $120,000 or more but less than $400,000?

c. Can you determine what percentage of homes sold for more than $130,000?

d. In what range of prices is the median price of the homes sold?

48. Our company produces metal parts that must have holes of a certain diameter punched for later use. As long as the center of each hole is within 1.5 centimeters (cm) of a particular spot on the metal part, the part will be acceptable to the buyer. The distance from the center of the hole punched to the desired center is called the error (errors can be positive or negative depending on their direction). Figure 2.32 shows a MINITAB histogram of the errors for a sample of the last 50 parts to come off the punch line. Managers have decided that as long as no more than 2 parts in a sample of 50 are in error by more than 1.5 cm, the punch machine is operating acceptably and no adjustments will be made.

   a. How many of the parts from the sample of 50 have an error of 1.5 cm or more?

   b. What percentage of parts in this sample has an error of 1.5 cm or more?

   c. Based on the histogram, what is management's decision?

49. Figure 2.33 is a MINITAB time-series plot of the errors for the 50 parts examined in Exercise 48. The errors are plotted in the order that the parts were handled by the punch machine. Based on the histogram in Exercise 48, management decided that the punch machine was operating correctly and needed no adjustment. After examining the time-series plot in addition to the histogram, do you agree or disagree with management's conclusion? Justify your answer.

50. The owner of a construction company makes bids on contracts for jobs. The company describes the probability distribution of X = "the number of jobs it is awarded per year" as shown:

x	p(x)
2	0.05
3	0.15
4	0.20
5	0.35
6	0.25

FIGURE 2.32
Histogram of Punch Errors.

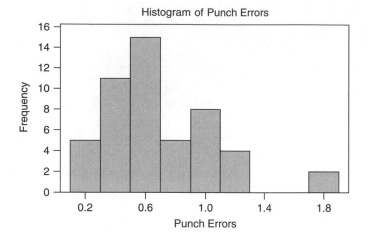

FIGURE 2.33 Time-Series Plot of Punch Errors.

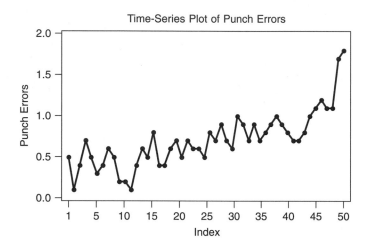

a. What is the expected or mean number of contracts that will be awarded per year?

b. Find the probability that more than three contracts will be awarded in a given year.

c. What is the most likely number of contracts that will be awarded in a given year?

51. An insurance company insures homes in Ft. Worth, TX. Roof damage due to hailstorms is always a persistent problem in this area. Suppose we simplify the range of damages in a given year to include just four possible values. We define a discrete random variable, X, to represent the dollar loss per year due to hail and assign the probabilities shown in the following distribution. For example, the probability is 0.9 that there will be no loss; it is 0.02 that there will be a $1000 loss, etc. Note that these losses are per household.

x	$p(x)$
0	0.90
1000	0.02
2000	0.04
3000	0.04

If the insurance company only wants to break even in the long run (dumb company), what should it charge annually for a policy for each household? Assume that the policy insures only against hail damage.

52. The filling machine for gallons of milk to be used by school districts in Texas can be set so that it discharges an average of μ ounces of milk per gallon. It is impossible to put exactly 1 gallon of milk in each gallon container due to natural variability in the dispensing machine. The amount of milk discharged is known to have a normal distribution with standard deviation equal to 0.4 ounce. When school officials discover that containers regularly have less than 1 gallon of milk in them, the state gets very upset. So we want to find a setting for μ so that only 1% of the containers of milk will have less than 1 gallon (128 ounces). What value should be used for μ?

53. A computer shop builds PCs from shipments of parts it receives from various suppliers. The number of defective hard drives per shipment is to be modeled as a random variable X. The random variable is assumed to have the following distribution:

x	$P(x)$
0	0.55
1	0.15
2	0.10
3	0.10
4	0.05
5	0.05

  a. What is the expected number of defective hard drives per shipment?

  b. What is the probability that a shipment will have more than two defective hard drives?

  c. What is the probability that a shipment will have four or fewer defective hard drives?

  d. What is the probability that a shipment will have more than two but less than five defective hard drives?

  e. What is the most likely number of defectives in a shipment?

54. Proctor and Gamble (P&G) has developed a new brand of toothpaste and plan to begin marketing the new toothpaste next month. From test market studies, P&G estimates that demand for the new toothpaste will average 200,000 tubes nationally with a standard deviation of 15,000 tubes during the first year. The *break-even point* for the toothpaste is the amount of sales necessary for P&G to begin making a profit. Management estimates that

P&G will need to sell 180,000 tubes to break even in the first year. Assume that demand for the new toothpaste follows a normal distribution with mean 200,000 and standard deviation 15,000. What is the probability that P&G will sell at or above the break-even point during the first year?

55. High-Tech, Inc. produces an electronic component, GS-7, that has an average life span of 1000 hours. The life span is normally distributed with a standard deviation of 25 hours. The company is considering how long a warranty to place on the component. If it wants to replace no more than 10% of the components for free, how many hours should it guarantee the components would last under the warranty?

56. A bank reports that the population of its demand deposit balances has a population mean of $7500 and a population standard deviation of $1000. An auditor refuses to certify the bank's claim and takes a random sample of 100 account balances. He will certify the bank's report only if the sample mean is no more than $125 above or below the bank's stated population mean.

  a. Assuming the bank's reported figures are correct, what is the probability that the auditor will certify the bank's report?

  b. What would be the probability that the auditor will certify the bank's report if the reported population mean were $9000 rather than $7500 (assuming the standard deviation is the same)?

57. Researchers at American Airlines and Sabre have developed a system to coordinate scheduling, yield management, and pricing decisions (*ORMS Today*, August 2000, pp. 36–44). In one of their examples they set up the following scenario: "Consider the case of a single flight leg with one fare class. For this example, we assume that the passenger demand follows a normal distribution with a mean of 125 passengers and a coefficient of variation of 0.3. The capacity of the leg is assumed to be 150 seats." The coefficient of variation is defined as the standard deviation divided by the mean. What is the probability that demand will exceed the flight capacity?

58. A city engineer recorded the number of vehicles passing through a certain intersection for each day of last year. One objective of this study was to determine the percentage of days that more than 425 vehicles used the intersection. Suppose the mean for the data was 375 vehicles per day and the standard deviation was 25 vehicles. What can you say about the percentage of days that more than 425 vehicles used the intersection? Assume the distribution of the data is a normal distribution.

59. Our company manufactures a certain electronic component for use in the new F-22 fighter being built by Lockheed. We want to test the reliability of the components, and we think testing a random sample of 50 components will be sufficient. The population standard deviation of the lives of the components is known to be 10 hours. If we use a random sample of 50 components, what is the probability that our sample mean will be within (plus or minus) 2 hours of the true population average lifetime of the components?

60. **Process Capability.** Specification limits for a particular characteristic of a product indicate the values at which the product will operate properly. Specification limits are set by engineering and are not determined by the data. The data must be examined to determine whether the product is within specifications. If the product is within specification limits, the process generating the product is often referred to as capable. For example, assume we have a product with an upper specification limit of USL = 4.25 and a lower specification limit of LSL = 3.0. As long as our product falls within these limits our buyers will be happy. We now investigate the process used to manufacture the product and find that the process mean is 4 and the process standard deviation is 0.2. Assuming the process is normally distributed, what percentage of items will fall outside the specification limits? Should we make changes in the process? If so, what would you suggest?

61. By law, a manufacturer of a food product is required to list Food and Drug Administration (FDA) estimates of the contents of the packaged product. Suppose the FDA wants to estimate the mean sugar content (by weight) in 16-ounce boxes of "Disney Chocolate Mud and Bugs" cereal. The FDA randomly selects 200 boxes of Disney Chocolate Mud and Bugs, measures the sugar content in each, and computes a 95% confidence interval estimate of the average sugar content to be (3.2, 4.5).

The manufacturer plans to use the interval to claim that 95% of all boxes of Disney Chocolate Mud and Bugs cereal have sugar content weights between 3.2 and 4.5 ounces. Is this a correct interpretation of the interval? Justify your answer.

62. Suppose a sample of $n = 100$ items is randomly chosen. The following hypotheses are to be tested:

$$H_0: \mu \leq 10$$
$$H_a: \mu > 10$$

A p value of 0.0409 for the test is obtained. If the population standard deviation is 5, what would the value of the sample mean have to be in order to achieve the stated p value?

USING THE COMPUTER

The Using the Computer section in each chapter describes how to perform the computer analyses in the chapter using Excel, MINITAB, and SAS. For further detail on Excel, MINITAB, and SAS, see Appendix C.

EXCEL

Descriptive Statistics

TOOLS: DATA ANALYSIS: HISTOGRAM

Click Tools, then Data Analysis, and then Histogram. Histogram creates a frequency distribution and a histogram. In the histogram dialog box (see Figure 2.34), fill in the input range of the variable to be graphed. In the language of worksheets, the range of the data indicates the cells in which the data are contained. For example, if you have

FIGURE 2.34 Excel Histogram Dialog Box.

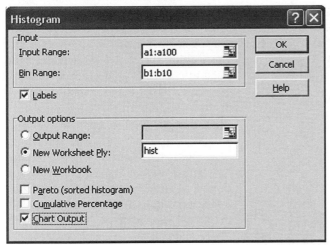

100 observations to be included in your frequency distribution/histogram and you have typed these 100 numbers into cells A1 through A100, you specify the range as A1:A100 in the Input Range box.

You can specify the bins you want Excel to use. If you enter nothing in this box, Excel picks the bin limits for your frequency distribution/histogram. To specify the bins for Excel to use, just put the numbers you want to use as bin limits in a column. Then indicate the range of this column in the Bin Range box.

If the first row of your column of data contains a label, check the Label box. The next three options determine where you want the frequency distribution to appear. If you click Output Range, you can tell Excel a specific spot in the spreadsheet to put your frequency distribution and histogram. For example, if you check the Output Range button and put C3 in this box, the frequency distribution starts in cell C3. Check the New Worksheet Ply button and Excel puts your frequency distribution/histogram on a new ply in this same workbook. You can name the worksheet ply if you want. Check New Workbook and Excel puts your frequency distribution/histogram in a completely new workbook.

The next three options determine exactly what kind of output you want from Excel. If you want only a frequency distribution, do not check any of these options. The Pareto option constructs a histogram with the bins arranged from biggest to smallest. This option is often not very useful when quantitative data are used. It is more useful when qualitative data are used and the order of importance of certain categories is of interest. A Pareto chart is often used in quality control situations. For example, if customer complaints are being monitored and the most frequent complaints are of primary interest, a Pareto chart emphasizes those complaints by listing them first. The Cumulative Percentage option constructs a histogram but superimposes a line on the histogram that represents the cumulative percentage (sometimes called an *ogive*). Chart Output requests Excel to construct a histogram in addition to a frequency distribution.

Once you have the options set as you want, click OK.

`TOOLS: DATA ANALYSIS: DESCRIPTIVE STATISTICS`

Click Tools, then Data Analysis, and then Descriptive Statistics. This procedure generates a variety of descriptive statistics (see Figure 2.6). In the Descriptive Statistics dialog box (see Figure 2.35), fill in the input range for the variable for which descriptive statistics are desired. Typically, this variable is in a column, but the option is available to indicate whether it is in a column or row. Choose the output option and click Summary statistics. Click Confidence Level for Mean to produce a 95% error bound. (The 95% error bound is the default value and can be set at any desired level). Click OK.

FIGURE 2.35 Excel Descriptive Statistics Dialog Box.

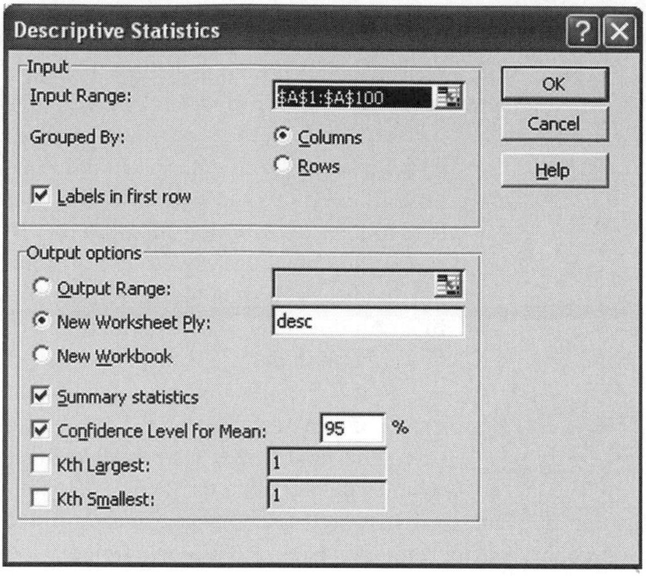

Descriptive Statistics ? X

┌─ Input ──────────────────────────────────
Input Range: $A$1:$A$100

Grouped By: ⊙ Columns
                     ○ Rows

☑ Labels in first row

┌─ Output options ─────────────────────────
○ Output Range:
⊙ New Worksheet Ply: desc
○ New Workbook

☑ Summary statistics
☑ Confidence Level for Mean: 95   %
☐ Kth Largest: 1
☐ Kth Smallest: 1

OK
Cancel
Help

Confidence Interval Estimate of μ

The descriptive statistics option on the Data Analysis toolpack produces a value for the sample mean and a 95% (or any other desired) error bound, which can be used to construct a confidence interval. See Descriptive Statistics and Figure 2.35 for information on this procedure. The confidence interval can be constructed as sample mean ± error bound.

Alternatively, a formula can be constructed to compute the upper and lower confidence interval bounds. For example, suppose we have data in cells A1 through A100 and want to construct a 95% confidence interval. The formulas for the upper (UCL) and lower (LCL) confidence interval limits are

$$\text{UCL} = \text{average(A1:A100)}+\text{tinv(0.05,99)*stdev(A1:A100)/sqrt(100)}$$

$$\text{LCL} = \text{average(A1:A100)}-\text{tinv(0.05,99)*stdev(A1:A100)/sqrt(100)}$$

Note that tinv(0.05,99)*stdev(A1:A100)/sqrt(100) is the 95% error bound produced by the descriptive statistics procedure. The function tinv(0.05,99) returns the t value that puts a combined probability of 0.05 in the upper and lower tails of the t distribution with 99 degrees of freedom.

Hypotheses Tests About μ

Suppose we want to test the hypotheses $H_0: \mu = k$ versus $H_a: \mu \neq k$, where k represents the hypothesized value in the problem. Let's say the data are in cells A1 through A100. The t statistic for this test can be constructed using the formula

$$= \text{(average(A1:A100)} - k)/\text{(stdev(A1:A100)/sqrt(100))}$$

This value can be compared to the appropriate t critical value chosen with 99 degrees of freedom (or a z value since we have lots of degrees of freedom here) using a two-tailed decision rule.

To compute the p value associated with this test statistic requires use of the tdist function. Suppose we build the formula for the t statistic in cell C10. Then the p value is computed as

$$= \text{tdist(abs(C10),99,2) or}$$

$$= \text{tdist(test statistic, degrees of freedom, number of tails)}$$

Note that the number of tails is 2 because this is a two-tailed test. You have to make sure that the numeric value provided to the tdist function is positive so the absolute value function is applied to C10.

For a one-tailed test, the test statistic is computed in exactly the same way, but a one-tailed decision rule is used. The computation of the p value is a little trickier.

Upper-tailed test:	If the test statistic in C10 is positive, use = tdist(abs(C10),99,1)
	If the test statistic in C10 is negative, use = 1 − tdist(abs(C10),99,1)
Lower-tailed test:	If the test statistic in C10 is positive, use = 1 − tdist(abs(C10),99,1)
	If the test statistic in C10 is negative, use = tdist(abs(C10),99,1)

Hypotheses Tests About $\mu_1 - \mu_2$
Confidence Interval Estimate of $\mu_1 - \mu_2$

To construct confidence intervals or test hypotheses about the difference between two population means, use either t-Test: Two-Sample Assuming Equal Variances or t-Test: Two-Sample Assuming Unequal Variances, as shown in Figure 2.36. The t-Test: Paired Two Sample for Means option is used when samples are matched rather than independent. The procedure for matched-sample tests was not discussed in this book. The z-Test: Two Sample for Means option is used when the population standard deviations are known (a fairly unlikely situation). The dialog boxes for the test procedures assuming either equal or unequal variances are identical. The dialog box from the unequal variance option is shown in Figure 2.37. Fill in the Variable Range 1 and 2 boxes with the ranges of the two independent samples. The Hypothesized Mean Difference is typically set to zero. This is the case discussed in this text, although tests for nonzero differences can be performed if these make sense. The Alpha level should be adjusted to the level desired for your test. This is because Excel prints out the critical value to compare with the standardized test statistic and needs to know the level of significance to do this. Specify the output option and click OK.

FIGURE 2.36 Excel Data Analysis Dialog Box with Two-Sample Test Options.

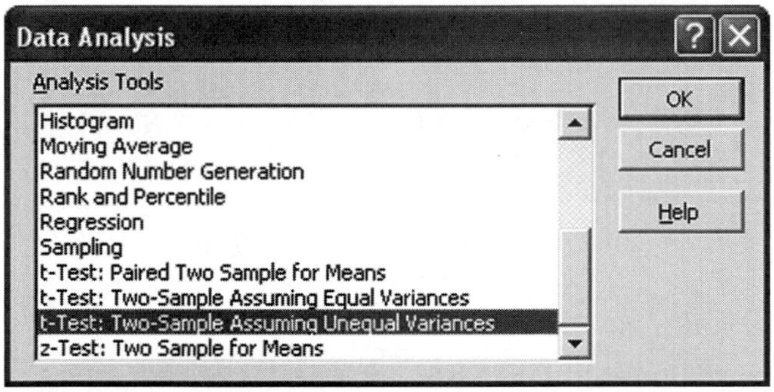

FIGURE 2.37 Excel Two-Sample Test Dialog Box with Unequal Variances Option.

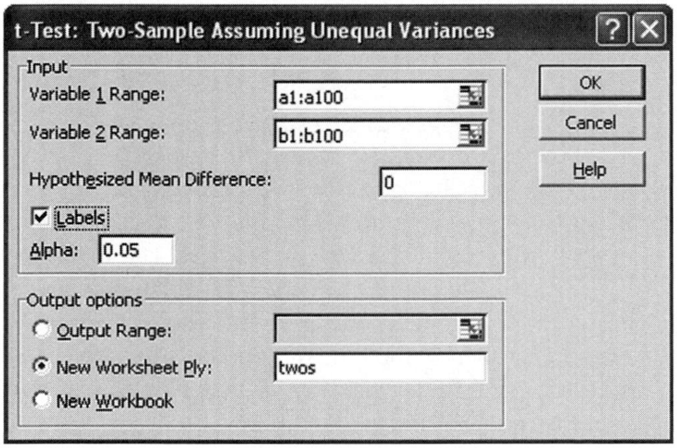

MINITAB
Descriptive Statistics

GRAPH: HISTOGRAM

Creates a histogram. Click Graph, then Histogram. Choose the type of histogram desired (See Figure 2.38). In the histogram dialog box (see Figure 2.39), fill in the variables to be graphed and click OK. There are a variety of options available, which are described in the MINITAB Help facility.

USING THE COMPUTER

FIGURE 2.38 Choose Type of Histogram Desired in **MINITAB**.

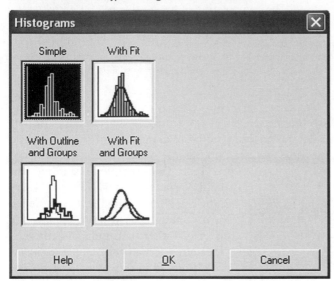

FIGURE 2.39 MINITAB Histogram Dialog Box.

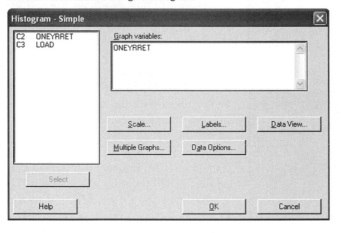

STAT: BASIC STATISTICS: DISPLAY DESCRIP-
TIVE STATISTICS

Click on Stat, then on Basic Statistics, then on Display Descriptive Statistics. This procedure generates a variety of descriptive statistics (see Figure 2.5). In the Display Descriptive Statistics dialog box (see Figure 2.40), fill in the variables for which descriptive statistics are desired and click OK.

Confidence Interval Estimate of μ

STAT: BASIC STATISTICS: 1-SAMPLE t

Click Stat, then Basic Statistics, then 1-Sample t. This procedure can be used to construct a confidence interval estimate of the population mean, μ. Fill in the name of the variable in the 1-Sample t dialog box and then click on Options (See Figures 2.41 and 2.42). Indicate the confidence level desired (the default value is 95%) and choose the not equal alternative for a two-sided confidence interval. One-sided intervals can also be constructed but were not discussed in this text. Choosing the alternative less than will provide an upper bound on the population mean, and the alternative greater than will provide a lower bound.

Hypotheses Tests About μ

STAT: BASIC STATISTICS: 1-SAMPLE t

Click Stat, then Basic Statistics, and then 1-Sample t. This procedure can be used to produce the output necessary to test the hypotheses $H_0: \mu = k$ versus $H_a: \mu \neq k$, where k represents the hypothesized value. The Options window is used to specify the type of hypothesis structure desired: "less than" for a lower-tailed test, "greater than" for an upper-tailed test, and "not equal" for a two-tailed (see Figures 2.43 and 2.44).

Confidence Interval Estimate of $\mu_1 - \mu_2$

STAT: BASIC STATISTICS: 2-SAMPLE T

Click Stat, then Basic Statistics, and then 2-Sample t. This procedure is used to construct confidence interval estimates of the difference between two population means, $\mu_1 - \mu_2$. Sample data from two different populations with means μ_1 and μ_2 are assumed to be available. These data can be in either one or two columns. The 2-Sample t dialog box is illustrated in Figure 2.45. If the data from both

FIGURE 2.40 MINITAB Descriptive Statistics Dialog Box.

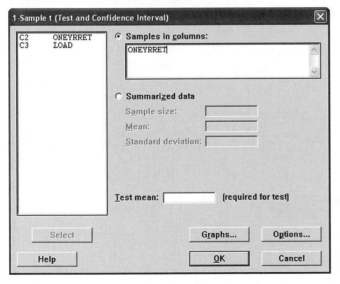

FIGURE 2.41 MINITAB 1-Sample t Dialog Box for Confidence Interval.

samples are contained in one column, check the "Samples in one column" button. If you use this option, a second column must contain numbers indicating from which sample each data point came (number the items from the first sample 0 and the items from the second sample 1, for example). If the data from each sample are in a different column, check the "Samples in different columns" button. Then indicate the two columns containing your data. The choice between these two options is based purely on how your data are arranged.

The assumption that the population variances (or standard deviations) are equal can be included by checking the "Assume equal variances" box in the dialog box. The Options window allows the user to choose a level of confidence and the hypothesized difference (Test difference). The default value for the hypothesized difference is zero because this is what is appropriate for a confidence interval for the difference between the two means. A two-sided confidence interval would be created if the Alternative is "not equal." For a one-sided interval use either "less than" for an upper bound on the difference or "greater than" for a lower bound. See Figure 2.46.

Hypotheses Tests About $\mu_1 - \mu_2$

STAT: BASIC STATISTICS: 2-SAMPLE t

Click Stat, then Basic Statistics, and then 2-Sample t. This procedure is used to produce the output necessary to test $H_0: \mu_1 - \mu_2 = 0$ versus $H_a: \mu_1 - \mu_2 \neq 0$ (see Figure 2.45). The same dialog box is used for constructing a confidence interval for the difference between two means and testing hypotheses about the difference between two means. See the previous section for a description of how data should be arranged and how the equal variance option is handled.

The Options window allows the user to choose the type of hypothesis structure desired ("less than" for a lower-tailed test, "greater than" for an upper-tailed test, and "not equal" for a two-tailed test) and the hypothesized difference (Test difference).

USING THE COMPUTER

FIGURE 2.42 MINITAB 1-Sample t Options Window Set for a Two-Sided Confidence Interval.

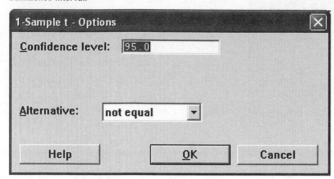

FIGURE 2.43 MINITAB 1-Sample t Dialog Box for Hypothesis Test.

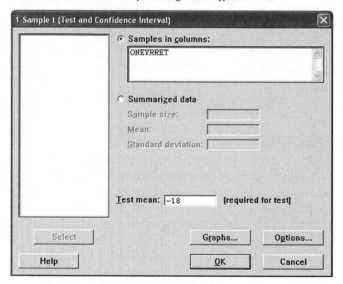

FIGURE 2.44 MINITAB 1-Sample t Options Window Set for a Lower-Tail Test.

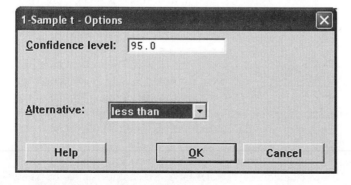

The default value for the hypothesized difference is zero because this is what is typically used (and is the only case discussed in this text). See Figure 2.46.

SAS

Hypothesis Tests About $\mu_1 - \mu_2$

```
PROC TTEST;
CLASS V1;
VAR V2;
```

Produces output to conduct a test of the difference between two population means. V1 is a variable used to separate the observations of the variable V2 into two groups. The two group means can be called μ_1 and μ_2. The t statistic produced can be used to conduct either a one- or two-tailed test. The p values in the Prob |T| column are designed specifically for two-tailed tests. Two versions of the test statistic are produced: the variances-equal version and the variances-unequal version. The appropriate statistic depends on the assumptions deemed correct for the population variances.

FIGURE 2.45 MINITAB 2-Sample *t* Dialog Box.

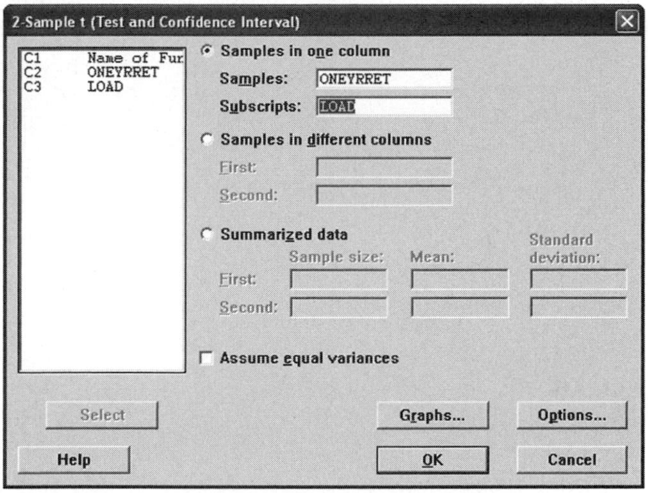

FIGURE 2.46 MINITAB 2-Sample *t* Options Window.

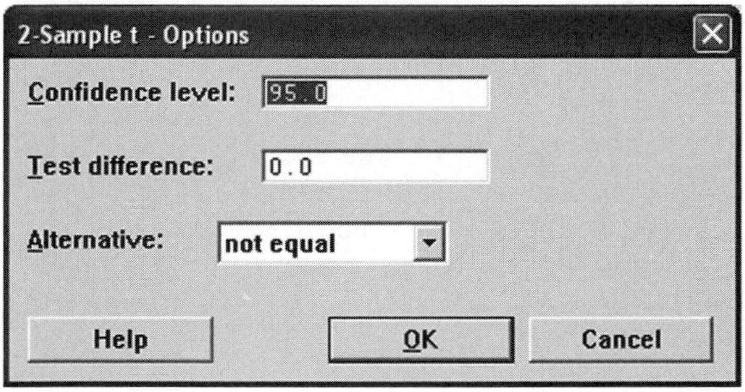

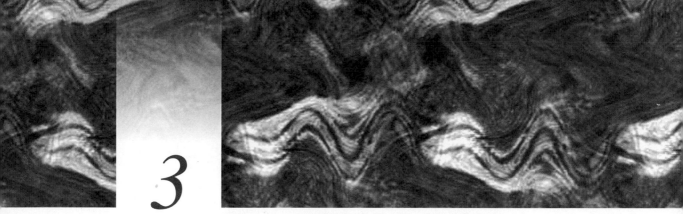

3

Simple Regression Analysis

3.1 USING SIMPLE REGRESSION TO DESCRIBE A LINEAR RELATIONSHIP

Regression analysis is a statistical technique used to describe relationships among variables. The simplest case to examine is one in which a variable *y*, referred to as the *dependent* variable, may be related to one variable *x*, called an *independent* or *explanatory* variable. If the relationship between *y* and *x* is believed linear, then the equation expressing this relationship is the equation for a line:

$$y = b_0 + b_1 x$$

If a graph of all the (x, y) pairs is constructed, then b_0 represents the *y intercept,* the point where the line crosses the vertical (y) axis, and b_1 represents the *slope* of the line.

Consider the data shown in Table 3.1. A graph of the (x, y) pairs would appear as shown in Figure 3.1. Regression analysis is not needed to obtain the equation expressing the relationship between these two variables. In equation form,

$$y = 1 + 2x$$

This is an *exact* or *deterministic* linear relationship. Exact linear relationships are sometimes encountered in business environments. For example, from accounting:

$$\text{assets} = \text{liabilities} + \text{owner equity}$$

$$\text{total costs} = \text{fixed costs} + \text{variable costs}$$

TABLE 3.1 Example Data: Exact Relationship

x	1	2	3	4	5	6
y	3	5	7	9	11	13

FIGURE 3.1 Graph of an Exact Linear Relationship (data in Table 3.1).

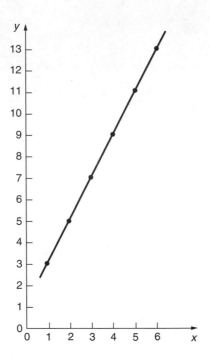

Other exact relationships may be encountered in various science courses (for example, physics or chemistry). In the social sciences (for example, psychology or sociology) and in business and economics, exact linear relationships are the exception rather than the rule. Data encountered in a business environment are more likely to appear as in Table 3.2. These data graph as shown in Figure 3.2.

It appears that x and y may be linearly related, but it is not an exact relationship. Still, it may be of use to describe the relationship in equation form. This can be done by drawing what appears to be the "best-fitting" line through the points and guessing what the values of b_0 and b_1 are for this line. This has been done in Figure 3.2. For the line drawn, a good guess might be the following equation:

$$\hat{y} = -1 + 2.5x$$

The notation $\hat{y}$ is used here to indicate that we do not expect this to be an exact relationship. Given a value of x, the equation will produce an estimate of y, which is denoted $\hat{y}$.

TABLE 3.2 Example Data: Not an Exact Relationship

x	1	2	3	4	5	6
y	3	2	8	8	11	13

FIGURE 3.2 Graph of a Relationship That Is Not Exact (data in Table 3.2).

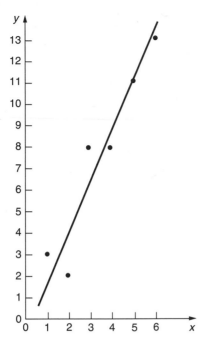

The drawbacks to this method of fitting the line should be clear. For example, if the (x, y) pairs graphed in Figure 3.2 were given to two people, each would probably guess different values for the intercept and slope of the best-fitting line. Furthermore, there is no way to assess who would be more correct. To make line fitting more precise, a definition of what it means for a line to be the "best" is needed. The criterion for a best-fitting line that we will use might be called the "minimum sum of squared errors" criterion or, as it is more commonly known, the least-squares criterion.

In Figure 3.3, the (x, y) pairs from Table 3.2 have been plotted and an arbitrary line drawn through the points. Consider the pair of values denoted (x^*, y^*). The actual y value is indicated as y^*; the value predicted to be associated with x^* if the line shown were used is indicated as $\hat{y}^*$. The difference between the actual y value and the predicted y value at the point x^* is called a *residual* and represents the "error" involved. This error is denoted $y^* - \hat{y}^*$. If the line is to fit the data points as accurately as possible, these errors should be minimized. This should be done not just for the single point (x^*, y^*), but for all the points on the graph. There are several possible ways to approach this task.

1. Use the line that minimizes the sum of the errors, $\sum_{i=1}^{n}(y_i - \hat{y}_i)$. The problem with this approach is that for any line that passes through the point ($\bar{x}$, $\bar{y}$), $\sum_{i=1}^{n}(y_i - \hat{y}_i) = 0$, so there are an infinite number of lines satisfying this criterion, some of which obviously do not fit the data well. For example, in Figure 3.4, lines A and B have both been constructed so that $\sum_{i=1}^{n}(y_i - \hat{y}_i) = 0$. But line A obviously fits the data better than line B; that is, it keeps the distances $y^* - \hat{y}^*$ small.

FIGURE 3.3
Motivation for the
Least-Squares
Regression Line.

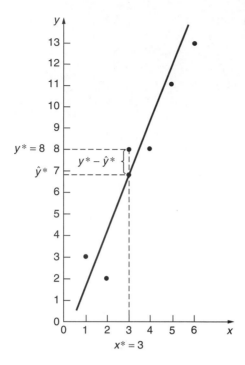

FIGURE 3.4 Lines *A*
and *B* Both Satisfy the
Criterion

$$\sum_{i=1}^{n}(y_i - \hat{y}_i) = 0.$$

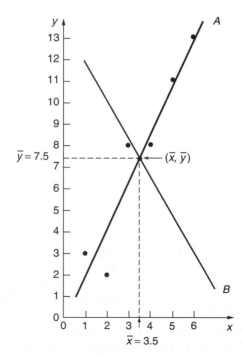

As mentioned previously, any line that passes through the point represented by the means of x and y ($\bar{x}, \bar{y}$) has errors that sum to zero. The line passes through the points in such a way that positive and negative errors cancel each other out. Because a criterion is desired that makes the errors small regardless of whether they are positive or negative, some method of removing negative signs is required. One such method uses absolute values of the errors; another squares the errors. Each method provides another possible criterion.

2. Use the line that minimizes the sum of the absolute values of errors, $\sum_{i=1}^{n} |y_i - \hat{y}_i|$. This is called the *minimum sum of absolute errors criterion*. The resulting line is called the *least absolute value* (LAV) regression line. Although use of this criterion is gaining popularity in many situations, it is not the one that we use in this text. Finding the line that satisfies the minimum sum of absolute errors criterion requires solving a fairly complex problem by a technique called *linear programming*. This is a difficult problem by hand, and the LAV procedure is not readily available in most statistical software packages. Furthermore, there may be no unique LAV regression line.

3. Use the line that minimizes the sum of the squared errors, $\sum_{i=1}^{n} (y_i - \hat{y}_i)^2$. This is called the *least-squares criterion*, and the resulting line is called the *least-squares regression line*. Applying the least-squares (LS) criterion results in a unique least-squares regression line. Its advantages over the LAV line include computational simplicity and the wide availability of statistical packages that contain easily implemented least-squares regression routines.

Now that a criterion has been established, the next question is: Can convenient computational formulas for the values of b_0 and b_1 that minimize

$$\sum_{i=1}^{n} (y_i - \hat{y}_i)^2 \tag{3.1}$$

be developed? The answer is "yes" and the resulting equations are

$$b_1 = \frac{\sum_{i=1}^{n} (x_i - \bar{x})(y_i - \bar{y})}{\sum_{i=1}^{n} (x_i - \bar{x})^2} \tag{3.2}$$

$$b_0 = \bar{y} - b_1 \bar{x} \tag{3.3}$$

A computationally simpler form of Equation (3.2) is

$$b_1 = \frac{\sum_{i=1}^{n} x_i y_i - \frac{1}{n} \sum_{i=1}^{n} x_i \sum_{i=1}^{n} y_i}{\sum_{i=1}^{n} x_i^2 - \frac{1}{n} \left(\sum_{i=1}^{n} x_i \right)^2} \tag{3.4}$$

EXAMPLE 3.1   As an example of the use of these formulas, consider again the data in Table 3.2. The intermediate computations necessary for finding b_0 and b_1 are shown in Figure 3.5. The slope, b_1, can now be computed using the formula in Equation (3.4):

$$b_1 = \frac{196 - \dfrac{1}{6}(21)(45)}{91 - \dfrac{1}{6}(21)^2} = \frac{38.5}{17.5} = 2.2$$

The intercept, b_0, is computed as in Equation (3.3):

$$b_0 = 7.5 - 2.2(3.5) = -0.2$$

because

$$\bar{x} = \frac{21}{6} = 3.5 \text{ and } \bar{y} = \frac{45}{6} = 7.5$$

The least-squares regression line for these data is

$$\hat{y} = -0.2 + 2.2x$$

There is no longer any guesswork associated with computing the best-fitting line once a criterion has been stated that defines "best." Using the criterion of minimum sum of squared errors, the regression line we computed provides the best description of the relationship between the variables x and y. Any other values used for b_0 and b_1 result in larger sum of squared errors. For example, Figure 3.6(a) shows the computation of the sum of squared errors for the original "guessed" line $\hat{y} = -1 + 2.5x$, and Figure 3.6(b) shows the same computation for the least-squares line $\hat{y} = -0.2 + 2.2x$. As expected, the sum of squared errors for the line in (a) is larger than that for the least-squares line.

FIGURE 3.5
Computations for
Finding b_0 and b_1.

i	x_i	y_i	$x_i y_i$	x_i^2
1	1	3	3	1
2	2	2	4	4
3	3	8	24	9
4	4	8	32	16
5	5	11	55	25
6	6	13	78	36
Sums	21	45	196	91

FIGURE 3.6

(a) Computation of Sum of Squared Errors for: $\hat{y} = -1 + 2.5x$

x	y	$\hat{y}$	$y - \hat{y}$	$(y - \hat{y})^2$
1	3	1.5	1.5	2.25
2	2	4.0	−2.0	4.00
3	8	6.5	1.5	2.25
4	8	9.0	−1.0	1.00
5	11	11.5	−0.5	0.25
6	13	14.0	−1.0	1.00

$$\sum_{i=1}^{6}(y_i - \hat{y}_i)^2 = 10.75$$

(b) Computation of Sum of Squared Errors for: $\hat{y} = -0.2 + 2.2x$

x	y	$\hat{y}$	$y - \hat{y}$	$(y - \hat{y})^2$
1	3	2.0	1.0	1.00
2	2	4.2	−2.2	4.84
3	8	6.4	1.6	2.56
4	8	8.6	−0.6	0.36
5	11	10.8	0.2	0.04
6	13	13.0	0.0	0.00

$$\sum_{i=1}^{6}(y_i - \hat{y}_i)^2 = 8.80$$

EXERCISES

Exercises 1 and 2 should be done by hand.

1. **Flexible Budgeting.** A budget is an expression of management's expectations and goals concerning future revenues and costs. To increase their effectiveness, many budgets are flexible, including allowances for the effect of variation in uncontrolled variables. For example, the costs and revenues of many production plants are greatly affected by the number of units produced by the plant during the budget period, and this may be beyond a plant manager's control. Standard cost-accounting procedures can be used to adjust the direct-cost parts of the budget for the level of production, but it is often more difficult to handle overhead. In many cases, statistical methods are used to estimate the relationship between overhead (y) and the level of production (x) using historical data. As a simple example, consider the historical data for a certain plant:

Production
(in 10,000)
units: 5   6   7   8   9   10   11

Overhead
costs
(in $1000): 12   11.5   14   15   15.4   15.3   17.5

   a. Construct a scatterplot of y versus x.

   b. Find the least-squares regression line relating overhead costs to production.

   c. Graph the regression line on the scatterplot.

2. **Central Company.** The Central Company manufactures a certain specialty item once a month in a batch production run. The number of items produced in each run varies from month to month as demand fluctuates. The company is interested in the relationship between the size of the production run (x) and the number of hours of labor (y) required for the run. The company has collected the following data for the ten most recent runs:

Number
of items: 40   30   70   90 50   60 70   40   80   70

Labor
(hours):   83   60 138 180 97 118 140 75 159 144

a. Construct a scatterplot of y versus x.

b. Find the least-squares line relating hours of labor to number of items produced.

c. Graph the regression line on the scatterplot.

3.2 EXAMPLES OF REGRESSION AS A DESCRIPTIVE TECHNIQUE

EXAMPLE 3.2   **Pricing Communications Nodes**   In recent years the growth of data communications networks has been amazing. The convenience and capabilities afforded by such networks are appealing to businesses with locations scattered throughout the United States and the world. Using networks allows centralization of an information system with access through personal computers at remote locations.

The cost of adding a new communications node at a location not currently included in the network was of concern for a major Fort Worth manufacturing company. To try to predict the price of new communications nodes, data were obtained on a sample of existing nodes. The installation cost and the number of ports available for access in each existing node were readily available information. These data are shown in Table 3.3 and a scatterplot of cost (y = COST) versus number of ports (x = NUMPORTS) is shown in Figure 3.7. (See the file COMNODE3 on the CD.)

Using a statistical package, the equation relating the price of the new communications node to the number of access ports to be included at the node was computed to be

$$COST = 16,594 + 650 NUMPORTS$$

This equation could be used to help predict the cost of installing new communications nodes based on the number of access ports to be included.

TABLE 3.3 Cost and Number of Ports for Communications Nodes Example*

COST	NUMPORTS
52,388	68
51,761	52
50,221	44
36,095	32
27,500	16
57,088	56
54,475	56
33,969	28
31,309	24
23,444	24
24,269	12
53,479	52
33,543	20
33,056	24

*These data have been modified as requested by the company to provide confidentiality.

FIGURE 3.7
Scatterplot of Cost
Versus Number of
Ports for the
Communications
Nodes Example.

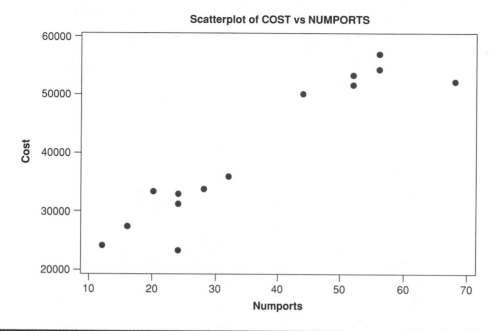

Scatterplot of COST vs NUMPORTS

EXAMPLE 3.3   **Estimating Residential Real Estate Values**   The Tarrant County Appraisal District must appraise properties for the entire county. The appraisal district uses data such as square footage of the individual houses as well as location, depreciation, and physical condition of an entire neighborhood to derive individual appraisal values on each house. This avoids labor-intensive inspections each year.

Regression can be used to establish the weight assigned to various factors used in assessing values. For example, Table 3.4 shows the value and size in square feet for a sample of 100 Tarrant County homes (these data are from 1990). A scatterplot of value (y = VALUE) versus size (x = SIZE) is shown in Figure 3.8. (See the file REALEST3 on the CD.)

Using a statistical package, the regression equation relating value to size can be determined as

$$\text{VALUE} = -50{,}035 + 72.8\text{SIZE}$$

If size were the only factor thought to be of importance in determining value, this equation could be used by the appraisal district. But obviously, other factors need to be considered. Developing an equation that includes more than one important factor (explanatory variable) is discussed in Chapter 4.

EXAMPLE 3.4   **Forecasting Housing Starts**   Forecasts of various economic measures are important to the U.S. government and to various industries throughout the United States. The construction industry is concerned with the number of housing starts in a given year. Accurate forecasts can help with plans for expansion or cutbacks within the industry.

TABLE 3.4 VALUE and SIZE for Residential Real Estate Value Example

VALUE	SIZE	VALUE	SIZE	VALUE	SIZE
23,974	1442	12,001	783	21,536	1404
24,087	1426	37,650	1874	24,147	1676
16,781	1632	27,930	1242	17,867	1131
29,061	910	16,066	772	21,583	1397
37,982	972	20,411	908	15,482	888
29,433	912	23,672	1155	24,857	1448
33,624	1400	24,215	1004	17,716	1022
27,032	1087	22,020	958	224,182	2251
28,653	1139	52,863	1828	182,012	1126
33,075	1386	41,822	1146	201,597	2617
17,474	756	45,104	1368	49,683	966
33,852	1044	28,154	1392	60,647	1469
29,046	1032	20,943	1058	49,024	1322
20,715	720	17,851	1375	52,092	1509
19,461	734	16,616	648	55,645	1724
21,377	720	38,752	1313	51,919	1559
52,881	1635	44,377	1780	55,174	2133
43,889	1381	43,566	1148	48,760	1233
45,134	1372	38,950	1363	45,906	1323
47,655	1349	44,633	1262	52,013	1733
53,088	1599	12,372	840	56,612	1357
38,923	1171	12,148	840	69,197	1234
57,870	1966	19,852	839	84,416	1434
30,489	1504	20,012	852	60,962	1384
29,207	1296	20,314	852	47,359	995
44,919	1356	22,814	974	56,302	1372
48,090	1553	24,696	1135	88,285	1774
40,521	1142	23,443	1170	91,862	1903
43,403	1268	35,904	960	242,690	3581
38,112	1008	21,799	1052	296,251	4343
27,710	1120	28,212	1296	107,132	1861
27,621	960	27,553	1282	77,797	1542
22,258	920	15,826	916		
29,064	1259	18,660	864		

Table 3.5 shows data on the number of housing starts for the years 1963 to 2002. Also shown are data on home mortgage rates for new home purchases (U.S. average) for the same years. (See the file HSTARTS3 on the CD.) These data were obtained from the Department of Commerce and the Federal Housing Finance Board. A scatterplot of housing starts ($y =$ STARTS) versus mortgage rates ($x =$ RATES) is shown in Figure 3.9. Note that the relationship appears to be considerably "weaker" than in the other scatterplots presented in this section. Intuitively, we might expect the relationship between housing starts and mortgage rates to be a strong one. But from the data, this does not appear to be the case. Perhaps there are other variables that might be more strongly related to housing starts that could be used to provide accurate forecasts for future years. From viewing the scatterplot, mortgage rates alone do not appear to be particularly helpful.

Cross-sectional data are gathered on a number of different individual units at approximately the same point in time. Examples 3.2 and 3.3 use cross-sectional

FIGURE 3.8
Scatterplot of VALUE
Versus SIZE for
Residential Real
Estate Value Example.

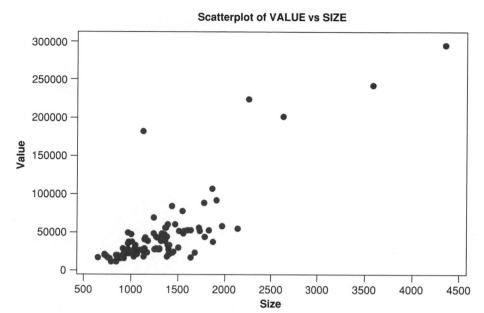

Scatterplot of VALUE vs SIZE

TABLE 3.5 Annual Housing Starts and Mortgage Rates for 1963–2002

Year	STARTS	RATES	Year	STARTS	RATES
1963	1,603.2	5.89	1983	1,703.0	12.57
1964	1,528.8	5.83	1984	1,749.5	12.38
1965	1,472.8	5.81	1985	1,741.8	11.55
1966	1,164.9	6.25	1986	1,805.4	10.17
1967	1,291.6	6.46	1987	1,620.5	9.31
1968	1,507.6	6.97	1988	1,488.1	9.19
1969	1,466.8	7.81	1989	1,376.1	10.13
1970	1,433.6	8.45	1990	1,192.7	10.05
1971	2,052.2	7.74	1991	1,013.9	9.32
1972	2,356.6	7.60	1992	1,199.7	8.24
1973	2,045.3	7.96	1993	1,287.6	7.20
1974	1,337.7	8.92	1994	1,457.0	7.49
1975	1,160.4	9.00	1995	1,354.1	7.87
1976	1,537.5	9.00	1996	1,476.8	7.80
1977	1,987.1	9.02	1997	1,474.0	7.71
1978	2,020.3	9.56	1998	1,616.9	7.07
1979	1,745.1	10.78	1999	1,640.9	7.04
1980	1,292.2	12.66	2000	1,568.7	7.52
1981	1,084.2	14.70	2001	1,602.7	7.00
1982	1,062.2	15.14	2002	1,704.9	6.43

data. The data examined in Example 3.4 are called *time-series data* because they are gathered over a time sequence (years in this case).

Most of the techniques discussed in this and subsequent chapters can be applied to either time-series or cross-sectional data. There are certain special techniques available when working with time-series data that may be helpful in developing forecasts. These techniques will be discussed throughout subsequent chapters where

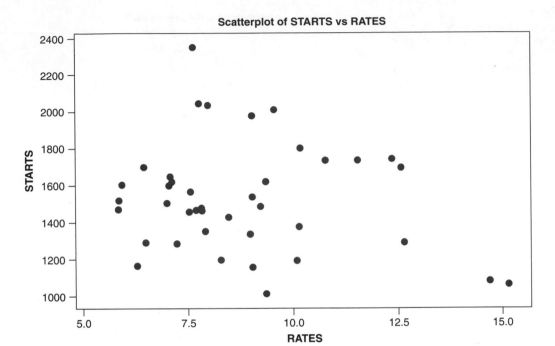

FIGURE 3.9 Scatterplot of Housing Starts Versus Mortgage Rates for Housing Starts Example.

appropriate. When special techniques apply to only one type of data, this will be mentioned.

EXERCISES

3. For each of the data sets discussed in this section, use a computer to access the data, construct a scatterplot of y versus x, and produce the regression output relating y to x.

a. Estimating Residential Real Estate Values: The file name is REALEST3.

b. Pricing Communications Nodes: The file name is COMNODE3.

c. Forecasting Housing Starts: The file name is HSTARTS3.

3.3 INFERENCES FROM A SIMPLE REGRESSION ANALYSIS

3.3.1 ASSUMPTIONS CONCERNING THE POPULATION REGRESSION LINE

Thus far, regression analysis has been viewed as a way to describe the relationship between two variables. The regression equation obtained can be viewed in this manner simply as a descriptive statistic. However, the power of the technique of least-squares regression is not in its use as a descriptive measure for one particular sample, but in its ability to draw inferences or generalizations about the relationship for the entire population of values for the variables x and y.

To draw inferences from a sample regression equation, we must make some assumptions about how x and y are related in the population. These initial assumptions describe an "ideal" situation. Later, each of these assumptions is relaxed and we demonstrate modifications to the basic least-squares approach that provide a model that is still suitable for statistical inference.

Assume that the relationship between the variables x and y is represented by a population regression line. The equation of this line is written as

$$\mu_{y|x} = \beta_0 + \beta_1 x \tag{3.5}$$

where $\mu_{y|x}$ is the *conditional mean* of y given a value of x, β_0 is the y intercept for the population regression line, and β_1 is the slope of the population regression line. Examples of possible relationships are shown in Figure 3.10.

The use of $\mu_{y|x}$ requires some additional explanation. Suppose the y variable represents the cost of installing a new communications node as discussed in Example 3.3, and x represents the number of access ports (NUMPORTS) to be included. It is possible that these two variables are related. Now consider all possible communications nodes with 30 access ports. If the costs were known, the average value for all communications nodes with 30 access ports could be calculated. This is the conditional mean of y given $x = 30$:

$$\mu_{y|x=30}$$

FIGURE 3.10
Examples of Possible Population Regression Lines.

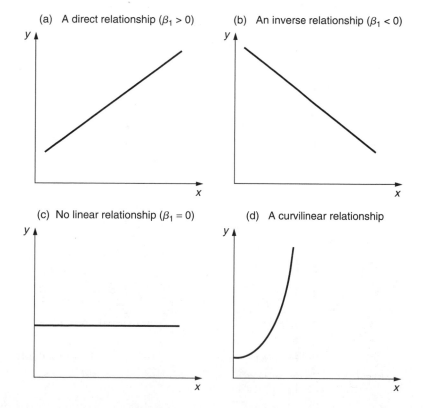

(a) A direct relationship ($\beta_1 > 0$)

(b) An inverse relationship ($\beta_1 < 0$)

(c) No linear relationship ($\beta_1 = 0$)

(d) A curvilinear relationship

Suppose this computation could be done for a number of x values and the resulting conditional means plotted as in Figure 3.11. The population regression line is the line passing through the conditional means. The relationship between y and x is linear if all conditional means lie on a straight line (or nearly so).

For a given number of ports (say, 30) costs vary; that is, not every communications node has a cost equal to the mean of y given $x = 30$. The actual cost is distributed around the point $\mu_{y|x=30}$ or around the regression line. Thus, in a sample of communications nodes with 30 ports, the costs are expected to differ from points on the population regression line (see Figure 3.12).

Because of this variation of the y values around the regression line, it is convenient to rewrite the equation representing an individual response as

$$y_i = \beta_0 + \beta_1 x_i + e_i \tag{3.6}$$

where β_0 and β_1 have the same interpretation as they did in Equation (3.5). The term e_i represents the difference between the true cost for communications node i and the conditional mean of all costs for nodes with that number of ports:

$$e_i = \text{COST}_i - \mu_{y|x} = y_i - (\beta_0 + \beta_1 x_i)$$

The e_i are called *disturbances*. These disturbances keep the relationship from being an exact one. If the e_i were all equal to zero, then there would be an exact linear relationship between COST and NUMPORTS. The effects of all factors other than the x variable that influence COST are included in the disturbances. To allow statistical inference from a sample to the population, some assumptions about the population regression line are necessary.

FIGURE 3.11 The Population Regression Line Passes Through the Conditional Means.

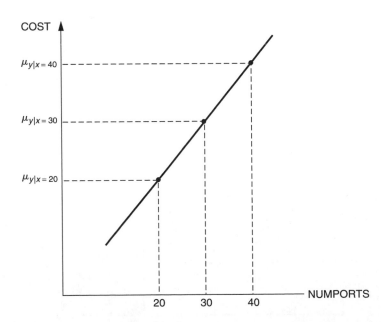

FIGURE 3.12
Distribution of Costs
Around the Regression
Line.

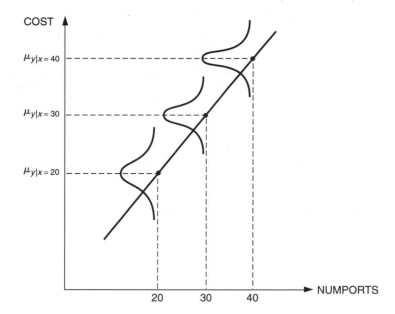

1. The expected value of the disturbances is zero: $E(e_i) = 0$. This implies that the regression line passes through the conditional means of the x variable. For our purposes, we interpret this assumption as: The population regression equation is linear in the explanatory variable.[1]

2. The variance of each e_i is equal to σ_e^2. Referring to Figure 3.12, this assumption means that each of the distributions along the regression line has the same variance regardless of the value of x.

3. The e_i are normally distributed.

4. The e_i are independent. This is an assumption that is most important when data are gathered over time. When the data are cross-sectional (that is, gathered at the same point in time for different individual units), this is typically not an assumption of concern.

   These assumptions allow inferences to be made about the population regression line from a sample regression line. The first inferences considered will be those made about β_0 and β_1, the intercept and slope, respectively, of the population regression line.

   The previous assumptions define an ideal case for linear regression. In Chapters 3 and 4, we examine regression procedures designed for this ideal case. In Chapter 6, we examine how violations of each of the assumptions might be detected and how corrections for these violations can be made.

---

[1] Our assumption here is that the population regression equation is linear in the x variable. In Chapter 5, we relax this assumption and find that we can fit curves by allowing equations that are not linear in the x variables. Throughout this text, however, we always assume that the equations are linear in the parameters. This means that equations such as $y = \beta_0\beta_1^2x + e$, for example, are not considered. These types of equations are beyond the scope of this text.

3.3.2 INFERENCES ABOUT β_0 AND β_1

The point estimates of β_0 and β_1 were previously justified by saying that b_0 and b_1 minimize the sum of squared errors for the sample. With the assumptions made concerning the random disturbances of the model, additional justification for the use of b_0 and b_1 can be made by stating certain properties these estimators possess. To fully discuss these properties, some characteristics of the sampling distributions of b_0 and b_1 must first be established.

Recall that a statistic is any value calculated from a sample. Thus, b_0 and b_1 are statistics. Because statistics are random variables, they have probability distributions called *sampling distributions*. Some characteristics of the sampling distributions of b_0 and b_1 are given here.

Sampling Distribution of b_0

1. $E(b_0) = \beta_0$ (3.7)

2. $Var(b_0) = \sigma_e^2 \left(\dfrac{1}{n} + \dfrac{\bar{x}^2}{\displaystyle\sum_{i=1}^{n}(x_i - \bar{x})^2} \right) = \sigma_e^2 \left(\dfrac{1}{n} + \dfrac{\bar{x}^2}{(n-1)s_x^2} \right)$ (3.8)

where $s_x^2 = \displaystyle\sum_{i=1}^{n}(x_i - \bar{x})^2/(n-1)$

3. The sampling distribution of b_0 is normally distributed.

Sampling Distribution of b_1

1. $E(b_1) = \beta_1$ (3.9)

2. $Var(b_1) = \dfrac{\sigma_e^2}{\displaystyle\sum_{i=1}^{n}(x_i - \bar{x})^2} = \dfrac{\sigma_e^2}{(n-1)s_x^2}$ (3.10)

3. The sampling distribution of b_1 is normally distributed.

The sampling distributions of both b_0 and b_1 are centered at the true parameter values of β_0 and β_1, respectively. Because the means of the sampling distributions are equal to the parameter values to be estimated, b_0 and b_1 are called unbiased estimators of β_0 and β_1 [see Figure 3.13(a)]. The variances of the sampling distributions are given in Equations (3.8) and (3.10). The standard deviations of the sampling distributions are obtained by taking the square roots of the variances:

$$\sigma_{b_0} = \sigma_e \sqrt{\frac{1}{n} + \frac{\bar{x}^2}{(n-1)s_x^2}} \tag{3.11}$$

$$\sigma_{b_1} = \sigma_e \sqrt{\frac{1}{(n-1)s_x^2}} \tag{3.12}$$

If the disturbances are normally distributed, the sampling distributions of b_0 and b_1 are also normally distributed regardless of the sample size.

FIGURE 3.13
Properties of b_0 and b_1
as Estimators of β_0
and β_1 (illustrated for
b_1).

(a) *Unbiased Estimators:* The mean of the sampling distribution is equal to the population parameter being estimated.

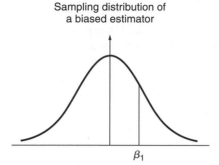

Sampling distribution of
a biased estimator

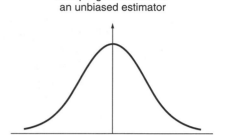

Sampling distribution of
an unbiased estimator

β_1 β_1

(b) *Consistent Estimators:* As *n* increases, the probability that the estimator will be close to the true parameter increases.

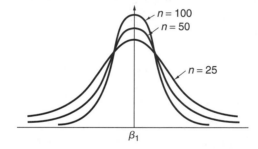

$n = 100$
$n = 50$
$n = 25$

β_1

(c) *Minimum Variance Estimators:* The variance of b_1 is smaller than the variance of any other linear unbiased estimator of β_1, say b_1^*.

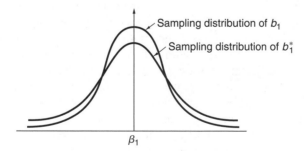

Sampling distribution of b_1
Sampling distribution of b_1^*

β_1

The estimators, b_0 and b_1, possess certain other properties that make them desirable as estimators of β_0 and β_1. Although these properties do not have a direct bearing on the work in this text, they are stated here for completeness. Each is a consistent estimator of its population counterpart. Using b_1 as an example, this means that as sample size increases, the probability increases that b_1 is "close" to β_1. Another way to view this property is by considering the standard deviation of the sampling distribution of b_1. As the sample size increases, the standard deviation of the sampling distribution decreases (as it did for the sample mean when viewed as an estimator of the population mean in Chapter 2). When this happens, the probability under the curve representing the sampling distribution becomes more concentrated near the center, β_1, and less concentrated in the extreme tails [see Figure 3.13(b)].

A final property of b_0 and b_1 can be illustrated by considering all other possible estimators of β_0 and β_1 that are unbiased. The standard deviations of the sampling distributions of b_0 and b_1 are smaller than those of any of the other unbiased estimators.

This minimum variance property can be restated by saying that b_0 and b_1 have smaller sampling errors than any other unbiased estimator. Of course, this says nothing about estimators that are biased. It does specify that the least-squares estimators are best within a certain class of estimators [see Figure 3.13(c)].

Using the properties of the sampling distributions of b_0 and b_1, inferences about the population parameters β_0 and β_1 can be made. This is analogous to what was done in Chapter 2 when properties of the sampling distribution of the sample mean, $\bar{y}$, were used to make inferences about the population mean, μ.

First, however, an estimate of one other unknown parameter in the regression model is needed: an estimate of σ_e^2, the variance around the regression line. This estimate of σ_e^2 is given by

$$s_e^2 = \frac{\displaystyle\sum_{i=1}^{n}(y_i - \hat{y}_i)^2}{n-2} = \frac{SSE}{n-2} = MSE$$

where $\hat{y}_i = b_0 + b_1 x_i$. The term s_e^2 represents an estimate of the variance around the regression line. Recall that in Figure 3.3, $y_i - \hat{y}_i$ represented the distance from the ith sample y value to the point on the regression line associated with the ith sample x value. The sum of the squares of these distances (SSE) is divided by the degrees of freedom ($n-2$) and is used as an estimator of the variance around the regression line. MSE stands for mean square error. A *mean square* is any sum of squares divided by its degrees of freedom. In any regression problem, the number of degrees of freedom associated with SSE is n (the sample size) minus the number of regression coefficients to be estimated. In a simple regression, there are two regression coefficients, β_0 and β_1, so there are $n-2$ degrees of freedom.

The square root of s_e^2, denoted s_e, is an estimate of the standard deviation around the regression line. It is often referred to as the *standard error of the regression*.

Now it is possible to discuss inferences about the population regression coefficients. Point estimates of β_0 and β_1 are simply the least-squares estimates b_0 and b_1. Point estimates are single numbers. Often it is more desirable to state a range of values in which the parameter is thought to lie rather than a single number. These ranges are called *confidence interval* estimates. They are less precise than point estimates because they cover a range of values rather than a single number. There is a trade-off, however, between the precision of an estimate and confidence in it.

Think back to Chapter 2 and the construction of estimates for the population mean, μ. A point estimate of μ was given by the sample mean, say, $\bar{y} = 40$. This is a very precise estimate, but the population mean will not be equal to 40. This value was obtained from one of a large number of possible samples, each of which has a different sample mean. The chances of one of these sample means being exactly equal to the population mean is so small that a 0% level of confidence is assigned to the possibility that μ is exactly 40. When constructing a confidence interval estimate of μ, however, the situation changes. For example, using the fact that the sampling distribution of $\bar{y}$ is normal (or approximately normal), 95% error bounds could be determined using a value from the standard normal table times the standard deviation of the sampling distribution. Putting these numbers into interval form, a 95% confidence interval for the population mean is

$$\left(\bar{y} - 1.96\frac{\sigma}{\sqrt{n}}, \bar{y} + 1.96\frac{\sigma}{\sqrt{n}} \right)$$

Although the interval estimate is less precise, confidence that the true population mean falls between the interval limits is considerably increased. In constructing interval estimates, a high level of confidence in the estimates is desired (90%, 95%, or 99% are commonly used), but the interval also should be precise enough to be practically useful. Telling the boss, "I'm 90% confident that our average monthly sales will be between $10,000 and $15,000" is probably better than "I'm 100% confident that our average monthly sales will be between $0 and $100,000." These same considerations in constructing interval estimates of the population mean also apply to interval estimates of population regression coefficients.

To construct a confidence interval estimate for β_1, the slope of the regression line, an estimate of the standard deviation of the sampling distribution is needed. This estimate is obtained by substituting s_e for σ_e in Equation (3.12):

$$s_{b_1} = s_e\sqrt{\frac{1}{(n-1)s_x^2}}$$

When sample sizes are small (say, $n \leq 30$), the t distribution is used to construct the interval estimate. A $(1 - \alpha)$ 100% confidence interval for β_1 is given by

$$\left(b_1 - t_{\alpha/2,n-2}\, s_{b_1},\, b_1 + t_{\alpha/2,n-2}\, s_{b_1} \right)$$

The value $t_{\alpha/2,n-2}$ is a number chosen from the t table to ensure the appropriate level of confidence. For example, for a 90% confidence interval estimate, $\alpha = 0.10$, so that $(1 - \alpha)$ 100% = 90%, and a t value with $\alpha/2 = 0.05$ probability in each tail of the t distribution with $n - 2$ degrees of freedom is used.

For β_0, the $(1 - \alpha)$ 100% confidence interval is

$$\left(b_0 - t_{\alpha/2,n-2}\, s_{b_0},\, b_0 + t_{\alpha/2,n-2}\, s_{b_0} \right)$$

where

$$s_{b_0} = s_e\sqrt{\frac{1}{n} + \frac{\bar{x}^2}{(n-1)s_x^2}}$$

The estimated standard deviations of the sampling distribution of b_0 and b_1 are sometimes referred to as *standard errors of the coefficients*. Thus, s_{b_0} is the standard error of b_0, and s_{b_1} is the standard error of b_1.

Hypothesis tests about β_0 and β_1 also can be performed. The most common hypothesis test in simple regression is

$$H_0: \beta_1 = 0$$
$$H_a: \beta_1 \neq 0$$

where H_0 represents the null hypothesis and H_a is the alternative hypothesis. The null hypothesis states that the slope of the population regression line is zero. This means that there is no linear relationship between y and x and that knowledge of x does not help explain the variation in y. The alternative hypothesis states that the slope of the population regression line is not equal to zero; that is, x and y are linearly related. Knowledge of the value of x does provide information concerning the associated value of y.

To test this hypothesis, a t statistic is used:

$$t = \frac{b_1}{s_{b_1}}$$

If the null hypothesis is true, then the t statistic has a t distribution with $n - 2$ degrees of freedom, and it should be small in absolute value. If the null hypothesis is false, then the t statistic should be large in absolute value.

To decide whether or not to reject the null hypothesis, a level of significance, α, must first be chosen. The level of significance is the probability of a Type I error; that is, α is equal to the probability of rejecting the null hypothesis if the null hypothesis is really true. Typical values for α are 0.01, 0.05, and 0.10. The decision rule for the test can be stated as:

Reject H_0 if $t > t_{\alpha/2, n-2}$ or $t < -t_{\alpha/2, n-2}$
Do not reject H_0 if $-t_{\alpha/2, n-2} \leq t \leq t_{\alpha/2, n-2}$

The value $t_{\alpha/2, n-2}$ is called a *critical value* and is chosen from the t table to ensure that the test is performed with the stated level of significance. A t value with probability $\alpha/2$ in each tail of the t distribution with $n - 2$ degrees of freedom is used.

Although the test of the null hypothesis $H_0: \beta_1 = 0$ is the most common and important test in simple regression analysis, tests of whether β_1 is equal to any value are possible. The general hypotheses can be stated as

$$H_0: \beta_1 = \beta_1^*$$
$$H_a: \beta_1 \neq \beta_1^*$$

where β_1^* is any number chosen as the hypothesized value. The decision rule is

Reject H_0 if $t > t_{\alpha/2, n-2}$ or $t < -t_{\alpha/2, n-2}$
Do not reject H_0 if $-t_{\alpha/2, n-2} \leq t \leq t_{\alpha/2, n-2}$

and the test statistic is

$$t = \frac{b_1 - \beta_1^*}{s_{b_1}}$$

When the null hypothesis is true, t should be small in absolute value because b_1 (the estimate of β_1) should be close to β_1^*, making the numerator, $b_1 - \beta_1^*$, close to zero. When the null hypothesis is false, b_1 should be different in value from the hypothesized value, β_1^*, and the difference $b_1 - \beta_1^*$ should be large in absolute value, resulting in a large absolute value for the t statistic.

Tests for hypotheses about β_0 proceed in a similar fashion. To test

$$H_0: \beta_0 = \beta_0^*$$
$$H_a: \beta_0 \neq \beta_0^*$$

the test statistic is

$$t = \frac{b_0 - \beta_0^*}{s_{b_0}}$$

where β_0^* is any hypothesized value. The decision rule for the test is

Reject H_0 if $t > t_{\alpha/2, n-2}$ or $t < -t_{\alpha/2, n-2}$
Do not reject H_0 if $-t_{\alpha/2, n-2} \leq t \leq t_{\alpha/2, n-2}$

Note that tests about β_0 do not provide information about the existence of a relationship between x and y. Testing whether the slope coefficient is equal to zero tells you if there is a relationship between x and y; testing whether the intercept is equal to zero does not.

An alternative method of reporting hypothesis-testing results also is available in many software packages. Consider again the hypotheses

$$H_0: \beta_1 = 0$$
$$H_a: \beta_1 \neq 0$$

The test statistic used is

$$t = \frac{b_1}{s_{b_1}}$$

Some computerized regression software routines perform this computation and then report the p value associated with the computed test statistic. The p value is the probability of obtaining a value of t at least as extreme as the actual computed value if the null hypothesis is true. Suppose a simple regression analysis is performed on a sample of 25 observations and the computed test statistic value is 2.50. The p value for the two-tailed test of $H_0: \beta_1 = 0$ is p-value $= P(t > 2.5 \text{ or } t < -2.5) = 0.02$

from the t table because there is a probability of 0.01 above 2.5 and 0.01 below –2.5 in the t distribution with $n - 2 = 23$ degrees of freedom.

The p value can be viewed as the minimum level of significance, α, that can be chosen for the test that results in rejection of the null hypothesis. Thus, a decision rule using p values can be stated as:

Reject H_0 if p value $< \alpha$

Do not reject H_0 if p value $\geq \alpha$

For a given level of significance, the same decision results regardless of which test procedure is used.

Using the previous example, we would reject H_0 if $\alpha = 0.05$, but would not reject H_0 if $\alpha = 0.01$. Reporting p values associated with hypothesis tests provides additional information beyond simply reporting whether the null hypothesis was rejected or not at a single level of significance. With a p value, readers can make their own decisions about the strength of the relationship by comparing the p value to any desired significance level.

Figure 3.14(a) shows the structure of the initial portion of the MINITAB output for a regression analysis. The estimated regression coefficients b_0 and b_1 are given along with the standard errors of the coefficients, s_{b_0} and s_{b_1}, and the t ratios for testing either $H_0: \beta_0 = 0$ or $H_0: \beta_1 = 0$. The last column reports the p values associated with the two-tailed test of the hypotheses $H_0: \beta_0 = 0$ and $H_0: \beta_1 = 0$.

Figure 3.14(b) shows the structure of the equivalent Excel output. Again, the coefficients, standard errors, t ratios, and p values are reported. Excel also prints out

(a) MINITAB

Predictor	Coef	SE Coef	T	P
Constant	b_0	s_{b_0}	b_0/s_{b_0}	p-value
x1 variable name	b_1	s_{b_1}	b_1/s_{b_1}	p-value

(b) Excel

	Coefficients	Standard Error	t stat	p value	Lower 95%	Upper 95%
Intercept	b_0	s_{b_0}	b_0/s_{b_0}	p-value	$b_0 - t_{\alpha/2, n-2} s_{b_0}$	$b_0 + t_{\alpha/2, n-2} s_{b_0}$
x1 variable name	b_1	s_{b_1}	b_1/s_{b_1}	p-value	$b_1 - t_{\alpha/2, n-2} s_{b_1}$	$b_1 + t_{\alpha/2, n-2} s_{b_1}$

(c) SAS

Parameter Estimates

Variable	DF	Parameter Estimate	Standard Error	t Value	$Pr > |t|$
Intercept	1	b_0	s_{b_0}	b_0/s_{b_0}	p-value
Variable name	1	b_1	s_{b_1}	b_1/s_{b_1}	p-value

FIGURE 3.14 Illustration of MINITAB, Excel and SAS Regression Outputs.

FIGURE 3.15
Hypothesis Structures and Their Associated Decision Rules for Hypotheses About β_0 and β_1.

Hypotheses		Decision Rules
$H_0: \beta_1 = \beta_1^*$	or $H_0: \beta_0 = \beta_0^*$	Reject H_0 if $t > t_{\alpha/2, n-2}$ or $t < -t_{\alpha/2, n-2}$
$H_a: \beta_1 \neq \beta_1^*$	$H_a: \beta_0 \neq \beta_0^*$	Do not reject H_0 if $-t_{\alpha/2, n-2} \leq t \leq t_{\alpha/2, n-2}$
$H_0: \beta_1 \geq \beta_1^*$	or $H_0: \beta_0 \geq \beta_0^*$	Reject H_0 if $t < -t_{\alpha, n-2}$
$H_a: \beta_1 < \beta_1^*$	$H_a: \beta_0 < \beta_0^*$	Do not reject H_0 if $t \geq -t_{\alpha, n-2}$
$H_0: \beta_1 \leq \beta_1^*$	or $H_0: \beta_0 \leq \beta_0^*$	Reject H_0 if $t > t_{\alpha, n-2}$
$H_a: \beta_1 > \beta_1^*$	$H_a: \beta_0 > \beta_0^*$	Do not reject H_0 if $t \leq t_{\alpha, n-2}$

the 95% confidence interval estimates for both β_0 and β_1. In addition, Excel allows the user to request confidence intervals for levels of confidence other than 95%.

Figure 3.14(c) shows the structure of the equivalent SAS output. The DF column indicates that 1 degree of freedom is lost for each parameter estimated. The remaining columns show the coefficients, standard errors, t ratios, and p values.

The p values for Excel, MINITAB, and SAS represent the appropriate values only for a two-tailed test. Note that the t ratios can be used for performing either one- or two-tailed tests as long as the hypothesized value is zero. The one-tailed tests simply require an adjustment in the decision rules. These are shown in Figure 3.15 for the more general hypothesis structures. Example 3.6 illustrates the use of the regression output.

EXAMPLE 3.5   Consider the data in Table 3.2 and the computations required to obtain the least-squares estimates in Figure 3.5. To compute s_{b_0} and s_{b_1}, the following quantities are needed:

$$n = 6$$

$$\bar{x}^2 = \left(\frac{\sum_{i=1}^{n} x_i}{n} \right)^2 = \left(\frac{21}{6} \right)^2 = 12.25$$

and

$$(n-1)s_x^2 = \sum_{i=1}^{n} x_i^2 - \frac{1}{n}\left(\sum_{i=1}^{n} x_i \right)^2 = 91 - \frac{1}{6}(21)^2 = 17.5$$

which can be determined using the sums obtained in Figure 3.5. In addition, the standard error of the regression must be computed:

$$s_e = \sqrt{\frac{\sum_{i=1}^{n}(y_i - \hat{y}_i)^2}{n-2}} = \sqrt{\frac{8.8}{4}} = 1.48$$

The error sum of squares, $\sum_{i=1}^{n}(y_i - \hat{y}_i)^2$, was computed in Figure 3.6(b). Using this information,

$$s_{b_0} = s_e\sqrt{\frac{1}{n} + \frac{\overline{x}^2}{(n-1)s_x^2}} = 1.48\sqrt{\frac{1}{6} + \frac{12.25}{17.5}} = 1.38$$

and

$$s_{b_1} = s_e\sqrt{\frac{1}{(n-1)s_x^2}} = 1.48\sqrt{\frac{1}{17.5}} = 0$$

The 95% confidence interval estimates for β_0 and β_1 now can be constructed.

For β_0: $[-0.2 - 2.776(1.38), -0.2 + 2.776(1.38)]$ or $(-4.03, 3.63)$

For β_1: $[2.2 - 2.776(0.35), 2.2 + 2.776(0.35)]$ or $(1.23, 3.17)$

EXAMPLE 3.6

Pricing Communications Nodes (continued) Table 3.3 shows the cost of installing a sample of communications nodes for a large manufacturing firm whose headquarters is based in Fort Worth, Texas, with branches throughout the United States. The number of access ports at each of the sampled nodes is also shown. The administrator of the network wants to develop an equation that is helpful in pricing the installation of new communications nodes on the network.

Figure 3.7 shows the scatterplot of cost versus number of ports. The MINITAB, Excel, and SAS regression results are shown in Figures 3.16, 3.17, and 3.18, respectively. Use the outputs to answer the following questions:

1. What is the sample regression equation relating NUMPORTS to COST?
   Answer: COST = 16,594 + 650NUMPORTS

FIGURE 3.16
MINITAB Regression for Communications Node Example.

```
The regression equation is
COST = 16594 + 650 NUMPORTS

Predictor     Coef      SE Coef          T        P
Constant     16594         2687       6.18    0.000
NUMPORTS    650.17        66.91       9.72    0.000

S = 4306.91     R-Sq = 88.7%     R-Sq(adj) = 87.8%

Analysis of Variance

Source             DF           SS           MS        F        P
Regression          1   1751268376   1751268376    94.41    0.000
Residual Error     12    222594146     18549512

Total              13   1973862522

Unusual Observations

Obs      NUMPORTS         COST        Fit    SE Fit    Residual    St Resid
  1          68.0        52388      60805      2414       -8417      -2.36R
 10          24.0        23444      32198      1414       -8754      -2.15R

R denotes an observation with a large standardized residual.
```

SUMMARY OUTPUT

Regression Statistics

Multiple R	0.942
R Square	0.887
Adjusted R Square	0.878
Standard Error	4306.914
Observations	14.000

ANOVA

	df	SS	MS	F	Significance F
Regression	1	1751268375.709	1751268375.709	94.410	0.000
Residual	12	222594145.791	18549512.149		
Total	13	1973862521.500			

	Coefficients	Standard Error	t Stat	P-value	Lower 95%	Upper 95%
Intercept	16593.647	2687.050	6.175	0.000	10739.068	22448.226
NUMPORTS	650.169	66.914	9.717	0.000	504.376	795.962

FIGURE 3.17 Excel Regression for Communications Node Example.

FIGURE 3.18 SAS Regression for Communications Node Example.

The REG Procedure

Model: MODEL1

Dependent Variable: COST

Analysis of Variance

Source	DF	Sum of Squares	Mean Square	F Value	Pr > F
Model	1	1751268376	1751268376	94.41	<.0001
Error	12	222594146	18549512		
Corrected Total	13	1973862522			

Root MSE	4306.91446	R-Square	0.8872	
Dependent Mean	40186	Adj R-Sq	0.8778	
Coeff Var	10.71758			

Parameter Estimates

Variable	DF	Parameter Estimate	Standard Error	t Value	Pr > |t|
Intercept	1	16594	2687.05000	6.18	<.0001
NUMPORTS	1	650.16917	66.91389	9.72	<.0001

2. Is there sufficient evidence to conclude that a linear relationship exists between COST and NUMPORTS?

Answer: To answer this question, the hypotheses

$$H_0: \beta_1 = 0$$

$$H_a: \beta_1 \neq 0$$

should be tested. Assuming that we want to use a 5% level of significance, here are two options to conduct the test.

(a) Using the standardized test statistic:

Decision Rule: Reject H_0 if $t > 2.179$ or $t < -2.179$
Do not reject H_0 if $-2.179 \leq t \leq 2.179$

Test Statistic: $t = 9.72$ from either Figure 3.16 (MINITAB), 3.17 (Excel), or 3.18 (SAS)

Decision: Reject H_0.

Conclusion: There is sufficient evidence to conclude that a linear relationship between COST and NUMPORTS does exist.

(b) Using the p value:

Decision Rule: Reject H_0 if p value < 0.05
Do not reject H_0 if p value ≥ 0.05

Test Statistic: p value $= 0.000$

Decision: Reject H_0.

3. Find a 95% confidence interval estimate of β_1.

Answer: $650.17 \pm (2.179)(66.91)$ or 650.17 ± 145.80

Our point estimate of the change in cost, on average, for each additional port is $650.17. A 95% error bound for this estimate is $145.80. We can now construct a range of values that can be used as an estimate of the average change in cost for each additional port. The range is 650.17 ± 145.80 or $504.37 to $795.97. We can be 95% confident that the true average change falls within this range. The upper and lower confidence limits provide an idea of the accuracy of our estimate of the cost of each additional port. (Note that Excel prints out this interval).

4. Test whether there is a direct (positive) relationship between COST and NUMPORTS.

Answer: To answer this question, the hypotheses

$$H_0: \beta_1 \leq 0$$
$$H_a: \beta_1 > 0$$

should be tested. The null hypothesis states that the slope of the population regression line is either zero (no relationship) or negative (an inverse relationship). The alternate hypothesis states that the slope of the population regression line is positive (a direct relationship). Choosing a 5% level of significance:

Decision Rule: Reject H_0 if $t > 1.782$

Do not reject H_0 if $t \leq 1.782$

Test statistic: $t = 9.72$ from either Figure 3.16 (MINITAB), 3.17 (Excel), or 3.18 (SAS)

Decision: Reject H_0

Conclusion: There is sufficient evidence to conclude that a direct (positive) linear relationship between COST and NUMPORTS does exist.

If the null hypothesis had not been rejected, the conclusion would be that either there is an inverse relationship or there is no relationship. The relationship is not direct, but failure to reject does not imply that no linear relationship exists as in the case of the two-tailed test. Care must be taken in interpreting the results of one-tailed tests.

Also note that the p values printed out by MINITAB, Excel, and SAS are not intended to be used for this test. The printed p values are computed for a two-tailed test.

5. A claim is made that each new access port adds at least $1000 to the installation cost of a communications node. To examine this claim, we test the hypotheses

$$H_0: \beta_1 \geq 1000$$

$$H_a: \beta_1 < 1000$$

using a 5% level of significance.

Answer:

Decision rule: Reject H_0 if $t < -1.782$

Do not reject H_0 if $t \geq -1.782$

Test statistic: $t = \dfrac{b_1 - \beta_1^*}{s_{b_1}} = \dfrac{650.17 - 1000}{66.91} = -5.23$

Decision: Reject H_0.

Conclusion: The slope of the line is not 1000 or more. In practical terms, the evidence does not support the claim that each port adds at least $1000 to the cost of installing a new communications node.

6. The intercept in this problem might be interpreted as a measure of the fixed cost of installing a new communications node. The slope would be a measure of the variable cost, that is, the cost of installation per port. If we view the intercept as a measure of the fixed cost, then it would be of interest to find a confidence interval estimate of this amount. Find a 95% confidence interval estimate of β_0.

Answer: $16{,}594 \pm 2.179(2687)$ or $16{,}594 \pm 5854.97$

7. The following test has no *practical* significance in this problem and is performed merely to illustrate the test for the intercept provided on most computer output. However, it might be of interest to test whether the intercept is equal to some value other than zero. If the intercept represents the fixed cost of installing a communications node, there might be a situation when we would want to know if the fixed cost would be more than $15,000, for example. In that case, we would construct the standardized test statistic in a similar fashion to Problem 5 in this example. To test whether the intercept is zero, as in this problem, the test statistic is provided on the computer output.

Test the hypotheses

$$H_0: \beta_0 = 0$$
$$H_a: \beta_0 \neq 0$$

using a 5% level of significance.

Answer:

(a) Using the standardized test statistic:

Decision rule: Reject H_0 if $t > 2.179$ or $t < -2.179$
                         Do not reject H_0 if $-2.179 \leq t \leq 2.179$

Test statistic: $t = 6.18$
Decision: Reject H_0
Conclusion: The population intercept is not equal to zero.

(b) Using the p value:

Decision Rule: Reject H_0 if p value < 0.05
                        Do not reject H_0 if p value ≥ 0.05

Test statistic: p value $= 0.000$
Decision: Reject H_0.
Conclusion: The population intercept is not equal to zero.

Note that rejection of the null hypothesis $H_0: \beta_0 = 0$ does not indicate that x and y are related. It merely makes a statement about the intercept of the population regression line.

EXERCISES

Exercises 4 and 5 should be done by hand.

4. Flexible Budgeting (continued) Refer to Exercise 1.

  a. Test the hypotheses $H_0: \beta_1 = 0$ versus $H_a: \beta_1 \neq 0$ at the 5% level of significance. State the decision rule, the test statistic value, and your decision.

  b. From the result in part a, are production and overhead costs linearly related?

  c. Test the hypotheses $H_0: \beta_1 = 1$ versus $H_a: \beta_1 \neq 1$ at the 5% level of significance. State the decision rule, the test statistic value, and your decision.

  d. From the result in part c, what can be concluded?

5. Central Company (continued) Refer to Exercise 2.

  a. Test the hypotheses $H_0: \beta_1 = 0$ versus $H_a: \beta_1 \neq 0$ at the 5% level of significance. State the decision rule, the test statistic value, and your decision.

  b. From the result in part a, are hours of labor and number of items linearly related?

  c. Test the hypotheses $H_a: \beta_0 = 0$ versus $H_a: \beta_0 \neq 0$ at the 5% level of significance. State the decision rule, the test statistic value, and your decision.

  d. From the result in part c, what can be concluded?

6. Dividends A random sample of 42 firms was chosen from the S&P 500 firms listed in the Spring 2003 Special Issue of *Business Week* (The Business Week Fifty Best Performers). The dividend yield (DIVYIELD) and the 2002 earnings per share (EPS) were recorded for these 42 firms. These data are in a file named DIV3.

Using dividend yield as the dependent variable and EPS as the independent variable, a regression was run. Use the results to answer the questions. The scatterplot of DIVYIELD and EPS is shown in Figure 3.19. The regression results are shown in Figure 3.20.

a. What is the sample regression equation relating dividends to EPS?

b. Is there a linear relationship between dividend yield and EPS? Use $\alpha = 0.05$. State the hypotheses to be tested, the decision rule, the test statistic, and your decision.

c. What conclusion can be drawn from the test result?

d. Construct a 95% confidence interval estimate of β_1.

e. Construct a 95% confidence interval estimate of β_0.

FIGURE 3.19
Scatterplot for
Dividends Exercise.

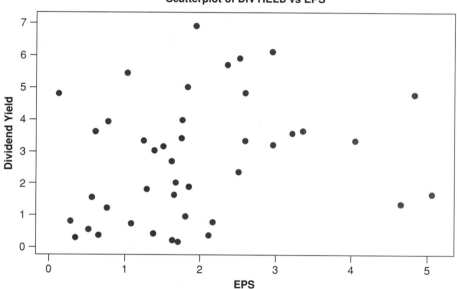

FIGURE 3.20
Regression Results for
Dividends Exercise.

Variable	Coefficient	Std Dev	T Stat	P Value
Intercept	2.0336	0.5405	3.76	0.001
EPS	0.3740	0.2395	1.56	0.126

Standard Error = 1.84975   R-Sq = 5.7%   R-Sq(adj) = 3.4%

Analysis of Variance

Source	DF	Sum of Squares	Mean Square	F Stat	P Value
Regression	1	8.345	8.345	2.44	0.126
Error	40	136.864	3.422		
Total	41	145.208			

7. Sales/Advertising The vice-president of marketing for a large firm is concerned about the effect
of advertising on sales of the firm's major product.
To investigate the relationship between advertising
and sales, data on the two variables were gathered
from a random sample of 20 sales districts. These
data are available in a file named SALESAD3.
(Sales and advertising are both expressed in hundreds of dollars. For example, 4250 represents
$425,000). The scatterplot is in Figure 3.21. The
results for the regression of sales (SALES) on
advertising (ADV) are shown in Figure 3.22.
Using the information given, answer the following
questions:

a. Is there a linear relationship between sales and
advertising? Use $\alpha = 0.05$. State the hypotheses to be tested, the decision rule, the test
statistic, and your decision.

b. What implications does this test result have
for the firm?

c. What is the sample regression equation relating sales to advertising?

d. If ADV increases by $1000, what would be
the resulting change in our prediction of
SALES?

e. Construct a 90% confidence interval estimate
of β_0.

FIGURE 3.21
Scatterplot for Sales
and Advertising
Exercise.

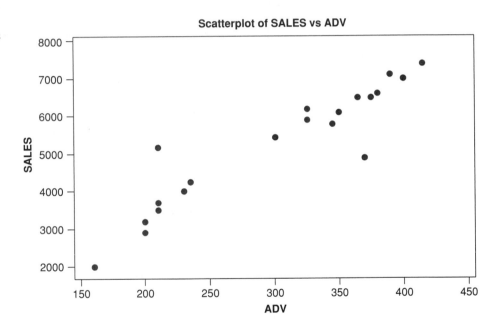

FIGURE 3.22
Regression Results for
Sales and Advertising
Exercise.

Variable	Coefficient	Std Dev	T Stat	P Value
Intercept	-57.281	509.750	-0.11	0.912
ADV	17.570	1.642	10.70	0.000

Standard Error = 594.808 R-Sq = 86.4% R-Sq(adj) = 85.7

Analysis of Variance

Source	DF	Sum of Squares	Mean Square	F Stat	P Value
Regression	1	40523671	40523671	114.54	0.000
Error	18	6368342	353797		
Total	19	46892014			

f. Construct a 95% confidence interval estimate of β_1.

g. Test the hypotheses

$$H_0: \beta_1 = 20$$
$$H_a: \beta_1 \neq 20$$

using a 5% level of significance. State the decision rule, the test statistic, and your decision.

h. What conclusion can be drawn from the result of the test in part g?

3.4   ASSESSING THE FIT OF THE REGRESSION LINE

3.4.1 THE ANOVA TABLE

In Example 3.6 (communications nodes), the goal might be to obtain the best possible prediction of the cost of a new node to be installed. Using a sample of n previously installed nodes, the sample mean of the n costs could be computed and used to predict the cost of any future node. But additional information on the number of access ports at each communications node might be used to obtain more accurate predictions of cost. This improvement in accuracy can be measured in terms of how much better the predictions are using the regression line instead of simply the mean of the y variable. If there is a significant improvement in prediction accuracy, then it is worthwhile to utilize the additional information.

In Figure 3.23, suppose that x^* represents the number of access ports at a particular communications node and y^* is the true cost of that node. If $\bar{y}$ is used to predict the cost of this node, then the prediction error is $y^* - \bar{y}$. But if the regression equation is used, the error is $y^* - \hat{y}$ thus reducing the error by $\hat{y} - \bar{y}$.

FIGURE 3.23
Partitioning the Variation in y.

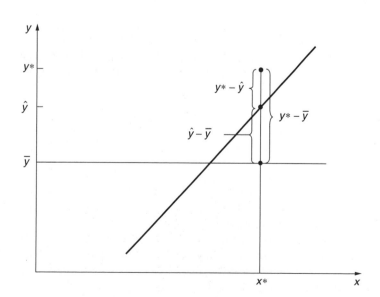

Note that

$$y* - \bar{y} = (y* - \hat{y}) + (\hat{y} - \bar{y}) \tag{3.13}$$

In words, the error in using the sample mean as our predictor of $y*$, $(y* - \bar{y})$, is equal to the error produced by using the regression line $(y* - \hat{y})$, plus the improvement over using the mean $(\hat{y} - \bar{y})$. Squaring both sides of Equation (3.13) gives

$$(y* - \bar{y})^2 = [(y* - \hat{y}) + (\hat{y} - \bar{y})]^2$$

or expanding the right-hand side,

$$(y* - \bar{y})^2 = (y* - \hat{y})^2 + 2(\hat{y} - \bar{y})(y* - \hat{y}) + (\hat{y} - \bar{y})^2 \tag{3.14}$$

The terms on either side of Equation (3.14) can be summed for all the individuals in the sample to obtain

$$\sum_{i=1}^{n}(y_i - \bar{y})^2 = \sum_{i=1}^{n}(y_i - \hat{y}_i)^2 + 2\sum_{i=1}^{n}(\hat{y}_i - \bar{y})(y_i - \hat{y}_i) + \sum_{i=1}^{n}(\hat{y}_i - \bar{y})^2$$

The term $\sum_{i=1}^{n}(\hat{y}_i - \bar{y})(y_i - \hat{y}_i)$ can be shown to equal zero so that

$$\sum_{i=1}^{n}(y_i - \bar{y})^2 = \sum_{i=1}^{n}(y_i - \hat{y}_i)^2 + \sum_{i=1}^{n}(\hat{y}_i - \bar{y})^2 \tag{3.15}$$

Each term in Equation (3.15) is a sum of squares and has a special interpretation in regression analysis. The term on the left-hand side of the equality,

$$SST = \sum_{i=1}^{n}(y_i - \bar{y})^2$$

is called the *total sum of squares (SST)*. This is the numerator of the fraction used to compute the sample variance of y and is interpreted as the total variation in y.

On the right-hand side of the equation are two additional sums of squares:

$$SSE = \sum_{i=1}^{n}(y_i - \hat{y}_i)^2$$

called the *error sum of squares (SSE)*, and

$$SSR = \sum_{i=1}^{n}(\hat{y}_i - \bar{y})^2$$

called the *regression sum of squares (SSR)*.

The terms in *SSE*, $y_i - \hat{y}_i$, are the prediction errors for the sample—that is, the differences between the true y values and the values predicted by using the regression line. *SSE* is often referred to as the "unexplained" sum of squares or as a measure of "unexplained" variation in y.

The terms in SSR, $\hat{y}_i - \bar{y}$, are measures of the improvement in using the regression line rather than the sample mean to predict. SSR, which is referred to as the "explained" sum of squares or as a measure of "explained" variation in y. The "explaining" is done through the use of the variable x.

The terms *explained* and *unexplained* should be interpreted with some caution. "Explained" does not necessarily mean that x causes y. It means that the variation in y around the regression line is smaller than the variation around the sample mean, $\bar{y}$. This caveat is explored in more detail in Section 3.7.

Figure 3.24 shows how MINITAB, Excel, and SAS report certain of the quantities discussed in this section. The tables shown are called *analysis of variance* (ANOVA) tables. This name refers to the partitioning of the total variation in the dependent variable into the regression and error sums of squares. These tables are typical of ANOVA tables presented in most statistical packages.

In the MINITAB ANOVA table in Figure 3.24(a), the three sources of variation in y are denoted Regression, Residual Error, and Total. In the Excel ANOVA table in Figure 3.24(b), the name Residual rather than Residual Error is used. In Figure 3.24(c), SAS uses the terms Model and Error in place of Regression and Residual Error or Residual. Total variation is referred to as Corrected Total. The quantities SSR, SSE, and SST are found in the SS column in both MINITAB and Excel and in the Sum of Squares column in SAS. In addition, the DF columns report the degrees of freedom associated with each of these sums of squares. SSR has 1 degree of freedom, SSE has $n - 2$ degrees of freedom, and SST has $n - 1$ degrees of freedom. Just as $SSR + SSE = SST$, it is also true that the regression and residual degrees of freedom always add up to the total degrees of freedom:

$$1 + (n - 2) = n - 1$$

FIGURE 3.24

Analysis of Variance Tables from MINITAB, Excel, and SAS Regressions.

(a) MINITAB

Source	DF	SS	MS	F	P
Regression	1	SSR	$MSR = SSR/1$	$F = MSR/MSE$	p value
Residual Error	$n - 2$	SSE	$MSE = SSE/(n - 2)$		
Total	$n - 1$	SST			

(b) Excel

	df	SS	MS	F	Significance F
Regression	1	SSR	$MSR = SSR/1$	$F = MSR/MSE$	p value
Residual	$n - 2$	SSE	$MSE = SSE/(n - 2)$		
Total	$n - 1$	SST			

(c) SAS

		Analysis of Variance			
Source	DF	Sum of Squares	Mean Square	F Value	Pr > F
Model	1	SSR	$MSR = SSR/1$	$F = MSR/MSE$	p value
Error	$n - 2$	SSE	$MSE = SSE/(n - 2)$		
Corrected Total	$n - 1$	SST			

Mean squares are shown in the MS column in MINITAB and Excel and in the Mean Square column in SAS. The mean squares are sums of squares divided by their degrees of freedom:

$$MSR = \frac{SSR}{1}$$

$$MSE = \frac{SSE}{n-2}$$

MSR is referred to as the *mean square due to regression,* and MSE is the *mean square due to error* (or more simply, "mean square regression" and "mean square error"). Note that MSR always equals SSR because the divisor is 1. This holds true in the case of simple regression, but will differ when multiple regression is discussed in the next chapter. MSE was used earlier in this chapter and was denoted s_e^2 to represent an estimate of the variance around the regression line. The square root of MSE, denoted s_e, was called the standard deviation around the regression line or the standard error of the regression.

The remaining columns in the ANOVA tables and further uses of MSR and MSE are discussed in the next sections.

3.4.2 THE COEFFICIENT OF DETERMINATION AND THE CORRELATION COEFFICIENT

In an exact or deterministic relationship, $SSR = SST$ and $SSE = 0$. A line could be drawn that passed through every sample point. But this is not the case in most practical business situations. A measure of how well the regression line fits the data is needed. In other words, "What proportion of the total variation has been explained?" This measure is provided through a statistic called the *coefficient of determination,* denoted R^2 (this is read "R squared"):

$$R^2 = \frac{SSR}{SST}$$

R^2 is computed by dividing the explained sum of squares by the total sum of squares. The result is the proportion of variation in y explained by the regression. R^2 falls between 0 and 1. The closer to 1 the value of R^2 is, the better the "fit" of the regression line to the data. An alternative formula for computing R^2 is to compute the proportion of variation unexplained by the regression and subtract this proportion from 1:

$$R^2 = 1 - \frac{SSE}{SST}$$

Most computerized regression routines report R^2, and some also report another quantity called the *correlation coefficient,* which is the square root of R^2 with an appropriate sign attached:

$$R = \pm\sqrt{R^2}$$

Note that this relationship holds in the case of simple regression, but not for multiple regression (discussed in Chapter 4). The sign of R is positive if the relationship is direct ($b_1 > 0$, or an upward-sloping line) and negative if the relationship is inverse ($b_1 < 0$, or a downward-sloping line). R ranges between -1 and 1.

Note that R^2 is referred to as a measure of fit of the regression line. It is not interpreted as a measure of the predictive quality of the regression equation even though the ANOVA decomposition was motivated using the concept of improved predictions. R^2 generally overstates the regression equation's predictive ability. This fact is discussed further in Section 3.5, and alternative measures of the regression's predictive ability are suggested. R^2 will continue to be referred to as a measure of fit.

3.4.3 THE F STATISTIC

An additional measure of how well the regression line fits the data is provided by the F statistic, which tests whether the equation $\hat{y} = b_0 + b_1 x$ provides a better fit to the data than the equation $\hat{y} = \overline{y}$. The F statistic is computed as

$$F = \frac{MSR}{MSE}$$

MSR is the mean square due to the regression, or the regression sum of squares divided by its degrees of freedom. MSE is the mean square due to error, or the error sum of squares divided by its degrees of freedom, so $MSE = SSE/(n-2)$. If the regression line fits the data well (that is, if the variation around the regression line is small relative to the variation around the sample mean, $\overline{y}$), then MSR should be large relative to MSE. If the regression line does not fit well, then MSR is small relative to MSE. Thus, large values of F support the use of the regression line, whereas small values suggest that x is of little use in explaining the variation in y.

To formalize the use of the F statistic, consider again the hypotheses

$$H_0: \beta_1 = 0$$
$$H_a: \beta_1 \neq 0$$

The F statistic can be used to perform this test. The decision rule is

Reject H_0 if $F > F(\alpha; 1, n-2)$

Do not reject H_0 if $F \leq F(\alpha; 1, n-2)$

where $F(\alpha; 1, n-2)$ is a critical value chosen from the F table for level of significance α. The F statistic has degrees of freedom associated with both the numerator and denominator sums of squares used in its computation. For a simple regression, there is 1 numerator degree of freedom and $n-2$ denominator degrees of freedom. These are the degrees of freedom associated with SSR and SSE, respectively. F tables are provided in Appendix B.

If the null hypothesis $H_0: \beta_1 = 0$ is rejected, then the conclusion is that x and y are linearly related. In other words, the line $\hat{y} = b_0 + b_1 x$ provides a better fit to the data than $\hat{y} = \overline{y}$.

The hypotheses tested by the F statistic also can be tested using the t test previously discussed. The decision made using either test is exactly the same. This is because the F statistic is equal to the square of the t statistic. Also, the $F(\alpha; 1, n - 2)$ critical value is the square of the $t_{\alpha/2,n-2}$ critical value:

$$F = \frac{MSR}{MSE} = t^2$$

and

$$F(\alpha; 1, n - 2) = t^2_{\alpha/2,n-2}$$

Because the two test procedures yield exactly the same decision, it does not matter which is used when testing $H_0: \beta_1 = 0$ versus $H_a: \beta_1 \neq 0$ (when testing $H_0: \beta_1 \leq 0$ versus $H_a: \beta_1 > 0$, or $H_0: \beta_1 \geq 0$ versus $H_a: \beta_1 < 0$ or any tests where $\beta_1^* \neq 0$, the t test should be used). The importance of the F statistic in multiple regression, however, makes it necessary to learn how to use this test. When there are two or more explanatory variables, the F test can be used to test hypotheses that cannot be tested using the t test.

Figure 3.25 shows the additional statistics provided by MINITAB [3.25(a)], Excel [3.25(b)], and SAS [3.25(c)]. In MINITAB, these statistics appear as shown directly above the ANOVA table. In Excel, the statistics appear as shown as the first entries in the regression output (called Regression Statistics). In SAS, the statistics appear directly below the ANOVA table. Figure 3.26 shows a representation of the complete MINITAB, Excel, and SAS outputs. The bracketed sections are discussed in later chapters.

In the MINITAB output, s is the standard deviation around the regression line or standard error of the regression. This was denoted s_e in the text. Also, R-Sq is the R^2 value. The Excel output labels s_e as *Standard Error* and R^2 as *R Square*.

FIGURE 3.25
Additional Statistics Provided on MINITAB, Excel, and SAS Regression Outputs.

(a) MINITAB

$S = s_e$ $\qquad$ R-Sq $= R^2$ $\qquad$ R-Sq(adj) $= R^2_{adj}$

(b) Excel

Regression Statistics
Multiple R	$= R$
R Square	$= R^2$
Adjusted R Square	$= R^2_{adj}$
Standard Error	$= s_e$
Observations	$= n$

(c) SAS

Root MSE	s_e	R-Square	R^2
Dependent Mean	$\bar{y}$	Adj R-Sq	R^2_{adj}
Coeff Var	$\dfrac{s_e}{\bar{y}} \times (100)$		

FIGURE 3.26
Complete Regression Outputs for MINITAB, Excel, and SAS.

(a) MINITAB

The regression equation is
$y = b_0 + b_1 x$

Predictor	Coef	SE Coef	T	P
Constant	b_0	s_{b_0}	b_0/s_{b_0}	p-value
x1 variable name	b_1	s_{b_1}	b_1/s_{b_1}	p-value

$S = s_e$ $\qquad$ R-Sq $= R^2$ $\qquad$ $\left[\text{R-Sq(adj)} = R^2_{adj}\right]^*$

Analysis of Variance

Source	DF	SS	MS	F	P
Regression	1	SSR	$MSR = SSR/1$	$F = MSR/MSE$	p value
Residual Error	$n-2$	SSE	$MSE = SSE/(n-2)$		
Total	$n-1$	SST			

Unusual Observations $\qquad\qquad\qquad\qquad\qquad\qquad\qquad\qquad\qquad\qquad\qquad$ *

Obs	X	Y	Fit	SE Fit	Residual	St Resid
Obs.No.	Value of X	Value of y	$\hat{y}$	s_m	$y - \hat{y}$	—

R denotes an observation with a large standardized residual.
X denotes an observation whose X value gives it large influence.

(b) Excel

Regression Statistics

Multiple R	$= R$
R Square	$= R^2$
Adjusted R Square	$= R^2_{adj}$
Standard Error	$= s_e$
Observations	$= n$

ANOVA

	df	SS	MS	F	Significance F
Regression	1	SSR	$MSR = SSR/1$	$F = MSR/MSE$	p value
Residual	$n-2$	SSE	$MSE = SSE/(n-2)$		
Total	$n-1$	SST			

	Coefficients	Standard Error	t stat	P-value	Lower 95%	Upper 95%
Intercept	b_0	s_{b_0}	b_0/s_{b_0}	p value	$b_0 - t_{\alpha/2,n-2}s_{b_0}$	$b_0 + t_{\alpha/2,n-2}s_{b_0}$
x1 variable name	b_1	s_{b_1}	b_1/s_{b_1}	p value	$b_1 - t_{\alpha/2,n-2}s_{b_1}$	$b_1 + t_{\alpha/2,n-2}s_{b_1}$

FIGURE 3.26
(Continued)

(c) SAS

Analysis of Variance

Source	DF	Sum of Squares	Mean Square	F Value	Pr > F
Model	1	SSR	$MSR = SSR/1$	$F = MSR/MSE$	p value
Error	$n-2$	SSE	$MSE = SSE/(n-2)$		
Corrected Total	$n-1$	SST			

Root MSE	s_e	R-Square	R^2
Dependent Mean	$\bar{y}$	Adj R-Sq	R^2_{adj}
Coeff Var	$\dfrac{s_e}{\bar{y}} \times (100)$		

Parameter Estimates

Variable	DF	Paramete Estimate	Standard Error	t Value	Pr > |t|
Intercept	1	b_0	s_{b_0}	b_0/s_{b_0}	p value
Variable name	1	b_1	s_{b_1}	b_1/s_{b_1}	p value

*Bracketed sections will be discussed in later chapters.

The SAS output calls the standard error *Root MSE* and refers to R^2 as *R-Square*. In addition, Excel shows the *Multiple R,* which is the positive square root of R^2, and the number of observations *(Observations).* All three outputs also show the *Adjusted R Square* [*R-Sq(adj)* in MINITAB and *Adj R-Sq* in SAS]. This quantity is discussed in Chapter 4. SAS also includes a number it calls the coefficient of variation *(Coeff Var)* which is computed by dividing the standard error of the regression by the mean of the y values and multiplying by 100. *Coeff Var* expresses the standard error of the regression in units of the mean of the dependent variable. The multiplication by 100 is merely a rescaling. *Coeff Var* is a number representing variability around the regression line and expresses this variation in unitless values. Thus, the coefficient of variation for two different regressions could be compared more readily than the standard errors because the influence of the units of the data has been removed.

EXAMPLE 3.7   To compute the R^2 for the data in Table 3.2, the quantities *SSE* and *SST* must be computed.

$$R^2 = 1 - \frac{SSE}{SST} = 1 - \frac{\sum\limits_{i=1}^{n}(y_i - \hat{y}_i)^2}{\sum\limits_{i=1}^{n}(y_i - \bar{y})^2}$$

SSE was computed in Figure 3.6 as $SSE = 8.8$. SST can be computed by the formula

$$\sum_{i=1}^{n} y_i^2 - \frac{1}{n}\left(\sum_{i=1}^{n} y_i\right)^2 = 431 - \frac{1}{6}(45)^2 = 93.5$$

The coefficient of determination or R^2 is

$$1 - \frac{8.8}{93.5} = 0.91$$

so 91% of the variation in y has been explained by the regression.

Note that the formula

$$R^2 = \frac{SSR}{SST}$$

could have been used here. But because SSE already had been computed and SSR had not, the alternative formula was used. If it were desired to compute SSR, this could be done by recalling that $SSR = SST - SSE = 93.5 - 8.8 = 84.7$.

The F statistic is computed as

$$F = \frac{MSR}{MSE} = \frac{SSR/1}{SSE/(n-2)} = \frac{84.7}{8.8/4} = 38.5$$

The hypotheses

$$H_0: \beta_1 = 0$$
$$H_a: \beta_1 \neq 0$$

can be tested using the F statistic. Using a 5% level of significance, the decision rule is

Reject H_0 if $F > F(0.05; 1,4) = 7.71$

Do not reject H_0 if $F \leq F(0.05; 1,4) = 7.71$

The test statistic was computed as $F = 38.5$, which results in a decision to reject H_0. In this case, the conclusion is that β_1 is not equal to zero and that the two variables x and y are linearly related.

EXAMPLE 3.8   **Pricing Communications Nodes (continued)**   Refer to Example 3.6 to complete the following problems using the regression output in Figure 3.16 (MINITAB), 3.17 (Excel), or 3.18 (SAS).

1. What percentage of the variation in COST is explained by the regression?

Answer: Using the R^2 value, 88.7% of the variation in COST has been explained by the regression.

2. Use the F test and a 5% level of significance to test the hypotheses

$$H_0: \beta_1 = 0$$
$$H_a: \beta_1 \neq 0$$

Answer:

(a) Using the standardized test statistic:

Decision rule: Reject H_0 if $F > 4.75$
Do not reject H_0 if $F \leq 4.75$

Test Statistic: $F = 94.41$
Decision: Reject H_0
Conclusion: There is evidence to conclude that COST and NUMPORTS are linearly related.

(b) Using the p value:

Decision rule: Reject H_0 if p value < 0.05
Do not reject H_0 if p value ≥ 0.05

Test Statistic: p value $= 0.000$
Decision: Reject H_0

EXERCISES

Exercises 8 and 9 should be done by hand.

8. Flexible Budgeting (continued) Refer to Exercise 1.

   a. Compute the coefficient of determination (R^2) for the regression of overhead costs on production.

   b. What percentage of the variation in overhead costs has been explained by the regression?

   c. Use the F test to test the hypotheses H_0: $\beta_1 = 0$ versus H_a: $\beta_1 \neq 0$ at the 5% level of significance. Be sure to state the decision rule, the test statistic value, and your decision.

   d. From the result in part c, are production and overhead costs linearly related?

9. Central Company (continued) Refer to Exercise 2.

   a. Compute the coefficient of determination (R^2) for the regression of the number of labor hours on number of items produced.

   b. What percentage of the variation in hours of labor has been explained by the regression?

   c. Use the F test to test the hypotheses H_0: $\beta_1 = 0$ versus H_a: $\beta_1 \neq 0$ at the 5%

level of significance. Be sure to state the decision rule, the test statistic value, and your decision.

   d. From the result in part c, are hours of labor and number of items produced linearly related?

10. Dividends (continued) Use the regression results in Figure 3.20 to help answer the questions.

   a. What percentage of the variation in dividend yield has been explained by the regression?

   b. Use the F test to test the hypotheses H_0: $\beta_1 = 0$ versus H_a: $\beta_1 \neq 0$ at the 5% level of significance. Be sure to state the decision rule, the test statistic value, and your decision.

11. Sales/Advertising (continued) Use the regression results in Figure 3.22 to help answer the questions.

   a. What percentage of the variation in sales has been explained by the regression?

   b. Use the F test to test the hypotheses H_0: $\beta_1 = 0$ versus H_a: $\beta_1 \neq 0$ at the 5% level of significance. Be sure to state the decision rule, the test statistic value, and your decision.

3.5 PREDICTION OR FORECASTING WITH A SIMPLE LINEAR REGRESSION EQUATION

One of the possible goals for fitting a regression line to data is to be able to use the regression equation to predict or forecast values of the dependent variable y. Given that a value of x has been observed, what is the best prediction of the response value, y? To discuss how to best predict y and how to make inferences using predictions based on a random sample, two cases that may arise in practice are considered.

3.5.1 ESTIMATING THE CONDITIONAL MEAN OF y GIVEN x

In Example 3.6, suppose that the network administrator wants to consider all possible nodes with 40 communications ports. The question to be answered is, "What will the cost be, on average, for nodes with 40 ports?"

The average cost of all nodes with 40 communications ports is to be estimated. Thus, an estimate of the conditional mean cost given $x = 40$, or an estimate of $\mu_{y|x=40}$, is required (see Figure 3.27). If a relationship between y and x does exist, the best estimate of this point on the population regression line is given by

$$\hat{y}_m = b_0 + b_1 x_m$$

where b_0 and b_1 are the least-squares estimates of β_0 and β_1, x_m is the number of ports for which an estimate is desired, and $\hat{y}_m$ is the estimate of the conditional mean cost. In this case, $\hat{y}_m$ represents the estimate of the point on the regression line corresponding to (or conditional on) $x = x_m$. Thus, it is the estimate of a population mean. The variance of this estimate can be shown to equal

$$\sigma_m^2 = \sigma_e^2 \left(\frac{1}{n} + \frac{(x_m - \bar{x})^2}{(n-1)s_x^2} \right) \tag{3.16}$$

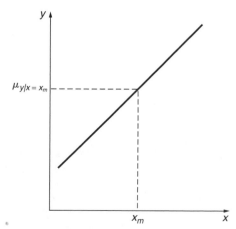

FIGURE 3.27
Estimating a Conditional Mean $\mu_{y|x=x_m}$.

Because σ_e^2 is unknown, s_e^2 is substituted to obtain an estimate of σ_m^2:

$$s_m^2 = s_e^2 \left(\frac{1}{n} + \frac{(x_m - \bar{x})^2}{(n-1)s_x^2} \right) \qquad (3.17)$$

The standard deviation or standard error of the estimate, s_m, is simply the square root of s_m^2.

The standard error of the estimate of the point on the regression line is affected by the distance of the value of x_m from the sample mean $\bar{x}$. The closer the value of x_m to the mean of the sample x values, the closer the term $(x_m - \bar{x})^2$ is to zero. If $(x_m = \bar{x})$, the term $(x_m - \bar{x})^2/[(n-1)s_x^2]$ equals zero, and the standard error equals $s_e/\sqrt{n}$. Thus, the closer the value x_m is to the sample mean, the smaller the standard error is or the more accurate the estimate is expected to be.

The reason is illustrated in Figure 3.28. Because all least-squares lines pass through the point $(\bar{x}, \bar{y})$, two least-squares lines have been drawn intersecting at the point $(\bar{x}, \bar{y})$. The two lines could represent least-squares lines fitted to two independent random samples taken from the same population. It is assumed that the two samples have means $\bar{x}$ and $\bar{y}$. Even though both lines pass through a common point, the slopes of the two estimated lines could be quite different, as illustrated. Thus, there is more certainty as to the value of a point on the regression line near the value $\bar{x}$ than at the extreme values of x. When estimating a point on the regression line for an extreme value of x, the greater uncertainty is reflected through a larger standard error.

It was previously stated that when the term $(x_m - \bar{x})^2/[(n-1)s_x^2]$ is zero, the standard error is $s_m = s_e/\sqrt{n}$. This happens when $(x_m = \bar{x})$. The quantity $s_e/\sqrt{n}$ is very much like the standard error associated with the sample mean $\bar{y}$, when it is used to estimate the (unconditional) population mean of the y values. A population mean is being estimated in the case of regression, so this is to be expected. The difference

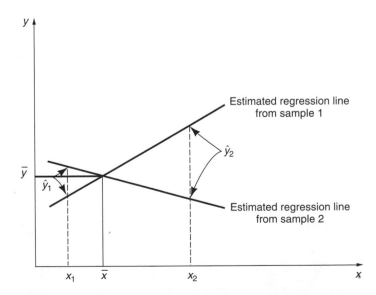

FIGURE 3.28 Effect on $\hat{y}_m$ of Variation in b_1 from Sample to Sample.

is that the mean in regression, $\mu_{y|x}$, is conditional on the value of x, rather than being unconditional.

Confidence intervals can be constructed for estimates of a conditional mean using

$$(\hat{y}_m - t_{\alpha/2,n-2}s_m, \hat{y}_m + t_{\alpha/2,n-2}s_m) \tag{3.18}$$

where $\hat{y}_m$ is the point estimate and $t_{\alpha/2,n-2}$ is chosen in the usual fashion from the t distribution with $n-2$ degrees of freedom.

Hypothesis tests also can be conducted. To test

$$H_0: \mu_{y|x_m} = \mu^*_{y|x_m}$$
$$H_a: \mu_{y|x_m} \neq \mu^*_{y|x_m}$$

where $\mu^*_{y|x_m}$ is a hypothesized value for the point on the population regression line, the decision rule is

Reject H_0 if $t > t_{\alpha/2,n-2}$ or $t < -t_{\alpha/2,n-2}$
Do not reject H_0 if $-t_{\alpha/2,n-2} \leq t \leq t_{\alpha/2,n-2}$

The test statistic, t, is computed as

$$t = \frac{\hat{y}_m - \mu^*_{y|x_m}}{s_m}$$

and it has a t distribution with $n-2$ degrees of freedom when H_0 is true.

One-tailed tests also can be performed provided the usual modifications are made in constructing the decision rule.

3.5.2 PREDICTING AN INDIVIDUAL VALUE OF y GIVEN x

Now suppose the network administrator is interested in a single communications node in a plant in Kansas City, Missouri, which will have 40 access ports. Predict the cost of installation for this particular node. With $x_p = 40$ access ports, the best prediction of the cost of this node is

$$\hat{y}_p = b_0 + b_1 x_p$$

which is exactly the same number that would be used to estimate the average cost for all nodes with 40 access ports. One can do no better in predicting cost for an individual node than to use the estimate of average cost for all nodes with the same number of access ports. This is because there is no additional information used in the regression that distinguishes this one node from all the others (see Figure 3.29).

The prediction for an individual value, however, is not as accurate as the estimate of a population mean for all individuals in a certain category. The variance of the prediction for an individual is

$$\sigma_p^2 = \sigma_e^2 \left(1 + \frac{1}{n} + \frac{(x_m - \bar{x})^2}{(n-1)s_x^2} \right) \tag{3.19}$$

FIGURE 3.29
Predicting an Individual y
Value.

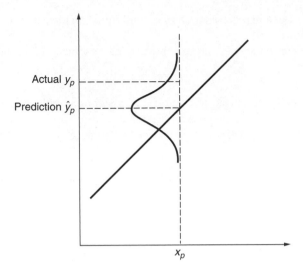

which can be estimated by replacing σ_e^2 by s_e^2:

$$s_p^2 = s_e^2\left(1 + \frac{1}{n} + \frac{(x_m - \bar{x})^2}{(n-1)s_x^2}\right) \tag{3.20}$$

To compare s_p^2 to the variance of the estimate of a conditional mean, write the prediction variance as

$$s_p^2 = s_e^2 + s_m^2$$

The variance of the prediction for an individual value is equal to the variance from estimating the point on the regression line for $x = x_p$, s_m^2, plus the estimate of the variation of the individual y values around the regression line, s_e^2. Even if the exact position of $\mu_{y|x_p}$ were known, y_p still would not be known. The individual y values are distributed around $\mu_{y|x_p}$ with standard deviation σ_e. Because $\mu_{y|x_p}$ is actually unknown, there is uncertainty associated with the estimation of this value (reflected in s_m or s_m^2) plus the uncertainty in predicting an individual value (reflected in s_e or s_e^2).

Interval estimation of y_p is accomplished by constructing prediction intervals. The term *prediction interval* is used rather than confidence interval because a population parameter is not being estimated in this case; instead, the response or performance of a single individual in the population is being predicted.

A $(1 - \alpha)100\%$ prediction interval for y_p is

$$(\hat{y}_p - t_{\alpha/2, n-2}s_p, \; \hat{y}_p + t_{\alpha/2, n-2}s_p) \tag{3.21}$$

where $\hat{y}_p$ is the predicted value and $t_{\alpha/2, n-2}$ is chosen from the t distribution with $n - 2$ degrees of freedom.

FIGURE 3.30
Confidence Interval
Limits Versus Prediction
Interval Limits.

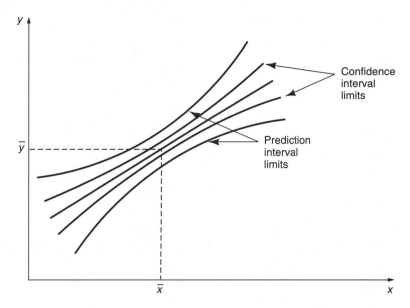

Figure 3.30 illustrates the difference between the confidence interval estimate of $\mu_{y|x}$ and the prediction interval for an individual. Both the confidence interval in Equation (3.18) and the prediction interval in Equation (3.21) are narrower (more precise) near $x = \bar{x}$ and wider at the extreme values of x. The prediction interval is always wider than the confidence interval because of the added uncertainty involved in predicting an individual response.

3.5.3 ASSESSING QUALITY OF PREDICTION

As noted in Section 3.4, the R^2 of the regression is a measure of the fit of the regression to the sample data. It is not generally considered an adequate measure of the regression equation's ability to estimate $\mu_{y|x}$ or to predict new responses. The standard R^2 overestimates the quality of future (or out-of-sample) predictions.

Two possible means of assessing prediction quality are presented in this section. The first is called *data splitting.* In this method, the data set is partitioned into two groups. One group of n_1 data points is used to estimate or fit possible equations used for forecasting. The second group of n_2 data points, called a *holdout sample* or *validation sample,* is used to assess predictive ability of the models estimated using the fitting sample. Any models that are considered possible candidates are estimated using the fitting sample. Predictions, $\hat{y}_i$, are then computed for these models using the explanatory variable values in the validation sample. For each candidate model, prediction errors, $y_i - \hat{y}_i$, are computed for all n_2 observations in the validation sample. A measure of forecast accuracy based on the forecast errors can then be computed. For example, the *mean square deviation (MSD)*

$$MSD = \frac{\sum_{i=1}^{n_2} (y_i - \hat{y}_i)^2}{n_2}$$

the *mean absolute deviation (MSD)*

$$MAD = \frac{\sum_{i=1}^{n_2} |y_i - \hat{y}_i|^2}{n_2}$$

and the *mean absolute percentage error (MAPE)*

$$MAPE = \frac{\sum_{i=1}^{n_2} \left(\frac{|y_i - \hat{y}_i|}{|y_i|} \right)}{n_2}$$

are three commonly computed measures.

Models with smaller *MSD*, *MAD*, or *MAPE* are better for prediction purposes. The advantage of using this approach is that the models are tested on data that were not used in the fitting or model estimation process. This provides an independent assessment of the models' predictive ability.

After an appropriate model has been chosen, the entire data set can be used to estimate the model parameters. This model is then used to produce future predictions.

A second means of assessing prediction quality is to use the *PRESS* statistic. *PRESS* stands for prediction sum of squares and is defined as

$$PRESS = \sum_{i=1}^{n} (y_i - \hat{y}_{i,-1})^2$$

In this formula, $\hat{y}_{i,-1}$ represents the prediction obtained from a model estimated with one of the sample observations deleted. If there are n observations in the sample, there are n different predictions, $\hat{y}_{i,-1}$. The prediction $\hat{y}_{i,-1}$ is obtained by evaluating the regression equation at x_i, but the data point (x_i, y_i) is not used in obtaining the estimated regression equation. Thus, as with the use of a validation sample, predictions are obtained from data that are not used to fit the model.

The quantities $y_i - \hat{y}_{i,-1}$ often are called *PRESS* residuals because they are similar to the actual regression residuals, $y_i - \hat{y}_i$. The prediction sum of squares also is similar to the error sum of squares, *SSE*. This suggests construction of an R^2-like statistic that might be called the prediction R^2:

$$R^2_{PRED} = 1 - \frac{PRESS}{SST}$$

Larger values of R^2_{PRED} (or smaller values of *PRESS*) suggest models of greater predictive ability.

EXAMPLE 3.9 **1.** Refer to the data in Table 3.2. Find an estimate of the conditional mean of y when $x = 6$ and find the standard deviation of this estimate.

Answer: Using the least-squares regression equation $\hat{y} = -0.2 + 2.2x$, an estimate of the point on the regression line when $x = 6$ is $\hat{y} = -0.2 + 2.2(6)$ $=13$. The standard deviation of the estimate of the point on the regression line is

$$s_m = s_e\sqrt{\frac{1}{n} + \frac{(x_m - \bar{x})^2}{(n-1)s_x^2}} = 1.48\sqrt{\frac{1}{6} + \frac{(6-3.5)^2}{17.5}} = 1.07$$

[For computation of s_e, $\bar{x}$, and $(n-1)s_x^2$, see Example 3.5.]

2. Find a prediction of the y value when $x = 6$ and find the standard deviation of the prediction.

Answer: Using the least-squares regression equation $\hat{y} = -0.2 + 2.2x$, the prediction of y when $x = 6$ is $\hat{y} = -0.2 + 2.2(6) = 13$. The standard deviation of the prediction is

$$s_p = s_e\sqrt{1 + \frac{1}{n} + \frac{(x_p - \bar{x})^2}{(n-1)s_x^2}} = 1.48\sqrt{1 + \frac{1}{6} + \frac{(6-3.5)^2}{17.5}} = 1.83$$

3. Find a 95% confidence interval and 95% prediction interval when $x = 6$.

Answer: The 95% confidence and prediction intervals are, respectively,

$$[13 - 2.776(1.07),\ 13 + 2.776(1.07)] \quad \text{or} \quad (10.03,\ 15.97)$$

and

$$[13 - 2.776(1.83),\ 13 + 2.776(1.83)] \quad \text{or} \quad (7.92,\ 18.08)$$

EXAMPLE 3.10   Pricing Communication Nodes (continued)

1. On average, how much do we expect communication nodes to cost if there are to be 40 access ports?

Answer: Figure 3.31 shows the MINITAB regression output using the option to request a prediction with 40 as the value of the x variable. Figure 3.32 shows similar output for SAS. Both outputs show the predicted value (*Fit* in MINITAB and *Predicted Value* in SAS), the standard error of the estimate of the point on the regression line, s_m (*SE Fit* in MINITAB and *Std Error Mean Predict* in SAS), a 95% confidence interval for the estimate of the conditional mean (*95% CI* in MINITAB and *95% CL Mean* in SAS), and a 95% prediction interval for each prediction (*95% PI* in MINITAB and *95% CL Predict* in SAS). Note that the SAS output provides this information for the observations used to estimate the equation as well as the observations for which predictions are desired.

A point estimate of cost for all nodes with 40 access ports is $42,600. A 95% confidence interval estimate is given by ($40,035, $45,166). Thus, we can say with 95% confidence that the average cost of all nodes with 40 access ports is expected to be between $40,035 and $45,166.

FIGURE 3.31
MINITAB Regression
Prediction Output for
Example 3.10.

```
Predicted Values for New Observations

New
Obs       Fit      SE Fit      95% CI            95% PI
  1      42600      1178    (40035, 45166)    (32872, 52329)


Values of Predictors for New Observations

New
Obs     NUMPORTS
  1        40.0
```

```
                         Output Statistics

       Dep Var  Predicted    Std Error
Obs     COST      Value    Mean Predict    95% CL Mean       95% CL Predict      Residual

 1     52388      60805       2414       55545    66065     50047    71563      -8417
 2     51761      50402       1559       47006    53799     40423    60382       1359
 3     50221      45201       1262       42452    47950     35423    54979       5020
 4     36095      37399       1186       34814    39984     27666    47132      -1304
 5     27500      26996       1780       23119    30874     16843    37150     503.6461
 6     57088      53003       1751       49189    56818     42873    63133       4085
 7     54475      53003       1751       49189    56818     42873    63133       1472
 8     33969      34798       1278       32015    37582     25010    44587    -829.3840
 9     31309      32198       1414       29116    35280     22321    42075    -888.7073
10     23444      32198       1414       29116    35280     22321    42075      -8754
11     24269      24396       1991       20057    28735     14057    34734    -126.6772
12     53479      50402       1559       47006    53799     40423    60382       3077
13     33543      29597       1585       26143    33051     19598    39596       3946
14     33056      32198       1414       29116    35280     22321    42075     858.2927
15        .       42600       1178       40035    45166     32872    52329         .


            Sum of Residuals                        0
            Sum of Squared Residuals        222594146
            Predicted Residual SS (PRESS)   345066019
```

FIGURE 3.32 SAS Regression Prediction Output for Example 3.10.

2. For an individual node with 40 access ports, find a prediction of cost.

Answer: The point prediction is again $42,600. A 95% prediction interval for the individual node is ($32,872, $52,329). Note that the prediction interval is considerably wider than the confidence interval, reflecting the additional uncertainty of predicting for an individual as opposed to estimating an average.

EXERCISES

Exercises 12 and 13 should be done by hand.

12. **Flexible Budgeting (continued)**, Refer to Exercise 1.

   a. Find a point estimate of the overhead costs, on average, for production runs of 80,000 units.

   b. Find a 95% confidence interval estimate of overhead costs, on average, for production runs of 80,000 units.

   c. Find a point prediction of the overhead costs for a single production run of 80,000 units.

   d. Find a 95% prediction interval for overhead costs for a single production run of 80,000 units.

   e. State why the prediction interval is wider than the confidence interval.

13. **Central Company (continued)**, Refer to Exercise 2.

   a. Find a point estimate for the number of hours of labor required, on average, when 60 units are produced.

   b. Find a 95% confidence interval estimate of hours of labor required, on average, when 60 units are produced.

   c. Find a point prediction of the number of hours of labor required for one run producing 60 units.

   d. Find a 95% prediction interval for the number of hours of labor required for one run producing 60 units.

14. **Dividends (continued)**, Consider the dividend-yield problem in Exercise 6 and the associated computer results in Figure 3.20. An analyst wants an estimate of dividend yield for all firms with earnings per share of $3. Does the equation developed provide a more accurate estimate than simply using the sample mean dividend yield for all 42 firms examined? State why or why not.

15. **Sales/Advertising (continued)**, Use the results in Figure 3.33 to help solve these problems. These results were obtained requesting a prediction with $x = 200, 250, 300$ and 350, respectively (representing $20,000, $25,000, $30,000 and $35,000).

   a. Find an estimate of average sales for all sales districts with advertising expenditures of $25,000. Find a point estimate and a 95% confidence interval estimate.

   b. Predict sales for individual districts having advertising expenditures of $20,000, $25,000, $30,000, and $35,000. Find point predictions as well as 95% prediction intervals.

FIGURE 3.33 Prediction Output for Exercise 15.	Predicted Value	SE Fit	95% Confidence Int.	95% Prediction Int.
	3457	211	(3013, 3900)	(2131, 4783)
	4335	156	(4007, 4663)	(3043, 5627)
	5214	133	(4934, 5493)	(3933, 6494)
	6092	157	(5763, 6421)	(4800, 7384)

3.6 FITTING A LINEAR TREND TO TIME-SERIES DATA

Data gathered on individuals at the same point in time are called *cross-sectional data. Time-series data* are data gathered on a single individual (person, firm, and so on) over a sequence of time periods, which may be days, weeks, months, quarters, years, or virtually any other measure of time. In a given problem, however, it is assumed that the data are gathered over only one interval of time (daily and weekly data are not combined, for example).

When dealing with time-series data, the primary goal often is to be able to produce forecasts of the dependent variable for future time periods. Two separate approaches to this problem can be identified. On the one hand, a researcher may identify variables that are related to the dependent variable in a causal manner and use these in developing a *causal regression model*. For example, when trying to forecast sales for a particular product, causal variables might include advertising expenditures and competitors' market share. Changes in these variables are felt to produce or cause changes in sales. Thus, the term *causal regression model* is used.

The researcher may, on the other hand, identify patterns of movement in past values of the dependent variable and extrapolate these patterns into the future using an *extrapolative regression model*. An extrapolative model uses explanatory variables, although they are not related to the dependent variable in a causal manner. They simply describe the past movements of the dependent variable so that these movements can be extended into future time periods. Variables that represent trend and seasonal components often are included in extrapolative models.

Both causal and extrapolative models have their benefits and drawbacks. Causal models require the identification of variables that are related to the dependent variable in a causal manner. Then data must be gathered on these explanatory variables to use the model. Furthermore, when forecasting for future time periods, the values of the explanatory variables in these periods must be known. In extrapolative models, only past values of the dependent variable are required, and thus variable selection and data gathering are simpler processes.

Whether a causal or extrapolative model performs better is determined to some extent by how far into the future the forecast refers. Forecasts often are classified as short-term (0 to 3 months), medium-term (3 months to 2 years), or long-term (2 years and longer). (Note that these cutoffs to classify forecast horizons are somewhat arbitrary and may not apply to all situations). Extrapolative models tend to perform well in the short term, but they can be reasonably accurate for medium-term forecasts. Often, extrapolative models can produce more accurate forecasts than causal models in the short term. Causal models often outperform extrapolative models when long-term forecasts are desired. In addition to being just as effective as causal models in the short term and often in the medium term, extrapolative models tend to be easier to develop and use.

The success of extrapolative models depends on the stability of the behavior of the time series. If past time-series patterns are expected to continue into the future, then an extrapolative model should be relatively successful in making accurate forecasts. If these past patterns are altered for some reason, and future movements differ in general from past movements, then extrapolative models do not perform well. Thus, an assumption when using an extrapolative model for forecasting is that past patterns of data movement are reflective of future patterns.

Causal models can respond, to some extent, to more drastic changes in patterns. Changes in the explanatory variables caused by changes in economic or market conditions should produce relatively accurate forecasts of changes in the dependent variable. Of course, this assumes that the changes in the explanatory variables will be known for future time periods. In addition, if the model itself changes (that is, if the way the variables are related changes in the future), the causal model is not capable of making accurate forecasts.

In this section, the use of a *linear trend model* for time-series data is examined. The linear trend model is a type of extrapolative model that may be useful in certain time-series applications. In subsequent chapters, other techniques useful in building extrapolative time-series models are examined.

A *trend* in time-series data is a tendency for the series to move upward or downward over many time periods. This movement may follow a straight line or a curvilinear pattern. Regression analysis can be used to model certain trends and to extrapolate these trends into future time periods.

The simplest form of a trend over time is a linear trend. The linear trend model can be written

$$y_i = \beta_0 + \beta_1 t + e_i$$

The explanatory variable simply indicates the time period ($x_i = t$). Usually, the variable t is constructed by using the integers, 1, 2, 3, . . . to indicate the time period.

FIGURE 3.34
Examples of Types of Trends.

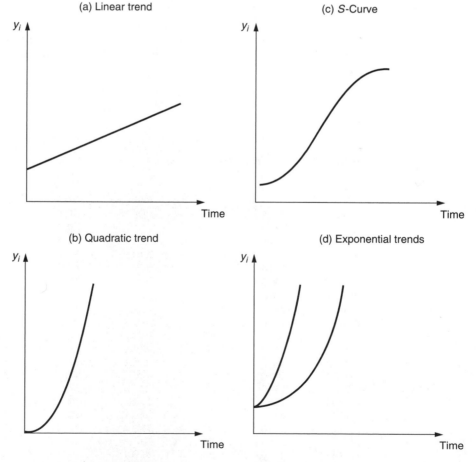

This is preferred to using the actual years (1980, 1981, 1982, . . .) because it reduces computational problems.

Forecasts are simple to compute when the linear trend model is used. Simply insert the appropriate value for the time period to be forecast into the regression equation. The time period T forecast can be written as

$$\hat{y}_T = b_0 + b_1 T$$

Many other types of trends can be modeled using regression. Some examples, including the linear trend, are shown in Figure 3.34. Note that the other types of trends are represented by curves. Equations to represent curvilinear trends are discussed in Chapter 5.

EXAMPLE 3.11   **ABX Company Sales**   The ABX Company sells winter sports merchandise including skis, ice skates, sleds, and so on. Quarterly sales (in thousands of dollars) for the ABX Company are shown in Table 3.6. The time period represented starts in the first quarter of 1994 and ends in the fourth quarter of 2003. (See the file ABXSALES3 on the CD.)

A time-series plot of the sales figures is shown in Figure 3.35. The time-series plot suggests a strong linear trend in the sales figures. A regression with a linear trend variable (labeled TIME) was estimated and the regression results are shown in Figure 3.36. The linear trend model estimated is

$$y_i = \beta_0 + \beta_1 t + e_i$$

TABLE 3.6 Data for ABX Company Sales Example

Year.Qtr	SALES	TIME	Year.Qtr	SALES	TIME
1994.1	221.0	1	1999.1	260.5	21
1994.2	203.5	2	1999.2	244.0	22
1994.3	190.0	3	1999.3	256.0	23
1994.4	225.5	4	1999.4	276.5	24
1995.1	223.0	5	2000.1	291.0	25
1995.2	190.0	6	2000.2	255.5	26
1995.3	206.0	7	2000.3	244.0	27
1995.4	226.5	8	2000.4	291.0	28
1996.1	236.0	9	2001.1	296.0	29
1996.2	214.0	10	2001.2	260.0	30
1996.3	210.5	11	2001.3	271.5	31
1996.4	237.0	12	2001.4	299.5	32
1997.1	245.5	13	2002.1	297.0	33
1997.2	201.0	14	2002.2	271.0	34
1997.3	230.0	15	2002.3	270.0	35
1997.4	254.5	16	2002.4	300.0	36
1998.1	257.0	17	2003.1	306.5	37
1998.2	238.0	18	2003.2	283.5	38
1998.3	228.0	19	2003.3	283.5	39
1998.4	255.0	20	2003.4	307.5	40

FIGURE 3.35
Time-Series Plot of ABX Company Sales.

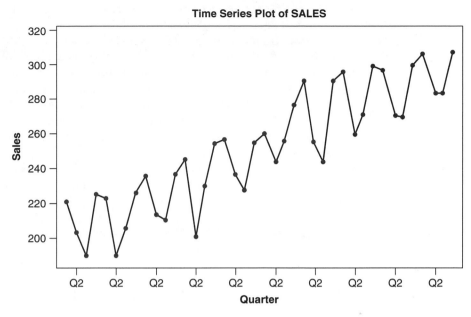

FIGURE 3.36
Regression Results for ABX Company Sales Example.

Variable	Coefficient	Std Dev	T Stat	P Value
Intercept	199.017	5.128	38.81	0.000
TIME	2.556	0.218	11.73	0.000

Standard Error = 15.9126   R-Sq = 78.3%   R-Sq(adj) = 77.8%

Analysis of Variance

Source	DF	Sum of Squares	Mean Square	F Stat	P Value
Regression	1	34818	34818	137.50	0.000
Error	38	9622	253		
Total	39	44440			

To test whether the linear trend component is useful in explaining the variation in sales, the following hypotheses should be tested:

$$H_0: \beta_1 = 0$$
$$H_a: \beta_1 \neq 0$$

Using a 5% level of significance, the decision rule is

Reject H_0 if $t > 1.96$ or $t < -1.96$

Do not reject H_0 if $-1.96 \leq t \leq 1.96$

The z value of 1.96 is used as a critical value because the number of degrees of freedom is large (38).

When using time-series data for forecasting, it is generally true that prediction intervals are more appropriate than confidence intervals for representing the uncertainty in predictions for future time periods. Figures 3.37 and 3.38 show the additional MINITAB and SAS output obtained when forecasts for the next four time periods are requested. Both outputs show the predicted value (*Fit* in MINITAB and *Predicted Value* in SAS), the standard error of the estimate of the point on the regression line (*SE Fit* in MINITAB and *Std Error Mean Predict* in SAS), a 95% confidence interval for the estimate of the conditional mean (*95% CI* in MINITAB and *95% CL Mean* in SAS) and a 95% prediction interval for each prediction (*95% PI* in MINITAB and *95% CL Predict* in SAS). Note that the SAS output provides this information for the observations used to estimate the equation as well as the observations for which predictions are desired.

Using the estimated linear trend equation, the point forecasts are determined by substituting values of the trend variable for the appropriate time period into the equation:

$$2004.1 \text{ sales} = 199.017 + 2.5559(41) = 303.81$$
$$2004.2 \text{ sales} = 199.017 + 2.5559(42) = 306.36$$
$$2004.3 \text{ sales} = 199.017 + 2.5559(43) = 308.92$$
$$2004.4 \text{ sales} = 199.017 + 2.5559(44) = 311.48$$

The prediction intervals in the outputs would be used as our interval predictions of sales in each of the four quarters. Thus, our interval prediction for sales in the first quarter of 2004 is $269,960 to $337,650.

If you look again at the time-series plot of sales in Figure 3.35, you may notice a pattern other than the trend. Note that the sales figures for the first and fourth quarters tend to be higher than the figures for the second and third quarters. This systematic variation among time periods from year to year is called *seasonal variation.* In Chapter 7, methods to account for seasonal variation are discussed.

FIGURE 3.37
MINITAB Output
Showing the Next Four
Periods' Forecasts.

```
Predicted Values for New Observations

New
Obs        Fit  SE Fit       95% CI             95% PI
  1     303.81    5.13   (293.43, 314.19)  (269.96, 337.65)
  2     306.36    5.32   (295.60, 317.13)  (272.40, 340.33)
  3     308.92    5.51   (297.76, 320.08)  (274.83, 343.01)
  4     311.48    5.71   (299.92, 323.03)  (277.25, 345.70)


Values of Predictors for New Observations

New
Obs        TIME
  1        41.0
  2        42.0
  3        43.0
  4        44.0
```

Obs	Dep Var sales	Predicted Value	Std Error Mean Predict	95% CL Mean		95% CL Predict		Residual
1	221.0000	201.5732	4.9391	191.5745	211.5719	167.8436	235.3027	19.4268
2	203.5000	204.1290	4.7528	194.5074	213.7507	170.5094	237.7487	-0.6290
3	190.0000	206.6849	4.5694	197.4347	215.9351	173.1696	240.2002	-16.6849
4	225.5000	209.2408	4.3891	200.3555	218.1260	175.8244	242.6571	16.2592
5	223.0000	211.7966	4.2123	203.2692	220.3241	178.4736	245.1196	11.2034
6	190.0000	214.3525	4.0396	206.1747	222.5303	181.1172	247.5878	-24.3525
7	206.0000	216.9083	3.8715	209.0709	224.7458	183.7552	250.0615	-10.9083
8	226.5000	219.4642	3.7085	211.9567	226.9718	186.3875	252.5409	7.0358
9	236.0000	222.0201	3.5515	214.8305	229.2097	189.0141	255.0261	13.9799
10	214.0000	224.5759	3.4012	217.6906	231.4612	191.6349	257.5170	-10.5759
11	210.5000	227.1318	3.2585	220.5353	233.7283	194.2499	260.0137	-16.6318
12	237.0000	229.6877	3.1245	223.3624	236.0129	196.8591	262.5162	7.3123
13	245.5000	232.2435	3.0004	226.1695	238.3176	199.4624	265.0246	13.2565
14	201.0000	234.7994	2.8875	228.9540	240.6448	202.0599	267.5389	-33.7994
15	230.0000	237.3553	2.7870	231.7133	242.9972	204.6514	270.0591	-7.3553
16	254.5000	239.9111	2.7004	234.4444	245.3778	207.2371	272.5851	14.5889
17	257.0000	242.4670	2.6291	237.1446	247.7894	209.8168	275.1172	14.5330
18	238.0000	245.0228	2.5743	239.8114	250.2343	212.3906	277.6551	-7.0228
19	228.0000	247.5787	2.5372	242.4425	252.7149	214.9584	280.1991	-19.5787
20	255.0000	250.1346	2.5184	245.0364	255.2327	217.5202	282.7490	4.8654
21	260.5000	252.6904	2.5184	247.5923	257.7886	220.0760	285.3048	7.8096
22	244.0000	255.2463	2.5372	250.1101	260.3825	222.6259	287.8666	-11.2463
23	256.0000	257.8022	2.5743	252.5907	263.0136	225.1699	290.4344	-1.8022
24	276.5000	260.3580	2.6291	255.0356	265.6804	227.7078	293.0082	16.1420
25	291.0000	262.9139	2.7004	257.4472	268.3806	230.2399	295.5879	28.0861
26	255.5000	265.4697	2.7870	259.8278	271.1117	232.7659	298.1736	-9.9697
27	244.0000	268.0256	2.8875	262.1802	273.8710	235.2861	300.7651	-24.0256
28	291.0000	270.5815	3.0004	264.5074	276.6555	237.8004	303.3626	20.4185
29	296.0000	273.1373	3.1245	266.8121	279.4626	240.3088	305.9659	22.8627
30	260.0000	275.6932	3.2585	269.0967	282.2897	242.8113	308.5751	-15.6932
31	271.5000	278.2491	3.4012	271.3638	285.1344	245.3080	311.1901	-6.7491
32	299.5000	280.8049	3.5515	273.6153	287.9945	247.7989	313.8109	18.6951
33	297.0000	283.3608	3.7085	275.8532	290.8683	250.2841	316.4375	13.6392
34	271.0000	285.9167	3.8715	278.0792	293.7541	252.7635	319.0698	-14.9167
35	270.0000	288.4725	4.0396	280.2947	296.6503	255.2372	321.7078	-18.4725
36	300.0000	291.0284	4.2123	282.5009	299.5558	257.7054	324.3514	8.9716
37	306.5000	293.5842	4.3891	284.6990	302.4695	260.1679	327.0006	12.9158
38	283.5000	296.1401	4.5694	286.8899	305.3903	262.6248	329.6554	-12.6401
39	283.5000	298.6960	4.7528	289.0743	308.3176	265.0763	332.3156	-15.1960
40	307.5000	301.2518	4.9391	291.2531	311.2505	267.5223	334.9814	6.2482
41	.	303.8077	5.1279	293.4269	314.1885	269.9629	337.6525	.
42	.	306.3636	5.3189	295.5961	317.1310	272.3982	340.3289	.
43	.	308.9194	5.5119	297.7612	320.0776	274.8282	343.0107	.
44	.	311.4753	5.7067	299.9227	323.0278	277.2530	345.6976	.

Sum of Residuals   0
Sum of Squared Residuals   9622.06053
Predicted Residual SS (PRESS)   10620

FIGURE 3.38 SAS Output Showing the Next Four Periods' Forecasts.

EXERCISES

16. Fort Worth Water Department. In 1990, the city of Fort Worth, Texas, conducted a study examining the level of water purity. One aspect helpful in maintaining water purity is monitoring the quality of water at storm drains that pour into the Trinity River. This river supplies drinking water for Fort Worth. Even though water from the river is filtered later, preventing contaminants from entering the river from storm drains is helpful in maintaining purity. Five of the variables the city monitored to test purity of water entering the river from storm drains are:

ODOR: determined by a sensory test

COLOR: no color is best—determined by comparison to a standard water sample

SCUM: floatable solids

hydrocarbon SHEEN: hydrocarbon (oil) sheen on surface of water

sewage BACTERIA: filamentous sewage bacteria

These are monthly data from January 1986 through December 1989. In all cases, lower numbers are better.

These data are in a file named WATER3 on the CD.

Your job is to use time-series plots and linear trend regression to examine the performance of the city's water department in improving the quality of storm drain water entering the Trinity River. Which of the variables show a significant decrease? Are there areas where the city might concentrate its efforts to achieve future improvements? Use a 5% level of significance in any tests.

3.7 SOME CAUTIONS IN INTERPRETING REGRESSION RESULTS

3.7.1 ASSOCIATION VERSUS CAUSALITY

A common mistake made when using regression analysis is to assume that a strong fit (high R^2) of a regression of y on x automatically means that "x causes y." This is not necessarily true. Some alternative explanations for the good fit include:

1. The reverse is true; y causes x. Linear regression computations pay no attention to the direction of causality. If x and y are highly correlated, a high R^2 value results even if the causal order of the variables is reversed.

2. There may be a third variable related to both x and y. It may be that neither x causes y nor y causes x. Both variables may be related to some third common cause. As an example, consider the price and gasoline mileage of automobiles. These two variables are inversely related. As mileage rises, price goes down (on average). But it is not the rise in mileage that "causes" the price to drop. A third variable, size of car, may be influencing both of the other two variables. As size increases, price increases and mileage drops. There are a variety of interesting examples in this category. For example, the mortality rate in countries is inversely related to the number of televisions. As the number of televisions increases, mortality rate decreases. I don't think this is a causal relationship.

To infer that x causes y requires that additional conditions be satisfied. A high R^2 for a regression of y on x might be considered supporting evidence for causality, but on its own, this is not enough to ensure that x causes y.

Note that the absence of causality is not necessarily a drawback in regression analysis. An equation showing a relationship between x and y can be important and useful even if it is recognized that x does not cause y.

3.7.2 FORECASTING OUTSIDE THE RANGE OF THE EXPLANATORY VARIABLE

When using an estimated regression equation to construct estimates of $\mu_{y|x}$ or to predict individual values of the dependent variable, some caution must be used if forecasts are outside the range of the x variable. Consider the communications nodes example. The explanatory variable was NUMPORTS, the number of access ports. The sample values ranged from 12 to 68. The estimated regression model can be expected to be reliable over this range of the x variable. If, however, a node is to be installed with 100 ports, there is some question as to how reliable the model will be. The relationship that holds over the range from 12 to 68 may differ from the relationship outside this range. Estimates of $\mu_{y|x}$ or predictions outside the range of the x variable require some caution for this reason.

There are often occasions where forecasts outside the range of the x variable must be made. One common example is when time-series data are used and forecasts for future time periods are desired. It may be that the values of the explanatory variables in future time periods are outside the range observed in the past, as, for example, when the linear trend model is used. In such cases, it must be recognized that the quality of the forecasts depends on whether the estimated relationship still holds for values of the explanatory variables that are outside the observed range.

EXERCISES

17. Sales/Advertising (continued). Use the results in Figure 3.22 to help answer the following questions.

a. Find a point estimate of average sales for all sales districts with advertising expenditures of $60,000. Are there any cautions that should be exercised regarding this estimate?

b. A district sales manager examines the model developed. The manager points out that $0 advertising expenditure results in sales of −$5700, which is impossible. She suggests that this means the model is of no use. Do you agree or disagree with her assessment? Explain why.

ADDITIONAL EXERCISES

18. Indicate whether the following statements are true or false:

a. If the hypothesis $H_0: \beta_1 = 0$ is rejected, then it can be safely concluded that x causes y.

b. Suppose a regression of y on x is run and the t statistic for testing $H_0: \beta_1 = 0$ versus $H_a: \beta_1 \neq 0$ has a p value of 0.0295 associated with it. Using a 5% level of significance, the null hypothesis should be rejected.

c. If the correlation between y and x is 0.9, then the R^2 value for a regression of y on x is 90%.

d. As long as the R^2 value is high for an estimated regression equation, it is safe to use the equation to predict for any value of x.

e. If the R^2 value for a regression of y on x is 75%, then the R^2 value for a regression of x on y is also 75%.

19. Suppose a regression analysis provides the following results:

$$b_0 = 1, \quad b_1 = 2, \quad s_{b_0} = 0.05,$$
$$s_{b_1} = 0.25, \quad SST = 117.2873, \quad SSE = 30.0$$

and $n = 24$. Use this information to solve the following problems.

a. Test the hypotheses

$$H_0: \beta_1 = 0$$
$$H_a: \beta_1 \neq 0$$

using a 5% level of significance. State the decision rule, the test statistic, and your decision. Use a t test.

b. Perform the same test as in part a using an F test. Use a 10% level of significance.

c. Compute the R^2 for the regression.

20. Suppose a regression analysis provides the following results:

$$b_0 = 4.0, \quad b_1 = 10.0, \quad s_{b_0} = 1.0,$$
$$s_{b_1} = 4.0, \quad SST = 67.36, \quad SSE = 50.0$$

and $n = 20$. Use this information to solve the following problems:

a. Test the hypotheses

$$H_0: \beta_0 = 0$$
$$H_a: \beta_0 \neq 0$$

using a 5% level of significance. State the decision rule, the test statistic, and your decision.

b. Test the hypotheses

$$H_0: \beta_1 \leq 0$$
$$H_a: \beta_1 > 0$$

using a 5% level of significance. State the decision rule, the test statistic, and your decision. What conclusion can be drawn from the test result?

c. Compute the R^2 for the regression.

21. Fill in the missing blanks on the following ANOVA table:

ANOVA Source	DF	SS	MS	F
Regression	1		1000	
Error (Residual)		800		
Total	81			

22. Fill in the missing blanks on the following ANOVA table:

ANOVA Source	DF	SS	MS	F
Regression			100	10
Error (Residual)	40			
Total				

23. Salary/Education. Data on beginning salary ($y = SALARY$) and years of education ($x = EDUC$) for 93 employees of Harris Bank Chicago in 1977 are provided in a data file named SALED3 on the CD. These data were obtained from an article by Daniel W. Schafer, "Measurement-Error Diagnostics and the Sex Discrimination Problem," *Journal of Business and Economic Statistics,* 5: 529–537, 1987. (Copyright 1987 by the American Statistical Association. Used with permission. All rights reserved.)

The scatterplot of salary verses education is shown in Figure 3.39. The regression results are shown in Figure 3.40. Use the results to answer the following questions:

a. Is there a linear relationship between salary and education? State the hypotheses to be tested, the decision rule, the test statistic, and your decision. Use a 10% level of significance.

b. What percentage of the variation in salary has been explained by the regression?

c. For an individual with 12 years of education, find a point prediction of beginning salary.

d. For all individuals with 12 years of education, find a point estimate of the conditional mean beginning salary.

e. What other factors, in addition to education, might be useful in helping to estimate beginning salary?

24. Cost Estimation. The file COSTEST3 on the CD contains data on production runs at a manufacturing plant. There are two columns of data:

$y = $ COST is the total cost of the production run.

$x = $ NUMBER is the number of items produced during that run.

Run the regression using COST as the dependent variable and NUMBER as the independent variable and use the result to help answer the following questions:

a. What is the estimated regression equation relating y to x?

b. What percentage of the variation in y has been explained by the regression?

c. Are y and x linearly related? Conduct a hypothesis test to answer this question and use a 5% level of significance. State the hypotheses to be tested, the decision rule, the test statistic,

FIGURE 3.39
Scatterplot for
Salary and Education
Exercise.

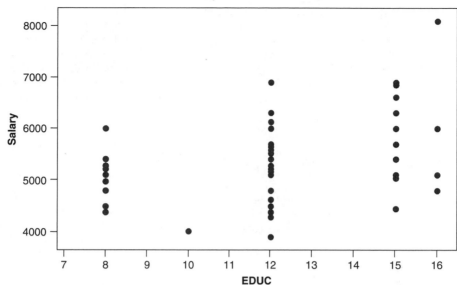

FIGURE 3.40
Regression Results for
Salary and Education
Exercise.

```
Variable        Coefficient    Std Dev    T Stat    P Value

Intercept         3818.6        377.4      10.12     0.000
EDUC               128.1         29.7       4.31     0.000


Standard Error = 650.112    R-Sq = 17.0%      R-Sq(adj) = 16.1%


Analysis of Variance

Source          DF    Sum of Squares   Mean Square    F Stat    P Value

Regression       1        7862534        7862534       18.60     0.000
Error           91       38460756         422646
Total           92       46323290
```

and your decision. What conclusion can be drawn from the result of the test?

d. Estimate the fixed cost involved in the production process. Find a point estimate and a 95% confidence interval estimate.

e. Estimate the variable cost involved in the production process. Find a point estimate and a 95% confidence interval estimate.

25. **Income/Consumption.** The following data are annual disposable income and total annual consumption for 12 families selected at random from a large metropolitan area. Regard annual disposable income as the explanatory variable and total annual consumption as the dependent variable. From the regression of y on x, answer the questions that follow. These data are in a file named INCONS3 on the CD.

Annual Disposable Income ($)	Total Annual Consumption ($)
INC	CONS
16,000	14,000
30,000	24,545
43,000	36,776
70,000	63,254

Annual Disposable Income ($)	Total Annual Consumption ($)
56,000	40,176
50,000	49,548
16,000	16,000
26,000	22,386
14,000	16,032
12,000	12,000
24,000	20,768
30,000	34,780

a. What is the estimated regression equation relating y to x?

b. What percentage of the variation in y has been explained by the regression?

c. Construct a 90% confidence interval estimate of β_1.

d. Use a t test to test the hypotheses $H_0: \beta_1 = 0$ versus $H_a: \beta_1 \neq 0$ at the 5% level of significance. State the decision rule, the test statistic, and your decision. What conclusion can be drawn from the result of the test?

e. Use an F test to test the hypotheses $H_0: \beta_1 = 0$ versus $H_a: \beta_1 \neq 0$ at the 5% level of significance. State the decision rule, the test statistic, and your decision.

f. Can the F test be used to test the hypotheses $H_0: \beta_1 \leq 0$ versus $H_a: \beta_1 > 0$?

g. Test the hypotheses $H_0: \beta_1 = 1$ versus $H_a: \beta_1 \neq 1$ at the 5% level of significance. State the decision rule, the test statistic, and your decision. What conclusion can be drawn from the result of the test?

26. **Apex Corporation.** The Apex Corporation produces corrugated paper. It has collected monthly data from January 2001 through March 2003 on the following two variables:

> y, total manufacturing cost per month (in thousands of dollars) (COST)
> x, total machine hours used per month (MACHINE)

The data are shown in Table 3.7 and are available in a file named APEX3 on the CD. Perform any analyses necessary to answer the following questions:

a. What is the estimated regression equation relating y to x?

b. What percentage of the variation in y has been explained by the regression?

c. Are y and x linearly related? Conduct a hypothesis test to answer this question and use a 5% level of significance. State the hypotheses to be tested, the decision rule, the test statistic, and your decision. What conclusion can be drawn from the result of the test?

d. Use the equation developed to estimate the average manufacturing cost in a month with 350 machine hours. Find a point estimate and a 95% confidence interval estimate. How reliable do you believe this forecast might be?

e. Use the equation developed to estimate the average manufacturing cost in a month with 550 machine hours. Find a point estimate and a 95% confidence interval estimate. How reliable do you believe this forecast might be?

27. **New Construction.** Our construction firm is interested in forecasting new construction in the United States for the years 2002 and 2003. We have data in billions of dollars for the years 1991 through 2001 from the Department of Commerce. These data are in the file NEWCON3 on the CD and are shown in Table 3.8.

a. Fit a linear trend to these data. What is the resulting regression equation?

b. What percentage of the variation in y has been explained by the regression?

c. Based on your answer in part b and on any other regression results you obtain, how well does the equation fit the data? Does a good fit ensure that forecasts for future years will be accurate?

d. Use the equation developed to predict new construction in both 2002 and 2003. Find a point prediction and a 95% prediction interval.

e. How reliable do you believe the forecast in part d might be? What factors might influence this accuracy?

28. **U.S. Population.** The data file USPOP3 on the CD contains the population of the United States for the years 1930 through 1999. Fit a linear trend to these data.

a. What is the resulting regression equation?

b. What percentage of the variation in y has been explained by the regression?

TABLE 3.7 Data for APEX Exercise

Date	COST	MACHINE	Date	COST	MACHINE
1/01	1102	218	3/02	1287	259
2/01	1008	199	4/02	1451	286
3/01	1227	249	5/02	1828	389
4/01	1395	277	6/02	1903	404
5/01	1710	363	7/02	1997	430
6/01	1881	399	8/02	1363	271
7/01	1924	411	9/02	1421	286
8/01	1246	248	10/02	1543	317
9/01	1255	259	11/02	1774	376
10/01	1314	266	12/02	1929	415
11/01	1557	334	1/03	1317	260
12/01	1887	401	2/03	1302	255
1/02	1204	238	3/03	1388	281
2/02	1211	246			

Source: These data were created by Professor Roger L. Wright, RLW Analytics, Inc., and are used (with modification) with his permission.

TABLE 3.8 Data for New Construction Exercise

YEAR	NEWCON
1991	432.6
1992	463.7
1993	491
1994	539.2
1995	557.8
1996	615.9
1997	653.4
1998	705.7
1999	765.9
2000	820.3
2001	842.5

c. Based on your answer in part b and on any other regression results you obtain, how well does the equation fit the data? Does a good fit ensure that forecasts for future years will be accurate?

d. Use the equation developed to predict the U.S. population in the years 2000 and 2001. Find a point prediction and a 95% prediction interval.

e. How reliable do you believe the forecast in part d might be? What factors might influence this accuracy?

29. **Wheat Exports.** The relationship between exchange rates and agricultural exports is of interest to agricultural economists. One such export of interest is wheat. The following data

   y, U.S. wheat export shipments (SHIPMENT)

x, the real index of weighted-average exchange rates for the U.S. dollar (EXCHRATE)

are available in a file named WHEAT3 on the CD.

These time-series data were observed monthly from January 1974 through March 1985. Perform any analyses necessary to answer the following questions:

a. What is the estimated regression equation relating *y* to *x*?

b. Are *y* and *x* linearly related? Conduct a hypothesis test to answer this question and use a 5% level of significance. State the hypotheses to be tested, the decision rule, the test statistic, and your decision. What conclusion can be drawn from the result of the test?

c. What percentage of the variation in y has been explained by the regression?

d. Construct a 95% confidence interval estimate of β_1.

(*Source*: Data are from D. A. Bessler and R. A. Babubla, "Forecasting Wheat Exports: Do Exchange Rates Really Matter?" *Journal of Business and Economic Statistics*, 5, 1987, pp. 397–406. Copyright 1987 by the American Statistical Association. Used with permission. All rights reserved.)

30. Major League Baseball Salaries. The owners of Major League Baseball (MLB) teams are concerned with rising salaries (as are owners of all professional sports teams). Table 3.9 provides the average salary (AVESAL) of the 30 MLB teams for the 2002 season. Also provided is the number of wins (WINS) for each team during the 2002 season. Is there evidence that teams with higher total payrolls tend to be more successful? Justify your answer. These data are available in a file named BBALL3 on the CD.

31. Computing Beta. In finance class you will discuss (or have discussed) the use of simple regression to estimate the relationship between the return on a stock and the market return. This relationship can be written as

$$y = \beta_0 + \beta_1 x + e$$

where y = return on the stock and x = the return on the market. The slope coefficient, β_1, is called the *beta coefficient* and is used to measure how responsive a stock's price is to movements in the market. The beta coefficient is used as a measure of a firm's systematic risk. In the file named BETA3 on the CD, the return on the stock of three companies is provided: Dell, Sabre, and Wal-Mart. Also provided is the return on the market (This is the value-weighted return computed by CRSP, the Center for Research on Security Prices). Five year's of monthly returns (January 1998 through December 2002) are used so there are a total of 60 observations for each company. Run the regression using the firm's return as the dependent variable and the market return as the independent variable for each of the three companies. Use the three regression results to answer the following questions:

a. What are the beta coefficients for each of the three companies?

b. Is there a relationship between the firm return and the market return for each of these three companies? Be sure to state the hypotheses to be tested, the decision rule, the test statistic, and your decision. Use a 5% level of significance.

c. The beta coefficient measures a security's responsiveness to movements in the market. For example, a beta of 2 would mean that a 1% increase (decrease) in the market return would result in, on average, a 2% increase (decrease) in the security's return. A beta of 1 would mean that movements in the market were matched, on average, by movements in the security's return. For each of the companies in the data file, test to see if the beta coefficient is equal to one or not. Be sure to state the decision rule, the test statistic, and your decision. Use a 5% level of significance.

d. Test to see if Dell's beta coefficient is greater than 1. Be sure to state the decision rule, the test statistic, and your decision. Use a 5% level of significance.

e. Test to see if Wal-Mart's beta coefficient is less than 1. Be sure to state the decision rule, the test statistic, and your decision. Use a 5% level of significance.

32. Major League Baseball Wins. What factor is most important in building a winning baseball team? Some might argue for a high batting average. Or it might be a team that hits for power as measured by the number of home runs. On the other hand, many believe that it is quality pitching as measured by the earned run average of the team's pitchers. The file MLB3 on the CD contains data on the following variables for the 30 major league baseball teams during the 2002 season:

$$\begin{aligned} \text{WINS} &= \text{ number of games won} \\ \text{HR} &= \text{ number of home runs hit} \\ \text{BA} &= \text{ average batting average} \\ \text{ERA} &= \text{ earned run average} \end{aligned}$$

Using WINS as the dependent variable, use scatterplots and regression to investigate the relationship of the other three variables to WINS. Which of the three possible explanatory

TABLE 3.9 Data for Major League Baseball Salaries Exercise

Team	WINS	AVESAL	Team	WINS	AVESAL
Anaheim Angels	99	2160054	Atlanta Braves	101	3166233
Baltimore Orioles	67	1855318	Chicago Cubs	67	2528398
Boston Red Sox	93	3633457	Cincinnati Reds	78	1658363
Chicago White Sox	81	1791286	Colorado Rockies	73	1848858
Cleveland Indians	74	2106591	Florida Marlins	79	1506567
Detroit Tigers	55	1562847	Houston Astros	84	2449680
Kansas City Royals	62	1832594	Los Angeles Dodgers	92	3396961
Minnesota Twins	94	1430068	Milwaukee Brewers	56	1338991
New York Yankees	103	4902777	Montreal Expos	83	1497309
Oakland Athletics	103	1746264	New York Mets	75	3192482
Seattle Mariners	93	3337435	Philadelphia Phillies	80	2086812
Tampa Bay Devil Rays	55	1131474	Pittsburgh Pirates	72	1370088
Texas Rangers	72	3123803	St. Louis Cardinals	97	2998072
Toronto Blue Jays	78	1868356	San Diego Padres	66	1292744
Arizona Diamondbacks	98	3199608	San Francisco Giants	95	3030571

Source: Reprinted courtesy of the *Fort Worth Star-Telegram*.

variables exhibits the strongest relationship to WINS? What might this suggest to managers of major league baseball teams?
(*Source*:Data courtesy of the *Fort Worth Star-Telegram*.)

33. **Work Orders.** During the construction phase of a nuclear plant, the number of corrective work orders open should gradually decline until reaching a steady state that would be present during the operational phase. The Nuclear Regulatory Commission has licensing requirements that the number of work orders open at licensing and for operational plants be less than 1000. (This was, of course, back in the days when nuclear plants were still being constructed in the United States.) This number is set to provide a goal indicating operational readiness. The number of work orders for a consecutive 120-working-day period during the construction phase of a nuclear plant are available in a file named WKORDER3 on the CD.

As a consultant to the plant, you have been asked to estimate how many days it will take to reach the operational level of 1000 work orders. In determining the number of days, state any assumptions you make and any caveats that might be in order.

34. **Fanfare.** Fanfare International, Inc., designs, distributes, and markets ceiling fans and lighting fixtures. The company's product line includes 120

basic models of ceiling fans and 138 compatible fan light kits and table lamps. These products are marketed to over 1000 lighting showrooms and electrical wholesalers that supply the remodeling and new construction markets. The product line is distributed by a sales organization of 58 independent sales representatives.

In the summer of 1994, Fanfare decided it needed to develop forecasts of future sales to help determine future salesforce needs, capital expenditures, and so on. The file named FAN3 on the CD contains monthly sales data and data on three additional variables for the period July 1990 through May 1994. The variables are defined as follows

SALES = total monthly sales in thousands of dollars

ADEX = advertising expense in thousands of dollars

MTGRATE = mortgage rate for 30-year loans (%)

HSSTARTS = housing starts in thousands of units

The data file contains the four variables as shown, plus columns for year and month. The sales data have been transformed to provide confidentiality.

As a consultant to Fanfare, your job is to find the best single variable to forecast future sales. Try each of the three variables in a simple regression and decide which is the best to create a forecasting model for Fanfare. Justify your choice. What problems do you see with using each of the three possible variables to help forecast sales?

35. College Graduation Rates. *Kiplinger's Personal Finance* provides information on the best public and private college values. Some of the variables included in this issue are as follows. All are based on the most recent available data.

GRADRATE4	the percentage of students who earned a bachelor's degree in four years (expressed as a percentage)
ADMISRATE	admission rate expressed as a percentage
SFACRATIO	student faculty ratio
AVGDEBT	average debt at graduation

This information is included in a file on the CD named COLLEGE3 for 195 schools. Only schools with complete information on all the categories listed are included. Using the graduation rate as the dependent variable, examine simple regressions using the independent variables provided. Which variable appears to do the best job of explaining graduation rate? How might you go about determining which of the possible variables used in a simple regression provides the best equation for predicting graduation rates?

(*Source*: Used by permission from the November and December 2003 issues of *Kiplinger's Personal Finance*. Copyright © 2003 The Kiplinger Washington Editors, Inc. Visit our website at www.kiplingers.com for further information.)

36. Retail Furniture Sales. The file FURNSALES3 on the CD contains monthly sales data (in millions of dollars) for retail furniture stores from January 1992 through December 2002. The data file contains a column with the year, the month (coded 1 = Jan through 12 = Dec) and SALES.

a. Use a linear trend model to forecast sales for each month of 2003.

b. Are there patterns present in this time series besides the trend in the data? If so, what are they?

37. Cubs Attendance. The Chicago Cubs baseball organization is interested in examining the relationship between attendance and the number of wins during the season. One possible hypothesized model is

$$\text{ATTENDANCE} = \beta_0 + \beta_1 \text{WINS} + e$$

They plan to use the equation to forecast future attendance and have annual data from 1972 through 1999. These data are shown in Table 3.10 and are in the file named CUBSWIN3 on the CD.

a. Find the regression equation using ATTENDANCE as the dependent variable and WINS as the explanatory variable.

TABLE 3.10 Cubs Attendance and Wins

YEAR	ATTENDANCE	WINS	YEAR	ATTENDANCE	WINS
1972	1299163	85	1986	1859102	70
1973	1351705	77	1987	2035130	76
1974	1015378	66	1988	2089034	77
1975	1034819	75	1989	2491942	93
1976	1026217	75	1990	2243791	77
1977	1439834	81	1991	2314250	77
1978	1525311	79	1992	2126720	78
1979	1648587	80	1993	2653763	84
1980	1206776	64	1994	1845208	49
1981	565637	38	1995	1918265	73
1982	1249278	73	1996	2219110	76
1983	1479717	71	1997	2190368	68
1984	2107655	96	1998	2623194	90
1985	2161534	77	1999	2813854	67

b. The Cubs organization wants to use the regression to forecast attendance next season: If we win 110 games next year, that means our forecast for attendance will be almost 2,600,000. With Sammy Sosa back, and if Kerry Wood is healthy, I think we've got a good chance at 110 wins. Do you see any problems with using the forecast for attendance specifically when the number of wins is 110?

USING THE COMPUTER

The Using the Computer section in each chapter describes how to perform the computer analyses in the chapter using Excel, MINITAB, and SAS. For further detail on Excel, MINITAB, and SAS, see Appendix C.

EXCEL

Plotting Data

FIGURE 3.41 Chart Wizard Dialog Box Showing Types of Scatterplots.

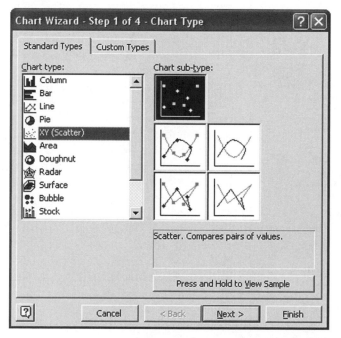

Use the Chart Wizard to create a scatterplot (or XY plot). Click the Chart Wizard button. A window opens showing chart types. Click on XY (Scatter), pick the type of scatterplot you want (see Figure 3.41), click Next> and follow the directions to create the scatterplot. To create a time-series plot, use the Line Plot feature in the Chart Wizard.

Regression

TOOLS: DATA ANALYSIS: REGRESSION

To perform a simple regression in Excel, use the Regression procedure on the Data Analysis menu. The Regression dialog box is shown in Figure 3.42. Fill in the Input Y Range and Input X Range with the cells containing the Y variable and X variable, respectively. Click the Labels box if your variables have labels in the first row. You can request an alternate level of confidence (instead of 95%) for confidence intervals for the regression coefficients by clicking the Confidence Level box and filling in the desired level. Choose the desired Output option and click OK.

Creating a Trend Variable

To put the numbers 1 through n in column B (for example), type 1 in B1, 2 in B2, then select these two cells, put the cursor on the rectangle at the bottom right-hand corner of cell B2, and drag through cell n.

FIGURE 3.42 Excel Regression Dialog Box.

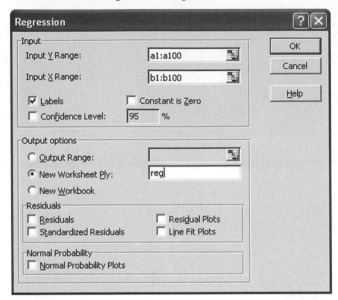

FIGURE 3.43 MINITAB Request for Type of Scatterplot Desired.

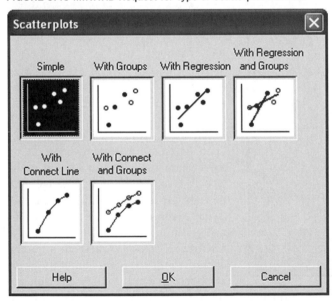

MINITAB

Plotting Data
GRAPH: SCATTERPLOT

Click Scatterplot and the dialog box in Figure 3.43 allows you to choose the type of plot you want. Click OK and the next dialog box (Figure 3.44) asks you indicate which columns represent the Y and X variables. When you click OK, MINITAB will plot the Y-data on the vertical axis and the X-data on the horizontal axis to create a scatterplot. There are a variety of other options available.

GRAPH: TIME-SERIES PLOT

Click Time Series Plot and the dialog box in Figure 3.45 allows you to choose the type of plot you want. Click OK and the next dialog box (Figure 3.46) asks you to indicate which column represents the data to be plotted. When you click OK, MINITAB plots the data values on the vertical axis versus a time indicator on the horizontal axis. MINITAB assumes the data are entered in a column with the most recent time period as the last entry in the column. There are a variety of other options available. For example, the use of month, quarter, and so on is chosen using the Time/Scale option.

Regression
STAT: REGRESSION: REGRESSION

See the Regression dialog box in Figure 3.47. Fill in the response (Y) and predictor (X) variable and click OK. A variety of options are available. For example, click Options and the PRESS, and predicted R^2 can be requested by checking the appropriate box. Other options are discussed in later chapters.

FIGURE 3.44 MINITAB Scatterplot Dialog Box.

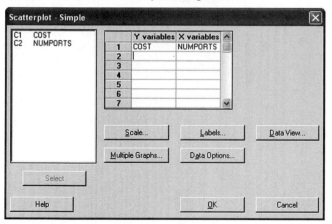

FIGURE 3.45 MINITAB Request for Type of Time-Series Plot Desired.

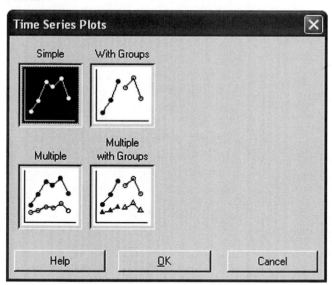

Forecasting with the Regression Equation

See the Regression–Options dialog box in Figure 3.48. Click OPTIONS in the Regression dialog box (Figure 3.47) to get to this screen. To generate forecasts and appropriate intervals, put the value of the *x* variable for which a forecast is desired in the line labeled Prediction intervals for new observations. If you want forecasts for several values, type those values in a column starting in row one and indicate that column in the Prediction intervals for new observations line.

Creating a Trend Variable

CALC: MAKE PATTERNED DATA: SIMPLE SET OF NUMBERS

See the Simple Set of Numbers dialog box in Figure 3.49. Type in the column number for the trend variable. Then enter a first value of 1 and a last value of *n* (the number of time periods) in steps of 1. Make sure "List each value" and "List the whole sequence" are set at 1. Then click OK.

SAS

Plotting Data

Plots in SAS are generated using the following command sequence:

```
PROC GPLOT;
PLOT COST*NUMPORTS;
```

or

```
PROC PLOT;
PLOT COST*NUMPORTS;
```

The variable to be plotted on the vertical axis (COST) is listed first, with the variable to be plotted on the horizontal axis (NUMPORTS) second. PLOT produces character plots. GPLOT produces high-resolution plots.

FIGURE 3.46 MINITAB Time-Series Plot Dialog Box.

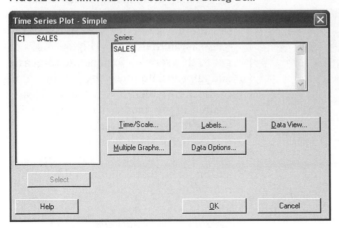

FIGURE 3.47 MINITAB Regression Dialog Box.

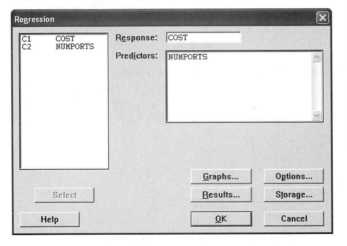

FIGURE 3.48 MINITAB Regression-Options Dialog Box.

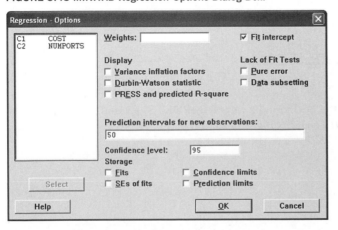

Regression

The following command sequence produces a regression with COST as the dependent variable and NUMPORTS as the independent variable:

```
PROC REG;
MODEL COST=NUMPORTS;
```

Forecasting with the Regression Equation

Forecasts in SAS are generated using an "appended" data set. To the values of the independent variable in the original data set, add the values for which predictions are desired. Then add to the values of the dependent variable the SAS symbol for missing data, a period, because we do not know those values. Now rerun the regression as follows:

```
PROC REG;
MODEL COST=NUMPORTS/P CLM CLI;
```

The option P requests forecasts (or predicted values), CLM requests upper and lower confidence interval limits for the estimate of the conditional mean, and CLI requests upper and lower prediction interval limits for an individual prediction.

Creating a Trend Variable

In the data input phase in SAS, use the command

```
TREND=_N_;
```

to create a trend variable. The command TREND=_N_ sets the variable TREND

FIGURE 3.49 MINITAB Dialog Box to Create a Trend Variable.

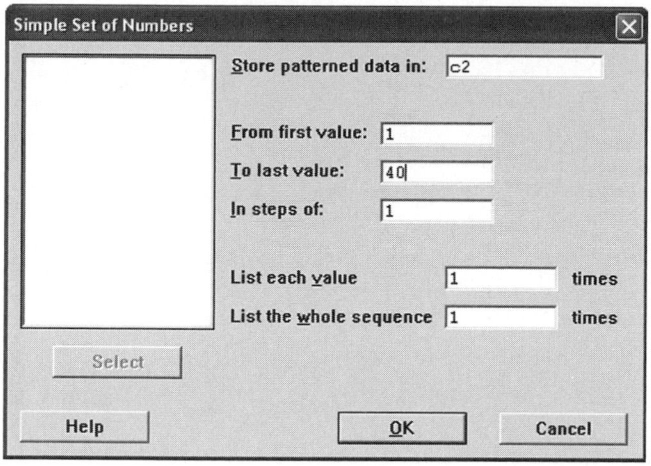

equal to the integers 1 through N, where N is the total number of observations in the data set. To do a time-series plot of the variable SALES, use the commands

```
PROC GPLOT;
PLOT SALES*TREND;
```

To fit the linear trend model, use the commands

```
PROC REG;
MODEL SALES=TREND;
```

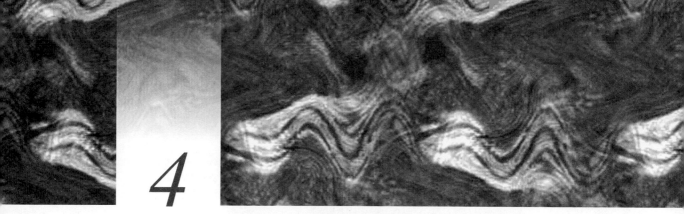

4 Multiple Regression Analysis

4.1 USING MULTIPLE REGRESSION TO DESCRIBE A LINEAR RELATIONSHIP

In Chapter 3, the method of least squares was used to develop the equation of a line that best described the relationship between a dependent variable y and an explanatory variable x. In business and economic applications, however, there may be more than one explanatory variable that is useful in explaining variation in the dependent variable y or obtaining better predictions of y. An equation of the form

$$\hat{y} = b_0 + b_1 x_1 + b_2 x_2$$

where x_1 and x_2 are the explanatory variables and b_1 and b_2 are estimates of the population regression coefficients may be desired. The relationship is still "linear"; each term on the right-hand side of the equation is additive, and the regression coefficients do not enter the equation in a nonlinear manner (such as $b_1^2 x_1$). The graph of the relationship is no longer a line, however, because there are three variables involved.

Graphing the equation thus requires the use of three dimensions rather than two, and the equation graphs as a plane passing through the three-dimensional space. Figure 4.1 shows how this graph might appear. The x_1 axis and y axis are drawn as before; the x_2 axis can be thought of as moving toward you to imitate the three-dimensional space. Because of the difficulty of drawing graphs in more than two dimensions on paper, the usefulness of graphical methods such as scatterplots is somewhat limited.

FIGURE 4.1 Graph
Showing Regression
"Plane."

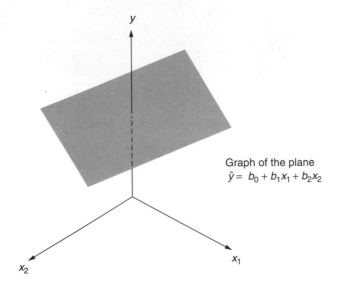

Graph of the plane
$\hat{y} = b_0 + b_1 x_1 + b_2 x_2$

Still, when two or more explanatory variables are involved, two-dimensional scatterplots between the dependent variable and each explanatory variable can provide an initial indication of the relationships present. The relationship involving more than one explanatory variable may differ, however, from that involving each explanatory variable individually. The least-squares method can still be used to develop regression equations involving more than one explanatory variable. These equations are referred to as *multiple regression equations*. As discussed, the equations no longer graph as lines, but the terms *linear regression* and even regression *line* (when perhaps regression *surface* might be more appropriate) still are used.

As the number of explanatory variables increases, the formulas for computing the estimates of the regression coefficients become increasingly complex. The availability of computerized regression routines precludes the need for hand computation of the estimates. The equations for the coefficient estimates when there are two or more explanatory variables are not presented in this text. There is a convenient method for writing the equations for the least-squares estimates for any number of explanatory variables, but it requires using matrices and matrix algebra. Because this text attempts to avoid as much mathematical detail as possible and concentrate on the use of computer regression output, the matrix presentation has been avoided; however, Appendix D does contain a brief introduction to the topic. A more advanced treatment of multiple regression that utilizes the matrix presentation is found, for example, in *Classical and Modern Regression with Applications* by R. Myers and in *Regression Analysis: Concepts and Applications* by F. Graybill and H. Iyer.[1]

---

[1] See References for complete publication information.

The concepts involved in producing least-squares coefficient estimates for a multiple regression equation are very similar to those for simple regression. An equation that "best" describes the relationship between a dependent variable y and K explanatory variables $x_1, x_2, \ldots, x_K$ can be written

$$\hat{y} = b_0 + b_1 x_1 + b_2 x_2 + \cdots + b_K x_K$$

where $b_0, b_1, b_2, \ldots, b_K$ are the least-squares coefficients. The case $K = 1$ is simple regression. The criterion for "best" is the same as it was for a simple regression; the difference between the true values of y and the values predicted by the multiple regression equation, $\hat{y}$, should be as small as possible. As before, this is accomplished by choosing $b_0, b_1, b_2, \ldots, b_K$ so that the sum of squares of the differences between the y and $\hat{y}$ values, $\sum_{i=1}^{n}(y_i - \hat{y}_i)^2$, is minimized. The optimizing values, $b_0, b_1, b_2, \ldots, b_K$ are the least-squares coefficients printed out by regression routines such as those available in Excel, MINITAB, and SAS.

EXAMPLE 4.1   **Meddicorp Sales**   Meddicorp Company sells medical supplies to hospitals, clinics, and doctors' offices. The company currently markets in three regions of the United States: the South, the West, and the Midwest. These regions are each divided into many smaller sales territories. Data for Meddicorp is contained in the MEDDICORP4 file on the CD.

Meddicorp's management is concerned with the effectiveness of a new bonus program. This program is overseen by regional sales managers and provides bonuses to salespeople based on performance. Management wants to know if the bonuses paid in 2003 were related to sales. (Obviously, if there is a relationship here, the managers expect it to be a direct—positive—one.) In determining whether this relationship exists, they also want to take into account the effects of advertising. The variables to be used in the study include:

> y, Meddicorp's sales (in thousands of dollars) in each territory for 2003 (SALES)
>
> x_1, the amount Meddicorp spent on advertising in each territory (in hundreds of dollars) in 2003 (ADV)
>
> x_2, the total amount of bonuses paid in each territory (in hundreds of dollars) in 2003 (BONUS)

Data for a random sample of 25 of Meddicorp's sales territories are shown in Table 4.1.

Figures 4.2 and 4.3 show the scatterplots of SALES versus ADV and BONUS, respectively. Figures 4.4, 4.5, and 4.6, respectively, show the MINITAB, Excel, and SAS regression output obtained relating SALES(y) to ADV(x_1) and BONUS(x_2). These outputs provide the multiple regression equation, which is used in this example, as well as additional information that will be used in later examples.

After rounding, the multiple regression equation describing the relationship between sales and the two explanatory variables may be written

$$\hat{y} = -516.4 + 2.47x_1 + 1.86x_2$$

or

$$SALES = -516.4 + 2.47ADV + 1.86BONUS$$

TABLE 4.1 Data for Meddicorp Example

Territory	SALES (in thousand $)	ADV (in hundred $)	BONUS (in hundred $)
1	963.50	374.27	230.98
2	893.00	408.50	236.28
3	1057.25	414.31	271.57
4	1183.25	448.42	291.20
5	1419.50	517.88	282.17
6	1547.75	637.60	321.16
7	1580.00	635.72	294.32
8	1071.50	446.86	305.69
9	1078.25	489.59	238.41
10	1122.50	500.56	271.38
11	1304.75	484.18	332.64
12	1552.25	618.07	261.80
13	1040.00	453.39	235.63
14	1045.25	440.86	249.68
15	1102.25	487.79	232.99
16	1225.25	537.67	272.20
17	1508.00	612.21	266.64
18	1564.25	601.46	277.44
19	1634.75	585.10	312.25
20	1159.25	524.56	292.87
21	1202.75	535.17	268.27
22	1294.25	486.03	309.85
23	1467.50	540.17	291.03
24	1583.75	583.85	289.29
25	1124.75	499.15	272.55

FIGURE 4.2
Scatterplot of SALES versus ADV.

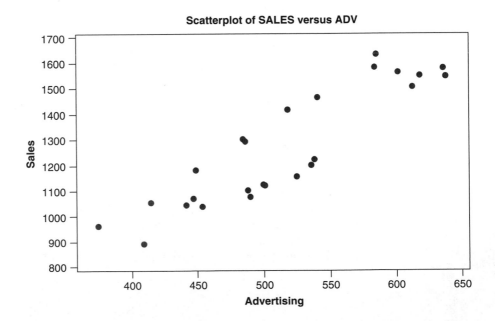

FIGURE 4.3
Scatterplot of SALES
versus BONUS.

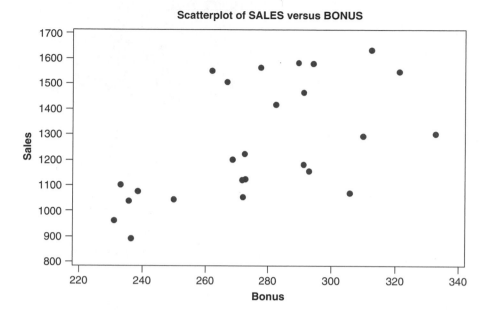

Scatterplot of SALES versus BONUS

FIGURE 4.4 MINITAB
Regression of SALES on
ADV and BONUS for
Meddicorp Example.

```
The regression equation is

SALES = -516 + 2.47 ADV + 1.86 BONUS

Predictor      Coef      SE Coef         T         P
Constant     -516.4        189.9     -2.72     0.013
ADV          2.4732       0.2753      8.98     0.000
BONUS        1.8562       0.7157      2.59     0.017


S = 90.7485      R-Sq = 85.5%      R-Sq(adj) = 84.2%

Analysis of Variance

Source         DF          SS        MS        F       P
Regression      2     1067797    533899    64.83   0.000
Residual Error 22      181176      8235
Total          24     1248974


Source         DF      Seq SS
ADV             1     1012408
BONUS           1       55389
```

This equation can be interpreted as providing an estimate of mean sales for a given level of advertising and bonus payment. Moreover, if advertising is held fixed, the equation shows that mean sales tends to rise by $1860 (1.86 thousands of dollars) for each unit increase in BONUS. Also, if bonus payment is held fixed, it shows that mean sales tends to rise by $2470 (2.47 thousands of dollars) for

SUMMARY OUTPUT

Regression Statistics

Multiple R	0.925
R Square	0.855
Adjusted R Square	0.842
Standard Error	90.749
Observations	25.000

ANOVA

	df	SS	MS	F	Significance F
Regression	2	1067797.321	533898.660	64.831	0.000
Residual	22	181176.419	8235.292		
Total	24	1248973.740			

	Coefficients	Standard Error	t Stat	P-value	Lower 95%	Upper 95%
Intercept	-516.444	189.876	-2.720	0.013	-910.223	-122.666
ADV	2.473	0.275	8.983	0.000	1.902	3.044
BONUS	1.856	0.716	2.593	0.017	0.372	3.341

FIGURE 4.5 Excel Regression of SALES on ADV and BONUS for Meddicorp Example.

FIGURE 4.6 SAS Regression of SALES on ADV and BONUS for Meddicorp Example.

The REG Procedure

Model: MODEL1

Dependent Variable: SALES

Analysis of Variance

Source	DF	Sum of Squares	Mean Square	F Value	Pr > F
Model	2	1067797	533899	64.83	<.0001
Error	22	181176	8235.29179		
Corrected Total	24	1248974			

Root MSE	90.74851	R-Square	0.8549
Dependent Mean	1269.02000	Adj R-Sq	0.8418
Coeff Var	7.15107		

Parameter Estimates

Variable	DF	Parameter Estimate	Standard Error	t Value	Pr > |t|
Intercept	1	-516.44428	189.87570	-2.72	0.0125
ADV	1	2.47318	0.27531	8.98	<.0001
BONUS	1	1.85618	0.71573	2.59	0.0166

each unit increase in ADV. (Note that a "unit" increase in either BONUS or ADV represents a $100 increase.) Clearly, such information provides a useful summary of the data.

4.2 INFERENCES FROM A MULTIPLE REGRESSION ANALYSIS

4.2.1 ASSUMPTIONS CONCERNING THE POPULATION REGRESSION LINE

In general, a population regression equation involving K explanatory variables can be written as

$$\mu_{y|x_1,x_2,\ldots,x_K} = \beta_0 + \beta_1 x_1 + \beta_2 x_2 + \cdots + \beta_K x_K$$

This equation says that the conditional mean of y given $x_1, x_2, \ldots, x_K$ is a point on the regression surface described by the terms on the right-hand side of the equation.

An alternative way of writing the relationship is

$$y_i = \beta_0 + \beta_1 x_{1i} + \beta_2 x_{2i} + \cdots + \beta_K x_{Ki} + e_i$$

where i denotes the ith observation and e_i is a random error or disturbance. Thus, y_i is related to the explanatory variables $x_{1i}, x_{2i}, \ldots, x_{Ki}$, although the relationship is not an exact one. The random error e_i shows that, given the same values for $x_{1i}, x_{2i}, \ldots, x_{Ki}$, each point y_i will not be exactly on the regression surface. Rather, the individual y_i values are distributed around the regression surface in the manner discussed for a simple regression line in Chapter 3. The following assumptions about the e_i are made:

1. The expected value of the disturbances is zero: $E(e_i) = 0$. This implies that the regression line passes through the conditional means of the y variable for each set of x variables. For our purposes, we interpret this assumption as: The population regression equation is linear in the explanatory variables.[2]

2. The variance of each e_i is equal to σ_e^2.

3. The e_i are normally distributed.

4. The e_i are independent. This is an assumption that is most important when data are gathered over time. When the data are cross-sectional (that is, gathered at the same point in time for different individual units), this is typically not an assumption of concern.

These assumptions allow inferences to be made about the population multiple regression line from a sample multiple regression line. The first inferences to be considered are those made about the individual population regression coefficients, $\beta_1, \beta_2, \ldots, \beta_K$. The effects of violations of the assumptions are considered in Chapter 6. In this chapter, each assumption is assumed to hold so that an ideal situation exists for the use of least-squares inference procedures.

4.2.2 INFERENCES ABOUT THE POPULATION REGRESSION COEFFICIENTS

This section considers estimates of the population regression coefficients and tests of hypotheses about the population coefficients. Much of the information required

---

[2] As pointed out in Chapter 3, the assumption here is that the population regression equation is linear in the x variables. In Chapter 5, we relax this assumption and find that we can fit curves by allowing equations that are not linear in the x variables. Throughout this text, however, we always assume that the equations are linear in the parameters. This means that equations such as $y = \beta_0 + \beta_1^2 x_1 + e$, are not considered. These types of equations are beyond the scope of this text.

to construct estimates and perform tests of hypotheses can be found in standard multiple regression output. For example, Figure 4.7 shows, in general, what information is provided by the multiple regression output for MINITAB, Excel, and SAS.

The least-squares estimates $b_0, b_1, b_2, \ldots, b_K$ are unbiased estimators of the corresponding population regression coefficients. A $(1 - \alpha)100\%$ confidence interval estimate of the population regression coefficient, β_k, is

$$b_k \pm t_{\alpha/2, n-K-1} s_{b_k}$$

Here, k refers to the kth regression coefficient, $k = 0, 1, \ldots, K$. The value $t_{\alpha/2, n-K-1}$ is a number chosen from the t table to ensure the appropriate level of confidence, and s_{b_k} is the standard deviation of the sampling distribution of b_k. The number of degrees of freedom used in determining the t value is $n - (K + 1)$, where $K + 1$ is the number of regression coefficients to be estimated (K coefficients corresponding to the K explanatory variables and one intercept or constant). Note that $n - (K + 1) = n - K - 1$ and is written in this manner throughout the text.

Hypothesis tests about the individual β_k also can be performed. The general form of two-tailed hypotheses about the individual β_k is as follows:

$$H_0: \beta_k = \beta_k^*$$
$$H_a: \beta_k \neq \beta_k^*$$

where β_k^* is any number chosen as the hypothesized value of the kth regression coefficient.

(a) MINITAB

The regression equation is
$y = b_0 + b_1 x_1 + b_2 x_2 + \cdots + b_K x_K$

Predictor	Coef	SE Coef	T	P
Constant	b_0	s_{b_0}	b_0/s_{b_0}	p value
x1 variable name	b_1	s_{b_1}	b_1/s_{b_1}	p value
x2 variable name	b_2	s_{b_2}	b_2/s_{b_2}	p value
.	.	.	.	.
.	.	.	.	.
.	.	.	.	.
xK variable name	b_K	s_{b_K}	b_K/s_{b_K}	p value

$S = s_e$ R-Sq $= R^2$ R-Sq(adj) $= R_{adj}^2$

Analysis of Variance

Source	DF	SS	MS		F	P
Regression	K	SSR	$MSR = SSR/K$		$F = MSR/MSE$	p value
Residual Error	$n - K - 1$	SSE	$MSE = SSE/(n - K - 1)$			
Total	$n - 1$	SST				

Unusual Observations

Obs.		X1	Y	Fit	SE Fit	Residual	St Resid
Obs.No.		Value of X1	Value of y	$\hat{y}$	s_m	$y - \hat{y}$	—

R denotes an observation with a large standardized residual
X denotes an observation whose X value gives it large influence

FIGURE 4.7 Illustration of MINITAB, Excel, and SAS Multiple Regression Output.

(b) Excel

Regression Statistics
Multiple R $= R$
R Square $= R^2$
Adjusted R Square $= R^2_{adj}$
Standard Error $= s_e$
Observations $= n$

ANOVA	df	SS	MS		F	Significance F
Regression	K	SSR	$MSR = SSR/K$		$F = MSR/MSE$	p value
Residual	$n - K - 1$	SSE	$MSE = SSE/(n - K - 1)$			
Total	$n - 1$	SST				

	Coefficients	Standard Error	t stat	P-value	Lower 95%	Upper 95%
Intercept	b_0	s_{b_0}	b_0/s_{b_0}	p value	$b_0 - t_{\alpha/2,n-K-1}s_{b_0}$	$b_0 + t_{\alpha/2,n-K-1}s_{b_0}$
x1 variable name	b_1	s_{b_1}	b_1/s_{b_1}	p value	$b_1 - t_{\alpha/2,n-K-1}s_{b_1}$	$b_1 + t_{\alpha/2,n-K-1}s_{b_1}$
x2 variable name	b_2	s_{b_2}	b_2/s_{b_2}	p value	$b_2 - t_{\alpha/2,n-K-1}s_{b_2}$	$b_2 + t_{\alpha/2,n-K-1}s_{b_2}$
.	.	.	.	.	.	.
.	.		.	.	.	.
.	.		.		.	.
xK variable name	b_K	s_{b_K}	b_K/s_{b_K}	p value	$b_K - t_{\alpha/2,n-K-1}s_{b_K}$	$b_K + t_{\alpha/2,n-K-1}s_{b_K}$

(c) SAS

Analysis of Variance

Source	DF	Sum of Squares	Mean Square	F Value	Pr > F
Model	K	SSR	$MSR = SSR/K$	$F = MSR/MSE$	p value
Error	$n - K - 1$	SSE	$MSE = SSE/(n - K - 1)$		
Corrected Total	$n - 1$	SST			

Root MSE	s_e	R-Square	R^2
Dependent Mean	$\bar{y}$	Adj R-Sq	R^2_{adj}
Coeff Var	$\dfrac{s_e}{\bar{y}} \times (100)$		

Parameter Estimates

Variable	DF	Parameter Estimate	Standard Error	t Value	Pr > |t|
Intercept	1	b_0	s_{b_0}	b_0/s_{b_0}	p value
x1 variable name	1	b_1	s_{b_1}	b_1/s_{b_1}	p value
x2 variable name	1	b_2	s_{b_2}	b_2/s_{b_2}	p value
.	.	.	.	.	
.	.	.	.	.	
.	.		.	.	
xK variable name	1	b_K	s_{b_K}	b_K/s_{b_K}	p value

FIGURE 4.7 (*Continued*).

The decision rule for this test is

$$\text{Reject } H_0 \text{ if } t > t_{\alpha/2,n-K-1} \text{ or } t < -t_{\alpha/2,n-K-1}$$
$$\text{Do not reject } H_0 \text{ if } -t_{\alpha/2,n-K-1} \leq t \leq t_{\alpha/2,n-K-1}$$

where α is the probability of a Type I error.

The standardized test statistic is

$$t = \frac{b_k - \beta_k^*}{s_{b_k}}$$

When the null hypothesis is true, the standardized test statistic, t, should be small in absolute value because the estimate, b_k, is close to the hypothesized value β_k^*, making the numerator $b_k - \beta_k^*$ close to zero. When the null hypothesis is false, the difference between b_k and β_k^* is large in absolute value, leading to a large absolute value of the test statistic and resulting in the decision to reject H_0.

The most common hypothesis test encountered in multiple regression analysis is

$$H_0: \beta_k = 0$$
$$H_a: \beta_k \neq 0$$

as in simple regression. This test is typically most important when β_k refers to the coefficient of the explanatory variable x_k rather than the intercept.

If the null hypothesis $H_0: \beta_k = 0$ is not rejected, then the conclusion is that, once the effects of all other variables in the multiple regression are included, x_k is not linearly related to y. In other words, adding x_k to the regression equation is of no help in explaining any additional variation in y left unexplained by the other explanatory variables.

On the other hand, if the null hypothesis is rejected, then there is evidence that y and x_k are linearly related and that x_k does help explain some of the variation in y not accounted for by the other explanatory variables.

Figure 4.7 shows that the test statistic for testing $H_0: \beta_k = 0$ is printed out on the regression output. The test statistic is

$$t = \frac{b_k}{s_{b_k}}$$

and is found in the column labeled "T" for MINITAB, the column labeled "t stat" for Excel, and the column labeled "t Value" for SAS. Also note that the p values for testing whether each population regression coefficient is equal to zero are found in the column labeled "P" in MINITAB, the column labeled "P-value" in Excel, and "Pr > |t|" in SAS.

EXAMPLE 4.2 **Meddicorp (Continued)**   Refer again to the MINITAB output in Figure 4.4, the Excel output in Figure 4.5, or the SAS output in Figure 4.6.

1. Use the regression output to test the following hypotheses:

$$H_0: \beta_1 = 0$$
$$H_a: \beta_1 \neq 0$$

where β_1 is the coefficient of ADV. Use a 5% level of significance. What conclusion can be drawn from the result of the test?

Answer 1: Using the standardized test statistic:

Decision Rule:	Reject H_0 if $t > 2.074$ or $t < -2.074$
	Do not reject H_0 if $-2.074 \leq t \leq 2.074$

Note: The t value with 22 degrees of freedom is 2.074 for a two-tailed test with a 5% level of significance.

Test Statistic:	8.98
Decision:	Reject H_0
Conclusion:	ADV is related to SALES (even when the effect of BONUS is taken into account).

Answer 2: Using the p value:

Decision Rule:	Reject H_0 if p value < 0.05
	Do not reject H_0 if p value ≥ 0.05
Test Statistic:	p value $= 0.000$
Decision:	Reject H_0

2. Use the regression output to test the following hypotheses:

$$H_0: \beta_2 = 0$$
$$H_a: \beta_2 \neq 0$$

where β_2 is the coefficient of BONUS. Use a 5% level of significance. What conclusion can be drawn from the result of the test?

Answer 1: Using the standardized test statistic:

Decision Rule:	Reject H_0 if $t > 2.074$ or $t < -2.074$
	Do not reject H_0 if $-2.074 \leq t \leq 2.074$.

Note: The t value with 22 degrees of freedom is 2.074 for a two-tailed test with a 5% level of significance.

Test Statistic:	2.59
Decision:	Reject H_0
Conclusion:	BONUS is related to SALES (even when the effect of ADV is taken into account).

Answer 2: Using the p value:

Decision Rule:	Reject H_0 if p value < 0.05
	Do not reject H_0 if p value ≥ 0.05
Test Statistic:	p value $= 0.017$
Decision:	Reject H_0

3. Use the regression output to test the following hypotheses:

$$H_0: \beta_2 \leq 0$$
$$H_a: \beta_2 > 0$$

where β_2 is the coefficient of BONUS. Use a 5% level of significance. What conclusion can be drawn from the result of the test?

Answer: Decision Rule: Reject H_0 if $t > 1.717$

Do not reject H_0 if $t \leq 1.717$

Note: The t value with 22 degrees of freedom is 1.717 for a one-tailed test with a 5% level of significance.

Test Statistic: $t = 2.59$

Decision: Reject H_0

Conclusion: BONUS is directly related to SALES (even when the effect of ADV is taken into account).

4. Construct a 95% confidence interval estimate of β_1, the coefficient of ADV.

Answer: $(2.4732 - (2.074 * 0.2753), 2.4732 + (2.074 * 0.2753))$ or $(1.9022, 3.0442)$

4.3 ASSESSING THE FIT OF THE REGRESSION LINE

4.3.1 THE ANOVA TABLE, THE COEFFICIENT OF DETERMINATION, AND THE MULTIPLE CORRELATION COEFFICIENT

As with simple regression, the variation in the dependent variable y in a multiple regression can be written as follows:

$$SST = SSE + SSR$$

The total variation in y is given by the total sum of squares:

$$SST = \sum_{i=1}^{n}(y_i - \bar{y})^2$$

The error sum of squares represents the variation in y left "unexplained" by the regression:

$$SSE = \sum_{i=1}^{n}(y_i - \hat{y}_i)^2$$

The regression sum of squares represents the variation in y "explained" by the regression:

$$SSR = \sum_{i=1}^{n}(\hat{y}_i - \bar{y})^2$$

In SSE and SSR, the $\hat{y}_i$ values are the predicted or fitted values from the multiple regression equation. These three sums of squares can be interpreted as in a simple regression context.

Figure 4.7 shows the format of the MINITAB, Excel, and SAS multiple regression output. Each of the three sums of squares is listed in the analysis of variance

(ANOVA) table as shown in the output. Also listed in the ANOVA table are the numbers of degrees of freedom associated with each of the sums of squares. For *SSR*, the number of degrees of freedom is equal to the number of explanatory variables, *K*. For *SSE*, the number of degrees of freedom is $n - K - 1$. As in the simple regression ANOVA table, the mean squares are also shown. These are computed by dividing the sums of squares by the appropriate number of degrees of freedom.

SST, SSR, and *SSE* can be used to evaluate how well the regression equation is explaining the variation in *y*. One measure of the goodness of fit of the regression is the coefficient of determination, R^2. The R^2 value, as for a simple regression, is computed by dividing *SSR* by *SST*:

$$R^2 = \frac{SSR}{SST}$$

Thus, R^2 represents the proportion of the variation in *y* explained by the regression. As before, R^2 ranges between 0 and 1. The closer to 1 the value of R^2 is, the better the fit of the regression equation to the data. If R^2 is multiplied by 100, it represents the percentage of the variation in *y* explained by the regression.

Although R^2 has a nice interpretation, there is a drawback to its use in multiple regression. As more explanatory variables are added to the regression model, the value of R^2 will never decrease, even if the additional variables are explaining an insignificant proportion of the variation in *y*. The addition of these unnecessary explanatory variables is not desirable. An alternative measure of the goodness of fit that is useful in multiple regression is R^2 adjusted for degrees of freedom (or simply, adjusted R^2). Recall that another way of writing R^2 is as

$$R^2 = 1 - \frac{SSE}{SST}$$

SSE/SST can be interpreted as the unexplained proportion of the total variation in *y*. Because the addition of explanatory variables to the model causes *SSE* to decrease, R^2 gets increasingly closer to 1. This happens even if the added explanatory variables have little significant relationship to *y*.

The adjusted R^2 does not suffer from this limitation. The adjusted R^2 is denoted as R^2_{adj}. It is computed by

$$R^2_{adj} = 1 - \frac{SSE/(n - K - 1)}{SST/(n - 1)}$$

Note that the sums of squares have been divided by (adjusted by) their degrees of freedom before they are used in computing R^2_{adj}. Now, suppose an explanatory variable is added to the regression model that produces only a very small decrease in *SSE*. The divisor, $n - K - 1$, also decreases because the number of explanatory variables, *K*, has been incremented by 1. It is possible that $SSE/(n - K - 1)$ may increase if the decrease in *SSE* from the addition of an explanatory variable is very small, because there is also a decrease in the size of the divisor. Thus, R^2_{adj} may decrease when the added explanatory variable adds little to the ability of the model to explain the variation in *y*. It is also possible that negative R^2_{adj} values may occur. This is not a mistake, but a result of a model that fits the data very poorly. (In

MINITAB, when negative R^2_{adj} values occur, they are printed as 0.0. Excel and SAS will print the actual value.)

R^2_{adj} no longer represents the proportion of variation in y explained by the regression, but it can be useful when comparing two regressions with different numbers of explanatory variables (say, a two-variable model with a three- or more variable model). A decrease in R^2_{adj} from the addition of one or more explanatory variables signals that the added variable(s) was of little importance in the regression equation. R^2_{adj} is purely a descriptive measure, however. The t test discussed previously can be used to compare two regressions that differ by just one variable. Comparing two regressions that differ by more than one variable will be discussed in more detail in Section 4.4.

MINITAB, SAS, and Excel print out the R^2 value for the regression and the value of R^2_{adj}. For regression routines that do not print out R^2_{adj}, it can be computed using the equation

$$R^2_{adj} = 1 - \frac{SSE/(n - K - 1)}{SST/(n - 1)}$$

or by using the relationship

$$R^2_{adj} = 1 + \frac{(n - 1)}{(n - K - 1)}(R^2 - 1)$$

Excel also prints a measure called the multiple correlation coefficient, R, which is the positive square root of R^2. The multiple correlation coefficient is equal to the simple correlation between the predicted y values, $\hat{y}$, and the true y values. Thus, it represents a measure of how closely associated the true values of y are with the points on the regression line. R^2 may be a preferable measure of goodness of fit because of its interpretation as percentage of variance explained.

4.3.2 THE F STATISTIC

Another measure of how well the multiple regression equation fits the data is the F statistic:

$$F = \frac{MSR}{MSE}$$

MSR is the mean square due to the regression, or the regression sum of squares divided by its degrees of freedom:

$$MSR = \frac{SSR}{K}$$

Note that SSR has K degrees of freedom, where K is the number of explanatory variables in the model. MSE is the mean square due to error, or the error sum of squares divided by its degrees of freedom:

$$MSE = \frac{SSE}{n - K - 1}$$

SSE has $n - K - 1$ degrees of freedom.

The F statistic is used to test the hypotheses

$$H_0: \beta_1 = \beta_2 = \cdots = \beta_K = 0$$
$$H_a: \text{At least one coefficient is not equal to zero}$$

The decision rule for the test is:

Reject H_0 if $F > F(a; K, n - K - 1)$

Do not reject H_0 if $F \leq F(a; K, n - K - 1)$

where $F(a; K, n - K - 1)$ is a value chosen from the F table for the appropriate level of significance, a. The critical value depends on the number of degrees of freedom associated with the numerator of the F statistic, K, and the number of degrees of freedom associated with the denominator, $n - K - 1$.

Failing to reject the null hypothesis implies that the explanatory variables in the regression equation are of little or no use in explaining the variation in the dependent variable, y. Rejection of the null hypothesis implies that *at least one* of the explanatory variables helps explain the variation in y. Rejection does not mean that all the population regression coefficients are different from zero (although this *may* be the case). Rejection does mean that the regression equation is useful, however.

If the hypothesis that all the population regression coefficients are zero is rejected, the t test discussed previously can be used to determine which of the individual variables contribute significantly to the model's ability to explain the variation in y. If the null hypothesis is not rejected, there is no need to perform the individual t tests. The F test can be thought of as a global test designed to assess the overall fit of the regression.

The information necessary to perform F tests is typically included in computer regression output in the ANOVA table. The Excel, SAS, and MINITAB outputs provide the computed value of the F statistic and the p value associated with the statistic (as shown in Figure 4.7).

EXAMPLE 4.3 **Meddicorp (Continued)**   Refer again to the MINITAB output in Figure 4.4, the Excel output in Figure 4.5, or the SAS output in Figure 4.6.

1. What percentage of the variation in sales has been explained by the regression?

Answer: 0.855, or 85.5%

2. What is the adjusted R^2?

Answer: 0.842

3. Conduct the F test for overall fit of the regression. Use a 5% level of significance.

$$H_0: \beta_1 = \beta_2 = 0$$
$$H_a: \text{At least one coefficient is not equal to zero}$$

Answer 1: Using the standardized test statistic:

Decision Rule: Reject H_0 if $F > F(0.05;2,22) = 3.44$

Do not reject H_0 if $F \leq F(0.05;2,22) = 3.44$

Test Statistic: $F = 64.83$

Decision: Reject H_0

Answer 2: Using the p value:

Decision Rule: Reject H_0 if p value < 0.05

Do not reject H_0 if p value ≥ 0.05

Test Statistic: p value $= 0.000$

Decision: Reject H_0

4. What conclusion can be drawn from the result of the F test for overall fit?

Answer: At least one of the coefficients (β_1, β_2) is not equal to zero. In other words, at least one of the variables (x_1, x_2) is important in explaining the variation in y.

EXERCISES

1. **Cost Control.** Ms. Karen Ainsworth is an employee of a well-known accounting firm's management services division. She is currently on a consulting assignment to the Apex Corporation, a firm that produces corrugated paper for use in making boxes and other packing materials. Apex called in consulting help to improve its cost control program, and Ms. Ainsworth is analyzing manufacturing costs to understand more fully the important influences on these costs. She has assembled monthly data on a group of variables, and she is using regression analysis to help assess how these variables are related to total manufacturing cost. The variables Ms. Ainsworth has selected to study, the data for which are contained in the file COST4 on the CD, are

- y, total manufacturing cost per month in thousands of dollars (COST)
- x_1, total production of paper per month in tons (PAPER)
- x_2, total machine hours used per month (MACHINE)
- x_3, total variable overhead costs per month in thousands of dollars (OVERHEAD)
- x_4, total direct labor hours used each month (LABOR)

The data shown in Table 4.2 refer to the period January 2001 through March 2003. Ms. Ainsworth wants to use a cost function developed by means of

regression analysis that initially includes all four of the explanatory variables. Use the regression results in Figure 4.8 to help answer the following questions:

a. What is the equation that is determined using all four explanatory variables?

b. Conduct the F test for overall fit of the regression. State the hypotheses to be tested, the decision rule, the test statistic, and your decision. Use a 5% level of significance. What conclusion can be drawn from the result of the test?

c. In the cost accounting literature, the sample regression coefficient corresponding to x_k is regarded as an estimate of the true marginal cost of output associated with the variable x_k. Find a point estimate of the true marginal cost associated with total machine hours per month. Also, find a 95% confidence interval estimate of the true marginal cost associated with total machine hours.

d. Test the hypothesis that the true marginal cost of output associated with total production of paper is 1.0. Use a 5% level of significance and a two-tailed test procedure. State the hypotheses to be tested, the decision rule, the test statistic, and your decision. What conclusion can be drawn from the result of the test?

e. What percentage of the variation in y has been explained by the regression?

TABLE 4.2 Data for Cost Control Exercise

COST	PAPER	MACHINE	OVERHEAD	LABOR
1102	550	218	112	325
1008	502	199	99	301
1227	616	249	126	376
1395	701	277	143	419
1710	838	363	191	682
1881	919	399	210	751
1924	939	411	216	813
1246	622	248	124	371
1255	626	259	127	383
1314	659	266	135	402
1557	740	334	181	546
1887	901	401	216	655
1204	610	238	117	351
1211	598	246	124	370
1287	646	259	127	387
1451	732	286	155	433
1828	891	389	208	878
1903	932	404	216	660
1997	964	430	233	694
1363	680	271	129	405
1421	723	286	146	426
1543	784	317	158	478
1774	841	376	199	601
1929	922	415	228	679
1317	647	260	126	378
1302	656	255	117	380
1388	704	281	142	429

Source: These data were created by Dr. Roger L. Wright, RLW Analytics, Inc., Sonoma, CA, and are used (with modification) with his permission.

FIGURE 4.8 Regression Results for Cost Control Exercise.

Variable	Coefficient	Std Dev	T Stat	P Value
Intercept	51.72	21.70	2.38	0.026
PAPER	0.95	0.12	7.90	0.000
MACHINE	2.47	0.47	5.31	0.000
OVERHEAD	0.05	0.53	0.09	0.927
LABOR	−0.05	0.04	−1.26	0.223

Standard Error = 11.0756   R-Sq = 99.9%   R-Sq(adj) = 99.9%

Analysis of Variance

Source	DF	Sum of Squares	Mean Square	F Stat	P Value
Regression	4	2271423	567856	4629.17	0.000
Error	22	2699	123		
Total	26	2274122			

FIGURE 4.9 Regression Results for Salaries Exercise.

Variable	Coefficient	Std Dev	T Stat	P Value
Intercept	3179.5	383.4	8.29	0.000
EDUC	139.6	27.7	5.04	0.000
EXPER	1.5	0.7	2.13	0.036
TIME	20.6	6.2	3.35	0.001

Standard Error = 602.728   R-Sq = 30.2%   R-Sq(adj) = 27.9%

Analysis of Variance

Source	DF	Sum of Squares	Mean Square	F Stat	P Value
Regression	3	13991247	4663749	12.84	0.000
Error	89	32332043	363281		
Total	92	46323290			

f. What is the adjusted R^2 for this regression?

g. Based on the regression equation, what actions might be taken to control costs?

2. Salaries. The file on the CD named HARRIS4 contains values of the following four variables for 93 employees of Harris Bank Chicago in 1977:

> y, beginning salary in dollars (SALARY)
> x_1, years of schooling at the time of hire (EDUC)
> x_2, number of months of previous work experience (EXPER)
> x_3, number of months after January 1, 1969, that the individual was hired (TIME)

The regression results for the regression of SALARY on the three explanatory variables are shown in Figure 4.9. Use the results to help answer the following questions:

a. What is the estimated regression equation relating SALARY to EDUC, EXPER, and TIME?

b. Conduct the F test for overall fit of the regression. Use a 5% level of significance. State the hypotheses to be tested, the decision rule, the test statistic, and your decision. What conclusion can be drawn from the result of the test?

c. Is education linearly related to beginning salary (after taking into account the effect of experience and time)? Perform the hypothesis test necessary to answer this question. State the hypotheses to be tested, the decision rule, the test statistic, and your decision. Use a 5% level of significance.

d. What percentage of the variation in salary has been explained by the regression?

4.4 COMPARING TWO REGRESSION MODELS

4.4.1 FULL AND REDUCED MODEL COMPARISONS USING SEPARATE REGRESSIONS

Thus far, two types of hypothesis tests for multiple regression models have been considered:

1. A test of the overall fit of the regression:

$$H_0: \beta_1 = \beta_2 = \cdots = \beta_K = 0$$

H_a: At least one coefficient is not equal to zero

2. A test of the significance of each individual regression coefficient:

$$H_0: \beta_k = 0$$
$$H_a: \beta_k \neq 0$$

In multiple regression models, it also may be useful to test whether subsets of coefficients are equal to zero. In this section, a *partial F* test to test whether any subset of coefficients in a multiple regression equals zero is considered.

To set up this hypothesis test, consider the following regression model:

$$y = \beta_0 + \beta_1 x_1 + \beta_2 x_2 + \cdots + \beta_L x_L + \beta_{L+1} x_{L+1} + \cdots + \beta_K x_K + e$$

Testing whether the variables $x_{L+1}, \ldots, x_K$ are useful in explaining any variation in y after taking account of the variation already explained by $x_1, \ldots, x_L$ can be viewed as a comparison of two regression models to determine whether it is worthwhile to include the additional variables. The two models for comparison are called the *full* and *reduced* models.

Full Model

$$y = \beta_0 + \beta_1 x_1 + \beta_2 x_2 + \cdots + \beta_L x_L + \beta_{L+1} x_{L+1} + \cdots + \beta_K x_K + e$$

This is called the full model because all K explanatory variables of interest are included.

Reduced Model

$$y = \beta_0 + \beta_1 x_1 + \beta_2 x_2 + \cdots + \beta_L x_L + e$$

This is called the reduced model because the variables $x_{L+1}, \ldots, x_K$ have been removed.

The question to be answered is, "Is the full model significantly better than the reduced model at explaining the variation in y?" This question can be formalized by setting up the following null and alternative hypotheses:

H_0: $\beta_{L+1} = \cdots = \beta_K = 0$

H_a: At least one of the coefficients $\beta_{L+1}, \ldots, \beta_K$ is not equal to zero

If the null hypothesis is not rejected, choose the reduced model; if the null hypothesis is rejected, at least one of $x_{L+1}, \ldots, x_K$ is contributing to the explanation of the variation in y, and the full model is chosen as superior to the reduced.

To test the hypotheses (that is, to compare the full and reduced models), an F statistic is used. The F statistic can be written:

$$F = \frac{(SSE_R - SSE_F)/(K - L)}{SSE_F/(n - K - 1)}$$

where the subscript F stands for full model and the subscript R stands for reduced model.

Now consider what is being computed in the F statistic. If the full and reduced models are estimated, the regression output includes the error sum of squares for each of these regressions. In the F statistic, SSE_F refers to the error sum of squares from the full model output using all K explanatory variables. SSE_R refers to the error sum of squares from the reduced model output using only L explanatory variables. Recall that the error sum of squares represents the variation in y unexplained by the regression. Also, the reduced model error sum of squares can never be less than the full model error sum of squares, so the difference $SSE_R - SSE_F$ is always greater than or equal to zero. This difference represents the additional amount of the variation in y explained by adding $x_{L+1}, \ldots, x_K$ to the regression model. This measure of improvement is then divided by the number of additional variables to be added to the model, $K - L$. The numerator thus represents the additional variation in y explained per additional variable used. Note that the numerator degrees of freedom, $K - L$, is equal to the number of coefficients included in the null hypothesis or, equivalently, to the difference in the number of explanatory variables in the full and reduced models.

The mean square error for the full regression model is used in the denominator:

$$MSE_F = \frac{SSE_F}{n - K - 1}$$

If the measure of improvement is large relative to the mean square error for the full model, then the F statistic is large. If the improvement measure is small relative to MSE_F, then the value of the F statistic is small. The decision rule for the test is

Reject H_0 if $F > F(\alpha; K - L, n - K - 1)$

Do not reject H_0 if $F \leq F(\alpha; K - L, n - K - 1)$

Here, α is the probability of a Type I error, and $F(\alpha; K - L, n - K - 1)$ is a value chosen from the F table for level of significance α, $K - L$ numerator degrees of freedom, and $n - K - 1$ denominator degrees of freedom. This test is referred to as a partial F test and can be performed with any statistical package by running both the full and reduced model regressions.

EXAMPLE 4.4   **Meddicorp (continued)**   Management of Meddicorp believes that, in addition to advertising and bonus, two other explanatory variables may be important in explaining the variation in sales. These variables are

x_3, market share currently held by Meddicorp in each territory (MKTSHR)

x_4, largest competitor's sales in each territory (COMPET)

These two additional variables are shown in Table 4.3 for each territory and in the file MEDDICORP4 on the CD.

The regression results for the regression of SALES on ADV, BONUS, MKTSHR, and COMPET are shown in Figure 4.10. This is the full model output. The hypothesized population regression model is

$$y = \beta_0 + \beta_1 x_1 + \beta_2 x_2 + \beta_3 x_3 + \beta_4 x_4 + e$$

Consider the test of the hypotheses

$H_0: \beta_3 = \beta_4 = 0$

$H_a:$ At least one of the coefficients (β_3, β_4) is not equal to zero

TABLE 4.3 Additional Data for Meddicorp Example

Territory	MKTSHR (percentage)	COMPET (in thousand $)
1	33	202.22
2	29	252.77
3	34	293.22
4	24	202.22
5	32	303.33
6	29	353.88
7	28	374.11
8	31	404.44
9	20	394.33
10	30	303.33
11	25	333.66
12	34	353.88
13	42	262.88
14	28	333.66
15	28	232.55
16	30	273.00
17	29	323.55
18	32	404.44
19	36	283.11
20	34	222.44
21	31	283.11
22	32	242.66
23	28	333.66
24	27	313.44
25	26	374.11

FIGURE 4.10 Results for the Regression of SALES on ADV, BONUS, MKTSHR, and COMPET.

Variable	Coefficient	Std Dev	T Stat	P Value
Intercept	-593.537	259.196	-2.29	0.033
ADV	2.513	0.314	8.00	0.000
BONUS	1.906	0.742	2.57	0.018
MKTSHR	2.651	4.636	0.57	0.574
COMPET	-0.122	0.372	-0.32	0.749

Standard Error = 93.7697   R-Sq = 85.9%   R-Sq(adj) = 83.1%

Analysis of Variance

Source	DF	Sum of Squares	Mean Square	F Stat	P Value
Regression	4	1073119	268280	30.51	0.000
Error	20	175855	8793		
Total	24	1248974			

The reduced model output is in Figure 4.4 (MINITAB), 4.5 (Excel), or 4.6 (SAS). The F statistic can be computed as

$$F = \frac{(181{,}176 - 175{,}855)/2}{175{,}855/20} = 0.303$$

(Note that $K = 4$ and $L = 2$, so $K - L = 2$.) If a 5% level of significance is used, the decision rule is

Reject H_0 if $F > 3.49$

Do not reject H_0 if $F \leq 3.49$

where 3.49 is the 5% F critical value with 2 numerator and 20 denominator degrees of freedom.

We do not reject H_0 and so the conclusion is that both coefficients β_3 and β_4 are equal to zero. Thus, the variables x_3 and x_4 are not useful in explaining any of the remaining variation in y.

SAS allows the user to request that the partial F test be conducted for any group of x variables desired (see Using the Computer at the end of this chapter). Figure 4.11 shows the result of requesting a partial F test for the variables MKTSHR and COMPET. The value of the F statistic is shown as 0.3. A p value of 0.7422 is also provided. Using the p value decision rule, the null hypothesis would not be rejected at reasonable levels of significance.

4.4.2 FULL AND REDUCED MODEL COMPARISONS USING CONDITIONAL SUMS OF SQUARES[3]

Another way to view partial F tests is through the use of conditional or sequential sums of squares. For a regression model with two explanatory variables,

$$\hat{y} = b_0 + b_1 x_1 + b_2 x_2$$

the standard ANOVA table appears as in Figure 4.12(a). In Figure 4.12(b), an alternative ANOVA table is presented. In this figure, the regression sum of squares has been decomposed into two parts. The first, $SSR(x_1)$, is the sum of squares explained by x_1 if it were the only explanatory variable. The second $SSR(x_2|x_1)$, is called a *conditional* or *sequential sum of squares*. It represents the sum of squares explained by x_2 in addition to that explained by x_1. That is, given that x_1 has explained a certain amount of variation in y, $SSR(x_2|x_1)$ shows how much of the remaining variation x_2 explains. Note that

$$SSR = SSR(x_1) + SSR(x_2|x_1)$$

Here, SSR is the variation explained by both x_1 and x_2.

To test the hypotheses

$$H_0: \beta_2 = 0$$
$$H_a: \beta_2 \neq 0$$

FIGURE 4.11 SAS Output Showing the Partial F Test for the MKTSHR and COMPET Variables.

Test 1 Results for Dependent Variable SALES

Source	DF	Mean Square	F Value	Pr > F
Numerator	2	2660.61069	0.30	0.7422
Denominator	20	8792.75990		

[3] Optional section.

FIGURE 4.12 ANOVA Tables for Two Explanatory Variable Regressions.

(a) Standard ANOVA Table

Source of Variation	DF	SS
Regression	2	SSR
Error	$n - 3$	SSE
Total	$n - 1$	SST

(b) ANOVA with Conditional Sums of Squares Explained by Each Explanatory Variable

Source of Variation	DF	SS
Regression		
x_1	1	$SSR(x_1)$
$x_2\|x_1$	1	$SSR(x_2\|x_1)$
Error	$n - 3$	SSE
Total	$n - 1$	SST

an F statistic can be constructed using the conditional sum of squares.

$$F = \frac{SSR(x_2|x_1)/1}{SSE/(n - 3)}$$

The numerator of F is the conditional sum of squares for x_2, given that x_1 is in the model, divided by its degrees of freedom. The conditional sum of squares has 1 degree of freedom because it represents the sum of squares explained by only one variable. The denominator is the error sum of squares for the full model, the model with both x_1 and x_2, divided by its degrees of freedom. If x_2 explains little of the additional unexplained variation in y, then $SSR(x_2|x_1)$ is small as is the F statistic. The more variation explained by x_2, the bigger the F statistic is. The decision rule for the test is

Reject H_0 if $F > F(\alpha; 1, n - 3)$

Do not reject H_0 if $F \leq F(\alpha; 1, n - 3)$

where $F(\alpha; 1, n - 3)$ is chosen from an F table for level of significance α, 1 is the numerator degree of freedom, and $n - 3$ is the denominator degrees of freedom.

Of course, this hypothesis could be tested with a two-tailed t test as described in Section 4.2 because it involves only one coefficient. It can be shown that an F statistic with 1 numerator degree of freedom is equal to the square of a t statistic and that the F critical value equals the square of a t critical value for appropriately chosen levels of significance and degrees of freedom, denoted df:

$$F(\alpha;1, df) = t^2_{\alpha/2,df}$$

Therefore, the decision made is the same regardless of which test is used. Since t statistics routinely appear on regression output, the t test is typically used when testing hypotheses about individual coefficients. The t test has additional advantages over the F test. The t test can be used to perform one-tailed hypothesis tests, whereas the F is restricted to the two-tailed test. It is also easier to test whether a coefficient is equal to some value other than zero using a t test than it is using an F test.

The F test gains its advantage when testing whether a subset of coefficients are all equal to zero—for example, to test

H_0: $\beta_{L+1} = \cdots = \beta_K = 0$

H_a: At least one of the coefficients $\beta_{L+1}, \ldots, \beta_K$ is not equal to zero

for the general model presented earlier. In this case, the t test cannot be used. Even performing individual t tests on each coefficient may not provide as much information as performing the F test on the coefficients as a group.

To test whether $\beta_{L+1}, \ldots, \beta_K$ are all zero, the following F statistic is used:

$$F = \frac{SSR(x_{L+1}, \ldots, x_K | x_1, x_2, \ldots, x_L)/(K - L)}{SSE/(n - K - 1)}$$

$SSR(x_{L+1}, \ldots, x_K | x_1, x_2, \ldots, x_L)$ is the additional variation in y explained by $x_{L+1}, \ldots, x_K$, given that $x_1, \ldots, x_L$ are already in the model. The number of degrees of freedom associated with this conditional sum of squares is $K - L$, the number of coefficients to be included in the test. SSE is the error sum of squares from the model with all the variables included and is divided by its degrees of freedom, $n - K - 1$. The conditional sum of squares can be computed as

$$SSR(x_{L+1}, \ldots, x_K | x_1, x_2, \ldots, x_L) = SSR(x_{L+1} | x_1, x_2, \ldots, x_L)$$
$$+ SSR(x_{L+2} | x_1, x_2, \ldots, x_{L+1}) + \ldots + SSR(x_K | x_1, x_2, \ldots, x_{K-1})$$

The regression output from certain statistical packages contains the necessary information to compute the conditional sums of squares. For example, in MINITAB, an additional sum of squares breakdown is provided as in Figure 4.13. The table provides the conditional sums of squares for each of the variables individually, given that the previous variables are in the model. By adding these sums of squares for appropriate individual variables, the conditional sum of squares for x_{L+1} through x_K is obtained. In the MINITAB output, the conditional sums of squares are denoted SEQ SS for sequential sums of squares. This term is used to indicate that the sums of squares explained by each of the x variables when entered sequentially are represented.

The order in which the variables enter the regression is very important when using the conditional sums of squares to construct a partial F statistic. In MINITAB, for example, the explanatory variables whose coefficients are included in the null hypothesis must be the last ones in the list of variables in the regression dialog box (see Using the Computer at the end of this chapter). This ensures that the conditional sums of squares are computed appropriately for the hypothesis to be tested.

A more extensive look at the use of these conditional sums of square is provided in the following example.

EXAMPLE 4.5   Meddicorp (continued)   Consider again the problem posed in Example 4.4. In addition to BONUS and ADV, Meddicorp wants to consider the possibility that MKTSHR and COMPET are important in explaining the variation in sales. The

FIGURE 4.13
Breakdown of SSR into Its Conditional Components as Provided by MINITAB Regression Output.

SOURCE	DF	SEQ SS
x_1	1	$SSR(x_1)$
x_2	1	$SSR(x_2|x_1)$
.	.	.
.	.	.
.	.	.
x_{L-1}	1	$SSR(x_{L-1}|x_1, \ldots, x_{L-2})$
x_L	1	$SSR(x_L|x_1, \ldots, x_{L-1})$
x_{L+1}	1	$SSR(x_{L+1}|x_1, \ldots, x_L)$
.	.	.
.	.	.
.	.	.
x_K	1	$SSR(x_K|x_1, \ldots, x_{K-1})$

MINITAB regression of SALES on ADV, BONUS, MKTSHR, and COMPET is shown in Figure 4.14.

The hypothesized population regression model is

$$y = \beta_0 + \beta_1 x_1 + \beta_2 x_2 + \beta_3 x_3 + \beta_4 x_4 + e$$

Consider the test of the hypotheses

$$H_0: \beta_3 = \beta_4 = 0$$

H_a: At least one of the coefficients (β_3, β_4) is not equal to zero

The MINITAB regression output gives the conditional sums of squares explained by each variable. The regression sum of squares for the full regression is $SSR = 1{,}073{,}119$. The conditional sums of squares are as follows:

$$SSR(x_1) = 1{,}012{,}408$$
$$SSR(x_2|x_1) = 55{,}389$$
$$SSR(x_3|x_1, x_2) = 4394$$
$$SSR(x_4|x_1, x_2, x_3) = 927$$

The decision rule for the test is

Reject H_0 if $F > 3.49$

Do not reject H_0 if $F \leq 3.49$

where 3.49 is the F critical value using a 5% level of significance with 2 and 20 degrees of freedom.

The test statistic, F, is

$$F = \frac{SSR(x_3, x_4|x_1, x_2)/2}{SSE/20} = \frac{[SSR(x_4|x_1, x_2, x_3) + SSR(x_3|x_1, x_2)]/2}{SSE/20}$$

$$= \frac{(927 + 4394)/2}{175{,}855/20} = 0.303$$

FIGURE 4.14
MINITAB Regression of
SALES on ADV, BONUS,
MKTSHR, and COMPET.

```
The regression equation is

SALES = - 594 + 2.51 ADV + 1.91 BONUS + 2.65 MKTSHR - 0.121 COMPET


Predictor              Coef      SE Coef        T        P
Constant             -593.5        259.2    -2.29    0.033
ADV                  2.5131       0.3143     8.00    0.000
BONUS                1.9059       0.7424     2.57    0.018
MKTSHR                2.651        4.636     0.57    0.574
COMPET              -0.1207       0.3718    -0.32    0.749


S = 93.7697    R-Sq = 85.9%    R-Sq(adj) = 83.1%


Analysis of Variance
Source            DF         SS          MS        F        P
Regression         4    1073119      268280    30.51    0.000
Residual Error    20     175855        8793
Total             24    1248974


Source        DF      Seq SS
ADV            1     1012408
BONUS          1       55389
MKTSHR         1        4394
COMPET         1         927


Unusual Observations
Obs   ADV    SALES      Fit   SE Fit   Residual   St Resid
 20   525   1159.3   1346.2     39.4     -187.0      -2.20R
R denotes an observation with a large standardized residual.
```

The null hypothesis cannot be rejected. The variables x_3 and x_4 do not significantly improve the model's ability to explain sales.   ▧

EXERCISES

3. **Cost Control (continued).** Consider again the cost data from Exercise 4.1 and the regression results in Figure 4.8. Consider this output to be for the full model:

$$y = \beta_0 + \beta_1 x_1 + \beta_2 x_2 + \beta_3 x_3 + \beta_4 x_4 + e$$

where y, x_1, x_2, x_3, and x_4 were defined in the first exercise.

Now consider the reduced model:

$$y = \beta_0 + \beta_1 x_1 + \beta_2 x_2 + e$$

Conduct the test to compare these two models. State the hypotheses to be tested, the decision

rule, the test statistic, and your decision. What conclusion can be drawn from the result of the test? The regression results for the reduced model can be found in Figure 4.15. Use a 5% level of significance.

4. **Salaries (continued).** Consider again the salary data from Exercise 4.2 and the regression results in Figure 4.9. Consider this output to be for the full model:

$$y = \beta_0 + \beta_1 x_1 + \beta_2 x_2 + \beta_3 x_3 + e$$

where y, x_1, x_2, and x_3 were defined in Exercise 4.2.

FIGURE 4.15
Regression Results for the Reduced Model in the Cost Control Exercise.

Variable	Coefficient	Std Dev	T Stat	P Value
Intercept	59.4318	19.6388	3.03	0.006
PAPER	0.9489	0.1101	8.62	0.000
MACHINE	2.3864	0.2101	11.36	0.000

Standard Error = 10.9835   R-Sq = 99.9%   R-Sq(adj) = 99.9%

Analysis of Variance

Source	DF	Sum of Squares	Mean Square	F Stat	P Value
Regression	2	2271227	1135613	9413.48	0.000
Error	24	2895	121		
Total	26	2274122			

FIGURE 4.16
Regression Results for the Reduced Model in the Salaries Exercise.

Variable	Coefficient	Std Dev	T Stat	P Value
Intercept	3818.56	377.44	10.12	0.000
EDUC	128.09	29.70	4.31	0.000

Standard Error = 650.112   R-Sq = 17.0%   R-Sq(adj) = 16.1%

Analysis of Variance

Source	DF	Sum of Squares	Mean Square	F Stat	P Value
Regression	1	7862534	7862534	18.60	0.000
Error	91	38460756	422646		
Total	92	46323290			

Now consider the reduced model:

$$y = \beta_0 + \beta_1 x_1 + e$$

Conduct the test to compare these two models. State the hypotheses to be tested, the decision rule, the test statistic, and your decision. What conclusion can be drawn from the result of the test? The regression results for the reduced model can be found in Figure 4.16. Use a 5% level of significance.

4.5 PREDICTION WITH A MULTIPLE REGRESSION EQUATION

As with simple regression, one of the possible goals of fitting a multiple regression equation is using it to predict values of the dependent variable. The two cases considered here are the same as in simple regression.

4.5.1 ESTIMATING THE CONDITIONAL MEAN OF Y GIVEN $X_1, X_2, \ldots, X_K$

In this case, the goal is to estimate the point on the regression surface for specific values of the explanatory variables. For example, in the Meddicorp example (Example 4.1), consider the population regression equation

$$\mu_{y|x_1, x_2} = \beta_0 + \beta_1 x_1 + \beta_2 x_2$$

where x_1 is ADV and x_2 is BONUS. The estimated regression equation from Figure 4.4 (MINITAB), 4.5(Excel), or 4.6(SAS) is

$$\hat{y} = -516.4 + 2.47x_1 + 1.86x_2$$

A point estimate of the conditional mean of y given x_1 and x_2 can be written as

$$\hat{y}_m = b_0 + b_1x_1 + b_2x_2$$

In the Meddicorp problem, the point estimate of the conditional mean of y given $x_1 = 500$ and $x_2 = 250$ is

$$\hat{y}_m = -516.4 + 2.47(500) + 1.86(250) = 1183.6$$

This is an estimate of average sales for *all* territories with advertising 500 and bonus 250. Confidence interval estimates can also be constructed. The formula for s_m, the standard deviation of $\hat{y}_m$, is omitted here due to its complexity.

Figure 4.17 shows the MINITAB output for this example, including the predicted value (*Fit*), the standard error of the estimate of the point on the regression line, s_m, (*SE Fit*), a 95% confidence interval for the estimate of the conditional mean (95% *CI*), and a 95% prediction interval for each prediction (95% *PI*). (The difference in $\hat{y}_m$ computed by MINITAB and by hand is due to rounding. The MINITAB forecasts are more accurate and therefore are preferred.)

4.5.2 PREDICTING AN INDIVIDUAL VALUE OF Y GIVEN $X_1, X_2, \ldots, X_K$

Write the population regression equation for a single individual as

$$y_i = \beta_0 + \beta_1x_{1i} + \beta_2x_{2i} + e_i$$

where e_i is the random disturbance. Denote the predicted value of y for an individual as $\hat{y}_p$. To predict the value of a dependent variable for a single individual, the point on the regression surface is used:

$$\hat{y}_p = b_0 + b_1x_1 + b_2x_2$$

As in simple regression, the point estimate of $\mu_{y|x_1, x_2, \ldots, x_K}$ and the point prediction for an individual are the same. However, the standard error of the prediction, s_p, is larger than the standard error of the forecast, s_m. Thus, the prediction interval is wider than the confidence interval, reflecting the greater uncertainty in predicting for individuals than in estimating a conditional mean.

FIGURE 4.17
Prediction in the Meddi-corp Example Using MINITAB.

```
Predicted Values for New Observations

New
Obs          Fit       SE Fit        95% CI              95% PI
  1       1184.2        25.2     (1131.8, 1236.6)   (988.8, 1379.5)


Values of Predictors for New Observations

New
Obs      ADV      BONUS
  1      500        250
```

MINITAB and SAS produce prediction intervals when requested. The forecast standard error, s_m, also is printed in both outputs. If the prediction standard error is desired, it can be computed using the relationship $s_p^2 = s_m^2 + s_e^2$, where s_e^2 is the *MSE* of the regression.

4.6 MULTICOLLINEARITY: A POTENTIAL PROBLEM IN MULTIPLE REGRESSION

4.6.1 CONSEQUENCES OF MULTICOLLINEARITY

For a regression of y on K explanatory variables $x_1, x_2, \ldots, x_K$, it is hoped that the explanatory variables are highly correlated with the dependent variable. A relationship is sought that explains a large portion of the variation in y. At the same time, however, it is not desirable for strong relationships to exist among the explanatory variables. When explanatory variables are correlated with one another, the problem of *multicollinearity* is said to exist. How serious the problem is depends on the degree of multicollinearity. Low correlations among the explanatory variables generally do not result in serious deterioration of the quality of the least-squares estimates. But high correlations may result in highly unstable least-squares estimates of the regression coefficients.

The presence of a high degree of multicollinearity among the explanatory variables results in the following problems:

1. The standard deviations of the regression coefficients are disproportionately large. As a result, the t values computed to test whether the population regression coefficients are zero are small. The null hypothesis that the coefficients are zero may not be rejected even when the associated variable is important in explaining variation in y.

2. The regression coefficient estimates are unstable. Because of the high standard errors, reliable estimates of the regression coefficients are difficult to obtain. Signs of the coefficients may be the opposite of what is intuitively reasonable. Dropping one variable from the regression or adding a variable causes large changes in the estimates of the coefficients of other variables.

4.6.2 DETECTING MULTICOLLINEARITY

Numerous ways have been suggested in the literature to help detect multicollinearity. These are listed here with some recommendations on their usefulness:

1. **Pairwise Correlations:** Compute the pairwise correlations between the explanatory variables. Because multicollinearity exists when explanatory variables are highly correlated, these correlations should help identify any highly correlated pairs of variables. One rule of thumb suggested by some researchers is that multicollinearity may be a serious problem if any pairwise correlation is bigger than 0.5.

Limitations: There are two limitations to this approach. First, the correlation cutoff of 0.5 is somewhat arbitrary and not always effective in identifying serious pairwise multicollinearity problems. Second, only relationships between two explanatory variables can be investigated. For example, if there are three explanatory variables in the model, x_1, x_2, and x_3, the pairwise correlations can be computed between x_1 and x_2, x_1 and x_3, and x_2 and x_3. But the relationships resulting in the multicollinearity may be more complex than simple pairwise correlations. The variable x_1 may not be highly correlated with x_2 or x_3 individually, but may be highly correlated with some linear combination of the two variables. That is, x_1 may be highly correlated with $a_1x_2 + a_2x_3$.

Another suggested rule of thumb is that multicollinearity may be a serious problem if any of the pairwise correlations among the x variables is larger than the largest of the correlations between the y variable and the x variables. Although this rule does not suffer from an arbitrary cutoff point (such as 0.5), it does suffer from the same limitations concerning more complex relationships among the x variables.

2. **Large F, small t:** An indication of multicollinearity is a large overall F statistic but small t statistics. As mentioned, multicollinearity results in large standard deviations of the regression coefficients and small t ratios. Thus, the test for whether the individual regression coefficients are equal to zero may fail to reject the null hypotheses $H_0: \beta_k = 0$ even when the variables included in the regression are important in explaining the variation in y. The overall F statistic is typically not affected by the multicollinearity, however. If the variables are important, the F statistic should be large, indicating a good overall fit even if the t statistics appear to be saying that none of the variables is important.

Limitations: This method of detecting multicollinearity is not always effective because multicollinearity may result in some, but not all, of the t values being small. The question of whether the variable is unimportant or whether it just appears so because of multicollinearity cannot be answered by looking at the output. Although this approach may be helpful in pointing out that multicollinearity exists in some instances, it does not provide any information on which of the explanatory variables are highly correlated with others.

3. **Variance Inflation Factors (VIFs):** Let x_1, x_2, ..., x_K be the K explanatory variables in a regression. Perform the regression of x_j on the remaining $K - 1$ explanatory variables and call the coefficient of determination from this regression R_j^2. The *VIF* for the variable x_j is

$$VIF_j = \frac{1}{1 - R_j^2}$$

A variance inflation factor can be computed for each explanatory variable. It is a measure of the strength of the relationship between each explanatory variable and all other explanatory variables in the regression. Thus, pairwise correlations are taken into account as well as more complex relationships with two or more of the other variables. The value R_j^2 measures the strength of the relationship between x_j and the other $K - 1$ explanatory variables. If there is no relationship (an ideal case), then $R_j^2 = 0.0$ and $VIF_j = 1/(1 - 0) = 1$. As R_j^2 increases, VIF_j increases also. For example, if $R_j^2 = 0.9$, then $VIF_j = 1/(1 - 0.9) = 10$; if $R_j^2 = 0.99$, then $VIF_j = 1/(1 - 0.99) = 100$. Large values of VIF_j suggest that x_j may be highly related to other explanatory variables and, thus, multicollinearity may be a problem. How large the *VIF*s must be to suggest a serious problem with multicollinearity is not completely clear. Some suggested guidelines are as follows:

a. Any individual VIF_j larger than 10 indicates that multicollinearity may be influencing the least-squares estimates of the regression coefficients.

b. If the average of the VIF_j, $\overline{VIF} = \sum_{j=1}^{K} VIF_j/K$, is considerably larger than 1, then serious problems may exist. $\overline{VIF}$ indicates how many times larger the error sum of squares for the regression is due to multicollinearity than it is if the variables are uncorrelated.

c. *VIF*s also need to be evaluated relative to the overall fit of the model. Freund and Wilson[4] note that whenever the *VIF*s are less than $1/(1 - R^2)$ where R^2 is the coefficient of determination for the model with all x variables included, multicollinearity is not strong enough to affect the coefficient estimates. In this case the independent variables are more strongly related to the y variable than they are to each other.

4.6.3 CORRECTION FOR MULTICOLLINEARITY

One obvious solution to the multicollinearity problem is to remove those variables that are highly correlated with others and thus eliminate the problem. One obvious drawback of this approach is that no information is obtained on the omitted variables. In addition, the omission of one variable causes changes in the estimates of the regression coefficients of variables left in the equation. It is important for the researcher to make a careful selection of potential explanatory variables to include in the regression. If it is found that some of these variables result in multicollinearity problems, it may be necessary to exclude certain of the variables in favor of others. This is where the insight and expertise of the researcher come into play in developing the most useful regression equation.

In certain cases, adding more data can break the pattern of multicollinearity. But this solution is not always possible, especially in many business and economics situations. It also does not always work even when it is possible.

When multicollinearity is present, it affects the regression coefficient estimates in the ways noted earlier. However, it does not affect the ability to obtain a good fit of the regression (high R^2). Nor does it affect the quality of forecasts or predictions from the regression (as long as the pattern of multicollinearity continues for those observations where forecasts are desired). Thus, if the regression model is to be used strictly for forecasting, corrections may be unnecessary. Even when developing a

---

[4] Freund, R. J., and Wilson, W. J., *Regression Analysis: Statistical Modeling of a Response Variable*, San Diego: Academic Press, 1998.

model for forecasting, however, it is often desirable to test whether individual variables are contributing significantly to the explanatory power of the model. In this case, multicollinearity remains a problem.

Finally, several other statistical procedures have been proposed as possible remedial measures with multicollinearity. These include ridge regression and principal components regression. These techniques are beyond the scope of this text. The interested reader is referred to Raymond H. Myers, *Classical and Modern Regression with Applications*, pp. 243–263.

4.7 LAGGED VARIABLES AS EXPLANATORY VARIABLES IN TIME-SERIES REGRESSION

When using time-series data, it is possible to relate values of the dependent variable in the current time period to explanatory variable values in the current time period. For example, sales for a firm in the current month can be related to advertising expenditures in the current month. It may be, however, that sales in the current month are not affected as much by advertising expenditures in the current month as by advertising expenditures from the previous month or from 2 months ago. This fact can be incorporated into a time-series regression. To illustrate, let y_i represent sales in time period i, x_i represent advertising expenditures in time period i, x_{i-1} represent advertising expenditures in time period $i - 1$, and so on. Then a possible model for sales could be written

$$y_i = \beta_0 + \beta_1 x_i + \beta_2 x_{i-1} + \beta_3 x_{i-2} + e_i$$

Here the variable sales is modeled as a function of advertising expenditures in the current month and the 2 previous months.

The variables x_{i-1} and x_{i-2} are called *lagged variables*. Any lags felt to be appropriate may be used. Here the one- and two-period lags are used. Some caution must be exercised, however. Because lagged variables are likely to be highly correlated with each other, including several such variables may result in multicollinearity problems discussed in the previous section.

When lagged variables are used, a certain number of data points in the initial time periods are lost. This is illustrated in Table 4.4. Note that no value can be computed for x_{i-1} in time period 1. No prior time period exists from which to take this value. For the same reason, no value for the first or second time period can be computed for x_{i-2}. These time periods have to be omitted from the analysis, reducing the effective sample size from eight to six in the example.

Lagged values of the dependent variable also can be used as explanatory variables. Consider again the sales example. Now, however, assume that no information on advertising is available. Sales in the current month are modeled simply as a function of sales in the previous month:

$$y_i = \beta_0 + \beta_1 y_{i-1} + e_i$$

The data are illustrated in Table 4.5. Of course, further lags can be used if desired. One observation is lost for each lag.

TABLE 4.4 Creation of Lagged Values of the Explanatory Variables

i	x_i	x_{i-1}	x_{i-2}
1	4	*	*
2	7	4	*
3	8	7	4
4	10	8	7
5	11	10	8
6	9	11	10
7	15	9	11
8	16	15	9

*Indicates a missing value.

TABLE 4.5 Creation of Lagged Values of the Dependent Variable

i	y_i	y_{i-1}
1	22	*
2	24	22
3	27	24
4	35	27
5	38	35
6	42	38
7	47	42
8	50	47

*Indicates a missing value.

Note that the model with the lagged value of the dependent variable as an explanatory variable may be viewed as an extrapolative time-series model (introduced in Chapter 3). We are using only past information from the series itself to help describe the behavior of the series and to forecast future values.

It should be noted, however, that regressions that include lagged dependent variable values are still sometimes interpreted as causal relationships. For example, the one-period lagged value of sales might be included along with the current and lagged values of advertising:

$$y_i = \beta_0 + \beta_1 y_{i-1} + \beta_2 x_i + \beta_3 x_{i-1} + e_i$$

If it is believed that last month's sales might help to generate new sales in the current month or to maintain sales, then this model could be justified as causal.

In the economics literature, a model with lagged values of the dependent variable (and possible lagged values of other explanatory variables) might be called an adaptive expectations model or a partial adjustment model (see, for example, G. Judge et al., *The Theory and Practice of Econometrics,* pp. 379–380). For alternatives to regression analysis useful in analyzing time-series data, see B. Bowerman and R. O'Connel, *Forecasting and Time Series: An Applied Approach.*[5]

---

[5] See References for complete publication information.

EXAMPLE 4.6   **Unemployment Rate**   The file on the CD named UNEMP4 contains monthly unemployment rates from January 1983 until December 2002. (These data were obtained from the web site www.economagic.com and were obtained from the St. Louis Federal Reserve Bank.) The data have been seasonally adjusted.

In Figure 4.18, a time-series plot of unemployment rate is shown. The horizontal axis in this plot is an index that numbers each month from 1 to 240.

In Figure 4.19, the regression of unemployment rate (UNEMP) on the one-period lagged unemployment rate is shown. The one-period lagged variable has been called UNEMPL1. Note that in any analyses involving UNEMPL1, there is one less observation than the total time-series length because missing cases are not used.

The regression of the monthly unemployment rate on the previous month's rate obviously produces a good fit. The t value for UMEMPL1 is 134.43 (p value $= 0.000$), resulting in rejection of the hypothesis H_0: $\beta_1 = 0$. Also, 98.7% of the variation in unemployment rate has been explained by the regression.

Now consider adding a two-period lagged variable, UNEMPL2, to the equation. The regression of UNEMP on UNEMPL1 and UNEMPL2 is shown in Figure 4.20. The regression model can now be written

$$y_i = \beta_0 + \beta_1 y_{i-1} + \beta_2 y_{i-2} + e_i$$

To test whether the two-period lag is of any importance in the model, the following hypotheses should be tested:

$$H_0: \beta_2 = 0$$
$$H_a: \beta_2 \neq 0$$

The test statistic value is 1.23 (p value $= 0.218$). At a 5% level of significance, the decision rule for the test is

Reject H_0 if $t > 1.96$ or $t < -1.96$

Do not reject H_0 if $-1.96 \leq t \leq 1.96$

The z value of 1.96 is used since a large number of degrees of freedom are available (235). The null hypothesis is not rejected, suggesting that the two-period lagged variable is not useful in explaining any of the additional variation in unemployment rates. A three-period lagged variable was also tried, but it, too, was found to be of no additional help in the regression.

Obviously, this process of lagging the dependent variable can be continued for additional lags if desired. However, the use of continued lagged variables as explanatory variables is questionable in this problem. With an R^2 of 98.7% for the one-period lag model, it is obvious that not much room for improvement is available. Even if an additional variable were found to be significant, it still might be best for the sake of simplicity to use the one-period lagged variable model.

Also, if additional lags are examined, problems can arise. For example, the lagged variables are highly correlated among themselves, which could result in

FIGURE 4.18
Time-Series Plot
of Unemployment
Rates.

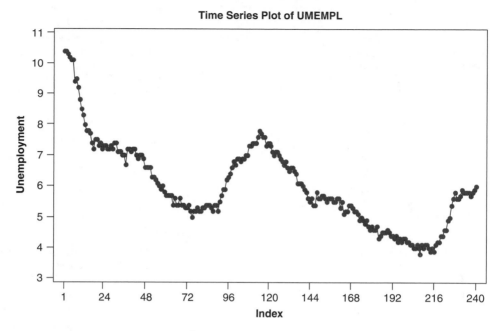

FIGURE 4.19 Re-
gression of Unemploy-
ment on One-Period
Lagged Unemployment.

```
Variable        Coefficient      Std Dev        T  Stat      P Value

Intercept       0.153195         0.044597          3.43      0.001
UNEMPL1         0.971495         0.007227        134.43      0.000


Standard Error = 0.151529       R-Sq = 98.7%   R-Sq(adj) = 98.7%


Analysis of Variance


Source       DF      Sum of Squares      Mean Square      F Stat      P Value

Regression    1          414.92             414.92        18070.47    0.000
Error       237            5.44               0.02
Total       238          420.36
```

multicollinearity problems. Also, for each additional lag used, one data point is lost. Since there were originally 240 monthly observations on unemployment, this loss is not a substantial part of the data set. However, in smaller data sets, the loss could be significant.

FIGURE 4.20
Regression of Unemployment on One- and Two-Period Lagged Unemployment.

Variable	Coefficient	Std Dev	T Stat	P Value
Intercept	0.16764	0.04565	3.67	0.000
UNEMPL1	0.89032	0.06497	13.70	0.000
UNEMPL2	0.07842	0.06353	1.23	0.218

Standard Error = 0.151381   R-Sq = 98.7%   R-Sq(adj) = 98.6%

Analysis of Variance

Source	DF	Sum of Squares	Mean Square	F Stat	P Value
Regression	2	395.55	197.77	8630.30	0.000
Error	235	5.39	0.02		
Total	237	400.93			

EXERCISES

5. **Mortgage Rates.** The file on the CD named MRATES4 contains monthly 30-year conventional mortgage rates from January 1985 to December 2002. (These figures are found on the web site www.economagic.com and are obtained from the Federal Home Mortgage Corporation).

Figure 4.21 provides a time-series plot of the rates. The following model is to be used to forecast mortgage rates:

$$y_i = \beta_0 + \beta_1 y_{i-1} + e_i$$

where y_i is the mortgage rate at time i and y_{i-1} is the rate at time $i - 1$. The regression results are shown in Figure 4.22. Note that RATEL1 in the output represents the one-period lagged variable, y_{i-1}. Use the output to answer the following questions:

a. What is the estimated regression equation?

b. Is there a relationship between current and previous period mortgage rates? State the hypotheses to be tested, the decision rule, the test statistic, and your decision. Use a 5% level of significance.

FIGURE 4.21
Time-Series Plot of Mortgage Rates.

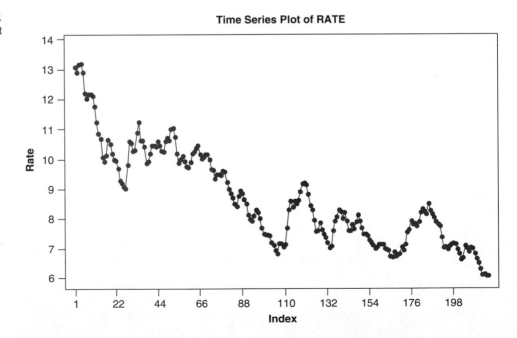

Time Series Plot of RATE

FIGURE 4.22
Regression Results for Mortgage Rates Exercise.

Variable	Coefficient	Std Dev	T Stat	P Value
Intercept	0.16077	0.08854	1.82	0.071
RATEL1	0.97774	0.01002	97.60	0.000

Standard Error = 0.235473   R-Sq = 97.8%   R-Sq(adj) = 97.8%

Analysis of Variance

Source	DF	Sum of Squares	Mean Square	F Stat	P Value
Regression	1	528.13	528.13	9524.82	0.000
Error	213	11.81	0.06		
Total	214	539.94			

c. What percentage of the variation in mortgage rates has been explained by the regression?

d. Use the estimated equation to produce a forecast of the mortgage rate in January 2003. Find out what the actual rate was and compare it to the forecast. How well did the equation do? Repeat this process for the remainder of 2003. Discuss any difficulties you encounter in forecasting more than 1 month ahead.

e. Test whether the intercept of the equation is equal to zero. State the hypotheses to be tested, the decision rule, the test statistic, and your decision. Use a 5% level of significance.

f. Test whether the slope of the equation is equal to 1. State the hypotheses to be tested, the decision rule, the test statistic, and your decision. Use a 5% level of significance.

ADDITIONAL EXERCISES

6. Fill in the missing blanks on the following ANOVA table:

ANOVA

Source	DF	SS	MS	F
Regression	2		250	
Error (Residual)		400		
Total	82			

7. Fill in the missing blanks on the following ANOVA table:

ANOVA

Source	DF	SS	MS	F
Regression		300	100	10
Error (Residual)	27			
Total				

8. Wheat Exports. The relationship between exchange rates, prices, and agricultural exports is of interest to agricultural economists. One such export of interest is wheat. The file named WHEAT4 on the CD contains data on the following variables:

y, U.S. wheat export shipments (SHIPMENT)
x_1, the real index of weighted-average exchange rates of the U.S. dollar (EXCHRATE)
x_2, the per-bushel real price of no. 1 red winter wheat (PRICE)

The dependent variable is U.S. wheat export shipments. The explanatory variables are exchange rate and price. The data are observed monthly from January 1974 through March 1985. The regression results are shown in Figure 4.23. Use the output to help answer the following questions:

a. What is the estimated regression equation relating SHIPMENT to EXCHRATE and PRICE?

b. Test the overall fit of the regression. State the hypotheses to be tested, the decision rule, the test statistic, and your decision. Use a 5% level of significance. What conclusion can be drawn from the result of the test?

c. After taking account of the effect of PRICE, are SHIPMENT and EXCHRATE related? Conduct a

FIGURE 4.23
Regression Results for
Wheat Export Exercise.

Variable	Coefficient	Std Dev	T Stat	P Value
Intercept	3361.932	633.194	5.31	0.000
EXCHRATE	1.869	4.223	0.44	0.659
PRICE	-2413.837	846.480	-2.85	0.005

Standard Error = 798.260   R-Sq = 8.8%   R-Sq(adj) = 7.4%

Analysis of Variance

Source	DF	Sum of Squares	Mean Square	F Stat	P Value
Regression	2	8117338	4058669	6.37	0.002
Error	132	84112922	637219		
Total	134	92230260			

hypothesis test to answer this question and use a 5% level of significance. State the hypotheses to be tested, the decision rule, the test statistic, and your decision. What conclusion can be drawn from the result of the test?

d. What percentage of the variation in the dependent variable has been explained by the regression?

e. Construct a 95% confidence interval estimate for the population regression coefficient of PRICE.

f. What is the value of the R^2 adjusted for degrees of freedom? What, if any, is the advantage of this number over the coefficient of determination?
(*Source*: Data are from D. A. Bessler and R. A. Babubla, "Forecasting Wheat Exports: Do Exchange Rates Really Matter?" *Journal of Business and Economic Statistics,* 5, 1987, pp. 397–406. Copyright 1987 by the American Statistical Association. Used with permission. All rights reserved.)

9. Mortgage Rates. The regression in Figure 4.24 is an attempt to develop an equation to forecast mortgage rates. The explanatory variables include prime rate and the one through six-period lagged values of prime rate. The forecaster initially believed that prime rate in some of the past time periods and possibly the current time period may have an effect on mortgage rates. After examining the regression, the forecaster notes the small *t* statistics (large *p* values) for all the variables included in the regression and concludes that the regression is basically worthless: None of the variables included—current or lagged—are helpful in predicting mortgage

rates. Do you agree or disagree? Justify your position.

10. Dividends. A random sample of 42 firms was chosen from the S&P 500 firms listed in the Spring 2003 Special Issue of *Business Week* (The Business Week Fifty Best Performers). The indicated dividend yield (DIVYIELD), the earnings per share (EPS), and the stock price (PRICE) were recorded for these 42 firms. These data are available on the CD in a file named DIV4. Run a regression using DIVYIELD as the dependent variable and EPS and PRICE as the independent variables. Use the output to answer the following questions:

a. What is the sample regression equation relating DIVYIELD to PRICE and EPS?

b. What percentage of the variation of DIVYIELD has been explained by the regression?

c. Test the overall fit of the regression. Use a 10% level of significance. State the hypotheses to be tested, the decision rule, the test statistic, and your decision.

d. What conclusion can be drawn from the test result?

e. Is it necessary to test each coefficient individually to see if either PRICE or EPS is related to DIVYIELD? Why or why not?
(*Source*: Copyright 2003, *Business Week*. Visit us at our Web site at www.businessweek.com for additional information.)

11. Fuel Consumption. The data file FUELCON4 on the CD contains the following variables for all 50 states plus the District of Columbia:

FIGURE 4.24
Mortgage Rate and
Prime Rate Regression.

Variable	Coefficient	Std Dev	T Stat	P Value
Intercept	5.3188	0.5203	10.22	0.000
PRIMERT	0.6266	0.5047	1.24	0.218
PRIMEM1	-0.1148	0.8281	-0.14	0.890
PRIMEM2	-0.2286	0.8286	-0.28	0.783
PRIMEM3	-0.3137	0.8295	-0.38	0.706
PRIMEM4	-0.2468	0.8286	-0.30	0.767
PRIMEM5	-0.0368	0.8262	-0.04	0.965
PRIMEM6	0.6845	0.4947	1.38	0.170

Standard Error = 0.7587   R-Sq = 29.7%   R-Sq(adj) = 23.7%

Analysis of Variance

Source	DF	Sum of Squares	Mean Square	F Stat	P Value
Regression	7	19.9288	2.8470	4.95	0.000
Error	82	47.1977	0.5756		
Total	89	67.1265			

FUELCON: Per capita fuel consumption in gallons

DRIVERS: The ratio of licensed drivers to private and commercial motor vehicles registered

HWYMILES: The number of miles of federally funded highways

GASTAX: The tax per gallon of gasoline in cents

INCOME: The average household income in dollars

Run the regression with FUELCON as the dependent variable and the other four variables as independent variables. Use the output to help answer the following questions:

a. What is the estimated regression equation?

b. Test the overall fit of the regression. Use a 5% level of significance. Be sure to state your decision rule, test statistic value, and your decision. What conclusion can you draw from the result of the test?

c. What percentage of the variation in FUEL-CON has been explained by the regression?

d. Are there any variables that appear to be unnecessary in the regression? Justify your answer.

e. Are there variables that have been omitted from the regression that you believe might be useful in explaining more of the variation in FUELCON? If so, what variables?

(*Source*: Data are from the U.S. Department of Transportation, Federal Highway Administration. Most data were found in the online publication Highway Statistics 2001 at www.fhwa.dot. gov/ohim/hs01/index.htm)

12. Pricing Communications Nodes. The cost of adding a new communications node at a location not currently included on the network was of concern to a major Fort Worth manufacturing company. To try to predict the price of new communications nodes, data were obtained on a sample of existing nodes. The installation cost (COST) and the number of ports (NUMPORTS) available for access in each existing node were readily available. Data on two additional characteristics of communications nodes were also obtained: bandwidth (BANDWIDTH) and port speed (PORTSPEED). These data are shown in Table 4.6 and are available on the CD in a file named COMNODE4.

The network administrator wants to develop a method of estimating the cost of new nodes in a quick and fairly accurate manner. You have been asked to help in this project. Using the data available, develop an equation to help in the

TABLE 4.6 Data for Communications Nodes Exercise

Cost	Number of Ports	Bandwidth	Port Speed
52,388	68	58	653
51,761	52	179	499
50,221	44	123	422
36,095	32	38	307
27,500	16	29	154
57,088	56	141	538
54,475	56	141	538
33,969	28	48	269
31,309	24	29	230
23,444	24	10	230
24,269	12	56	115
53,479	52	131	499
33,543	20	38	192
33,056	24	29	230

pricing of new communications nodes. Justify your choice of equation.

Start with all three explanatory variables in the equation. Do you encounter any problems when you estimate this equation? Request variance inflation factors for the variables included in the regression. What do the VIFs tell you?

13. **Prime Rate.** The file named PRIME4 on the CD contains monthly prime rates for the time period from January 1988 through December 2002. (These data are from the web site www.econo-magic.com and are from the Federal Reserve Bank of St. Louis.) Develop an extrapolative model to forecast the prime rate for each month in 2003.

Find the actual rates for each month in 2003 and compare them to your forecasts. How well did your model do? (How will you measure the accuracy of your forecasts?)

14. **Graduation Rates.** *Kiplinger's Personal Finance* provides information on the best public and private college values. Some of the variables included in this issue are as follows. All are based on the most recent available data.

GRADRATE4	the percentage of students who earned a bachelor's degree in four years (expressed as a percentage)
ADMISRATE	admission rate expressed as a percentage
SFACRATIO	student faculty ratio
AVGDEBT	average debt at graduation

This information is included in a file named COL-LEGE4 on the CD for 195 schools. Only schools with complete information on all the categories listed are included. Using the graduation rate as the dependent variable, use the techniques discussed in this chapter to develop an equation to explain four year graduation rates.

(*Source*: Used by permission from the November and December 2003 issues of *Kiplinger's Personal Finance*. Copyright © 2003 The Kiplinger Washington Editors, Inc. Visit our website at www.kiplingers.com for further information).

15. **Absenteeism.** The ABX Company is interested in conducting a study of the factors that affect absenteeism among its production employees. After reviewing the literature on absenteeism and interviewing several production supervisors and a number of employees, the researcher in charge of the project defined the variables shown in Figure 4.25. Then a sample of 77 employees was randomly selected, and the data contained in the file on the CD named ABSENT4 was collected. The dependent variable is absenteeism. The other variables are considered possible explanatory variables.

Use the procedures discussed in Chapters 3 and 4 to identify factors that may be related to absenteeism. Write down your final model and justify your choice of variables in the model. Check to see if your choice of variables and the coefficient estimates make intuitive sense. How much variation in absenteeism has been explained? What does this tell you? Does your model give you some sense of which

FIGURE 4.25
Absenteeism Study
Variables.

	Variable	Description
1.	Absenteeism (ABSENT):	The number of distinct occasions that the worker was absent during 2003. Each occasion consists of one or more consecutive days of absence.
2.	Job Complexity (COMPLX):	An index ranging from 0 to 100.
3.	Base Pay (PAY):	Base hourly pay rate in dollars.
4.	Seniority (SENIOR):	Number of complete years with the company on December 31, 2003.
5.	Age (AGE):	Employee's age on December 31, 2003.
6.	Dependents (DEPEND):	Determined by employee response to the question: "How many individuals other than yourself depend on you for most of their financial support?"

Source: These data were created by Dr. Roger L. Wright, RLW Analytics, Inc., Sonoma, CA, and are used (with modification) with his permission.

employees might be absent most often? If so, which ones? What might be done to reduce absenteeism?

16. **Fanfare.** Fanfare International, Inc. designs, distributes, and markets ceiling fans and lighting fixtures. The company's product line includes 120 basic models of ceiling fans and 138 compatible fan light kits and table lamps. These products are marketed to over 1000 lighting showrooms and electrical wholesalers that supply the remodeling and new construction markets. The product line is distributed by a sales organization of 58 independent sales representatives.

In the summer of 1994, Fanfare decided it needed to develop forecasts of future sales to help determine future sales force needs, capital expenditures, and so on. The data file named FAN4 on the CD contains data on the following variables:

SALES	=	total monthly sales in thousands of dollars
ADEX	=	advertising expense in thousands of dollars
MTGRATE	=	mortgage rate for 30-year loans (%)
HSSTARTS	=	housing starts in thousands of units

The data are monthly and cover the period from July 1990 through May 1994. (*Note:* These data have been modified as requested by the company to provide confidentiality.)

As a consultant to Fanfare, your job is to find a causal regression model to forecast future sales. Use the techniques discussed in Chapters 3 and 4

to help you decide which variables you should include in the equation and which should be omitted. Justify your choices. How well do you believe the equation you developed will do at forecasting future sales? What additional analyses might you use to examine forecasting ability?

Now use the techniques discussed in Chapters 3 and 4 to build an extrapolative model to forecast sales. Generate forecasts from both the causal model and the extrapolative model. What are the benefits and drawbacks of each of the models? How could you compare the forecasting ability of the two models?

17. **Major League Baseball.** What factor is most important in building a winning baseball team? Some might argue for a high batting average. Or it might be a team that hits for power as measured by the number of home runs. On the other hand, many believe that it is quality pitching as measured by the earned run average of the team's pitchers. The file MLB4 on the CD contains data on the following variables for the 30 major league baseball teams during the 2002 season:

WINS	=	number of games won for each team
HR	=	number of home runs hit by each team
BA	=	average batting average for each team
ERA	=	earned run average for each team

Using WINS as the dependent variable, use scatterplots and regression to investigate the relationship of the other variables to WINS. Use the variables to build a multiple regression model to explain WINS. Interpret what your model tells you about a successful baseball team.

(*Source*: Courtesy of the *Fort Worth Star-Telegram*.)

18. NBA. The following data were obtained from the *Fort Worth Star-Telegram* and refer to the 2002–2003 National Basketball Association (NBA) season. The data are included in a file on the CD named NBA4 for all 29 NBA teams:

wins (WINS)
field goals attempted (FGA)
field goals made (FGM)
field goals attempted for opponents (FGAOP)
field goals made for opponents (FGMOP)
three-point field goals attempted (TFGA)
three-point field goals made (TFGM)
three-point field goals attempted for opponents (TFGAOP)
three-point field goals made for opponents (TFGMOP)
offensive rebounds (OFFREB)
total rebounds (TOTREB)
offensive rebounds for opponents (OFFREBOP)
total rebounds for opponents (TOTREBOP)
assists (ASST)
assists for opponents (ASSTOP)
steals (STL)
steals for opponents (STLOP)
blocked shots (BLK)
blocked shots for opponents (BLKOP)

The dependent variable is the number of wins (WINS) for the season. The other variables are to be considered possible explanatory variables.

You have been hired by your favorite NBA team to try and determine what factors might be important in helping to achieve a winning season. Using the available data, determine which combination of the variables provides the best explanation of what makes a winning team.

Write a report with your results. Your report should consist of a letter/executive summary of your results for team management and a technical section with a description and justification of your regression equation. In the technical section you will want to discuss aspects such as your regression equation, the choice of variables, the strength of the relationship, and the practical usefulness of the results. What do your results tell you about winning teams?

(*Source*: Courtesy of the *Fort Worth Star-Telegram*.)

19. Multicollinearity. Consider the following regression model:

$$y_i = \beta_0 + \beta_1 x_{1i} + \beta_2 x_{2i} + \beta_3 x_{3i} + \beta_4 x_{4i} + e_i$$

Multicollinearity is suspected to exist among the four explanatory variables used in this regression. The analyst using the regression computes pairwise correlations between the four explanatory variables and the dependent variable. The following correlation matrix results:

	y	x_1	x_2	x_3	x_4
x_1	0.4	1.0			
x_2	0.3	0.2	1.0		
x_3	0.6	0.3	0.4	1.0	
x_4	0.7	0.3	0.5	0.3	1.0

Based on these correlations, the analyst concludes that there will be no problems with multicollinearity. Do these correlations provide sufficient evidence to conclude that multicollinearity will not be a problem? Justify your answer.

USING THE COMPUTER

The Using the Computer section in each chapter describes how to perform the computer analyses in the chapter using Excel, MINITAB, and SAS. For further detail on Excel, MINITAB, and SAS, see Appendix C.

EXCEL

Multiple Regression

TOOLS: DATA ANALYSIS: REGRESSION

FIGURE 4.26 Excel Regression Dialog Box.

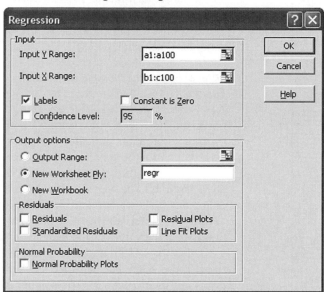

Figure 4.26 shows the Excel Regression dialog box. Regression is accessed in Excel by clicking on Tools and then Data Analysis. The Regression option is chosen from the Data Analysis menu. Put the range of the y variable in "Input Y Range." Put the range of the x variables in "Input X Range." Note that all x variables to be used in a multiple regression must be in adjacent columns. To accommodate this restriction, variables often must be moved around.

Click "Labels" if the variables have labels in the first row. Typically, "Constant is Zero" is not an option that is used. This option forces the constant or y intercept in a regression to be zero. It is seldom a good idea to use this option and can make interpretation of the regression results difficult.

Excel produces 95% confidence interval estimates of the population regression coefficients by default. If another level is required, click the "Confidence Level" box and insert the desired level.

Click the output option desired. The Residuals and Normal Probability options are discussed in Chapter 6.

Variance Inflation Factors

Excel does not provide options for automatically producing the variance inflation factors. Although these could be computed by creating formulas, this option is not discussed in this text.

A regression add-in is available that does compute a number of additional statistics. See the Excel section in Appendix C for more information on the add-in.

Creating a Lagged Variable

One way to create a lagged variable in Excel is simply to copy the necessary portion of the column to be lagged and paste it in the appropriate position. Figure 4.27 shows an example. Once the column is copied, the initial values with no matches are not used when running any regressions.

FIGURE 4.27 Creating a Lagged Variable with Excel.

	A	B	C	D	E	F
1	YEAR	MONTH	UNEMP	UNEMPL1	UNEMPL2	
2	1983	1	10.4			
3	1983	2	10.4	10.4		
4	1983	3	10.3	10.4	10.4	
5	1983	4	10.2	10.3	10.4	
6	1983	5	10.1	10.2	10.3	
7	1983	6	10.1	10.1	10.2	
8	1983	7	9.4	10.1	10.1	
9	1983	8	9.5	9.4	10.1	
10	1983	9	9.2	9.5	9.4	
11	1983	10	8.8	9.2	9.5	
12	1983	11	8.5	8.8	9.2	
13	1983	12	8.3	8.5	8.8	
14	1984	1	8	8.3	8.5	
15	1984	2	7.8	8	8.3	
16	1984	3	7.8	7.8	8	
17	1984	4	7.7	7.8	7.8	
18	1984	5	7.4	7.7	7.8	
19	1984	6	7.2	7.4	7.7	
20	1984	7	7.5	7.2	7.4	
21	1984	8	7.5	7.5	7.2	
22	1984	9	7.3	7.5	7.5	
23	1984	10	7.4	7.3	7.5	
24	1984	11	7.2	7.4	7.3	
25	1984	12	7.3	7.2	7.4	
26	1985	1	7.3	7.3	7.2	
27	1985	2	7.2	7.3	7.3	
28	1985	3	7.2	7.2	7.3	
29	1985	4	7.3	7.2	7.2	
30	1985	5	7.2	7.3	7.2	
31	1985	6	7.4	7.2	7.3	

MINITAB

Multiple Regression

STAT:REGRESSION:REGRESSION

To do a multiple regression in MINITAB, click the Stat menu and the Regression option. Then click the regression option on the subsequent menu. The Regression dialog box is shown in Figure 4.28. Fill in the Response variable and the Predictor variables. Then click OK.

Forecasting with a Multiple Regression Equation

STAT: REGRESSION: REGRESSION: OPTIONS

Click on Stat, then choose Regression from the Stat menu, and choose Regression again from the next menu. In the Regression dialog box (see Figure 4.28), click Options. The Regression–Options dialog box is shown in Figure 4.29. Fill in "Prediction intervals for new observations" with the values of the independent variables for which predictions are desired. These values can be single numbers or columns with numbers. The number of entries on this line must be the same as the number of independent variables in the regression equation—one entry for each variable.

Variance Inflation Factors

STAT: REGRESSION: REGRESSION: OPTIONS

Click Stat, then choose Regression from the Stat menu, and choose Regression again from the next menu. In the Regression dialog box (see Figure 4.28), click Options. The Regression–Options dialog box is shown in Figure 4.29. Click the box for variance inflation factors.

FIGURE 4.28 MINITAB Regression Dialog Box.

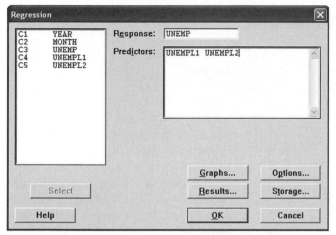

Creating a Lagged Variable

STAT: TIME-SERIES:LAG

To create a lagged variable in MINITAB, click on the Stat menu and then on Time Series. Choose Lag from the available options. Figure 4.30 shows the Lag dialog box. In "Series," put the column of the variable to be lagged. Put the location of the lagged variable in "Store lags in." In the "Lag" box, put the order of the lag desired: 1 for a one-period lag, 2 for a two-period lag, and so on.

SAS

Multiple Regression

The following command sequence produces a regression with dependent variable SALES and independent variables ADV and BONUS:

```
PROC REG;
MODEL SALES=ADV BONUS;
```

Partial F Tests in Multiple Regression

The TEST command in SAS produces the partial F statistic for testing whether several coefficients are equal to zero. The following command sequence illustrates the use of the TEST command:

```
PROC REG;
MODEL SALES=ADV BONUS MKTSHR COMPET;
TEST MKTSHR, COMPET;
```

FIGURE 4.29 MINITAB Regression—Options Dialog Box.

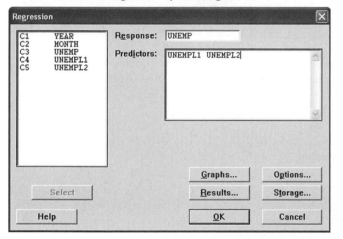

This sequence produces the F statistic to test whether the coefficients of MKTSHR and COMPET are equal to zero. When the SAS TEST command is used, the explanatory variables in the MODEL command do not have to be listed in any particular order. For example, the following command sequence produces the same test result as the previous sequence:

FIGURE 4.30 MINITAB Lag Dialog Box.

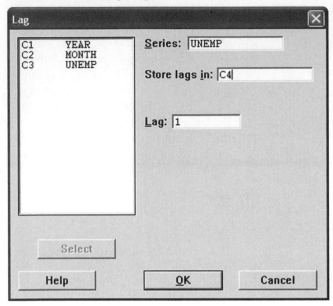

```
PROC REG;
MODEL SALES=ADV MKTSHR BONUS COMPET;
TEST MKTSHR, COMPET;
```

Forecasting with the Multiple Regression Equation

Forecasts in SAS for multiple regression are generated using an "appended" data set just as with simple regression. To the values of the independent variables in the original data set add the values for which predictions are desired. Then to the values of the dependent variable add the SAS symbol for mission data, which is a period, because we do not know those values. Now run the regression with the following options:

```
PROC REG;
MODEL SALES = ADV BONUS/P CLM CLI;
```

The option P requests forecasts (or predicted values), CLM requests upper and lower confidence interval limits for the estimate of the conditional mean, and CLI requests upper and lower prediction interval limits for an individual prediction.

Variance Inflation Factors

```
PROC REG;
MODEL SALES = ADV BONUS/VIF;
```

requests that variance inflation factors be printed.

Creating a Lagged Variable

In the data input phase in SAS, use the LAG_ command to create lagged variables. For example, to create a one-period lagged variable for the unemployment variable in Example 4.6, use

```
UNEMPL1 = LAG1(UNEMP);
```

For a two-period lagged variable, use

```
UNEMPL2 = LAG2(UNEMP);
```

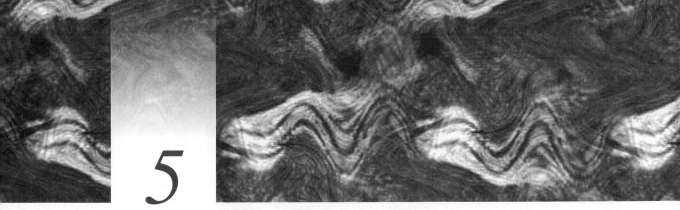

5 Fitting Curves to Data

5.1 INTRODUCTION

In Chapter 4, the multiple linear regression model was presented as

$$y_i = \beta_0 + \beta_1 x_{1i} + \beta_2 x_{2i} + \cdots + \beta_K x_{Ki} + e_i \qquad (5.1)$$

There we assumed that the true relationship was linear in the x variables. In this chapter, we find that this assumption need not be true to fit a regression equation to the data. We can fit *curvilinear* as well as linear relationships. This is accomplished through transformations of the variables in the model. The equation $y_i = \beta_0 + \beta_1 x_i + e_i$ represents a straight-line relationship between y and x. But the equation $y_i = \beta_0 + \beta_1 x_i + \beta_2 x_i^2 + e_i$ represents a curve (a parabola). The same x variable is involved in the equation; the fitting of the curve is accomplished through the transformation of the x variable to x^2. There are many possible transformations that produce some type of curvilinear relationship. The most commonly used transformations in business and economic applications are discussed in this chapter.

When a curvilinear relationship is suspected, the appropriate transformation of the variables to produce the best-fitting curve for the data is not always obvious. The variables y and x are related in some curvilinear fashion, but there are many equations that describe curvilinear relationships. The idea behind the use of any equation of this sort is to transform the variables in such a way that a linear relationship is achieved. If x and y are related in a curvilinear fashion, then perhaps x^2 and y have a linear relationship.

In this text, the following four commonly used corrections are considered:

1. polynomial regression
2. reciprocal transformation of the x variable
3. log transformation of the x variable
4. log transformation of both the x and y variables

In this chapter, we suggest some ways to assess whether a good choice of transformations was used to fit a curve to the data. In Chapter 6, some additional methods of assessing the choice are considered. In some texts, the choice of linear or curvilinear model and the type of curvilinear model to be used is called selection of *functional form* of the model.

5.2 FITTING CURVILINEAR RELATIONSHIPS

5.2.1 POLYNOMIAL REGRESSION

A common correction when the linearity assumption is violated is to add powers of the explanatory variable that is viewed as the curvilinear component of the model. This type of model is called a *polynomial regression*. The *order* of the model is the highest power used for the explanatory variable. For example, a second-order polynomial regression in one variable is written as

$$y_i = \beta_0 + \beta_1 x_i + \beta_2 x_i^2 + e_i$$

Higher-order polynomial models may be developed by adding higher powers of x. A Kth-order polynomial regression model in one variable, x, is written as

$$y_i = \beta_0 + \beta_1 x_i + \beta_2 x_i^2 + \cdots + \beta_K x_i^K + e_i$$

In practice, the second-order model is often sufficient to describe curvilinear relationships encountered.

EXAMPLE 5.1   **Telemarketing**   A company that provides transportation services uses a telemarketing division to help sell its services. The division manager is interested in the time spent on the phone by the telemarketers in the division. Data on the number of months of employment and the number of calls placed per day (an average for 20 working days) is recorded for 20 employees. These data are shown in Table 5.1 and in the file TELEMARK5 on the CD.

The average number of calls for all 20 employees is 28.95. The division manager, however, suspects that there may be a relationship between time on the job and number of calls. As time on the job increases, the employee becomes more familiar with the calling system and the correct procedures to use on the phone and also begins to acquire more regular clients. Thus, the longer the time on the job, the greater the number of calls per day. The scatterplot of CALLS (y) versus MONTHS (x) is shown in Figure 5.1. Looking at this scatterplot helps to verify that the relationship

TABLE 5.1 Data for Telemarketing Example

MONTHS	CALLS
10	18
10	19
11	22
14	23
15	25
17	28
18	29
20	29
20	31
21	31
22	33
22	32
24	31
25	32
25	32
25	33
25	31
28	33
29	33
30	34

FIGURE 5.1
Scatterplot of CALLS versus MONTHS.

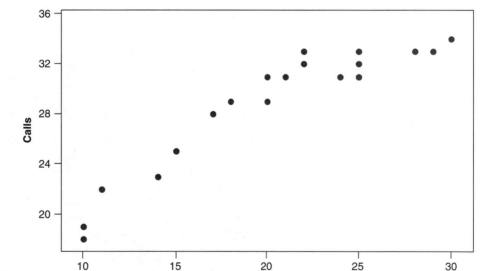

may not be linear. As the number of months on the job increases, the number of calls also increases. But the rate of increase begins to slow over time, thus resulting in a pattern that may be better modeled by a curve than a straight line.

In Figure 5.2, the plot of y versus x has been reproduced with a curve drawn in that approximates the relationship between the two variables.

FIGURE 5.2
Scatterplot of
CALLS versus
MONTHS with
Curve Drawn to
Represent the
Relationship.

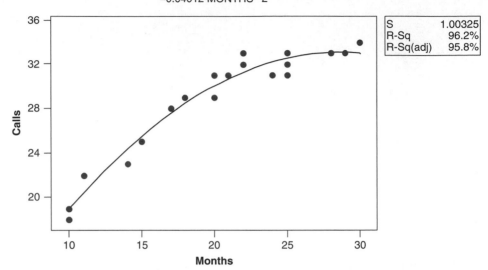

To model the curvilinear relationship in the telemarketing data, a second-order polynomial regression can be tried. The model can be written

$$CALLS = \beta_0 + \beta_1 MONTHS + \beta_2 XSQR + e$$

where XSQR is a variable created by squaring each value of the MONTHS variable.

For comparison purposes, the linear regression using CALLS as the dependent variable and MONTHS as the explanatory variable is shown in Figure 5.3. Figure 5.4 shows the regression estimates of the second-order model. The estimated regression is

$$CALLS = -0.14 + 2.31\ MONTHS - 0.04\ XSQR$$

The primary check that should be made at this point to determine whether the second-order model is preferred to the original linear model is to test whether the coefficient of the second-order term is significantly different from zero.

To determine whether the x^2 variable has significantly improved the fit of the regression, the following hypotheses can be tested:

$$H_0\colon \beta_2 = 0$$
$$H_a\colon \beta_2 \neq 0$$

where β_2 is the coefficient of x^2. The t test discussed in Chapter 4 can be used to conduct the test. For $\alpha = 0.05$, the decision rule is

Reject H_0 if $t > 2.11$ or $t < -2.11$

Do not reject H_0 if $-2.11 \leq t \leq 2.11$

FIGURE 5.3 Regression Results for Telemarketing Example Using Only MONTHS as an Explanatory Variable.

Variable	Coefficient	Std Dev	T Stat	P Value
Intercept	13.6708	1.4270	9.58	0.000
MONTHS	0.7435	0.0667	11.15	0.000

Standard Error = 1.78737   R-Sq = 87.4%   R-Sq(adj) = 86.7%

Analysis of Variance

Source	DF	Sum of Squares	Mean Square	F Stat	P Value
Regression	1	397.45	397.45	124.41	0.000
Error	18	57.50	3.19		
Total	19	454.95			

FIGURE 5.4 Regression Results for Telemarketing Example with Second-Order Term Added.

Variable	Coefficient	Std Dev	T Stat	P Value
Intercept	-0.14047	2.32226	-0.06	0.952
MONTHS	2.31020	0.25012	9.24	0.000
XSQR	-0.04012	0.00633	-6.33	0.000

Standard Error = 1.00325   R-Sq = 96.2%   R-Sq(adj)=95.8%

Analysis of Variance

Source	DF	Sum of Squares	Mean Square	F Stat	P Value
Regression	2	437.84	218.92	217.50	0.000
Error	17	17.11	1.01		
Total	19	454.95			

The test statistic value is $t = -6.33$. The null hypothesis is rejected. The x^2 term adds significantly to the ability of the regression to explain the variation in y. Thus, the term should remain in the equation. Note that the p value could also have been used to conduct this test ($p$value $= 0.000 < 0.05$, so reject H_0).

Once a decision is made to keep the second-order term in the model, the lower-order term is typically kept in the model regardless of the t test result on its coefficient. There are good statistical reasons for keeping lower-order terms in a polynomial regression when the higher-order terms are judged important (see "A Property of Well-Formulated Polynomial Regression Models," by J. L. Peixoto).[1] Other indicators that the regression has been improved by adding the x^2 term include the reduction in the standard error of the regression from 1.787 to 1.003 and the increase in adjusted R^2 from 86.7% to 95.8%.

---

[1]See References for complete publication information.

In our example, the second-order model is an improvement over the first-order model. Higher-order terms could be added to the model to see whether additional improvements are possible. Figure 5.5 shows the estimates of a third-order model. The explanatory variables are MONTHS, XSQR, and $X^\wedge 3$, the cube (third power) of the MONTHS variable. Note that the coefficient of the $X^\wedge 3$ variable is not significant at the 0.05 level, suggesting that the addition of this term is of little additional help in explaining the variation in CALLS. The third-order term is unnecessary in the model. Note that second-order term is not significant either in this model. This is a result of multicollinearity. The explanatory variables are highly correlated and this causes least-squares to have difficulty determining which of the variables (the second-order or the third-order term) is the important one. We will opt for the simpler second-order model in this case.

Table 5.2 summarizes the different measures that may be useful in determining the best model to use. The p values suggest that the second-order term is useful, but the third-order term is not. The R^2 increases from 87.4% for the linear model to 96.2% for the second-order model. The increase for the third-order model is very small, however. If R^2_{adj} is used, there is actually a decrease from the second-order to the third-order model. This is further verification that the third-order term is unnecessary. This is also reflected in the standard error, which decreases from 1.787 for the linear model to 1.003 for the second-order model but increases to 1.020 for the third-order model. (Recall that we want increases in R^2 and R^2_{adj} but decreases in the standard error.)

One caution should be observed in using higher-order polynomial regression models. Correlations between powers of a variable can result in *multicollinearity* problems, as discussed in Chapter 4. This problem was encountered in the telemarketing example when a third-order model was estimated. To reduce the possibility of computational difficulties, the use of explanatory variables that have been centered often is recommended. For example, instead of using the explanatory variables, x, x^2, x^3, use instead

$$x - \overline{x}, (x - \overline{x})^2, \text{ and } (x - \overline{x})^3$$

where $\overline{x}$ is the sample mean of the variable values. Using the centered variables helps avoid multicollinearity problems in polynomial regressions to some extent.

FIGURE 5.5 Regression Results for Telemarketing Example with Second-Order and Third-Order Terms Added.

Variable	Coefficient	Std Dev	T Stat	P Value
Intercept	-5.58003	8.38720	-0.67	0.515
MONTHS	3.25806	1.42526	2.29	0.036
XSQR	-0.09075	0.07518	-1.21	0.245
X^3	0.00085	0.00125	0.68	0.509

Standard Error = 1.01967   R-Sq = 96.3%   R-Sq(adj) = 95.7%

Analysis of Variance

Source	DF	Sum of Squares	Mean Square	F Stat	P Value
Regression	3	438.31	146.10	140.52	0.000
Error	16	16.64	1.04		
Total	19	454.95			

TABLE 5.2 Summary Measures for Linear, Second-Order, and Third-Order Models for Telemarketing Example

Model	p Value for Highest-Order Term	R^2	R^2_{adj}	s_e
Linear Model	0.000	87.4%	86.7%	1.787
Second-Order Model	**0.000**	96.2%	**95.8%**	**1.003**
Third-Order Model	0.509	96.3%	95.7%	1.020

Choosing which curvilinear model to use in a particular case is not always a simple matter. In the telemarketing example, I chose to use a second-order polynomial regression as my starting point. Why not use the logarithm of the x variable instead? This is another of the transformations to be discussed in this chapter. Familiarity with the look of certain curves can be helpful in choosing the right curvilinear model. The curve shown in Figure 5.2 is similar to a parabola (or half of a parabola, at any rate), and this led me to make the second-order model my first choice (since the second-order equation is the equation of a parabola). If you are not sure about what type of transformation is best, you can always try different ones and check the summary measures used in the example to help make the choice. Chapter 6 also presents some additional methods of assessing the validity of a curvilinear model and choosing the best transformation. ▨

5.2.2 RECIPROCAL TRANSFORMATION OF THE X VARIABLE

Other transformations may produce a linear relationship. A fairly common example is the reciprocal transformation:

$$y_i = \beta_0 + \beta_1\left(\frac{1}{x_i}\right) + e_i$$

In this equation, x and y are inversely related, but the inverse relationship is not a linear one. (Note that this transformation is not defined when $x = 0$.)

EXAMPLE 5.2 **MPG Versus HP**   A scatterplot of a possible curvilinear inverse relationship is shown in Figure 5.6. The variables are HWYMPG (y), which is the number of miles per gallon obtained by a car in highway driving, and HP (x), the horsepower of the car. This information is available for 147 cars listed in the *Road and Track* October 2002 issue and in the file MPGHP5 on the CD.

As HP increases, the mileage decreases, as would be expected. However, it appears that the rate of decrease may be slower as the cars get more powerful. The regression results for the linear regression of HWYMPG on HP are shown in Figure 5.7. Can we find a curvilinear model that better describes the relationship between these two variables? The scatterplot suggests that the following curvilinear inverse relationship might be appropriate:

$$HWYMPG = \beta_0 + \beta_1\left(\frac{1}{HP}\right) + e$$

FIGURE 5.6
Scatterplot of HWYMPG
Versus HP.

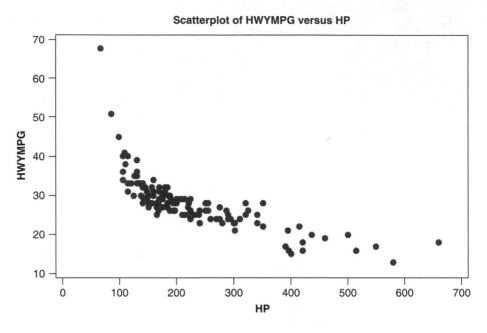

Scatterplot of HWYMPG versus HP

FIGURE 5.7 Results
for Regression of
HWYMPG on HP.

```
Variable          Coefficient      Std Dev        T Stat        P Value

Intercept         38.730875        0.803300       48.21         0.000
HP                -0.047732        0.003274       -14.58        0.000

Standard Error = 4.17503      R-Sq = 59.4%       R-Sq(adj) = 59.2%


Analysis of Variance

Source      DF     Sum of Squares     Mean Square     F Stat      P Value
Regression  1      3705.2             3705.2          212.57      0.000
Error       145    2527.5             17.4
Total       146    6232.7
```

An inverse relationship is one where y decreases as x increases. In a linear model, an inverse relationship results in a negative slope coefficient. If a relationship is expected to be inverse but curvilinear, then the reciprocal transformation of the x variable is often useful in representing this relationship.

Figure 5.8 shows the scatterplot of HWYMPG versus the transformed explanatory variable 1/HP (named HPINV).[2] This scatterplot appears linear. The regression results for the regression of HWYMPG on HPINV are shown in Figure 5.9.

---

[2] A graphical method to see if a transformation might be effective in modeling a curvilinear relationship is to plot the dependent variable versus the transformed x variable, as was done in this example. If the resulting graph looks linear, the transformation likely gives a good result. Note that this was not done in the first example using the second-order model. Because this model has two terms (x and x^2), examining the transformation graphically is more difficult.

FIGURE 5.8
Scatterplot of HWYMPG versus HPINV = 1/HP.

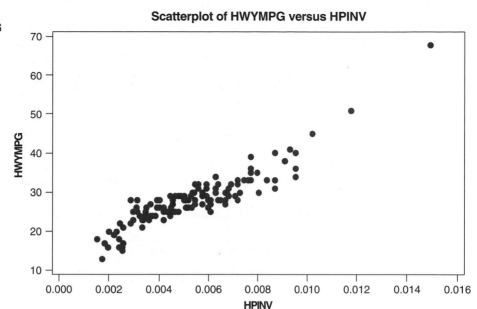

Scatterplot of HWYMPG versus HPINV

FIGURE 5.9
Results for Regression of HWYMPG on HPINV = 1/HP.

```
Variable          Coefficient       Std Dev        T Stat        P Value

Intercept            13.6310         0.6493         20.99         0.000
HPINV              2692.4675        11.7526         24.09         0.000

Standard Error = 2.93107      R-Sq = 80.0%   R-Sq(adj) = 79.9%

Analysis of Variance

Source           DF    Sum of Squares    Mean Square    F Stat     P Value
Regression        1       4987.0           4987.0       580.48     0.000
Error           145       1245.7              8.6
Total           146       6232.7
```

The R^2 value has increased from 59.4% for the linear model to 80.0% for the model with the transformed x variable. The standard error of the regression has decreased from 4.17503 to 2.93107. Both of these facts support the use of the curvilinear model. Table 5.3 summarizes the statistics for the two models.

TABLE 5.3 Summary Measures for the Linear Model and the Model Using the Reciprocal of HP for MPG Versus HP Example

Model	p Value for Highest-Order Term	R^2	R^2_{adj}	s_e
Linear Model	0.000	59.4%	59.2%	4.17503
Reciprocal Model	0.000	**80.0%**	**79.9%**	**2.93107**

5.2.3 LOG TRANSFORMATION OF THE *X* VARIABLE

Another useful curvilinear equation is

$$y_i = \beta_0 + \beta_1 \ln(x_i) + e_i$$

where $\ln(x)$ is the natural logarithm of x. It is assumed here that the x values are positive, because $\ln(x)$ is not defined for $x \leq 0$.

EXAMPLE 5.3 **Fuel Consumption** Table 5.4 shows the fuel consumption (FUELCON) in gallons per capita for each of the 50 states and Washington, DC. (*Source*: U.S. Department of Transportation, Federal Highway Administration; see the file FUELCON5 on the CD). The following variables are also shown: the population of the state (POP), the area of the state in square miles (AREA), and the population

TABLE 5.4 Data for Fuel Consumption Example

State	FUELCON	POP	AREA	DENSITY
Alabama	547.92	4,486,508	50750	88.4041
Alaska	440.38	643,786	570374	1.128709
Arizona	456.90	5,456,453	113642	48.0144
Arkansas	530.08	2,710,079	52075	52.04184
California	426.21	35,116,033	155973	225.1417
Colorado	474.78	4,506,542	103729	43.44534
Connecticut	432.44	3,460,503	4845	714.2421
Delaware	492.97	807,385	1955	412.9847
Florida	461.55	16,713,149	53997	309.52
Georgia	564.82	8,560,310	57919	147.798
Hawaii	336.97	1,244,898	6423	193.8188
Idaho	484.83	1,341,131	82751	16.20683
Illinois	406.99	12,600,620	55593	226.6584
Indiana	524.01	6,159,068	35870	171.7053
Iowa	532.39	2,936,760	55875	52.55946
Kansas	483.31	2,715,884	81823	33.19218
Kentucky	532.77	4,092,891	39732	103.0125
Louisiana	513.80	4,482,646	43566	102.8932
Maine	472.68	1,294,464	30865	41.93954
Maryland	463.46	5,458,137	9775	558.3772
Massachusetts	436.57	6,427,801	7838	820.0818
Michigan	504.95	10,050,446	56809	176.9164
Minnesota	532.52	5,019,720	79617	63.04834
Mississippi	541.06	2,871,782	46914	61.21375
Missouri	549.16	5,672,579	68898	82.333
Montana	549.35	909,453	145556	6.248131
Nebraska	503.10	1,729,180	76878	22.49252
Nevada	448.81	2,173,491	109806	19.79392
New Hampshire	541.67	1,275,056	8969	142.1626
New Jersey	465.52	8,590,300	7419	1157.878
New Mexico	504.77	1,855,059	121364	15.28508
New York	296.44	19,157,532	47224	405.6736
North Carolina	510.05	8,320,146	48718	170.7818
North Dakota	580.32	634,110	68994	9.190799

TABLE 5.4 (*Continued*)

State	FUELCON	POP	AREA	DENSITY
Ohio	458.31	11,421,267	40953	278.8872
Oklahoma	523.89	3,493,714	68679	50.87019
Oregon	439.09	3,521,515	96002	36.68168
Pennsylvania	417.36	12,335,091	44820	275.214
Rhode Island	382.82	1,069,725	1045	1023.66
South Carolina	557.53	4,107,183	30111	136.4014
South Dakota	577.84	761,063	75896	10.02771
Tennessee	506.30	5,797,289	41219	140.646
Texas	502.17	21,779,893	261914	83.15666
Utah	430.53	2,316,256	82168	28.18927
Vermont	555.78	616,592	9249	66.6658
Virginia	529.52	7,293,542	39598	184.1897
Washington	446.63	6,068,996	66581	91.15207
West Virginia	466.31	1,801,873	24087	74.80687
Wisconsin	466.08	5,441,196	54314	100.1804
Wyoming	715.55	498,703	97105	5.135709
Washington D.C.	289.99	570898	61	9358.984

density (DENSITY) defined as population/area. The object is to develop a regression equation to predict fuel consumption based on the population density. FUELCON is the dependent variable and DENSITY is the explanatory variable. The scatterplot of FUELCON versus DENSITY is shown in Figure 5.10. Looking at the scatterplot of FUELCON versus DENSITY, it is clear that this is not a linear

FIGURE 5.10
Scatterplot of
FUELCON versus
DENSITY.

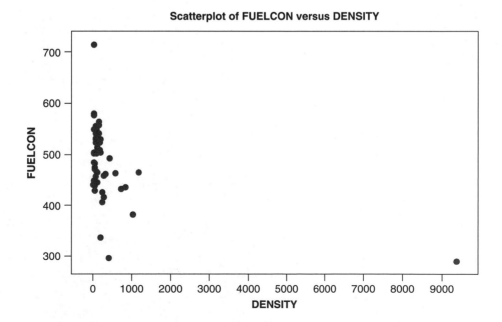

Scatterplot of FUELCON versus DENSITY

relationship. One thing to note about this plot is how the values spread out on the x axis. At the left-hand side of the x axis, the values are clumped together. Moving from left to right, the values become progressively more spread out. This suggests the use of a log transformation of DENSITY. The log transformation puts values on a different scale that compresses large distances so that they are more comparable to smaller distances. Table 5.5 shows the effect of applying the log to the base 10 to a series of numbers. (If we let q represent the log to the base 10 of a number x, then q is defined as the value that makes the following equation true: $10^q = x$. If we use log to the base 2, then the defining equation becomes $2^q = x$.) Note that the values of x in the table are successively more and more spread out; the distances between the values are becoming greater. The $\log_{10}(x)$ values do not exhibit this tendency. The log transformation evens out the successively larger distances between the values.

The scatterplot of FUELCON versus the log of DENSITY (LOGDENS) is shown in Figure 5.11. The natural log of DENSITY is used. The natural log uses the number called e (e is approximately equal to 2.718) as its base. It is common in business and economic applications to use natural logarithms, although the base used is usually not important. The relationship in Figure 5.11 appears to be linear.

TABLE 5.5 Effect of Applying Log to the Base 10 to a Set of Numbers

x	10	100	1000	10000
$\log_{10}(x)$	1	2	3	4

FIGURE 5.11
Scatterplot of
FUELCON versus
LOGDENS.

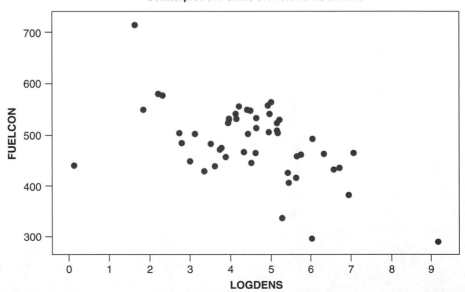

Fitting a line to the values in Figure 5.11 makes much more sense than trying to fit a line to the values in Figure 5.10.The log transformation is a good choice.

Figure 5.12 shows the linear regression results for the regression of FUELCON on DENSITY. These are used for comparison purposes. The regression results using the natural log of DENSITY as the explanatory variable are shown in Figure 5.13. The regression results indicate that using LOGDENS as the explanatory variable produces a better model fit than the regression using DENSITY. Table 5.6 provides summary statistics for the two models.

FIGURE 5.12
Regression of FUELCON on DENSITY.

Variable	Coefficient	Std Dev	T Stat	P Value
Intercept	495.628	9.481	52.28	0.000
DENSITY	−0.025	0.007	−3.56	0.001

Standard Error = 65.1675   R-Sq = 20.6%   R-Sq(adj) = 19.0%

Analysis of Variance

Source	DF	Sum of Squares	Mean Square	F Stat	P Value
Regression	1	53961	53961	12.71	0.001
Error	49	208093	4247		
Total	50	262054			

FIGURE 5.13
Regression of FUELCON on LOGDENS.

Variable	Coefficient	Std Dev	T Stat	P Value
Intercept	597.19	26.96	22.15	0.000
LOGDENS	−24.53	5.65	−4.34	0.000

Standard Error = 62.1561   R-Sq = 27.8%   R-Sq(adj) = 26.3%

Analysis of Variance

Source	DF	Sum of Squares	Mean Square	F Stat	P Value
Regression	1	72748	72748	18.83	0.000
Error	49	189306	3863		
Total	50	262054			

TABLE 5.6 Summary Measures for the Linear Model and the Model Using LOGDENS for Fuel Consumption Example

Model	p Value for Highest-Order Term	R^2	R^2_{adj}	S_e
Linear Model	0.001	20.6%	19.0%	65.17
Log x Model	0.000	**27.8%**	**26.3%**	**62.16**

5.2.4 LOG TRANSFORMATION OF BOTH THE X AND Y VARIABLES

It is also possible to transform the y variable in attempting to achieve a linear relationship. The natural logarithm of y is often used as the dependent variable with the natural logarithm of x as the explanatory variable:

$$\ln(y_i) = \beta_0 + \beta_1 \ln(x_i) + e_i$$

Some caution must be exercised if this model is chosen. First, all x and y values must be positive for the natural log transformation to be defined. Second, because $\ln(y)$ is used as the dependent variable, it becomes more difficult to compare this regression to any model using y as the dependent variable. The R^2 values of the two regressions cannot be compared, for example, because two different units of measurement are used for the dependent variable. (This applies as well to adjusted R^2 and the standard error.) Thus, increases in R^2 when the natural logarithm transformation is applied to y do not necessarily suggest an improved model. (Note that transformations of the explanatory variables do not create this type of problem. It is only when the y variable is transformed that comparison becomes more difficult.)

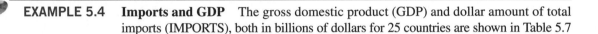

EXAMPLE 5.4   **Imports and GDP**   The gross domestic product (GDP) and dollar amount of total imports (IMPORTS), both in billions of dollars for 25 countries are shown in Table 5.7

TABLE 5.7 Data for Imports and GDP Example

Country	Imports	GDP
Argentina	20.300	391.000
Australia	68.000	528.000
Bolivia	1.500	21.400
Brazil	57.700	1340.000
Canada	229.000	923.000
Cuba	4.800	25.900
Denmark	47.900	155.500
Egypt	164.000	258.000
Finland	31.800	136.200
France	303.700	1540.000
Greece	31.400	201.100
Haiti	0.978	12.000
India	53.800	2660.000
Israel	30.800	122.000
Jamaica	3.100	9.800
Japan	292.100	3550.000
Liberia	0.170	3.600
Malaysia	76.900	200.000
Mauritius	2.000	12.900
Netherlands	201.100	434.000
Nigeria	13.700	105.900
Panama	6.700	16.900
Samoa	0.900	0.618
United Kingdom	330.100	1520.000
United States	1148.000	10,082.000

and contained in the file IMPGDP5 on the CD. These data were obtained from *The World Fact Book 2002* at www.odci.gov/cia/publications/factbook/index.html. The objective is to find an equation showing the relationship between IMPORTS (y) and GDP (x). The scatterplot of IMPORTS versus GDP in Figure 5.14 shows that this is not a linear relationship. One thing to note about this plot is how the values spread out on the x and y axes. At the left-hand side of the x axis and the bottom of the y axis, the values are clumped together. Moving from left to right on the x axis, the values become more spread out. The same thing happens when moving up the y axis —the values become progressively more spread out. This suggests the use of a log transformation for both the x and y variables. The motivation for using the log transformation is the same as in the Fuel Consumption example, but the transformation needs to be applied to both the x and y variables.

Figure 5.15 shows the scatterplot of the natural logarithm of imports (LOGIMP) versus the natural logarithm of GDP (LOGGDP). The relationship appears much closer to linear than in Figure 5.14. Figure 5.16 shows the results for the regression of LOGIMP on LOGGDP. The regression of IMPORTS on GDP is not shown for comparison purposes as in previous examples. As noted, since the dependent variable has been transformed, the usual comparisons are not valid. In Chapter 6, we find alternative methods for judging which of the functional forms of the model appears better. At this stage, the scatterplot strongly supports the use of the log transformation.

It is important to keep in mind that the type of transformation to correct for curvilinearity is not always obvious. If y and x are related in a curvilinear manner, the goal is to transform the variables in some manner to achieve a linear

FIGURE 5.14
Scatterplot of
IMPORTS versus GDP.

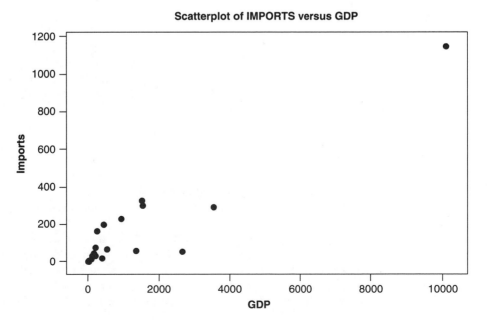

FIGURE 5.15
Scatterplot of Log
IMPORTS versus Log
GDP.

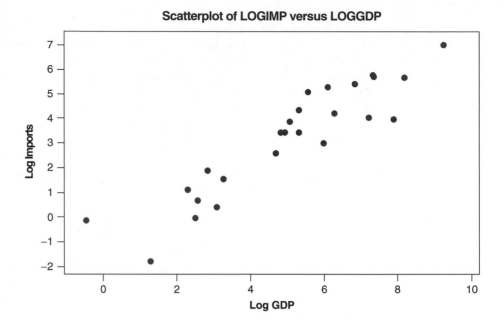

FIGURE 5.16
Regression of Log
IMPORTS on Log GDP.

Variable	Coefficient	Std Dev	T Stat	P Value
Intercept	-1.1275	0.4346	-2.59	0.016
LOGGDP	0.8670	0.0788	11.01	0.000

Standard Error = 0.914202   R-Sq = 84.0%   R-Sq(adj) = 83.4%

Analysis of Variance

Source	DF	Sum of Squares	Mean Square	F Stat	P Value
Regression	1	101.26	101.26	121.15	0.000
Error	23	19.22	0.84		
Total	24	120.48			

relationship. Different transformations may be tried (including transformations not discussed in this section) and the one that appears to do the best job chosen. There may be theoretical results as well that support the use of certain transformations in certain cases. As always, subject matter expertise is important in any analysis.

In deciding what type of transformation to use, look at the scatterplot showing the relationship between y and x. This may help identify the form of the relationship between the two variables. In Chapter 6, we discuss other methods of recognizing when a linear model is not the best choice, when a curvilinear model may be more appropriate, and which curvilinear model provides an improvement.

5.2.5 FITTING CURVILINEAR TRENDS

In Chapter 3, the linear trend model was presented:

$$y_i = \beta_0 + \beta_1 t + e_i$$

where t is simply a variable indicating time sequence, $t = 1, 2, \ldots, n$. Just as curvilinear patterns can be observed with regard to x variables as discussed in this chapter, so can curvilinear trends occur. It is possible to model certain curvilinear trends using regression. This is done in a manner very similar to the fitting of curves to data just discussed. A few basic curvilinear trend models are presented here.

A *quadratic trend* equation can be written

$$y_i = \beta_0 + \beta_1 t + \beta_2 t^2 + e_i$$

Examples of the linear and quadratic trends are shown in Figures 5.17(a) and 5.17(b), respectively.

An equation for a curve called an *S-curve* is given by

$$y_i = \exp\left(\beta_0 + \beta_1\left(\frac{1}{t}\right) + e_i \right)$$

where exp denotes the exponential operator: the value $e = 2.718$ (approximately) is raised to the power

$$\beta_0 + \beta_1\left(\frac{1}{t}\right) + e_i$$

The S-curve is shown in Figure 5.17(c). This type of trend might be used to model demand for certain products over their lifetime. Demand is slow initially until the product becomes better known. Then demand picks up until a saturation point is reached. At that time, demand levels off.

The S-curve equation cannot be estimated directly using least-squares. By taking natural logarithms of both sides of the equation, however, a new equation is obtained that can be estimated. Because $\ln(\exp(x)) = x$ for any x, taking natural logarithms of both sides of the equation produces

$$\ln(y_i) = \beta_0 + \beta_1\left(\frac{1}{t}\right) + e_i$$

Regressing $\ln(y_i)$ on $1/t$ produces estimates of β_0 and β_1. When forecasting with this model, care should be taken. For example, write $y_i' = \ln(y_i)$ and $t' = \frac{1}{t}$, and write the estimated regression equation as

$$\hat{y}_i' = b_0 + b_1 t'$$

The estimated equation provides the forecast for time period T:

$$\hat{y}_T' = b_0 + b_1\left(\frac{1}{T}\right)$$

Note that this is a forecast of y'_T or the natural logarithm of y_T. To obtain a forecast of the original dependent variable, y_T, the conversion back to the original units from logarithmic units must be made:

$$\exp(\hat{y}'_T) = \hat{y}_T$$

Exponential trends also are used in time-series applications. The equation for an exponential trend is

$$y_i = \exp(\beta_0 + \beta_1 t + e_i)$$

Again, to estimate β_0 and β_1, the equation is transformed using natural logarithms. Writing $y'_i = \ln(y_i)$, the transformed equation is

$$y'_i = \beta_0 + \beta_1 t + e_i$$

Regressing $y'_i = \ln(y_i)$ on t produces estimates of β_0 and β_1. As with the S-curve, exercise caution when using this equation for forecasting. The natural logarithm of $\hat{y}'_i$ must be transformed back to its original units to obtain the desired forecast. Examples of exponential trends are shown in Figure 5.17(d).

FIGURE 5.17
Examples of Types of Trends.

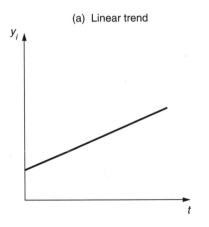

(a) Linear trend

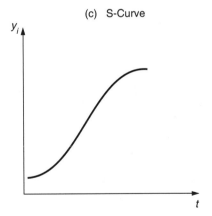

(c) S-Curve

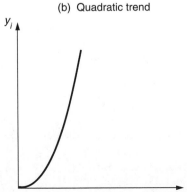

(b) Quadratic trend

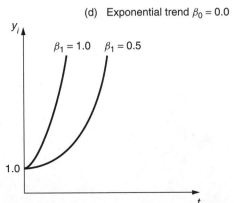

(d) Exponential trend $\beta_0 = 0.0$

EXERCISES

1. **Research and Development.** A company is interested in the relationship between profit on a number of projects and two explanatory variables. These variables are the expenditure on research and development for the project (RD) and a measure of risk assigned at the outset of the project (RISK). The file RD5 on the CD and Table 5.8

TABLE 5.8 Data for Research and Development Exercise

RD	RISK	PROFIT
132.580	8.5	396
81.928	7.5	130
145.992	10.0	508
90.020	8.0	172
114.408	7.0	256
53.704	7.5	32
76.244	7.0	102
71.680	8.0	102
151.592	9.5	536
74.816	7.5	102
108.752	6.0	214
92.372	8.5	200
92.260	7.0	158
60.732	6.5	32
78.120	7.5	116
90.000	5.5	120
105.532	9.0	270
111.832	8.0	270

show the data on the three variables PROFIT, RISK, and RD. PROFIT is measured in thousands of dollars and RD is measured in hundreds of dollars. The scatterplots of PROFIT versus RISK and PROFIT versus RD are shown in Figures 5.18 and 5.19, respectively. The regression of PROFIT on the two explanatory variables RISK and RD is shown in Figure 5.20.

Figure 5.21 shows the results of a regression using PROFIT as the dependent variable with RISK, RD, and RDSQR (the square of the RD variable) as explanatory variables. Choose the model you prefer for PROFIT and provide a justification for your choice.

2. **Piston Corporation (Part A).** Reginald Jackson was employed as a cost accountant by the Piston Corporation, a medium-size auto parts company located in the outskirts of Detroit. Kelly Jones, the controller for Piston, decided that she needed an assistant. Jackson was selected to fill that position. As part of his training program, Jackson was sent to night school to study quantitative applications in cost accounting.

Because the Piston Corporation's products were replacement parts, its sales were, fortunately, not as volatile as the new car market's. Piston had

FIGURE 5.18
Scatterplot of PROFIT versus RISK for Research and Development Exercise.

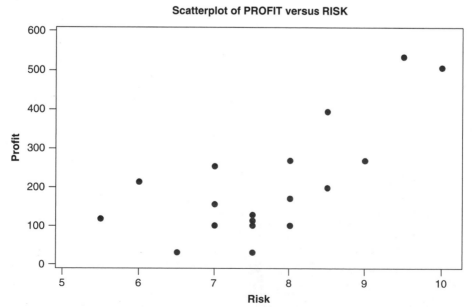

Scatterplot of PROFIT versus RISK

FIGURE 5.19
Scatterplot of PROFIT versus RD for Research and Development Exercise.

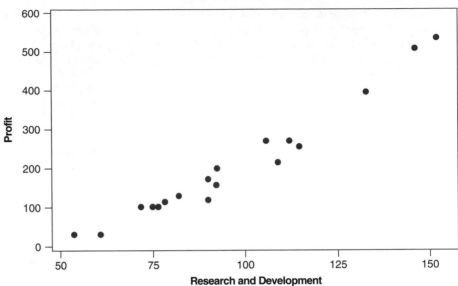

FIGURE 5.20
Regression of PROFIT on RISK and RD for Research and Development Exercise.

Variable	Coefficient	Std Dev	T Stat	P Value
Intercept	-453.1763	23.5061	-19.28	0.000
RISK	29.3090	3.6686	7.99	0.000
RD	4.5100	0.1538	29.33	0.000

Standard Error = 14.3420   R-Sq = 99.2%   R-Sq(adj) = 99.0%

Analysis of Variance

Source	DF	Sum of Squares	Mean Square	F Stat	P Value
Regression	2	361639	180820	879.08	0.000
Error	15	3085	206		
Total	17	364724			

FIGURE 5.21
Regression of PROFIT on RISK and First- and Second-Order Terms for RD for Research and Development Exercise.

Variable	Coefficient	Std Dev	T Stat	P Value
Intercept	-245.369584	14.811115	-16.57	0.000
RISK	23.249237	0.988413	23.52	0.000
RD	1.014314	0.232378	4.36	0.001
RDSQR	0.017567	0.001152	15.25	0.000

Standard Error = 3.53798   R-Sq = 100.0%   R-Sq(adj)=99.9%

Analysis of Variance

Source	DF	Sum of Squares	Mean Square	F Stat	P Value
Regression	3	364549	121516	9707.86	0.000
Error	14	175	13		
Total	17	364724			

experienced a rather stable growth in sales in recent years and had been required to increase its capacity regularly. It appeared to be time for another expansion, but with an uncertain stock market prevailing and uncertainty concerning interest rates, Jones was worried about obtaining funds at a reasonable cost. On the other hand, Piston's production manager had been complaining, more than usual, about various personnel, material handling, and scheduling bottlenecks that arose from the high level of output demanded of his present facilities.

The executive officers had been asked by Piston's directors to formulate a proposal for expansion and price adjustments. Jones asked Jackson what he could determine statistically about the effect of inflation and the level of production on unit costs.

By looking at old budgets, Jackson was able to obtain quarterly data on manufacturing costs per unit, production level (a percentage of the total capacity), and the index of direct material and direct labor costs for a five-year period (These data are in a file named PISTON5 on the CD). He immediately went to the computer and ran a regression of unit cost on production volume (PROD) and the cost index (INDEX). He began to wonder about the validity of modeling unit

costs as a linear function of production level and the cost index.

The scatterplots of the dependent variable (COST) versus each explanatory variable are shown in Figures 5.22 and 5.23. The regression results are shown in Figure 5.24.

After spending several days trying to improve his model, Jackson had several new solutions but was still unsure which one was best. That night after his quantitative accounting class, he asked his professor for advice. The professor suggested that Jackson first derive a theoretically plausible solution and then see if the data satisfied this relationship.

Jackson knew that the basic relationship with which he was dealing was:

total cost = variable cost per unit × volume + fixed cost

He also theorized that variable cost per unit was composed of a constant multiple of the index of direct materials and labor (x_2) and that fixed cost was simply the current capacity of the company times some constant.

Jackson realized that the basic equation could be rewritten as

total cost = $\beta_1$capacity + $\beta_2 x_2$volume

FIGURE 5.22
Scatterplot of COST versus PROD.

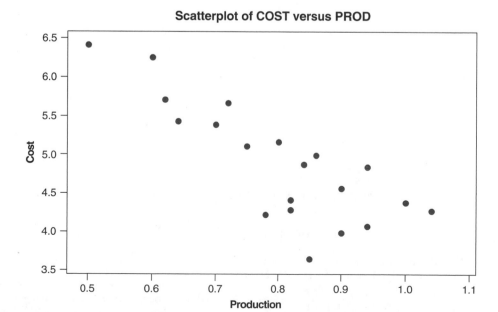

Scatterplot of COST versus PROD

FIGURE 5.23
Scatterplot of COST
versus INDEX.

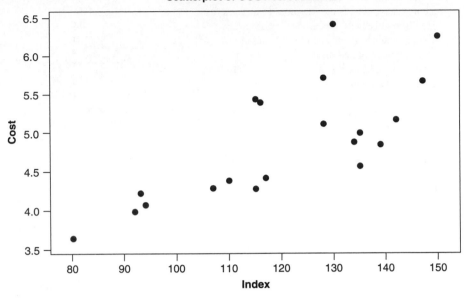

Scatterplot of COST versus INDEX

FIGURE 5.24
Regression of COST on
PROD and INDEX.

Variable	Coefficient	Std Dev	T Stat	P Value
Intercept	5.1829	0.5364	9.66	0.000
PROD	-3.4482	0.3961	-8.70	0.000
INDEX	0.0205	0.0028	7.33	0.000

Standard Error = 0.227957   R-Sq = 91.9%   R-Sq(adj) = 91.0%

Analysis of Variance

Source	DF	Sum of Squares	Mean Square	F Stat	P Value
Regression	2	10.0586	5.0293	96.78	0.000
Error	17	0.8834	0.0520		
Total	19	10.9420			

Then it dawned on him that he actually wanted cost per unit. The preceding equation could be divided by volume to get

total cost/volume = $\beta_1$capacity/volume + $\beta_2 x_2$

Capacity/volume, however, is simply the reciprocal of production level, so the new equation becomes

$$y = \beta_1\left(\frac{1}{x_1}\right) + \beta_2 x_2$$

where y is total cost/volume (or cost per unit), x_1 is production level and x_2 is cost index. Allowing for random error and allowing the equation to have an intercept term produces:

$$y = \beta_0 + \beta_1\left(\frac{1}{x_1}\right) + \beta_2 x_2 + e$$

Try Jackson's new model. How does this model compare with the original regression? In answering this question, use the R^2 values, the standard error of the regression, and any other information you feel might be useful in the comparisons. Which model do you prefer?

3. **Piston Corporation (Part B).** Use the model developed in Problem 2 (Piston Part A) to answer the following questions:

   a. A three-point rise in the cost index will cause what change in unit costs (assuming production level remains constant)?

   b. What is the marginal unit cost of a rise in production volume from 0.94 to 0.95 of capacity (marginal cost implies all other variables remain constant)?

   c. If forecasts of production level of 0.87 and cost index of 120 are obtained, find a prediction of the manufacturing cost per unit (use a point prediction).

   d. Construct a 95% prediction interval for manufacturing cost per unit under the conditions described in part c.

4. **Computer Repair.** A computer repair service is examining the time taken on service calls to repair computers. Data are obtained for 30 service calls. The data are in a file named COMPREP5 on the CD. Information obtained includes:

   x_1 = number of machines to be repaired (NUMBER)

   x_2 = years of experience of service person (EXPER)

   y = time taken (in minutes) to provide service (TIME)

Develop a polynomial regression model to predict average time on the service calls using EXPER and NUMBER as explanatory variables. Justify your model choice including transformations of any variables.

5. **Criminal Justice Expenditures.** The file named CRIMSPN5 on the CD contains the following data for each of the 50 states:

   total expenditures on a state's criminal justice system (in millions of dollars) (EXPEND)
   total number of police employed in the state (POLICE)

State governments must try to project spending in many areas. Expenditure on the criminal justice system is one area of continually rising cost. Your job is to build a model that can be used to forecast spending on a state's criminal justice system.

Once your model is complete, predict expenditures for a state that plans to hire 10,000 police personnel. Find a point prediction and a 95% prediction interval.

(*Source*: These data were obtained from the U.S. Department of Criminal Justice web site and are for the year 1999.)

6. **Predicting Movie Grosses.** The file named MOVIES5 on the CD contains data on movies released in the United States during the calendar year 1998. The two variables in this file are

   TDOMGROSS, the total domestic gross revenue
   WEEKEND, first weekend gross

We would like to find an equation to predict the total domestic gross revenue of movies based on their first weekend gross. People in the movie industry watch the first weekend gross revenues closely and use them to help make decisions about advertising, distribution, etc. We need to formalize the relationship between total domestic gross revenue and first weekend gross. Find an equation that represents this relationship. Use the scatterplot to guide in choosing the best model.

(*Source*: These data are discussed in the article "Predicting Movie Grosses: Winners and Losers, Blockbusters and Sleepers," by Jeffrey S. Simonoff and Ilana R. Sparrow. Chance, Vol. 13, 2000, 15—24, and were obtained from Dr. Simonoff's web site.)

7. **Kentucky Derby.** On the first Saturday in May, the granddaddy of horse races—the Kentucky Derby—is run at Churchill Downs in Louisville, Kentucky. The amount of money bet, in millions of dollars, on this race for the 66-year period from 1927 through 1992 is in the file named DERBY5 on the CD.

(*Source*: "How the Betting Went," Louisville *Courier-Journal*, May 3, 1992. Copyright 1992, Louisville *Courier-Journal*. Used with permission.).

Build an extrapolative model for the amount bet and provide a justification for the model. Use linear and/or nonlinear trends to build the model. Once you have chosen your preferred model, use it to forecast the amount bet in 1993 and 1994.

8. **Mileage and Weight.** The variables CITYMPG (y), which is the number of miles per gallon obtained by a car in city driving, and WEIGHT (x), the weight in pounds of the car, are in a file named MPGWT5 on the CD. This information is available for 147 cars listed in the *Road and Track* October 2002 issue.

Fit the linear regression using CITYMPG as the dependent variable and WEIGHT as the independent variable.

Examine a scatterplot of these two variables. Can you find a curvilinear model that better describes the relationship between these two variables? If so, what is the regression equation that describes this relationship? Justify your choice of equation.

USING THE COMPUTER

The Using the Computer section in each chapter describes how to perform the computer analyses in the chapter using Excel, MINITAB, and SAS. For further detail on Excel, MINITAB, and SAS, see Appendix C.

EXCEL

Variable Transformations

Variable transformations in Excel are accomplished through the use of formulas. Consider the screen shown in Figure 5.25. We want to create a new column containing the square of the numbers in column A. To do this, we create a formula in cell C2 to multiply the value in cell A2 by itself (=A2*A2). Then place the cursor on the lower right-hand corner of cell C2 and drag this cell to the last entry in column A (see Figure 5.26). Other transformations are created in a similar manner. If you are

FIGURE 5.25 Excel Screen Showing Formula Creation for Variable Transformation.

	A	B	C
1	MONTHS	CALLS	
2	10	18	=a2*a2
3	10	19	
4	11	22	
5	14	23	
6	15	25	
7	17	28	
8	18	29	
9	20	29	
10	20	31	
11	21	31	
12	22	33	
13	22	32	
14	24	31	
15	25	32	
16	25	32	
17	25	33	
18	25	31	
19	28	33	
20	29	33	
21	30	34	

FIGURE 5.26 Excel Screen Showing Formula Creation for Variable Transformation.

1	MONTHS	CALLS	
2	10	18	100
3	10	19	100
4	11	22	121
5	14	23	196
6	15	25	225
7	17	28	289
8	18	29	324
9	20	29	400
10	20	31	400
11	21	31	441
12	22	33	484
13	22	32	484
14	24	31	576
15	25	32	625
16	25	32	625
17	25	33	625
18	25	31	625
19	28	33	784
20	29	33	841
21	30	34	900

FIGURE 5.27 Excel Screen Showing How to Access Functions.

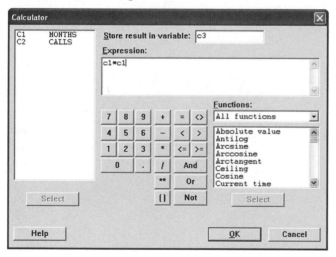

not sure of the form of a certain transformation, click the f_x button. Figure 5.27 shows an example using the natural log function. Most functions used to transform variables can be found in the Math & Trig category.

MINITAB

Variable Transformations

CALC: CALCULATOR

The CALCULATOR dialog box is shown in Figure 5.28. Any variable transformations can be performed using the calculator. As shown, the calculator is set up to multiply the numbers in column 1 by themselves (C1*C1) and place the result in C3. Thus, C3 will contain the square of the numbers in C1. Other transformations can be performed in similar fashion. Placing LOGE(C1) in the Expression box produces the natural logarithm of C1. Placing SQRT(C1) in the expression box produces the square root of C1. If you are not sure of the form of the function (LOGE, SQRT, and so on), just double click the desired function to the right of the keyboard and the appropriate expression appears in the Expression box.

FIGURE 5.28 MINITAB Dialog Box for Calculator.

SAS

Variable Transformations

In SAS, variable transformations are performed during the data input phase. Here are some typical examples:

Create the square of the variable MONTHS and call it XSQR:

```
XSQR = MONTHS**2;
```

Create the natural log of the variable MONTHS and call it LOGMONTH:

```
LOGMONTH = LOG(MONTH);
```

These transformed variables can then be used in PROC REG, PROC PLOT, and so on.

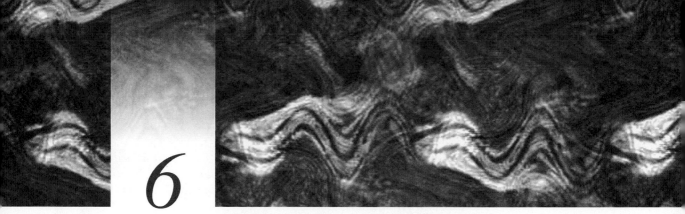

6

Assessing the Assumptions
of the Regression Model

6.1 INTRODUCTION

In Chapter 4, the multiple linear regression model was presented as

$$y_i = \beta_0 + \beta_1 x_{1i} + \beta_2 x_{2i} + \cdots + \beta_K x_{Ki} + e_i \qquad (6.1)$$

Certain assumptions were made concerning the disturbances, e_i, of this model. The e_i represent the differences between the true values of the dependent variable and the corresponding points on the population regression line. Because the true disturbances cannot be observed, they are modeled as realizations of a random variable about which certain assumptions are made. Under a set of ideal assumptions, the method of least squares provides the best possible estimates of the population regression coefficients. Certain assumptions are necessary for inference procedures (confidence interval estimates and hypothesis tests) to perform as desired. In this chapter, we consider the problems with estimation and inference that may arise if any of these assumptions are violated. Methods of assessing the validity of the assumptions also are discussed. Graphical procedures such as scatterplots and residual plots may be used to examine certain assumptions, and statistical tests are available for a more formal examination. Finally, we discuss appropriate techniques to correct for violated assumptions.

6.2 ASSUMPTIONS OF THE MULTIPLE LINEAR REGRESSION MODEL

The "ideal" conditions for estimation and inference in the multiple regression model are as follows:

a. The expected value of the disturbances is zero: $E(e_i) = 0$. This implies that the regression line passes through the conditional means of the y variable. For our purposes, we interpret this assumption as: The relationship is linear in the explanatory variables.

b. The disturbances have constant variance σ_e^2.

c. The disturbances are normally distributed.

d. The disturbances are independent.

The effects of violations of each of these assumptions on the least-squares estimates of the regression coefficients are examined in subsequent sections. Methods of assessing the validity of the assumptions are discussed and possible corrections for violations are offered. Because many of the methods of assessing assumption validity depend on the use of the residuals (the sample counterpart of the disturbances), the next section is devoted to a brief discussion of the computation and properties of the residuals.

6.3 THE REGRESSION RESIDUALS

The regression equation estimated from the sample data may be written

$$\hat{y}_i = b_0 + b_1 x_{1i} + b_2 x_{2i} + \cdots + b_K x_{Ki} \tag{6.2}$$

By substituting in the sample values for each explanatory variable, the predicted or fitted y value for each data point in the sample is obtained. The fitted y values are denoted as $\hat{y}_i$. The y values for each point in the sample are also available and are referred to as y_i. The differences between the true and fitted y values for the points in the sample are called the *residuals*. The residuals are denoted by $\hat{e}_i$:

$$\hat{e}_i = y_i - \hat{y}_i$$

They represent the distance that each dependent variable value is from the estimated regression line or the portion of the variation in y that cannot be "explained" with the data available. Because these "sample disturbances" approximate the population disturbances, they can be used to examine assumptions concerning the population disturbances.

After estimating a sample regression equation, it is highly recommended that some sort of analysis be conducted to assess the model assumptions. No regression analysis can be considered complete without such further examination. The residuals can be used to conduct such analyses through graphical techniques called *residual plots*. Often, violations of assumptions can be detected through the use of residual plots in combination with scatterplots without the use of statistical tests.

The use of graphical techniques, however, is not an exact science. It might, in fact, be considered an "art." It takes some experience at examining plots to become adept at determining which, if any, assumptions may be violated. Several examples are presented later to illustrate this art and to aid in mastering residual analysis.

First, consider some properties of the residuals.

Property 1: The average of the residuals is equal to zero. This property holds regardless of whether the assumptions are true or not and is a direct result of the way the least-squares method works. Least squares "forces" the mean of the residuals to be zero when it chooses the estimates of the regression coefficients.

Property 2: If assumptions a, b, and d of Section 6.2 are true, then the residuals should be randomly distributed about their mean (zero). There should be no systematic pattern in a residual plot.

Property 3: If assumptions a, b, and d are true and the disturbances are also normally distributed (assumption c), then the residuals should look like random numbers chosen from a normal distribution.

The residuals can be thought of as representing the variation in *y* that cannot be explained using the proposed regression model. Think of the process we are following as building a model for the data. We can write DATA = MODEL + ERROR. We have some DATA that we want to explain. We build a MODEL that we believe helps to explain patterns in the data. Any patterns in the DATA not included in the MODEL are accounted for in the ERROR term. These errors are represented by the residuals. Thus, if an assumption is violated, an indication of this violation appears as some type of pattern in the residuals. Identification of such patterns is a first step in correcting for the violation.

In a residual analysis, it is suggested that the following plots be used:

1. Plot the residuals versus each explanatory variable.
2. Plot the residuals versus the predicted or fitted values.
3. If the data are measured over time, plot the residuals versus some variable representing the time sequence.

As is shown in subsequent sections, each of these three types of plots plays a part in identifying violations of the basic assumptions.

If no assumptions are violated, then the residuals should be randomly distributed around their mean of zero and should look like numbers drawn randomly from a normal distribution. Figure 6.1 shows how a residual plot might appear when assumptions a through d are all true. There is no pattern visible in the scatter of residuals. For comparison, Figure 6.2 shows a residual plot with an obvious pattern to the residuals. Compare this to the random scatter of the residuals in Figure 6.1. A residual plot such as Figure 6.2 indicates that some assumption has been violated. (There are other patterns that could suggest violations, as is seen throughout this chapter.)

In most regression software packages (Excel, SAS, and MINITAB included), residual plots are easily constructed after a regression analysis has been performed.

FIGURE 6.1
Residual Plot Assuming
No Violation of
Assumptions a Through
d of Section 6.2.

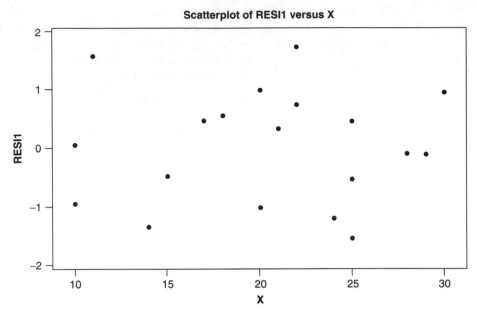

FIGURE 6.2
Residual Plot
Indicating That at
Least One of
Assumptions a
Through d of Section
6.2 Has Been Violated.

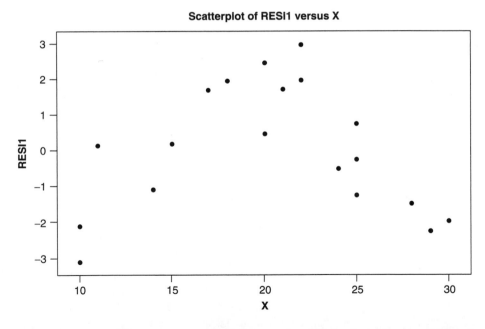

These plots may be constructed using the actual residuals, $\hat{e}_i$, or the standardized residuals. The *standardized residuals* are simply the residuals divided by their standard deviation. There is very little difference in the way residual plots with actual residuals or those with standardized residuals are used. To illustrate the difference in the two types of plots, compare Figure 6.1, a plot of actual residuals, to Figure 6.3, a plot of the same residuals after standardization. The residuals plotted in Figure 6.2

FIGURE 6.3 Plot of the Residuals from Figure 6.1 After Standardizing.

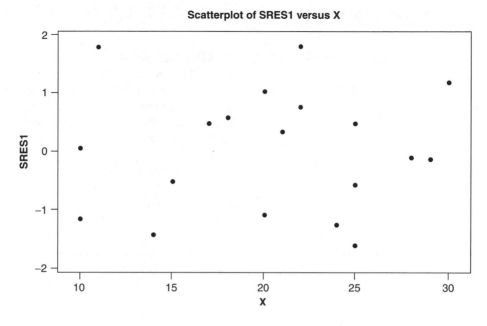

also have been standardized and plotted in Figure 6.4. The patterns in the actual and standardized plots are identical; only the scale has been changed. One advantage of using the standardized plots becomes more evident when the assumption of normality is discussed in Section 6.6. In this text, the residual plots shown are standardized plots unless otherwise indicated.

FIGURE 6.4 Plot of the Residuals from Figure 6.2 After Standardizing.

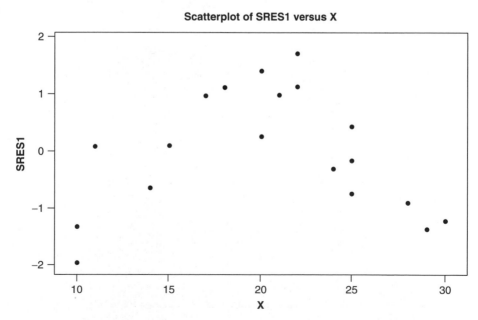

6.4 ASSESSING THE ASSUMPTION THAT THE RELATIONSHIP IS LINEAR

6.4.1 USING PLOTS TO ASSESS THE LINEARITY ASSUMPTION

The first assumption given in Section 6.2 was that the regression was linear in the explanatory variables. In Chapter 5, we saw that we can fit curvilinear as well as linear relationships using regression. In that chapter, we assumed that we could tell when a curvilinear relationship was needed simply by looking at the scatterplot. The scatterplots of y versus each of the explanatory variables may give an indication of whether the linearity assumption is an appropriate one, but this is not always the case. After performing a regression, this assumption can be checked visually through residual plots. Small deviations from linearity that are not evident in the scatterplots may show up clearly in the residual plots. The following example illustrates a violation of the linearity assumption.

EXAMPLE 6.1   **Telemarketing**   Consider again the telemarketing data from Example 5.1 (see the TELEMARK6 file on the CD). A company that provides transportation services uses a telemarketing division to help sell its services. The division manager is interested in the time spent on the phone by the telemarketers in the division. Data on the number of months of employment and the number of calls placed per day (an average for 20 working days) are recorded for 20 employees. These data are shown in Table 6.1.

The average number of calls for all 20 employees is 28.95. The division manager suspects, however, that there may be a relationship between time on the job and

TABLE 6.1 Data for Telemarketing Example

MONTHS	CALLS
10	18
10	19
11	22
14	23
15	25
17	28
18	29
20	29
20	31
21	31
22	33
22	32
24	31
25	32
25	32
25	33
25	31
28	33
29	33
30	34

number of calls. As time on the job increases, the employee becomes more familiar with the calling system and the correct procedures to use on the phone and also begins to acquire more regular clients. Thus, the longer the time on the job, the greater the number of calls per day. The scatterplot of CALLS versus MONTHS is in Figure 6.5 and the regression relating CALLS to MONTHS is shown in Figure 6.6.

Plots of the standardized residuals versus the fitted values and the explanatory variable MONTHS are shown in Figures 6.7 and 6.8, respectively. The standardized residuals have been labeled SRES1 in the plots. The fitted values are labeled FITS1.

A systematic pattern can be observed in both of the residual plots. The standardized residuals plot in a curvilinear pattern, suggesting a curvilinear component may be omitted from the equation expressing the relationship between CALLS and MONTHS. The plots of the standardized residuals versus the fitted values and versus MONTHS show identical patterns in this case. The plot versus the fitted values

FIGURE 6.5
Scatterplot of CALLS versus MONTHS for Telemarketing Example.

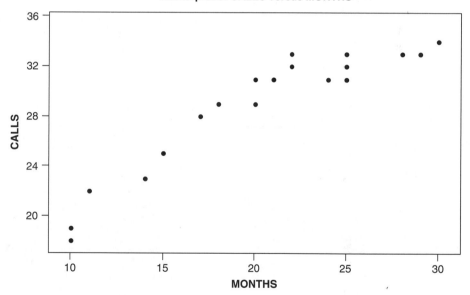

Scatterplot of CALLS versus MONTHS

FIGURE 6.6
Regression Results for Telemarketing Example.

Variable	Coefficient	Std Dev	T Stat	P Value
Intercept	13.6708	1.4270	9.58	0.000
MONTHS	0.7435	0.0667	11.15	0.000

Standard Error = 1.78737   R-Sq = 87.4%   R-Sq(adj) = 86.7%

Analysis of Variance

Source	DF	Sum of Squares	Mean Square	F Stat	P Value
Regression	1	397.45	397.45	124.41	0.000
Error	18	57.50	3.19		
Total	19	454.95			

FIGURE 6.7 Plot of Standardized Residuals versus Fitted Values for Telemarketing Example.

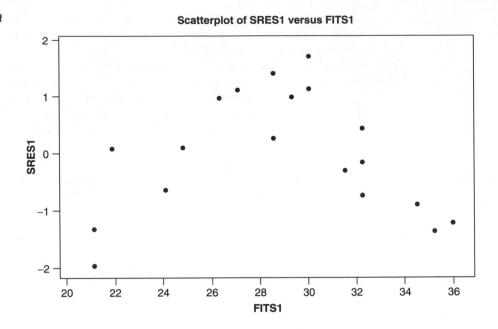

FIGURE 6.8 Plot of Standardized Residuals versus Explanatory Variable MONTHS for Telemarketing Example.

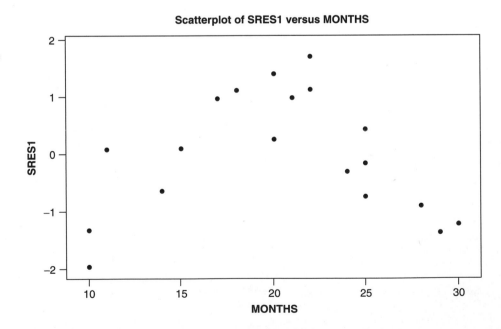

may differ from the plot versus one of the explanatory variables, especially in a multiple regression. The fitted values combine the effects of all the explanatory variables used in the regression. In a multiple regression, the plot of the standardized residuals versus the fitted values provides an overall picture, while the plots of the standardized residuals versus each explanatory variable may help identify any violations specifically related to an individual explanatory variable.

The systematic pattern observed in the residual plots suggests a violation of the linearity assumption. Looking at the scatterplot in Figure 6.5 also helps verify that the relationship may not be linear. As the number of months on the job increases, the number of calls also increases. But the rate of increase begins to slow over time, thus resulting in a pattern that may be better modeled by a curve than a straight line.

Note that the residual plots were used to determine whether the linearity assumption had been violated, although this violation might have been suspected from looking at only the scatterplot of CALLS versus MONTHS. In many cases, the violation of an assumption is not obvious from a scatterplot. The residual plot, however, is intended to magnify the consequences of any possible violation. Thus, the residual plot should be depended on to identify the violation. ◼

6.4.2 TESTS FOR LACK OF FIT[1]

MINITAB provides two tests to determine whether a curvilinear model might fit the data better than a linear model. These tests are referred to as *tests for lack of fit*. The first test is called the *pure error lack-of-fit test*. To perform this test, the error sum of squares is decomposed into two parts: the pure error component and the lack-of-fit component. These two components are used to construct an F statistic to test the hypotheses

H_0: The relationship is linear

H_a: The relationship is not linear

If H_0 is not rejected, the linear regression model is appropriate. If H_0 is rejected, the linear model does not fit the data well, and some other function may provide a better fit, although the test does not specify what that function is.

To conduct the F test, the decision rule is

Reject H_0 if $F > F(\alpha; c - K - 1, n - c)$

Do not reject H_0 if $F \leq F(\alpha; c - K - 1, n - c)$

where K is the number of explanatory variables and n is the sample size. The value c requires some additional explanation.

The pure error lack-of-fit test requires that there be repeated observations (replications) for at least one level of the x variables. In the telemarketing data in Table 6.1, there are replicates for $x = 10, 20, 22,$ and 25. The value c is the number of distinct levels of x. In the telemarketing example, there are 14 levels of (or distinct values of) the explanatory variable. The decision rule to perform this test on the telemarketing data is:

Reject H_0 if $F > F(\alpha; 12, 6)$

Do not reject H_0 if $F \leq F(\alpha; 12, 6)$

---

[1] This section refers specifically to MINITAB output, but similar tests could be available with other software.

(because $c = 14$, $K = 1$, and $n = 20$). The results are shown in Figure 6.9. From the output, the F statistic value is seen to be 5.25. If a 5% level of significance is used, the critical value for the test is $F(0.05; 12, 6) = 4.00$, and the decision is to reject H_0 and conclude that a curvilinear model may fit the data better than the linear model. Also, the p value (0.026) can be used in the usual manner to perform this test.

Note that this test cannot be performed unless there are replicates for at least one level of x. MINITAB does provide another test for lack of fit that does not require replicates. The *data subsetting test* actually involves a series of tests, and the results of several of these tests may be printed out on the output. For example, in Figure 6.10, results of tests examining curvilinearity in the variable MONTHS, lack of fit at the outer x values, and overall lack of fit are reported. These results are reported in terms of the p values, so the p value decision rule can be applied.

Reject H_0 if p value $< \alpha$

Do not reject H_0 if p value $\geq \alpha$

For $\alpha = 0.05$, the test result indicates possible curvature in the variable MONTHS and an overall lack of fit.

6.4.3 CORRECTIONS FOR VIOLATIONS OF THE LINEARITY ASSUMPTION

When the linearity assumption is violated, the appropriate correction is not always obvious. The violation of this assumption implies that y and x are related in some curvilinear fashion, but there are many equations that describe curvilinear relationships. The idea behind the use of any equation of this sort is to transform the variables in such a way that a linear relationship is achieved. If x and y are related in a curvilinear fashion, then perhaps x^2 and y have a linear relationship.

The violation of the linearity assumption was originally noted in the residual plots. If we have corrected for the violation, we should not see the same patterns in

FIGURE 6.9 MINITAB Output Showing Pure Error Lack-of-Fit Test for Telemarketing Example.

```
The regression equation is
CALLS = 13.7 + 0.744    MONTHS


Predictor      Coef  SE Coef       T      P
Constant     13.671    1.427    9.58  0.000
MONTHS      0.74351  0.06666   11.15  0.000


S = 1.78737   R-Sq = 87.4%     R-Sq(adj) = 86.7%


Analysis of Variance

Source          DF       SS       MS       F      P
Regression       1   397.45   397.45  124.41  0.000
Residual Error  18    57.50     3.19
  Lack of Fit   12    52.50     4.38    5.25  0.026
  Pure Error     6     5.00     0.83
Total           19   454.95


10 rows with no replicates
```

FIGURE 6.10
MINITAB Output Showing Data Subsetting Test for Telemarketing Example.

```
The regression equation is
CALLS = 13.7 + 0.744 MONTHS


Predictor      Coef   SE Coef      T      P
Constant     13.671     1.427   9.58  0.000
MONTHS      0.74351   0.06666  11.15  0.000


S = 1.78737   R-Sq = 87.4%   R-Sq(adj) = 86.7%


Analysis of Variance

Source         DF       SS      MS       F      P
Regression      1   397.45  397.45  124.41  0.000
Residual Error 18    57.50    3.19
Total          19   454.95


Lack of fit test
Possible curvature in variable MONTHS   (P-Value = 0.0001)

Possible lack of fit at outer X-values (P-Value = 0.097)
Overall lack of fit test is significant at P = 0.000
```

the residual plots from the corrected model. The residuals from a properly corrected model should be randomly scattered.

In Chapter 5, the following four commonly used corrections were considered:

1. polynomial regression
2. reciprocal transformation of the x variable
3. log transformation of the x variable
4. log transformation of both the x and y variables

After trying one of these transformations, check the new residual plots to see if the violation was effectively corrected. If not, try one of the other corrections. Refer to Chapter 5 for examples of the use of curvilinear models. The four corrections just listed are described in greater detail in Chapter 5 as well.

EXAMPLE 6.2   **Telemarketing (continued)**   In Chapter 5, a second-order polynomial regression was used to model the telemarketing data. The model can be written

$$\text{CALLS} = \beta_0 + \beta_1\text{MONTHS} + \beta_2\text{XSQR} + e$$

where XSQR is a variable created by squaring each value of the MONTHS variable. Figure 6.11 shows the regression estimates of the second-order model. The estimated regression is

$$\text{CALLS} = -0.14 + 2.31\text{MONTHS} - 0.04\text{XSQR}$$

Two checks should be made at this point to determine whether the second-order model is preferred to the original linear model: (a) test to see whether the coefficient

FIGURE 6.11
Regression Results for Telemarketing Example with Second-Order Term Added.

Variable	Coefficient	Std Dev	T Stat	P Value
Intercept	-0.14047	2.32226	-0.06	0.952
MONTHS	2.31020	0.25012	9.24	0.000
XSQR	-0.04012	0.00633	-6.33	0.000

Standard Error = 1.00325   R-Sq = 96.2%   R-Sq(adj) = 95.8%

Analysis of Variance

Source	DF	Sum of Squares	Mean Square	F Stat	P Value
Regression	2	437.84	218.92	217.50	0.000
Error	17	17.11	1.01		
Total	19	454.95			

of the second-order term is significantly different from zero and (b) check to see whether the new residual plots indicate an improved model. The *t* stat or the *p* value for the coefficient of XSQR can be used to verify that the coefficient is significant. Next, we should examine the new residual plots.

The test result on the second-order term by itself is not sufficient evidence to judge this model to be adequate. The goal in adding the second-order term was to correct for the curvilinear patterns noted in the original residual plots. To see if this has been accomplished, the residual plots from the new equation must be examined. The residual plots of the standardized residuals versus the fitted values, the MONTHS variable, and the XSQR variable are shown in Figures 6.12, 6.13, and 6.14, respectively.

FIGURE 6.12
Plot of Standardized Residuals versus Fitted Values for Second-Order Model.

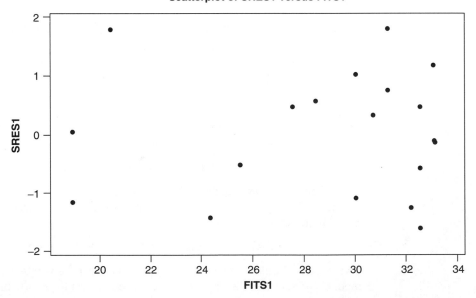

Scatterplot of SRES1 versus FITS1

FIGURE 6.13
Plot of
Standardized
Residuals versus
Explanatory
Variable MONTHS
for Second-Order
Model.

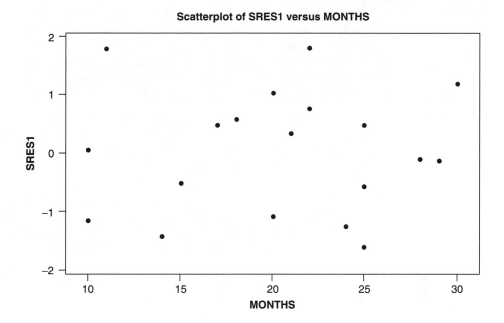

Scatterplot of SRES1 versus MONTHS

FIGURE 6.14
Plot of
Standardized
Residuals versus
Explanatory
Variable XSQR for
Second-Order
Model.

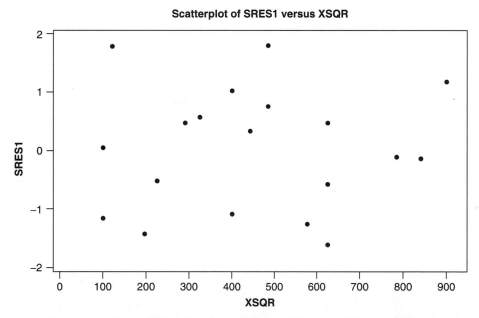

Scatterplot of SRES1 versus XSQR

Looking at the residual plots, no distinct patterns can be seen. Contrast this with the obvious patterns of Figures 6.7 and 6.8. The addition of the x^2 variable appears to have corrected for the curvilinearity. The second-order model is an improvement over the first-order model, and the regression assumptions appear to be satisfied.

Higher-order terms could be added to the model, but there appears to be little justification in doing so from looking at the second-order model regression output and residual plots.

Other indicators that the regression has been improved by adding the x^2 term (as discussed in Chapter 5) include the reduction in the standard error of the regression from 1.787 to 1.003 and the increase in adjusted R^2 from 86.7% to 95.8%. ◼

When curvilinear patterns appear in residuals plots, it is typically a sign that a linear model has been fit when a curvilinear model is more appropriate (or that an incorrect curvilinear model was fit). When this happens, a choice of the type of curvilinear model must be made. Some of the more common types of curvilinear models were discussed in Chapter 5. That discussion is not repeated again here in Chapter 6. The reader should be sure to review these models. When one of the models is chosen as a possible improvement, be sure to recheck the residual plots. A random scatter in the residual plots indicates that the correct model was fit to the data. If patterns in the residual plots persist, try a different correction.

EXERCISES

1. Parabola. Consider the following data:

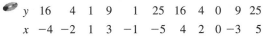

y	16	4	1	9	1	25	16	4	0	9	25
x	−4	−2	1	3	−1	−5	4	2	0	−3	5

Regard x as the explanatory variable and y as the dependent variable. Figure 6.15 shows the scatter-plot of y versus x. Figure 6.16 shows the regression output. These data are in a file named PARA6 on the CD.

a. Examine the scatterplot of y versus x. Does there appear to be a relationship between y and x?

FIGURE 6.15
Scatterplot of y
Versus x for
Parabola Exercise.

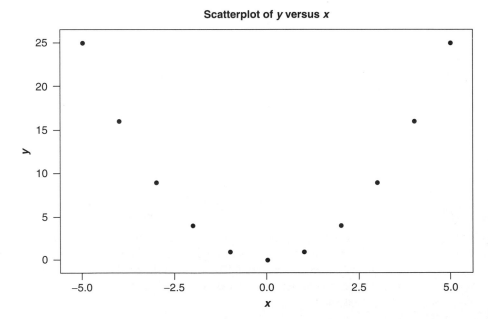

Scatterplot of y versus x

b. What is the estimated linear regression equation relating y to x?

c. Test the hypothesis $H_0: \beta_1 = 0$ against the alternate $H_a: \beta_1 \neq 0$ at the 1% level of significance. What conclusion can be drawn from the result of the test?

d. Despite the outcome of the test in part c, does there appear to be a "strong" or "weak" association between x and y? Express this association in the form of an equation.

2. Research and Development. A company is interested in the relationship between profit (PROFIT) on a number of projects and two explanatory variables. These variables are the expenditure on research and development for the project (RD) and a measure of risk assigned at the outset of the project (RISK). PROFIT is measured in thousands of dollars and RD is measured in hundreds of dollars. The scatterplots of PROFIT versus RISK and PROFIT versus RD are shown in Figures 6.17 and 6.18, respectively. The regression results are in

FIGURE 6.16
Regression of y on x for Parabola Exercise.

Variable	Coefficient	Std Dev	T Stat	P Value
Intercept	10.000	2.944	3.40	0.008
X	-0.000	0.931	-0.00	1.000

Standard Error = 9.76388   R-Sq = 0.0% R-Sq(adj) = 0.0%

Analysis of Variance

Source	DF	Sum of Squares	Mean Square	F Stat	P Value
Regression	1	0.00	0.00	0.00	1.000
Error	9	858.00	95.33		
Total	10	858.00			

FIGURE 6.17
Scatterplot of PROFIT versus RISK for Research and Development Exercise.

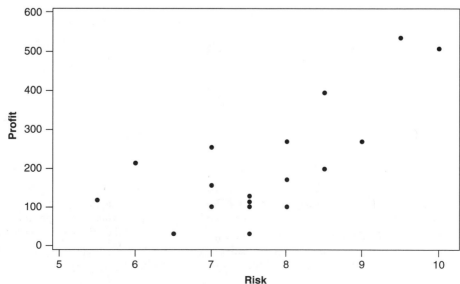

Scatterplot of PROFIT versus RISK

FIGURE 6.18
Scatterplot of PROFIT versus RD for Research and Development Exercise.

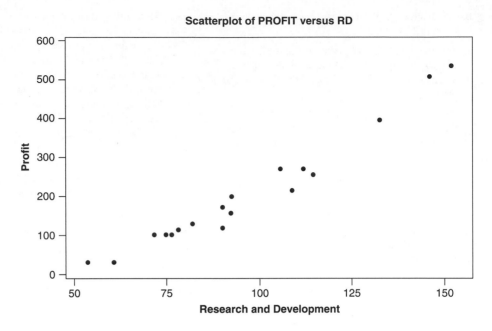

Scatterplot of PROFIT versus RD

FIGURE 6.19
Regression Results for Research and Development Exercise.

Variable	Coefficient	Std Dev	T Stat	P Value
Intercept	-453.1763	23.5061	-19.28	0.000
RISK	29.3090	3.6686	7.99	0.000
RD	4.5100	0.1538	29.33	0.000

Standard Error = 14.3420   R-Sq = 99.2% R-Sq(adj) = 99.0%

Analysis of Variance

Source	DF	Sum of Squares	Mean Square	F Stat	P Value
Regression	2	361639	180820	879.08	0.000
Error	15	3085	206		
Total	17	364724			

Figure 6.19. The residual plots of the standardized residuals versus the fitted values, RISK, and RD are shown in Figures 6.20, 6.21, and 6.22, respectively.

Using any of the given outputs, does the linearity assumption appear to be violated? Justify your answer. If you answered yes, state how the violation might be corrected. Then try your correction using a computer regression routine. Does your model appear to be an improvement over the original model? Justify your answer.

These data are available in a file named RD6 on the CD.

FIGURE 6.20
Plot of Standardized
Residuals versus
Fitted Values for
Research and
Development
Exercise.

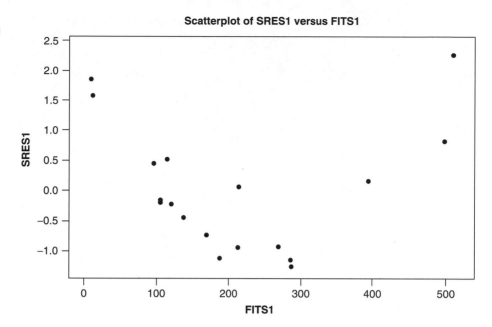

FIGURE 6.21
Plot of Standardized
Residuals versus
RISK for Research
and Development
Exercise.

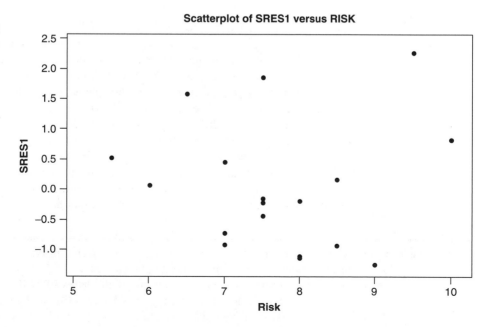

FIGURE 6.22
Plot of Standardized
Residuals versus RD
for Research and
Development
Exercise.

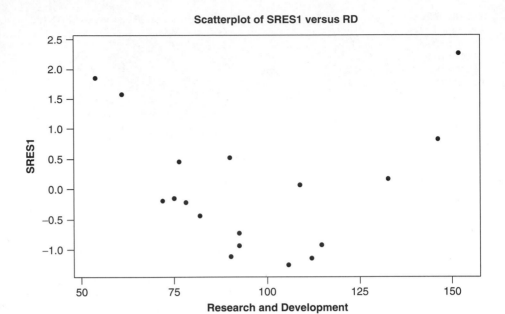

Scatterplot of SRES1 versus RD

6.5 ASSESSING THE ASSUMPTION THAT THE VARIANCE AROUND THE REGRESSION LINE IS CONSTANT

6.5.1 USING PLOTS TO ASSESS THE ASSUMPTION OF CONSTANT VARIANCE

Assumption b of Section 6.2 states that the disturbances in the population regression equation, e_i, have constant variance σ_e^2. In a residual plot of $\hat{e}_i$ versus an explanatory variable x, the residuals should appear scattered randomly about the zero line with no differences in the amount of variation in the residuals regardless of the value of x. If there appears to be a difference in variation (for example, if the residuals are more spread out for large values of x than for small values), then the assumption of constant variance may be violated. In a residual plot, nonconstant variance is often identified by a "cone-shaped" pattern, as shown in Figure 6.23. Again, the violation is indicated by a systematic pattern in the residuals. In many texts, the term *heteroskedasticity* is used in place of "nonconstant variance." Example 6.3 illustrates the use of plots to assess the constant variance assumption.

EXAMPLE 6.3   FOC Sales   Techcore is a high-tech company located in Fort Worth, Texas. The company produces a part called a fibre-optic connector (FOC) and wants to generate reasonably accurate but simple forecasts of the sales of FOCs over time. The company has weekly sales data for the past 265 weeks (in the file on the CD named FOC6). (The data have been disguised to provide confidentiality.) The time-series plot of FOC sales is shown in Figure 6.24. The regression of FOC sales using a

FIGURE 6.23
Cone-Shaped Pattern in
Residual Plot Suggesting
Nonconstant Variance.

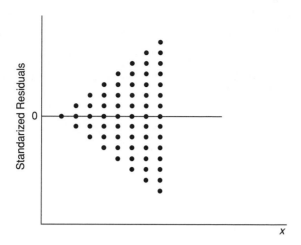

FIGURE 6.24
Time-Series Plot of
FOC Sales.

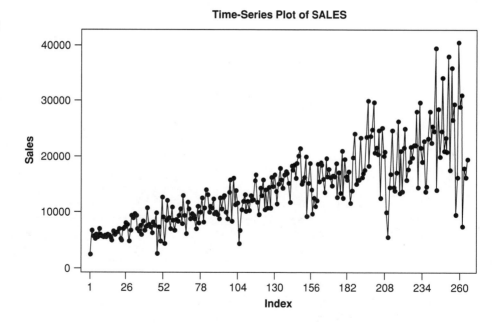

linear trend variable as the independent variable is shown in Figure 6.25. The plot of
the standardized residuals versus the fitted values is shown in Figure 6.26. The cone-
shaped pattern of residuals is obvious. Note that the variability of the residuals
increases over time. This pattern also can be seen in the time-series plot of FOC
sales. Such violations of assumptions, however, are typically magnified in the
residual plots, as in this case.

FIGURE 6.25
Regression Results for
FOC Sales Example.

```
Variable        Coefficient      Std Dev        T Stat        P Value

Intercept        4703.77          12.47          9.18          0.000
TIME               72.46           3.34         21.69          0.000


Standard Error = 4159.39      R-Sq = 64.2%   R-Sq(adj) = 64.0%


Analysis of Variance

Source          DF    Sum of Squares       Mean Square   F Stat   P Value

Regression       1       8142042913         8142042913   470.62   0.000
Error          263       4550042800           17300543
Total          264      12692085712
```

FIGURE 6.26 Plot of
Standardized Residuals
versus Fitted Values for
FOC Sales Example.

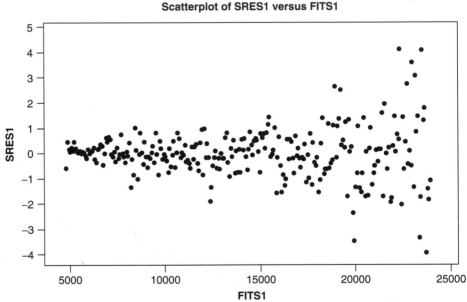

When the disturbance variance is not constant, the use of the least-squares method has two major drawbacks:

1. The estimates of the regression coefficients are no longer minimum variance estimates.

2. The estimates of the standard errors of the coefficients are biased.

The first drawback suggests that estimates of the coefficients with smaller sampling variability may exist. Because of the second drawback, hypothesis tests about the population regression parameters may provide misleading results.

6.5.2 A TEST FOR NONCONSTANT VARIANCE

Several tests are available for nonconstant variance, although a study by Griffiths and Surekha ("A Monte Carlo Evaluation of the Power of Some Tests for Heteroscedasticity"[2]) demonstrated that a test developed by J. Szroeter tends to be better at detecting nonconstant variance. The hypotheses to be tested are

H_0: Variance is constant

H_a: Variance is not constant

Szroeter's test statistic is

$$Q = \left(\frac{6n}{n^2 - 1} \right)^{1/2} \left(h - \frac{n + 1}{2} \right)$$

where n is the sample size,

$$h = \frac{\sum_{i=1}^{n} i\hat{e}_i^2}{\sum_{i=1}^{n} \hat{e}_i^2}$$

and $\hat{e}_i$ is the residual from the ith observation in the regression equation. The decision rule for the test is

Reject H_0 if $Q > z_\alpha$

Do not reject H_0 if $Q \leq z_\alpha$

where α is the level of significance for the test and z_α is chosen from the standard normal table with upper-tail area α.

Szroeter's test assumes that all the observations can be arranged in order of increasing variance. Typically, it is assumed that the variance increases as the value of one of the explanatory variables increases. Thus, the data need to be arranged according to the values of this explanatory variable. As a simple example, suppose the values of x and y in a simple regression are as follows:

x	3	2	7	9	4
y	6	4	16	15	8

Arranging the values of x in ascending order and maintaining the associated values of y results in the following arrangement of the data:

x	2	3	4	7	9
y	4	6	8	16	15

After reordering the data in this way, a regression is run and the residuals, $\hat{e}_i$, are saved and used to compute h. The value for h is then substituted into the equation to compute the test statistic Q.

---

[2] See References for complete publication information.

6.5.3 CORRECTIONS FOR NONCONSTANT VARIANCE

There are a number of possible corrections for nonconstant variance. All require a transformation of the dependent variable. This often makes comparison of the new regression equation with the old equation difficult. Some commonly used corrections for nonconstant variance are as follows:

1. In place of the dependent variable y, use the natural logarithm of y, $\ln(y)$. The natural logarithms of the y values are less variable than the original y values and may stabilize the variance. For example, consider the following numbers:

y	1	2	5	10	50
$\ln(y)$	0	0.69	1.61	2.30	3.91

   Note the difference in variation between the original y values and their natural logarithms. The natural logarithms are more equally spaced than are the original values. Note also that the natural logarithm transformation is only defined for positive numbers.

   The natural logarithm transformation is the appropriate transformation when the error standard deviation is proportional to the mean of the dependent variable.

2. In place of the dependent variable y, use the square root of y, $\sqrt{y}$. The square roots of the y values are less variable than the original y values and may stabilize the variance. Note that the square root transformation is not defined for negative numbers.

   The square root transformation is appropriate when the dependent variable is a count variable that follows a Poisson distribution.

3. There is a general class of transformations, called Box-Cox transformations, that can be used to transform the y variable when attempting to stabilize the variance. Box-Cox transformations can be defined as follows:

   Use y^p in place of y where $0 \leq p \leq 1$

   Thus, powers of the y variable are used in place of the original variable. The square root transformation defined in correction 2 is a special case of the Box-Cox transformation because $p = 0.5$ in that case. Note that the extreme values of p for this type of transformation are 0 and 1. When $p = 1$, the original y values are used. When $p = 0$, we define the transformation to be the natural log transformation discussed under correction 1. When $p = 1$, no compression of the data are necessary. When $p = 0$, we use the log transformation and obtain maximum compression of the data. Powers between 0 and 1 provide flexibility in the amount of compression obtained.

   (For more on the use of Box-Cox transformations, the interested reader is referred to Neter, Wasserman, and Kutner, *Applied Linear Statistical Models*, pp. 394–400.)

4. If the disturbance variance is thought to be proportional to some function of one of the x variables, the values of that variable can be used to stabilize the variance. For example, if

$$\sigma_{e_i}^2 = \sigma^2 x_i^2$$

is thought to express the relationship between the variance at each observation i and the associated value of the x variable, then dividing each variable in the regression by x stabilizes the variance. If the original equation is

$$y_i = \beta_0 + \beta_1 x_i + e_i$$

then, after dividing through by x_i, the transformed model becomes

$$\frac{y_i}{x_i} = \beta_0 \left(\frac{1}{x_i}\right) + \beta_1 + e_i'$$

where $e_i' = e_i/x_i$ is a new disturbance with constant variance. Note that the roles played by β_0 and β_1 have been reversed in the transformed equation. The transformation is not defined when x_i is zero.

EXAMPLE 6.4 **FOC Sales (continued)** The regression output for the regression of the natural log of sales (LOGSALES) on TIME is shown in Figure 6.27, and the residual plot of the standardized residuals versus the fitted values is in Figure 6.28. In the residual plot, the cone-shaped pattern has been greatly reduced. Using the log transformation stabilized the variance, so it is now relatively constant for all values of x. Caution should be exercised in interpreting and using the regression output, however. The output in Figure 6.27 shows results concerning the relationship of the log of sales to TIME. All information from the output must be interpreted in light of this fact. Thus, 68.7% of the variation in the log of sales has been explained. This value is not directly comparable to the R^2 from the regression of y on x in Figure 6.25 (64.2%). No conclusions can be drawn concerning which regression does "better" based on the R^2 because two different measures of the dependent variable are used.

Also keep in mind that if the equation from Figure 6.27 is used for forecasting, natural logs of the y values are forecasted, not the y values themselves. For example, what is the forecast of sales in week 300? Using the estimated regression equation

$$\text{LOGSALES} = 8.74 + 0.00537(300) = 10.351$$

FIGURE 6.27
Regression of
LOGSALES on TIME.

Variable	Coefficient	Std Dev	T Stat	P Value
Intercept	8.7390327	0.0342976	254.80	0.000
TIME	0.0053746	0.0002235	24.04	0.000

Standard Error = 0.278373   R-Sq = 68.7%   R-Sq(adj) = 68.6%

Analysis of Variance

Source	DF	Sum of Squares	Mean Square	F Stat	P Value
Regression	1	44.797	44.797	578.09	0.000
Error	263	20.380	0.077		
Total	264	65.177			

FIGURE 6.28
Plot of Standardized
Residuals versus
Fitted Values for
Transformed Model.

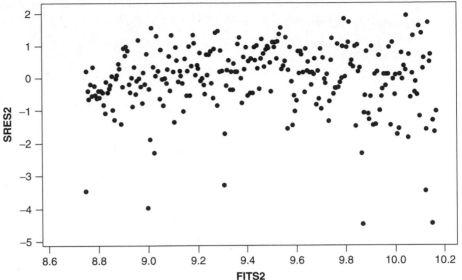

The forecast value for LOGSALES is 10.351. The resulting forecast value for y must be computed as

$$\text{SALES} = e^{10.351} = 31{,}288$$

When deciding whether the transformation has improved our results, comparing the R^2 is not always dependable, as noted. The residual plots can be used to help in this decision. If the pattern suggesting a violation in the original residual plots is no longer present in the plots from a transformed model, then the transformed model is preferable. If there is little or no improvement in the residual plots, then another transformation should be tried and the resulting residual plots examined to see whether they indicate an improved model. Choosing the correct transformation to produce an adequate model is thus an iterative process that may take several tries. And when all else fails, consult your neighborhood statistician!

EXERCISES

3. S&P 500 Index Prices. Data for the S&P 500 Stock Index for the time period January 1974 through December 2002 are in the file named SP5006 on the CD. These data were obtained from the web site *www.economagic.com*. The objective is to build a regression model relating the current price to the price in the previous month. The regression equation can be written as

$$y_i = \beta_0 + \beta_1 y_{i-1} + e_i$$

where y_i represents the S&P 500 index price in time period i.

Figure 6.29 is the output for the regression of the current S&P 500 index price on the previous month's price. Figure 6.30 is the plot of the standardized residuals versus the fitted values.

FIGURE 6.29
Regression of Current S&P 500 Index Price on Previous Month's Price.

Variable	Coefficient	Std Dev	T Stat	P Value
Intercept	8.806181	6.236058	1.41	0.159
S&P LAG1	0.998917	0.003629	275.27	0.000

Standard Error = 85.4760   R-Sq = 99.5%   R-Sq(adj) = 99.5%

Analysis of Variance

Source	DF	Sum of Squares	Mean Square	F Stat	P Value
Regression	1	553606990	553606990	75772.72	0.000
Error	345	2520622	7306		
Total	346	556127613			

FIGURE 6.30
Plot of Standardized Residuals versus Fitted Values for S&P 500 Index Exercise.

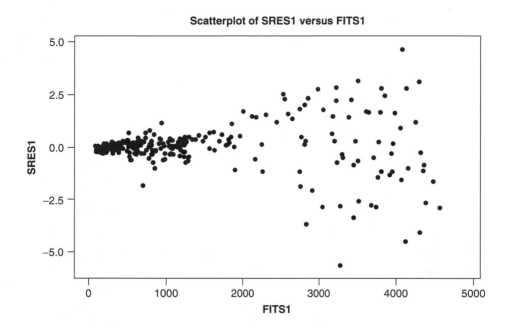

Is there evidence that the constant variance assumption has been violated? Justify your answer. Suggest a correction for the violation of the constant variance assumption for this example. Try the correction using a regression routine. Does the correction appear to have eliminated the problem of nonconstant variance? State why or why not.

6.6 ASSESSING THE ASSUMPTION THAT THE DISTURBANCES ARE NORMALLY DISTRIBUTED

6.6.1 USING PLOTS TO ASSESS THE ASSUMPTION OF NORMALITY

Residual plots of the standardized residuals versus the fitted values can be used to assess graphically whether the sample residuals have come from a normally distributed population. For normally distributed data, about 68% of the standardized residuals should be between −1 and +1, about 95% should be between −2 and +2, and about 99% should be between −3 and +3. Normal probability plots also can be a useful graphical technique in assessing normality.

EXAMPLE 6.5   **Communications Nodes (continued)**   Figure 6.31 shows the regression of cost (COST) on the number of ports (NUMPORTS) and the bandwidth (BANDWIDTH) for the communications nodes data discussed in several examples and exercises in Chapters 3 and 4. (The data can be found in the file named COMNODES6 on the CD.) The plot of the standardized residuals versus the fitted values is shown in Figure 6.32. A printout of the standardized residuals is shown in Table 6.2.

FIGURE 6.31
Regression of COST on NUMPORTS and BANDWIDTH.

Variable	Coefficient	Std Dev	T Stat	P Value
Intercept	17086.75	1865.41	9.16	0.000
NUMPORTS	469.03	66.98	7.00	0.000
BANDWIDTH	81.07	21.65	3.74	0.003

Standard Error = 2982.52   R-Sq = 95.0%   R-Sq(adj) = 94.1%

Analysis of Variance

Source	DF	Sum of Squares	Mean Square	F Stat	P Value
Regression	2	1876012662	938006331	105.45	0.000
Error	11	97849860	8895442		
Total	13	1973862522			

FIGURE 6.32 Plot of Standardized Residuals versus Fitted Values.

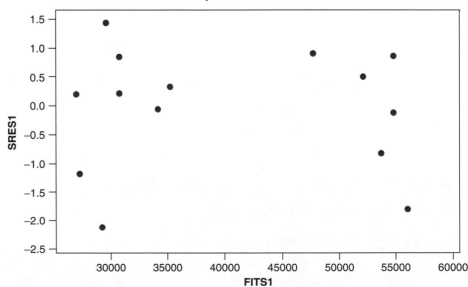

TABLE 6.2 List of Standardized Residuals for Regression
of COST on NUMPORTS and BANDWIDTH

Observation	Standardized Residuals
1	−0.82144
2	−1.80142
3	0.91087
4	0.32777
5	0.20567
6	0.85903
7	−0.11480
8	−0.04969
9	0.22068
10	−2.11672
11	−1.17904
12	0.50409
13	1.44088
14	0.84722

As can be seen from the plots and the printout, 13 of the 14 standardized residuals (93%) are between ±2. So there are about the number of residuals we would expect to see if they came from a normal distribution. From these plots, we conclude that the normality assumption seems reasonable.

When examining the normality assumption, concern should be placed on relatively large standardized residuals. Thus, an excessive number of residuals outside the ±2 limit or the ±3 limit might cause concern about this assumption. But remember to expect some values outside these limits, especially in large data sets.

The MINITAB regression output is shown in Figure 6.33. Note that MINITAB flags any observations with standardized residuals that are greater than or equal to 2 in absolute value. These values are shown with an R next to the value of the standardized residual in the table of Unusual Observations. Often, it is the observations with large standardized residuals with which we are concerned. This is why MINITAB takes the time to flag these observations. This does not mean that there is anything wrong with these data values or that they should be deleted. It simply means that these observations may be different from the others in our data set for some reason and may deserve special attention.

Figure 6.34 shows the normal probability plot for the residuals produced by MINITAB. In this plot, the observed residuals (horizontal axis) are plotted against the "normal scores." The normal scores can be thought of as the values expected if a sample of the same size as the one used (14 in this case) was selected from a normal distribution. The vertical scale shows the cumulative percentage (probability) at or below the normal scores rather than the normal scores themselves. The normal scores and the percentages are computed by MINITAB.

When the plot of the normal scores (cumulative probabilities) and the data is approximately a straight line, the normality assumption appears reasonable. The

FIGURE 6.33 MINITAB Regression of COST on NUMPORTS and BANDWIDTH.

```
The regression equation is
COST = 17086 + 469 NUMPORTS + 81.1 BANDWIDTH

Predictor       Coef     SE Coef         T         P
Constant       17086        1865      9.16     0.000
NUMPORTS      469.03       66.98      7.00     0.000
BANDWIDTH      81.07       21.65      3.74     0.003

S = 2982.52     R-Sq = 95.0%    R-Sq(adj) = 94.1%

Analysis of Variance

Source            DF          SS           MS          F         P
Regression         2  1876012662    938006331     105.45     0.000
Residual Error    11    97849860      8895442
Total             13  1973862522

Source           DF      Seq SS
NUMPORTS          1  1751268376
BANDWIDTH         1   124744286

Unusual Observations

Obs    NUMPORTS      COST      Fit    SE Fit    Residual    St Resid
  1        68.0     52388    53682      2532       -1294      -0.82 X
 10        24.0     23444    29153      1273       -5709      -2.12R

R denotes an observation with a large standardized residual.
X denotes an observation whose X value gives it large influence.
```

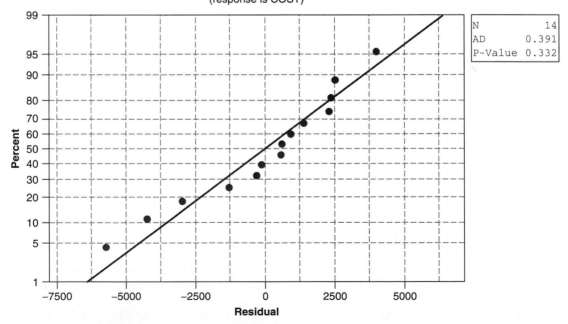

Normal Probability Plot of the Residuals
(response is COST)

N	14
AD	0.391
P-Value	0.332

FIGURE 6.34 MINITAB Normal Probability Plot and Test for Normality.

normal scores are numbers we expect to see from a sample from a normal distribution, so for the two to plot on a straight line, the two sets of data have to be similar. Thus, we reason that the data must also have come from a normal distribution. If the data did not come from a normal distribution, the plot will show curvature. (Note that whether the normal probability plot is linear or not has nothing to do with whether the relationship between y and x is linear. We are not assessing whether the relationship between the original variables is linear, but whether the disturbances come from a normal distribution.)

In this example, the plot of the normal scores versus the data is nearly linear (MINITAB has drawn in the line). This suggests that the standardized residuals could have come from a normal distribution. Thus, the normality assumption is supported.

6.6.2 TEST FOR NORMALITY

The plot of the residuals versus the normal scores in a normal probability plot should be approximately linear if the disturbances are normal. The "straightness" of the line in the normal probability plot can be measured by computing a variety of test statistics including the Kolmogorov–Smirnov (KS) statistic, the Anderson–Darling (AD) statistic, and the Ryan–Joiner (RJ) statistic. Regardless of the test statistic used, the hypotheses can be stated as

H_0: Disturbances are normal

H_a: Disturbances are nonnormal

In most statistical packages a p value will be printed for the appropriate test statistic (KS, AD, or RJ). The decision rule is

Reject H_0 if p value $< \alpha$

Do not reject H_0 if p value $\geq \alpha$

where α is the level of significance for the test.

EXAMPLE 6.6   **Communications Nodes (continued)**   In the MINITAB normal probability plot in Figure 6.34 the value of the Anderson–Darling (AD) statistic is AD = 0.391. The associated p value is 0.332. Using a 5% level of significance, the decision rule for the normality test is

Reject H_0 if p value < 0.05

Do not reject H_0 if p value ≥ 0.05

With a reported p value of 0.332, H_0 cannot be rejected. The conclusion is that the disturbances are normally distributed.

EXAMPLE 6.7   **Saving and Loan (S&L) Rate of Return**   In "Return, Risk, and Cost of Equity for Stock S&L Firms: Theory and Empirical Results," Lee and Lynge discuss methods of estimating the cost of equity capital for S&L associations. One aspect of their

analysis included an examination of the relationship between the rate of return of the S&L stocks (y) and two measures of the risk of the stocks: the beta coefficient (x_1), which is a measure of nondiversifiable risk, and the standard deviation of the security returns (x_2), which measures total risk. The data for their sample of 35 S&Ls is shown in Table 6.3 and is contained in the file SL6 on the CD. Scatterplots of y versus x_1 and y versus x_2 are shown in Figures 6.35 and 6.36, respectively. The regression output is in Figure 6.37. Figures 6.38 through 6.40 show the residual plots of the standardized residuals versus the fitted values, x_1 and x_2, respectively. These plots highlight the presence of one standardized residual that falls well above the +3 limit. The normal probability plot and results of the normal probability plot

TABLE 6.3 Data for S&L Rate of Return Example

Name	Time Period	RETURN	BETA	SIGMA
1. H.F. Ahnanson	1/78–12/82	2.29	1.2862	14.2896
2. Alamo Savings Bank	9/78–12/82	0.34	0.6254	10.9786
3. American Federal S&L	12/80–12/82	2.57	1.1706	9.7917
4. American S&L	1/78–12/82	2.91	1.6328	17.7219
5. Bell National Ind.	1/78–12/82	3.50	1.2492	18.4450
6. Beverly Hills S&L	11/78–12/82	0.47	1.1363	12.4691
7. Broadview Financial Ind.	1/78–12/82	−0.28	1.3585	12.3396
8. Buckeye Financial Ind.	10/80–12/82	0.40	1.5415	14.7000
9. Citizens Savings Financial	6/80–12/82	2.42	2.1457	18.9970
10. City Federal Bank	6/80–12/82	5.48	2.2701	18.0840
11. Danney S&L	1/78–12/82	1.67	1.4527	14.2785
12. Far West Financial	1/78–12/82	1.01	1.4532	13.8673
13. Financial Corp. of America	1/78–12/82	6.06	1.8826	18.1800
14. Financial Corp. of Santa Barbara	1/78–12/82	0.48	1.4493	14.7792
15. Financial Federation	1/78–12/82	0.96	1.5590	14.9088
16. First Charter Financial	1/78–12/82	1.24	1.2274	12.3504
17. First City Federal Ind.	5/81–12/82	6.39	2.2567	22.0455
18. First Financial S&L	1/81–12/82	2.39	1.7003	12.8343
19. First Lincoln Financial Bank	1/78–12/82	0.30	2.2226	1.8750
20. First Western Financial Bank	1/78–12/82	2.09	1.6535	16.5737
21. Freedom S&L Bank	5/80–12/82	2.46	1.3616	14.2680
22. Gibraltar Financial	1/78–12/82	2.10	1.9851	17.8500
23. Golden West Financial Corp.	1/78–12/82	2.76	1.4311	14.1036
24. Great Western Financial	1/78–12/82	2.06	1.3448	12.4630
25. Guarantee Financial Ind.	1/78–12/82	1.42	1.4560	13.9728
26. Homestead Financial Bank	1/78–12/82	4.12	1.5543	17.2628
27. Imperial Corp. of America	1/78–12/82	1.82	1.7280	15.2880
28. Land of Lincoln Ind.	12/79–12/82	1.59	1.3389	10.3032
29. Mercury Saving	1/78–12/82	13.05	1.2973	13.3110
30. Naples Federal	2/80–12/82	3.04	1.0945	10.5792
31. Palmetto Federal S&L	11/79–12/82	3.72	1.2051	12.9456
32. Prudential Federal S&L	1/78–12/82	0.75	1.0756	11.5200
33. Texas Federal Bank	7/81–12/82	1.00	1.9157	16.6000
34. Transohio Financial	1/78–12/82	−3.35	1.4456	11.7705
35. Western Financial Corp.	1/78–12/82	2.26	1.9128	16.8370

Source: Data are from Lee and Lynge, "Return, Risk, and Cost of Equity for Stock S&L Firms: Theory and Empirical Results," *Journal of the American Real Estate and Urban Economic Association*, 1985. Copyright © 1985 School of Business, Indiana University. Reprinted by permission.

FIGURE 6.35
Scatterplot of
RETURN versus
BETA for S&L Rate
of Return Example.

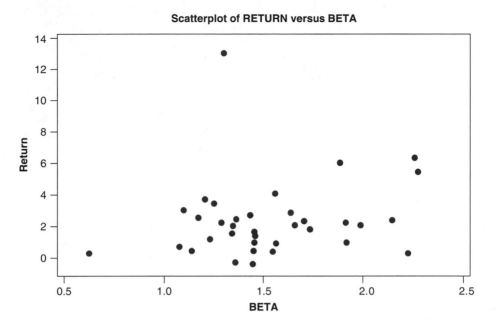

FIGURE 6.36
Scatterplot of
RETURN versus
SIGMA for S&L Rate
of Return Example.

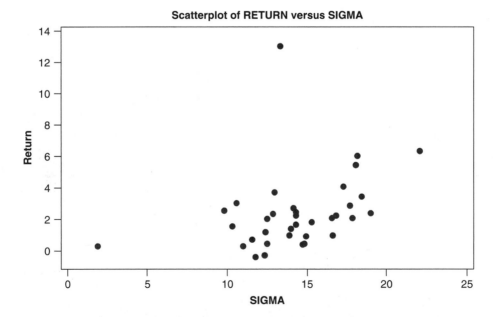

test for normality are shown in Figure 6.41. To perform the test for normal disturbances at the 5% level of significance, the decision rule is

Reject H_0 if $p$value < 0.05

Do not reject H_0 if $p$value ≥ 0.05

FIGURE 6.37
Regression Results for S&L Rate of Return Example.

Variable	Coefficient	Std Dev	T Stat	P Value
Intercept	−1.330	2.012	−0.66	0.513
BETA	0.300	1.198	0.25	0.804
SIGMA	0.231	0.126	1.84	0.075

Standard Error = 2.37707   R-Sq = 12.5%   R-Sq(adj) = 7.0%

Analysis of Variance

Source	DF	Sum of Squares	Mean Square	F	P
Regression	2	25.808	12.904	2.28	0.118
Error	32	180.815	5.650		
Total	34	206.624			

FIGURE 6.38
Plot of the Standardized Residuals versus the Fitted Values for S&L Rate of Return Example.

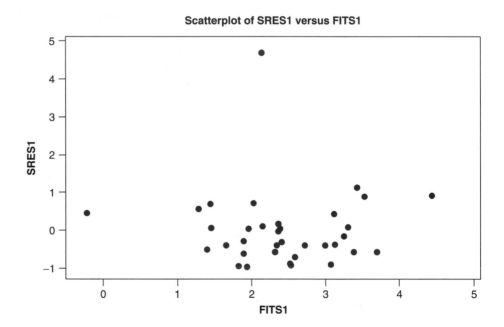

Scatterplot of SRES1 versus FITS1

The *p* value computed by MINITAB is less than 0.005, so the null hypothesis is rejected. The disturbances do not appear to be normally distributed.

6.6.3 CORRECTIONS FOR NONNORMALITY

The assumption of normally distributed disturbances is not necessary to use least-squares estimation to produce an estimated regression equation. However, for making inferences with small samples, it is necessary. In large samples, the assumption is not as important because the sampling distribution of the estimators of the regression coefficients is still approximately normal. Recall that when estimating a population mean, μ, the cutoff point for a large sample is $n = 30$. When $n \geq 30$, the

FIGURE 6.39
Plot of the
Standardized
Residuals versus the
Explanatory Variable
BETA for S&L Rate of
Return Example.

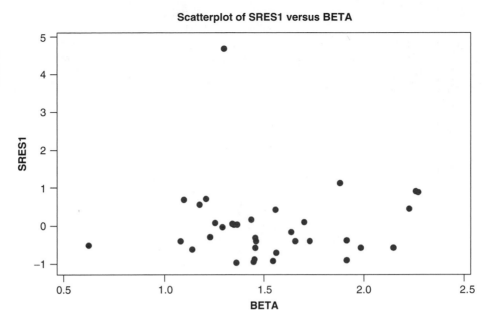

FIGURE 6.40
Plot of the
Standardized
Residuals versus
the Explanatory
Variable SIGMA for
S&L Rate of Return
Example.

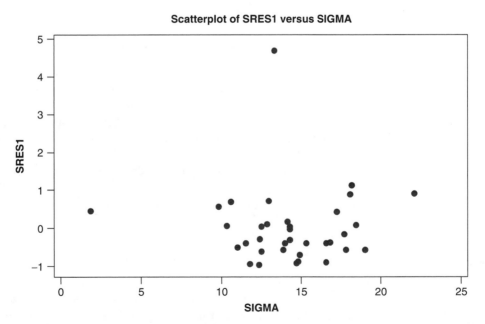

central limit theorem guarantees that the sampling distribution of the sample mean is approximately normal. A similar theorem operates in the regression context for the sampling distribution of the regression coefficients, b_k. The cutoff for a large sample may differ, however, because several coefficients may be estimated in a multiple regression context. It is uncertain exactly how many observations ensure

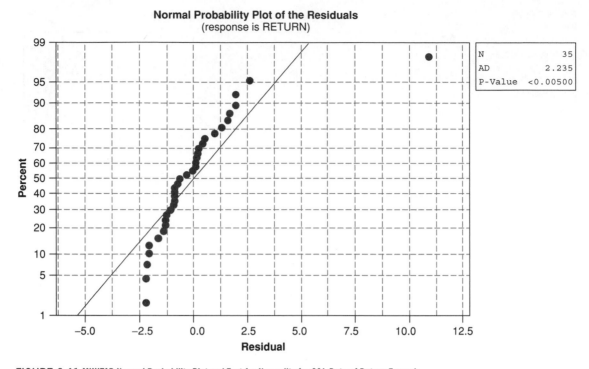

FIGURE 6.41 MINITAB Normal Probability Plot and Test for Normality for S&L Rate of Return Example.

normality of the sampling distributions in the multiple regression context. If the assumption of normal disturbances does not hold, additional observations are necessary for each additional explanatory variable. For a simple regression, 30 observations with 10 to 20 additional observations for *each* additional explanatory variable are commonly suggested.

When the normality assumption is violated and the sample size is too small to ensure normality of the sampling distributions, there are a variety of possible corrections. One type of correction is to transform the dependent variable using a Box-Cox transformation. (This was also a correction for nonconstant variance). Box-Cox transformations can be defined as follows:

$$\text{Use } y^p \text{ in place of } y \text{ where } 0 \leq p \leq 1$$

Thus, powers of the *y* variable are used in place of the original variable. Note that the extreme values of *p* for this type of transformation are 0 and 1. When $p = 1$, the original *y* values are used. When $p = 0$, we define the transformation to be the natural log transformation. When $p = 1$ is used, no correction is necessary; the disturbances are normal. The natural log transformation would provide the maximum amount of correction to the residuals to try to achieve normality. Powers of *y* between 0 and 1 can be used to adjust the amount of correction necessary. (For more on the use of Box-Cox transformations, the interested reader is referred to Neter, Wasserman, and Kutner, *Applied Linear Statistical Models*, pp. 394–400.)

When considering the normality assumption, be sure to correct for other violations before worrying about normality. A violation of the linearity or the constant variance assumption can introduce outliers (discussed in Section 6.7) into a data set that make the normality assumption appear to be violated also. Choosing the correct model by correcting for nonlinearity or nonconstant variance may eliminate the outliers, however. So check for violations of the linearity and constant variance assumptions before being too concerned with the normality assumption.

In some cases, the primary reason for the nonnormality of the disturbances may be the presence of one or a few data points that are much different from the remaining observations in the data set. Even in large samples, it is important to recognize such unusual observations because their presence may drastically alter results. In such instances, the sampling distributions of the estimated regression coefficients should not be assumed normal, even if the sample size is large. These cases, and some possible corrections, are discussed in the next section.

6.7 INFLUENTIAL OBSERVATIONS

6.7.1 INTRODUCTION

The method of least-squares estimation chooses the regression coefficient estimates so that the error sum of squares, *SSE*, is a minimum. In doing this, the sum of squared distances from the true y values, y_i, to the points on the regression line or surface, $\hat{y}_i$, are minimized. Least squares thus tries to avoid any large distances from y_i to $\hat{y}_i$. As shown in Figure 6.42, this can have an effect on the placement of the regression line. In Figure 6.42(a), the points are all clustered near the regression line. In Figure 6.42(b), one of the points has been moved so that its y value is much different from the y values of the remaining sample points. The effect of moving this one point on the placement of the estimated regression line is also shown. Note that the regression line has been pulled toward the point that was moved to the extreme position. This is a result of the requirement of the least-squares method to minimize the error sum of squares. Squaring the residuals gives proportionately more weight to extreme points in the error sum of squares. The least-squares regression line is often drawn toward such extreme points.

When a sample data point has a y value that is much different from the y values of the other points in the sample, it is called an *outlier*. Outliers can be either good or bad. They can provide information concerning the behavior of the process being studied that would be unavailable otherwise. In this sense, the presence of the outlier could be viewed as positive. On the other hand, the presence of an outlier can at times produce confusing results and mask important information that could otherwise be obtained from the regression. In either case, it is important to recognize an outlier when it is present. Outlier detection techniques are discussed in this section.

Now consider Figure 6.43. In both Figure 6.43(a) and 6.43(b), the same five initial points are used as in Figure 6.42(a). One additional point is added to each figure but is placed in a different position. The positioning of this point is seen to have a large effect on the estimated regression line. In Figure 6.43(a), the slope of the line appears to have changed little from its value with only the five original

FIGURE 6.42(a)
Scatterplot with Regression Line for Data Shown.

Data

x	1	2	3	4	5
y	1.1	2.2	2.4	2.8	3.3

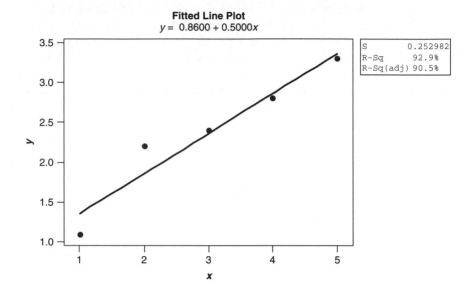

FIGURE 6.42(b)
Scatterplot Showing Effect on Regression Line of Outlier.

Data

x	1	2	3	4	5
y	1.1	2.2	5.0	2.8	3.3

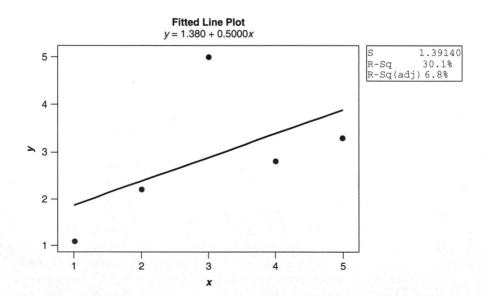

FIGURE 6.43(a)
Scatterplot Showing
Effect of Leverage Point
in Line with Other Data.

Data

x	1	2	3	4	5	10
y	1.1	2.2	2.4	2.8	3.3	6.0

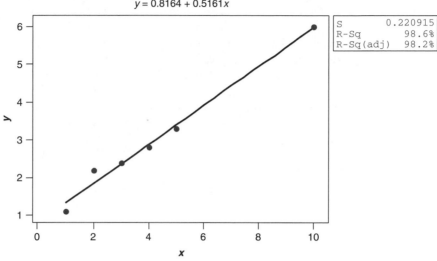

FIGURE 6.43(b)
Scatterplot Showing
Effect of Moving
Leverage Point to
Alternate Position.

Data

x	1	2	3	4	5	10
y	1.1	2.2	2.4	2.8	3.3	2.0

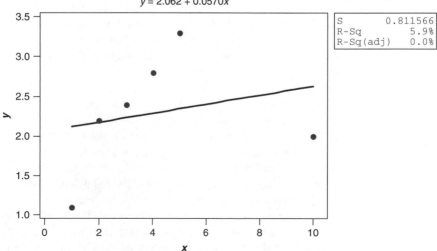

points [see Figure 6.42(a)]. In Figure 6.43(b), the placement of the additional point has drastically changed the slope of the line. In fact, the slope of the line appears to be determined almost entirely by this one point. This sixth observation is said to have high leverage and is referred to as a *leverage point*. The term "leverage point" means that the point is placed in such a way that it has the *potential* to affect the regression line.

In both Figure 6.43(a) and 6.43(b), the added point is a leverage point. This is due to its extreme placement on the x axis. The x value for this point is much different from the x values for the other sample points. The point, however, does not exert much influence in Figure 6.43(a) because it is in line with all the other points in terms of the position of its y value relative to its x value. It is important to recognize any leverage points due to their possible effect on the regression line. As with outliers, leverage points may be good or bad depending on whether they add information about the process under study or mask information that otherwise would be obtained.

Finally, it is important to note that an observation can be both a leverage point and an outlier.

6.7.2 IDENTIFYING OUTLIERS

The use of the standardized residuals has already been discussed. The standardized residuals are computed by dividing the raw residual, $\hat{e}_i = y_i - \hat{y}_i$, by the standard deviation of $\hat{e}_i$:

$$\hat{e}_{is} = \frac{\hat{e}_i}{\text{stdev}(\hat{e}_i)}$$

where $\hat{e}_{is}$ indicates the standardized residual. The variance of the standardized residuals is 1 (note that in this chapter it is necessary to distinguish between the raw residual $\hat{e}_i$ and the standardized residual $\hat{e}_{is}$).

If the residuals come from a normal distribution, then a standardized residual with an absolute value larger than 2 is expected only about 5% of the time. Thus, any observation with a standardized residual larger than 2 in absolute value might be classified as an outlier. MINITAB, for example, indicates any such observations in a table of Unusual Observations following the regression results.

Another measure sometimes used in place of the standardized residual is the standardized residual computed after deleting the *i*th observation. This measure is called the *studentized residual* or *studentized deleted residual*. To compute the studentized deleted residual, the residual, $\hat{e}_i$, is again standardized, but the divisor is different from that used to compute the standardized residual. The standard deviation of the *i*th residual is computed from the regression with the *i*th observation deleted. By doing this, the *i*th observation exerts no influence over the value of the standard deviation. If the *i*th observation's y value is unusual, this is reflected in the residual but not in its divisor. Thus, unusual y values should stand out. Also, because of the way they are computed, the studentized deleted residuals are known to follow a t distribution with $n - K - 1$ degrees of freedom. The studentized deleted residuals can be compared to a value chosen from the t table to determine whether they should be classified as outliers. (The standardized residuals do not follow a t distribution.) But this approach should not be used as a test of significance to determine whether the observation should be discarded. What to do about outliers once they are identified will be discussed later in this section.

6.7.3 IDENTIFYING LEVERAGE POINTS

Leverage was previously defined as the potential of an observation to affect the regression line. As shown in Figure 6.43, a point can possess leverage without significantly altering the position of the regression line. On the other hand, given sufficient leverage, a single point can significantly affect the slope of the regression line.

The leverage of the ith point in a sample is denoted h_i and is computed by some regression software packages. Leverage is a measure of how extreme the point is in terms of the values of the explanatory variables. Observations with extreme x values possess greater leverage than observations with x values that are similar to the other sample points. In Figure 6.42, the observation that was changed has smaller leverage than the observation that was changed in Figure 6.43. Note that the slope of the regression line is affected more by changes in the point with greater leverage.

MINITAB indicates certain observations with very high leverage. Any data value with leverage greater than $2(K + 1)/n$ (where K is the number of explanatory variables and n is the sample size) is indicated in a table of Unusual Observations following the regression results along with observations that have large standardized residuals.

6.7.4 COMBINING MEASURES TO DETECT OUTLIERS AND LEVERAGE POINTS

The effect of an observation on the regression line is determined both by the y value of the point and the x value(s). As shown in Figures 6.42 and 6.43, an observation with an unusual y value has a much greater effect on the regression line if it also has high leverage. Several statistics have been developed that consider extremity in both the y and x dimensions in an attempt to determine which points are highly influential on the regression line. Two of these measures, the DFITS statistic and Cook's D statistic, are discussed in this section.

Both of these measures combine information from the residuals and the leverage of each observation to try to pick out observations that may have a large influence on the regression line. As with the individual measures, unusual values of the DFITS statistic or Cook's D statistic are not indications that anything is wrong with the particular data value or that it should be discarded. It does indicate that the value with the unusual statistic is somehow different from the remaining values in the data set and should be given additional consideration before accepting the regression model.

The DFITS statistic can be written as

$$\text{DFITS}_i = \hat{e}_i \sqrt{\frac{n - K - 2}{SSE(1 - h_i) - \hat{e}_i^2}} \sqrt{\frac{h_i}{1 - h_i}}$$

where $\hat{e}_i$ is the residual for the ith observation, h_i is the leverage value for the ith observation, and SSE is the error sum of squares for the regression.

Cook's D statistic is computed as

$$D_i = \frac{\hat{e}_i^2}{MSE(K + 1)} \left(\frac{h_i}{(1 - h_i)^2} \right)$$

where $\hat{e}_i$ and h_i are as defined for the DFITS statistic and MSE is the mean square error for the regression.

As can be seen, both statistics use the residuals and the leverage of each individual point. Since they use the values in different ways, however, different information may be obtained from each of the statistics. Cook's D statistic is usually thought to represent the combined impact on all the regression coefficients of the *i*th observation. DFITS represents the combined impact on the fitted values of the *i*th observation.

There are two different schools of thought about how the DFITS statistic and Cook's D statistic should be used:

1. The values should be compared to some absolute cutoff. For example, the Cook's D value often is compared to an $F(\alpha; K + 1, n - K - 1)$ value. The DFITS value is compared to $2\sqrt{(K + 1)/n}$. Values bigger than either of the numbers should be further examined. But these are not to be viewed as statistical tests to reject or throw out observations.

2. Do not use absolute cutoffs. Simply pick out those observations whose Cook's D or DFITS values (if any) differ appreciably from most of the values.

The second approach is used in this text. Much recent research has shown that comparison to absolute cutoffs is not as effective in identifying influential observations as examining observations with unusually large DFITS or Cook's D values.

When specific cutoffs are not used, a method is needed to compare the different values of each statistic to determine which observations may be unusual. One way of doing this is simply to graph the DFITS (or Cook's D) values for each observation on the vertical axis and the number of the observation on the horizontal axis. This can also be accomplished by doing a time-series plot of the DFITS (or Cook's D) values, remembering that the index on the horizontal axis refers to the number of the observation and not necessarily to a time period. This method of presenting the DFITS and Cook's D statistics is demonstrated in Example 6.8.

6.7.5 WHAT TO DO WITH UNUSUAL OBSERVATIONS

As noted earlier, the fact that an observation has been classified as unusual does not mean that it is useless or that it should be deleted from the analysis. It is merely a flag to indicate that the observation deserves further examination. This is true regardless of which measure of "unusualness" has been used (standardized residual, DFITS, Cook's D statistic, and so on).

There are many reasons an observation may appear unusual. If there is a violation of the linearity or constant variance assumptions, this can cause certain observations to appear unusual until the violation has been corrected by choosing an appropriate transformation of the data.

If a data value has been typed in incorrectly, the value may be flagged as unusual. This is useful since incorrectly coded values should not be included in the data set. The true data value should be located and used to replace the incorrect value. If the true value cannot be found, then this is one case when it is almost always better to omit the incorrect data value before running the analysis.

If the unusual value is not due to the violation of an assumption or incorrect coding but is a correct value that is simply unusual with respect to most of the values in the data set, then the choice of what to do with it is more difficult. There are certain cases when deleting the observation from the data set is appropriate. These cases occur when the unusual observation is somehow very different from the observations included in the analysis. For example, if a production process is being examined, unusual observations may occur at the beginning of the process due to start-up problems. When the process reaches a steady state of performance, the data values generated may be quite different from those generated initially. The initial observations may be deleted if the process in its steady state is to be studied.

Consider another example. Suppose data on price and size of houses are obtained to develop an equation to help set prices for the houses. If interest lies in pricing houses with 1500 to 2500 square feet of space, then it is proper to exclude a house if it contains 4500 square feet. If the x values for the unusual observation fall outside the range of interest, it may be best to discard the observation and run the regression without that value.

In any case, before deleting an observation from an analysis, the observation should be studied carefully to see whether deletion is an appropriate option. If the observation is somehow so different from the others in the data set that it is inappropriate to include it in the analysis or if the value has been coded incorrectly and the true value is unknown, then deletion is a viable option. However, this option should not be used indiscriminately.

EXAMPLE 6.8   **S&L Rate of Return (continued)**   Figure 6.44 shows a plot of DFITS for the S&L rate of return data. Figure 6.45 shows a plot of the Cook's D statistic. Both of these plots were created by doing a time-series plot of the statistics. The index on the horizontal axis refers to the number of the observation from the sample. From each of these plots, the 29th observation stands out as unusual relative to the rest of the observations in the sample. In Table 6.3, the 29th observation is Mercury Saving. Note that the return for Mercury Saving is much higher than that for the other S&Ls. Mercury Saving has a very unusual y value resulting in a standardized residual of 4.69, which is far greater than would be expected if the disturbances were normally distributed.

This observation should definitely be examined further. The question at this point is what to do with it. In the case of Mercury Saving, it is unclear whether the return of 13.05% is correct. In the article from which these data were taken, however, it appears that Mercury Saving has been deleted from the analysis. Because the true return is not known, Mercury Saving is excluded from the data set and the analysis redone. It is important to emphasize that casual deletion of points from the data set is not recommended. If there is a good reason for not including a particular value, then it can be deleted. If the return for Mercury Saving is incorrect and the correct value is unknown, it is better to delete this observation than to use the incorrect information. Or if Mercury Saving is believed to be so different from the other S&Ls that it should not be included in the analysis, this is a good reason for deletion. For example, perhaps Mercury Saving had undertaken a particularly risky line of investments, different from those of the other S&Ls, which led to its very high return. Because the

FIGURE 6.44 Plot of
DFITS for S&L Rate of
Return Example.

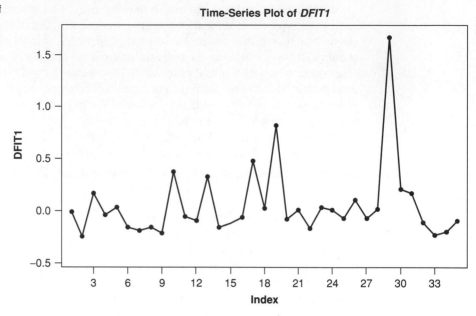

FIGURE 6.45 Plot of
Cook's D Statistic for
S&L Rate of Return
Example.

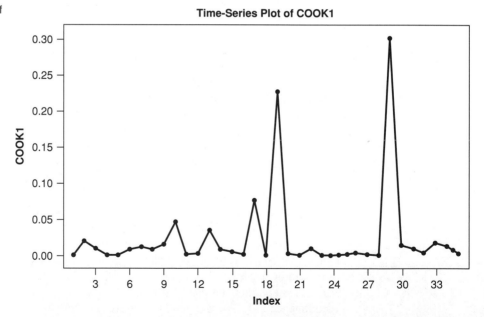

investment behavior of Mercury is extremely different from the other S&Ls in this case, there might be grounds for excluding it from the analysis. It is assumed here that there is sufficient reason for omitting Mercury Saving since that appears to be what was done in the original article. In a real application, it would be beneficial to study Mercury further to see what makes it so different from the other S&Ls.

In the following example, note the difference between the analyses with and without Mercury Saving.

EXAMPLE 6.9 **S&L Rate of Return Without Mercury Saving** Figures 6.46 and 6.47 show the scatterplots with Mercury Saving omitted. The regression results are shown in

FIGURE 6.46
Scatterplot of RETURN versus BETA with Mercury Saving Omitted.

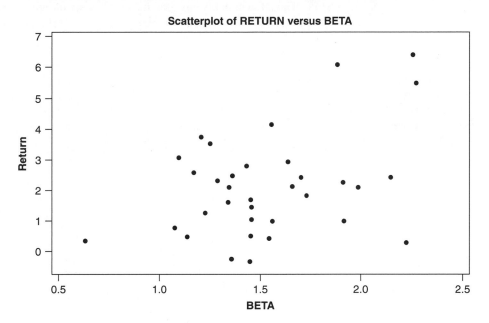

FIGURE 6.47
Scatterplot of RETURN versus SIGMA with Mercury Saving Omitted.

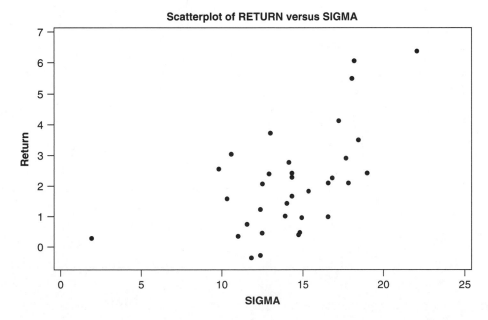

Figure 6.48. The residual plots are in Figures 6.49 to 6.51, and the plots of DFITS and Cook's D are in Figures 6.52 and 6.53.

The regression results differ considerably from the regression with Mercury Saving included. Compare the Figure 6.48 regression (without Mercury) to the Figure 6.37 regression (with Mercury). In the initial regression, neither variable is significant at the 5% level. In the regression without Mercury, SIGMA is now significant. Thus, the conclusions drawn from these two regressions are completely different because of the presence (or absence) of one influential observation. This

FIGURE 6.48
Regression for S&L Rate of Return Example with Mercury Saving Omitted.

Variable	Coefficient	Std Dev	T Stat	P Value
Intercept	-2.5103	1.1529	-2.18	0.037
BETA	0.8463	0.6843	1.24	0.225
SIGMA	0.2322	0.0714	3.25	0.003

Standard Error = 1.35165   R-Sq=37.2%   R-Sq(adj) = 33.1%

Analysis of Variance

Source	DF	Sum of Squares	Mean Square	F Stat	P Value
Regression	2	33.537	16.768	9.18	0.001
Error	31	56.635	1.827		
Total	33	90.172			

FIGURE 6.49 Plot of Standardized Residuals versus Fitted Values for S&L Rate of Return Example with Mercury Saving Omitted.

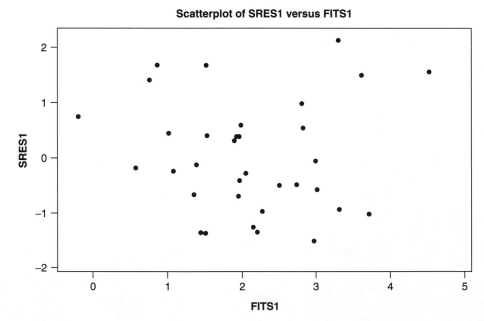

Scatterplot of SRES1 versus FITS1

FIGURE 6.50 Plot of Standardized Residuals versus BETA for S&L Rate of Return Example with Mercury Saving Omitted.

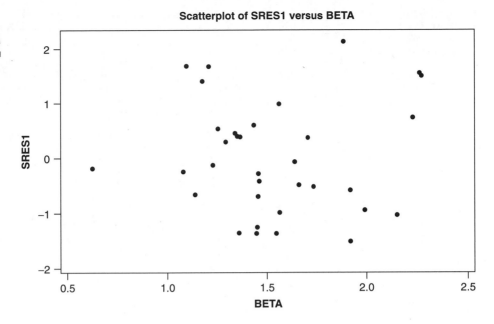

FIGURE 6.51 Plot of Standardized Residuals versus SIGMA for S&L Rate of Return Example with Mercury Saving Omitted.

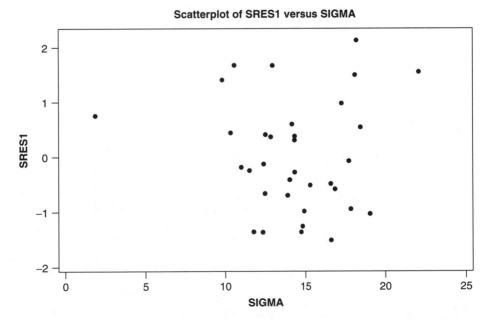

highlights the need to be careful when deleting influential observations. Knowing whether Mercury Saving belongs in our analysis is especially important because omitting it causes a reversal in our conclusion concerning the importance of the variable SIGMA.

FIGURE 6.52 Plot of DFITS for S&L Rate of Return Example with Mercury Saving Omitted.

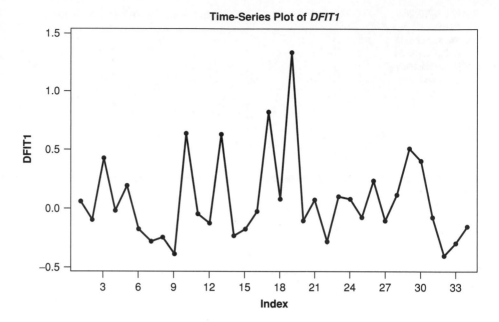

FIGURE 6.53 Plot of Cook's D Statistic for S&L Rate of Return Example with Mercury Saving Omitted.

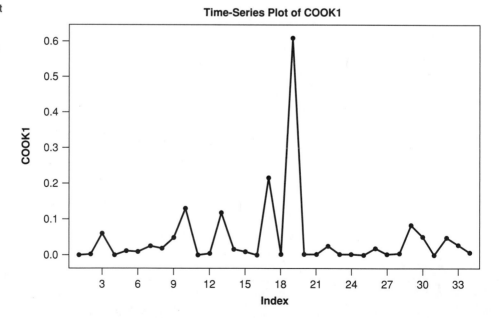

EXERCISES

4. Petroleum Imports. The data for U.S. monthly petroleum imports from 1991 through 1997 appear in Table 6.4. Figure 6.54 shows the regression results for the regression of monthly petroleum imports (in millions of barrels) on the one-period lagged imports. Figure 6.55 is a time-series plot of

TABLE 6.4 Monthly Petroleum Imports (in millions of barrels)

Date	PETROIMP	Date	PETROIMP	Date	PETROIMP
Jan-91	180.1	May-93	211.3	Sep-95	244.9
Feb-91	163.6	Jun-93	222.5	Oct-95	226.4
Mar-91	169.2	Jul-93	229.5	Nov-95	231.6
Apr-91	177.8	Aug-93	204.1	Dec-95	235.7
May-91	215.2	Sep-93	206.7	Jan-96	239.0
Jun-91	199.2	Oct-93	223.4	Feb-96	198.1
Jul-91	201.4	Nov-93	228.3	Mar-96	201.7
Aug-91	220.2	Dec-93	203.1	Apr-96	238.2
Sep-91	190.3	Jan-94	206.0	May-96	261.6
Oct-91	189.3	Feb-94	176.9	Jun-96	253.0
Nov-91	182.9	Mar-94	219.8	Jul-96	275.2
Dec-91	182.8	Apr-94	217.2	Aug-96	251.0
Jan-92	186.7	May-94	216.6	Sep-96	260.4
Feb-92	155.0	Jun-94	248.1	Oct-96	250.7
Mar-92	172.4	Jul-94	245.6	Nov-96	217.0
Apr-92	186.0	Aug-94	242.5	Dec-96	247.8
May-92	195.5	Sep-94	260.4	Jan-97	224.1
Jun-92	193.1	Oct-94	221.0	Feb-97	211.1
Jul-92	112.7	Nov-94	229.2	Mar-97	245.6
Aug-92	201.4	Dec-94	220.9	Apr-97	250.9
Sep-92	190.0	Jan-95	212.3	May-97	278.2
Oct-92	216.5	Feb-95	195.9	Jun-97	254.2
Nov-92	193.2	Mar-95	239.7	Jul-97	266.3
Dec-92	192.0	Apr-95	212.0	Aug-97	280.9
Jan-93	212.0	May-95	240.5	Sep-97	274.6
Feb-93	175.9	Jun-95	242.0	Oct-97	280.6
Mar-93	206.1	Jul-95	245.4	Nov-97	252.8
Apr-93	220.3	Aug-95	240.8	Dec-97	250.1

Source: *Business Statistics of the United States*. Lanham, MD: Berman Press.

FIGURE 6.54
Regression of Petroleum Imports on the Lagged Value of Petroleum Imports.

Variable	Coefficient	Std Dev	T Stat	P Value
Intercept	63.087	17.044	3.70	0.00
PETROIMPL1	0.716	0.077	9.30	0.00

Standard Error = 22.2218   R-Sq = 51.7%   R-Sq(adj) = 51.1%

Analysis of Variance

Source	DF	Sum of Squares	Mean Square	F Stat	P Value
Regression	1	42754	42754	86.58	0.000
Error	81	39999	494		
Total	82	82753			

FIGURE 6.55
Time-Series Plot of
Standardized
Residuals for
Petroleum Imports
Regression.

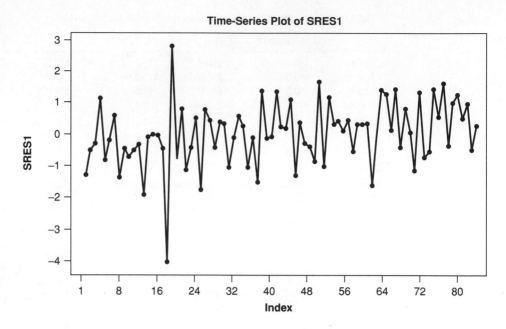

FIGURE 6.56
Plot of
Standardized
Residuals versus
Fitted Values for
Petroleum Imports
Regression.

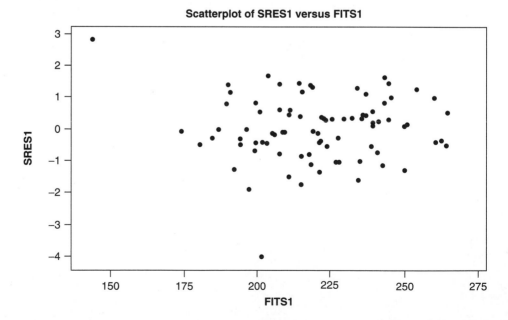

the standardized residuals from this regression. Figure 6.56 shows the plot of the standardized residuals versus the fitted values. Figures 6.57 and 6.58 show, respectively, plots of the DFITS and Cook's D statistics. These data are available in a file named PETRO6 on the CD.

FIGURE 6.57 Plot
of DFITS Statistic for
Petroleum Imports
Regression.

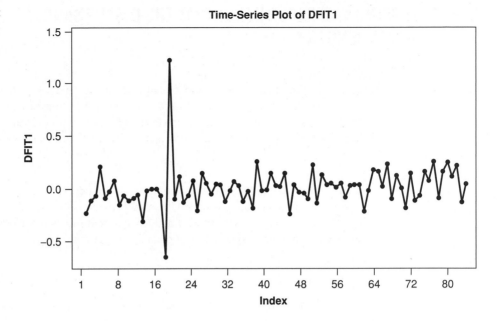

FIGURE 6.58 Plot
of Cook's D Statistic
for Petroleum
Imports Regression.

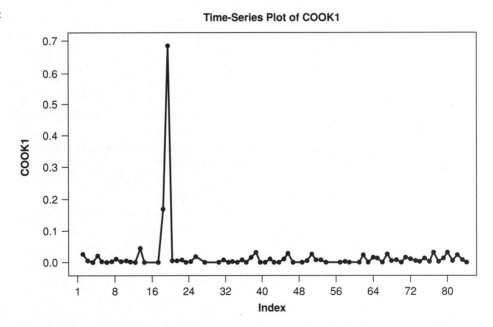

Are there any unusual observations that should be checked before accepting these regression results? If so, which observations? Can you determine what might be causing certain observations to appear unusual? Justify your answers.

6.8 ASSESSING THE ASSUMPTION THAT THE DISTURBANCES ARE INDEPENDENT

6.8.1 AUTOCORRELATION

One assumption that is frequently violated when using time-series data is that of independence of the disturbances, e_i. Disturbances in adjacent time periods are often correlated because an event in one time period may influence an event in the next time period. The relationship between disturbances in adjacent time periods is often represented by

$$e_i = \rho e_{i-1} + u_i$$

In this equation, e_i is the disturbance in the ith time period, e_{i-1} is the disturbance in the previous time period, ρ is called the *serial correlation coefficient* or *autocorrelation coefficient*, and u_i represents a disturbance that meets the assumption of independence. The regression relationship can be written as

$$y_i = \beta_0 + \beta_1 x_i + e_i$$

where $e_i = \rho e_{i-1} + u_i$.

When disturbances exhibit autocorrelation, least-squares estimates of the regression coefficients are unbiased, but the estimated standard errors of the coefficients are biased. As a result, confidence intervals and hypothesis tests do not perform as expected. The estimated regression coefficients also have larger sampling variance than certain other estimators that correct for autocorrelation.

The autocorrelation coefficient, ρ, determines the strength of the relationship between disturbances in successive time periods. Like any other correlation coefficient, it varies between -1 and $+1$, with values close to ± 1 indicating very strong relationships and values close to 0 indicating weak relationships. Ideally, a value of $\rho = 0$ is desired since this means that the disturbances are independent and the independence assumption has not been violated.

To determine whether autocorrelation is present, the Durbin–Watson test is used. This test is discussed in the next section.

6.8.2 A TEST FOR FIRST-ORDER AUTOCORRELATION

A well-known and widely used test for first-order autocorrelation is the *Durbin–Watson test*. When autocorrelation is present in business and economic data, it is typically positive autocorrelation ($\rho > 0$). For this reason, we test for positive autocorrelation.

The hypotheses to be tested may be written as follows:

$$H_0: \rho = 0$$
$$H_a: \rho > 0$$

where ρ is the first-order autocorrelation coefficient. If the null hypothesis is not rejected, the correlation, ρ, of adjacent disturbances is zero, and no problem of first-order autocorrelation exists. If the null hypothesis is rejected, the disturbances are correlated, and some correction for autocorrelation needs to be made.

The Durbin–Watson statistic is computed by first using least squares to estimate the regression equation and then by computing the residuals

$$\hat{e}_i = y_i - \hat{y}_i$$

where y_i represents one of the sample y values and $\hat{y}_i$ is the corresponding predicted y value. The residuals are used to compute the Durbin–Watson statistic, d:

$$d = \frac{\sum_{i=2}^{n}(\hat{e}_i - \hat{e}_{i-1})^2}{\sum_{i=1}^{n}\hat{e}_i^2}$$

When the disturbances are independent, d should be approximately equal to 2. When the disturbances are positively correlated, d tends to be smaller than 2.

The decision rule for the test is

Reject H_0 if $d < d_L(\alpha; n, K)$

Do not reject H_0 if $d > d_U(\alpha; n, K)$

Here, $d_L(\alpha; n, K)$ and $d_U(\alpha; n, K)$ are the critical values, which can be found in Table B.7 in Appendix B. The critical values depend on the level of significance of the test, α, the sample size, n, and the number of explanatory variables in the equation, K. For the Durbin–Watson test, there is a range of values for the test statistic where the test is said to be inconclusive. The inconclusive range of values for d is

$$d_L \le d \le d_U$$

If the test statistic d falls in this region, there is some question as to how to proceed.

Rejection of the null hypothesis suggests that a correction for autocorrelation is necessary. Failing to reject the null hypothesis means that no correction is necessary. But what should be done when d falls in the inconclusive region? There have been some additional procedures developed to further examine these cases, but they have not been incorporated into many computer regression routines. Without easy access to the additional procedures, one possibility is to treat values of d in the inconclusive region as if they suggested autocorrelation. If the regression results after correction for autocorrelation differ from those prior to the correction, then conclude that the correction was necessary. If the results are similar, then the correction was unnecessary, and the original uncorrected results can be used.[3]

 **EXAMPLE 6.10**   **Sales and Advertising**   Table 6.5 shows data on sales (in millions) and advertising (ADV) (in thousands) for the ABC Company (see the file SALESADV6 on the CD). These are annual data covering the period from 1967 to 2002. The regression of SALES on ADV is shown in Figure 6.59. The Durbin–Watson statistic is shown in

---

3 To test H_0: $\rho = 0$ versus H_a: $\rho < 0$, the decision rule would be: Reject H_0 if $d > 4 - d_L$; Do not reject H_0 if $d < 4 - d_U$; Inconclusive if $4 - d_U \le d \le 4 - d_L$. A two-sided test can be performed by using both the upper- and lower-tailed tests separately (adjusting for the desired level of significance).

TABLE 6.5 Sales and Advertising Data

Year	SALES	ADV
1967	381.0	5316.8
1968	383.9	5413.2
1969	384.4	5486.9
1970	370.5	5537.8
1971	396.4	5660.6
1972	421.8	5750.8
1973	379.2	5782.2
1974	390.9	5781.7
1975	420.9	5821.9
1976	408.8	5892.5
1977	407.2	5950.2
1978	408.4	6002.1
1979	444.2	6121.8
1980	437.2	6201.2
1981	376.1	6271.7
1982	454.6	6383.1
1983	459.2	6444.5
1984	478.2	6509.1
1985	492.8	6574.6
1986	541.2	6704.2
1987	512.0	6794.3
1988	562.0	6911.4
1989	590.1	6986.5
1990	617.7	7095.7
1991	629.3	7170.8
1992	653.9	7210.9
1993	698.6	7304.8
1994	707.8	7391.9
1995	735.9	7495.3
1996	748.3	7629.2
1997	755.4	7703.4
1998	762.0	7818.4
1999	794.3	7955.0
2000	815.5	8063.4
2001	840.9	8170.8
2002	820.8	8254.5

the output. A time-series plot of the standardized residuals is shown in Figure 6.60. To test for positive first-order autocorrelation, the following decision rule is used:

Reject H_0 if $d < d_L$ $(0.05; 36, 1) = 1.41$

Do not reject H_0 if $d > d_U$ $(0.05; 36, 1) = 1.52$

Inconclusive if $1.41 \leq d \leq 1.52$

Because $d = 0.47$, the null hypothesis of no autocorrelation is rejected. First-order autocorrelation is a problem and should be corrected before inferences or forecasts are made.

FIGURE 6.59
Regression of SALES on ADV Including Durbin–Watson Statistic.

Variable	Coefficient	Std Dev	T Stat	P Value
Intercept	−632.694476	47.276973	−13.38	0.000
ADV	0.177233	0.007045	25.16	0.000

Standard Error = 36.4920   R-Sq = 94.9%   R-Sq(adj) = 94.8%

Analysis of Variance

Source	DF	Sum of Squares	Mean Square	F Stat	P Value
Regression	1	842685	842685	632.81	0.000
Error	34	45277	1332		
Total	35	887961			

Durbin-Watson statistic = 0.467294

FIGURE 6.60 Time-Series Plot of Standardized Residuals for Sales and Advertising Example.

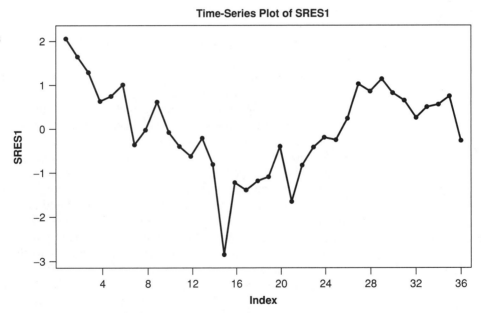

6.8.3 CORRECTION FOR FIRST-ORDER AUTOCORRELATION

When disturbances are autocorrelated, it may be due to the omission of an important variable from the regression. If the time-ordered effects of such a missing variable are positively correlated, then the disturbances in the regression tend to be positively correlated. The remedy to this problem is to locate the missing variable and include it in the regression equation, although this is easier said than done in many cases.

Another possible correction for first-order autocorrelation transforms the original time-series variables in the regression so that a regression using the transformed variables has independent disturbances. The original regression model for the time period i can be written

$$y_i = \beta_0 + \beta_1 x_i + e_i$$

where the disturbances, e_i, have first-order autocorrelation

$$e_i = \rho e_{i-1} + u_i$$

To remove the autocorrelation, the following transformations are used. Create new dependent and explanatory variable values y_i^* and x_i^* using

$$y_i^* = y_i - \rho y_{i-1} \text{ and } x_i^* = x_i - \rho x_{i-1}$$

for time periods $i = 2, 3, \ldots, n$.

In addition, transform the first time-period observation as

$$y_1^* = \sqrt{1 - \rho^2}\, y_1 \qquad \text{and} \qquad x_1^* = \sqrt{1 - \rho^2}\, x_1$$

The new regression can be written as

$$y_i^* = \beta_0 + \beta_1 x_i^* + u_i$$

The disturbances u_i are independent. Now regress y_i^* on x_i^* to obtain estimates of β_0 and β_1.

In practice, there are various refinements to this process. In choosing a statistical package to perform these transformations and run the regressions, one of the most important things to keep in mind is that the transformation should include the first observation (x_1^*, y_1^*). Many statistical packages incorporated routines that simply dropped the first observation rather than transforming it as shown and including it in the new regression. Recent research shows that dropping this observation results in the loss of important information and adversely affects the results of the new regression. One common estimation procedure that drops the first observation is called the Cochrane–Orcutt method. This method should be avoided. Procedures that incorporate the first observation include the Prais–Winsten method and full maximum likelihood. When using an automatic method from a statistical package to correct for autocorrelation, only those methods incorporating the initial observation should be used. Statistical packages such as SAS and SHAZAM have single commands that transform the data appropriately, rerun the regression on the transformed data, and print out the results so that correcting for autocorrelation in this case is a simple matter. MINITAB and Excel have not incorporated any of these automatic procedures, so this type of correction is difficult to perform.

When the Prais–Winsten transformation is used, forecasts are computed in a slightly different manner than shown for a regression without autocorrelation. The T-period-ahead forecast can be written in general as

$$\hat{y}_{n+T} = \hat{\beta}_0 + \hat{\beta}_1 x_{n+T} + \rho^T \hat{e}_n$$

where $\hat{\beta}_0$ and $\hat{\beta}_1$ are used to represent the estimates of β_0 and β_1 from the transformed regression model (rather than the least-squares estimates b_0 and b_1), x_{n+T} is the value of the explanatory variable in the period to be forecast, ρ is the autocorrelation coefficient, and $\hat{e}_n$ is the residual from the last sample time period. Note that only the last residual in the sample contains any information about the future. Also note that the value of this information declines as forecasts are generated further into the future. The term ρ^T can be seen to decrease (because $-1 < \rho < 1$) as T increases.

A third option to correct for autocorrelation is to add a lagged value of the dependent variable as an explanatory variable. This is a viable option especially if building an extrapolative model and if the number of observations is reasonably large (since one observation is lost because of the lagged variable).

EXAMPLE 6.11   **Sales and Advertising (again)**   A lagged value of the dependent variable will be introduced to try to correct for first-order autocorrelation in the model for sales. The new equation can be written

$$y_i = \beta_0 + \beta_1 y_{i-1} + \beta_2 x_i + e_i$$

where y_i is sales in time period i, y_{i-1} is the lagged value of sales, and x_i is ADV. The regression is shown in Figure 6.61. Evaluating whether this model is an improvement over the previous model is discussed in the next section.

6.8.4 H TEST FOR AUTOCORRELATION

When lagged values of the dependent variable are used as explanatory variables, the Durbin–Watson test is no longer appropriate. An alternative test typically recommended in this case is Durbin's h. The hypotheses to be tested are

$$H_0\colon \rho = 0$$
$$H_a\colon \rho > 0$$

The test statistic is

$$h = r\left(\frac{n}{1 - ns_{b_1}^2}\right)^{1/2}$$

FIGURE 6.61
Regression of SALES on ADV and a Lagged Dependent Variable.

Variable	Coefficient	Std Dev	T Stat	P Value
Intercept	-234.4752	78.0688	-3.00	0.005
SALESL1	0.6751	0.1123	6.01	0.000
ADV	0.0631	0.0202	3.12	0.004

Standard Error = 24.1223   R-Sq=97.8%   R-Sq(adj)=97.7%

Analysis of Variance

Source	DF	Sum of Squares	Mean Square	F Stat	P Value
Regression	2	841098	420549	722.74	0.000
Error	32	18620	582		
Total	34	859718			

Durbin-Watson statistic = 2.33302

where r is an estimate of the first-order autocorrelation coefficient, n is the sample size, and $s_{b_1}^2$ is the estimated variance of the regression coefficient of the lagged dependent variable (y_{i-1}). If the null hypothesis is true, h has a standard normal distribution. The decision rule to conduct the test is

Reject H_0 if $h > z_\alpha$

Do not reject H_0 if $h \leq z_\alpha$

Note that the test statistic h cannot be computed if $1 - ns_{b_1}^2 < 0$ because this results in the need to take the square root of a negative number. (Alternative test procedures are available in this case. The reader is referred to Judge et al., *The Theory and Practice of Econometrics*, pp. 326–327, for an example.) In practice, a quick way to estimate the autocorrelation correlation coefficient is to use $r = 1 - d/2$, where d is the Durbin–Watson statistic.

Finally, it should be noted that other procedures are available for analyzing time-series data. Certain of these procedures are designed especially for extrapolative models using lagged dependent variables, autocorrelated errors, or both. For more detail on such time-series forecasting methods, the reader is referred to Bowerman and O'Connell, *Forecasting and Time Series: An Applied Approach*.

EXAMPLE 6.12   **Sales and Advertising (again)**   In Example 6.10, a regression of SALES on ADV was run. The Durbin–Watson test indicated that the disturbances from this regression were not independent. In Example 6.11, a lagged dependent variable was added to the regression to try to correct for the first-order autocorrelation. The appropriate test to determine whether the introduction of the lagged variable has eliminated the autocorrelation is the h test. The quantities necessary to compute the h statistic are available in the regression output in Figure 6.61.

Using a 5% level of significance, the decision rule to conduct the test is

Reject H_0 if $h > 1.645$

Do not reject H_0 if $h \leq 1.645$

The estimate of the autocorrelation coefficient can be computed as

$$r = 1 - d/2 = 1 - 2.33/2 = -0.165$$

The sample size used in the regression is 35 (one of the original observations was lost because of the use of the lagged variable), and the standard deviation of the coefficient of the lagged variable is 0.1123, so the h statistic is

$$h = -0.165\sqrt{\frac{35}{1 - 35(0.1123)^2}} = -1.31$$

Our decision is that H_0 cannot be rejected. Conclude that first-order autocorrelation is not a problem. Adding the lagged variable proved effective in correcting for autocorrelation.   ◼

EXAMPLE 6.13   **Sales and Advertising (for the last time)**   Figure 6.62 shows the resulting output from SAS when PROC AUTOREG is used (see Using the Computer section for

FIGURE 6.62
SAS Regression of SALES on ADV Using Prais–Winsten Correction for Autocorrelation.

The AUTOREG Procedure

Dependent Variable   SALES

Ordinary Least Squares Estimates

SSE	45276.5625	DFE		34
MSE	1332	Root MSE		36.49197
SBC	366.263543	AIC		363.096505
Regress R-Square	0.9490	Total R-Square		0.9490
Durbin-Watson	0.4673			

Variable	DF	Estimate	Standard Error	t Value	Approx Pr > |t|
Intercept	1	-632.6945	47.2770	-13.38	<.0001
ADV	1	0.1772	0.007045	25.16	<.0001

Estimates of Autocorrelations

Lag	Covariance	Correlation	-1 9 8 7 6 5 4 3 2 1 0 1 2 3 4 5 6 7 8 9 1
0	1257.7	1.000000	| |*******************|
1	891.8	0.709089	| |************* |

Preliminary MSE   625.3

Estimates of Autoregressive Parameters

Lag	Coefficient	Standard Error	t Value
1	-0.709089	0.122745	-5.78

Algorithm converged.

The AUTOREG Procedure

Yule-Walker Estimates

SSE	19487.808	DFE		33
MSE	590.53964	Root MSE		24.30102
SBC	340.302604	AIC		335.552047
Regress R-Square	0.8057	Total R-Square		0.9781
Durbin-Watson	2.0797			

Variable	DF	Estimate	Standard Error	t Value	Approx Pr > |t|
Intercept	1	-577.8041	97.8730	-5.90	<.0001
ADV	1	0.1696	0.0145	11.70	<.0001

details). As used here, this is the same as the Prais–Winsten correction discussed in the text. SAS first estimates the relationship using least squares and reports the results (Ordinary Least Squares Estimates). The usual coefficient estimate, standard error, t value, and p value are reported along with a number of other statistics, some of which were discussed in this text (SSE, MSE, R-square, etc.) and some of which were not (AIC and SBC). The estimate of the first-order autocorrelation coefficient is $r = 0.709089$. The results under "The AUTOREG procedure: Yule-Walker estimates" show the model estimates after using the Prais–Winsten correction for first-order autocorrelation. The coefficient estimate, standard error, t value and p value are reported and are used as with least squares. In this example the revised equation would be written

$$SALES = -577.8041 + 0.1696ADV$$

In Section 6.8.3, the coefficient estimates after the Prais–Winsten transformation were written as

$$\hat{\beta}_0 = -577.8041 \text{ and } \hat{\beta}_1 = 0.1696$$

to distinguish them from the original least-squares estimates.

EXERCISES

5. **Cost Control (reconsidered).** The file COST6 on the CD contains data on three variables:

COST is total manufacturing cost per month in thousands of dollars (the dependent variable)

PAPER is total production of paper per month in tons

MACHINE is total machine hours used per month

The regression of COST on PAPER and MACHINE is shown in Figure 6.63. The Durbin–Watson statistic is included in the output. A time-series plot of the standardized residuals is shown in Figure 6.64.

Test whether the disturbances are autocorrelated. Use a 5% level of significance. Be sure to state the hypotheses to be tested, the decision rule, the test statistic, and your decision.

FIGURE 6.63
Regression of COST on PAPER and MACHINE Including Durbin–Watson Statistic.

Variable	Coefficient	Std Dev	T Stat	P Value
Intercept	59.4318	19.6388	3.03	0.006
PAPER	0.9489	0.1101	8.62	0.000
MACHINE	2.3864	0.2101	11.36	0.000

Standard Error = 10.9835   R-Sq = 99.9%   R-Sq(adj) = 99.9%

Analysis of Variance

Source	DF	Sum of Squares	Mean Squares	F Stat	P Value
Regression	2	2271227	1135613	9413.48	0.000
Error	24	2895	121		
Total	26	2274122			

Durbin-Watson statistic = 2.14197

FIGURE 6.64
Time-Series Plot of
Standardized
Residuals from
Regression of COST
on PAPER and
MACHINE.

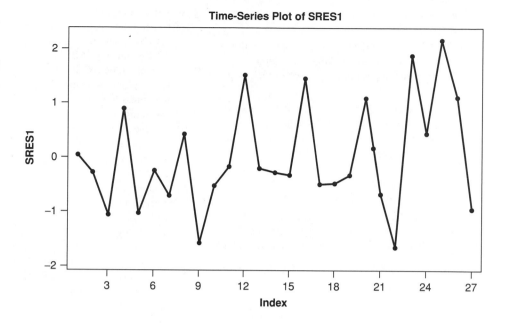

Time-Series Plot of SRES1

ADDITIONAL EXERCISES

6. Major League Baseball Salaries. The owners
of Major League Baseball (MLB) teams are
concerned with rising salaries (as are owners of
all professional sports teams). Table 3.9 provides
the average salary (AVESAL) of the 30 MLB
teams for the 2002 season. Also provided is the
number of wins (WINS) for each team during
the 2002 season. In Chapter 3 you were asked to
run the regression of WINS on AVESAL to
determine whether there is evidence that teams
with higher total payrolls tend to be more successful. Rerun this regression now. The data are
in the file BBALL6 on the CD. Examine the
standardized residuals. Are there any teams that
appear to be winning more games than expected
given the size of their payroll? Justify your
answer. Which team or teams, if any, did you
identify?
(*Source*: Data courtesy of the *Fort Worth Star-Telegram*.)

7. Imports. The gross domestic product (GDP) and
imports (IMPORTS) for 25 countries are available
in a file named IMPORTS6 on the CD. Construct
the scatterplot of IMPORTS versus GDP. Run the
regression using IMPORTS as the dependent
variable and GDP as the explanatory variable. Plot

the standardized residuals versus the fitted values
and the explanatory variable. Use the results to
help answer the following questions:

a. Do any of the assumptions of the regression
model appear to be violated? If so, which one
(or ones)? Justify your answer.

b. Construct the scatterplot and rerun the regression with the United States omitted. Construct
the residual plots for this new regression. Do
the results appear any different from the
original regression results? If so, how do they
differ? Do you prefer the original results or
the results with the United States omitted? On
what do you base your choice? Do there still
appear to be problems with this regression?

c. Try to develop a curvilinear model using the
original data (with the United States included)
that provides improved results over the linear
model. Be sure to examine the residual plots
from the curvilinear model to see if any regression assumptions are violated for this model.

8. Outlier. Consider the following time-series regression model:

$$y_i = \beta_0 + \beta_1 x_i + e_i$$

Suppose that the *y* value in one time period can be regarded as an outlier.

a. Indicate how you might be able to detect the presence of the outlier from any computer output you request or receive while analyzing the data.

b. Suppose the outlier is detected and the person in charge of the analysis decides to delete the point from the analysis and to rerun the regression without this observation. Are there alternate suggestions you would make prior to taking this course of action, or do you believe deletion of the observation is the best course of action?

9. Consumer Credit. Consider the time series of consumer installment credit (U.S.) in billions of dollars. Two different models are being considered:

$$\text{MODEL 1: CREDIT} = \beta_0 + \beta_1 \text{TREND} + e$$

and

$$\text{MODEL 2: } \log(\text{CREDIT}) = \beta_0 + \beta_1 \text{TREND} + e$$

where CREDIT is billions of dollars of consumer installment credit at the end of each month, TREND is a linear trend component, and log(CREDIT) is the natural logarithm of consumer installment credit. The regressions for these two models are shown in Figures 6.65 and 6.66, respectively. The researcher wishes to compare these two models and choose the best one. Can the choice be made on the basis of the two outputs shown? If yes, state how and state which model you would choose. If no, state why not and state what you would need to make the comparison.

10. ABX Company. The file named ABSENT6 on the CD contains data on two variables. The dependent variable is absenteeism among employees at the ABX Company (ABSENT). The explanatory variable to be considered is seniority (SENIOR), the employee's time with the company. Our goal is to express the relationship between absenteeism and seniority in the form of an equation. Start with the linear regression using absenteeism as the dependent variable and seniority as the independent variable.

Based on an examination of scatterplots and residual plots, do any assumptions of the linear regression model appear to be violated? If so, which one (or ones)? If any violations are detected, suggest possible corrections. Rerun the regression with the suggested corrections and compare your results to the original results. Be sure to do residual plots for the model using the suggested correction. Which model do you prefer and why?

11. Coal Mining Fatalities. The file CUTTING6 on the CD contains data on the following two variables:

> FATALS: the annual number of fatalities from gas and dust explosions in coal mines for the years 1915 to 1978
>
> CUTTING: the number of cutting machines in use

(*Source*: These data are from K. D. Lawrence and L. C. Marsh, "Robust Ridge Estimation Methods for Predicting U.S. Coal Mining Fatalities." *Communications in Statistics,* 13 (1984): pp. 139–149. Used by permission.)

Run the regression using FATALS as the dependent variable and CUTTING as the independent variable.

FIGURE 6.65
Regression Results for Consumer Credit Model 1.

Variable	Coefficient	Std Dev	T Stat	P Value
Intercept	737.933	4.650	158.71	0.000
TREND	9.763	0.165	59.10	0.000

Standard Error = 15.86　　　　R-Sq = 98.7%　　R-Sq(adj) = 98.7%

Analysis of Variance

Source	DF	Sum of Squares	Mean Square	F Stat	P Value
Regression	1	878056	878056	3492.70	0.000
Error	46	11564	251		
Total	47	889621			

FIGURE 6.66
Regression Results for Consumer Credit Model 2.

Variable	Coefficient	Std Dev	T Stat	P Value
Intercept	6.62890	0.00441	1504.82	0.000
TREND	0.01004	0.00016	64.14	0.000

Standard Error = 0.01502   R-Sq = 98.9%   R-Sq(adj) = 98.9%

Analysis of Variance

Source	DF	Sum of Squares	Mean Square	F Stat	P Value
Regression	1	0.92843	0.92843	4114.28	0.000
Error	46	0.01038	0.00023		
Total	47	0.93881			

Based on an examination of scatterplots and residual plots, do any assumptions of the linear regression model appear to be violated? If so, which one (or ones)? If any violations are detected, suggest possible corrections. Rerun the regression with the suggested corrections and compare your results to the original results. Be sure to do residual plots for the model using the suggested correction. Which model do you prefer and why?

12. **Piston Corporation (reconsidered).** Reexamine the data from the PISTON exercise in Chapter 5. The original variables in this problem were

COST, the dependent variable; a measure of cost for the company
PROD, production level
INDEX, a cost index for the industry

   The data for these variables are in the file PISTON6 on the CD.

Use the variables 1/PROD and INDEX as explanatory variables. Test to see whether the disturbances are positively autocorrelated. State the hypotheses to be tested, the decision rule, the test statistic, and your decision. On the basis of the test result, what action should be taken? Use a 0.05 level of significance.

13. **Computer Repair.** A computer repair service is examining the time taken on service calls. The data obtained for 30 service calls are in the file named COMPREP6 on the CD. Information obtained includes:

number of machines to be repaired (NUMBER)
years of experience of service person (EXPER)

time taken (in minutes) to provide service (TIME)

Develop a model to predict average time on the service calls using EXPER and NUMBER as explanatory variables. Use scatterplots and residual plots to determine whether any of the assumptions of the linear regression model have been violated.

If any of the assumptions have been violated, state which one or ones and suggest possible corrections. Try the new model to see if it is an improvement over the original one. Be sure to examine residual plots from the corrected model (or models) that you try. Indicate your choice for the best model.

14. **U. S. Population.** The data file USPOP6 on the CD contains the population of the United States for the years 1930 through 1999. Fit a linear trend to these data.

   a. What is the resulting regression equation?
   b. What percentage of the variation in y has been explained by the regression?
   c. Test to see whether the disturbances are positively autocorrelated. State the hypotheses to be tested, the decision rule, the test statistic, and your decision. On the basis of the test result, what action should be taken? Use a 0.05 level of significance.
   d. Add a lagged value of the dependent variable to the equation. Your equation will now have a term for linear trend as well as the lagged dependent variable. What is the resulting regression equation?
   e. Test to see whether the disturbances are positively autocorrelated. (What test should you

use for this model?) State the hypotheses to be tested, the decision rule, the test statistic, and your decision. Use a 0.05 level of significance. On the basis of the test result, what action should be taken?

f. Add a two-period lagged value of the dependent variable to the equation. Your equation will now have a term for linear trend as well as the one- and two-period lagged dependent variables. What is the resulting regression equation? (*Source*: U.S. Bureau of the Census.)

g. Test to see whether the disturbances are positively autocorrelated. (What test should you use for this model?) State the hypotheses to be tested, the decision rule, the test statistic, and your decision. Use a 0.05 level of significance. On the basis of the test result, what action should be taken?

h. Use your choice of the "best" model from among the three examined in this exercise to predict the U.S. population for 2000 and 2001. Do you encounter any problems with the use of this model to generate these forecasts? If so, what can you do to overcome these problems?

15. **Major League Baseball Wins.** What factor is most important in building a winning baseball team? Some might argue for a high batting average. Or it might be a team that hits for power as measured by the number of home runs. On the other hand, many believe that it is quality pitching as measured by the earned run average of the team's pitchers. The file MLB6 on the CD contains data on the following variables for the 30 major league baseball teams during the 2002 season:

WINS = number of games won
HR = number of home runs hit
BA = average batting average
ERA = earned run average

Using WINS as the dependent variable, run the regression relating the three explanatory variables to WINS. Examine the standardized residuals from this regression and any other regression diagnostics you believe might be useful. Do any of the regression assumptions appear to be violated? Justify your answer. (*Source*: Data courtesy of the *Fort Worth Star-Telegram*.)

16. **Byron Nelson Donations.** Since 1982, the year that PGA Tour officials began tracking charitable donations from its tournaments on an annual basis,

no event has contributed more to charities in its community than the Byron Nelson tournament. The amount donated each year from 1982 through 2002 (in millions of dollars) is provided in the file named DONATIONS6 on the CD. Fit a linear trend to these data.

a. What is the resulting regression equation?

b. What percentage of the variation in *y* has been explained by the regression?

c. Test to see whether the disturbances are positively autocorrelated. State the hypotheses to be tested, the decision rule, the test statistic, and your decision. Use a 0.05 level of significance. On the basis of the test result, what action should be taken?

d. Add a lagged value of the dependent variable to the equation. Your equation will now have a term for linear trend as well as the lagged dependent variable. What is the resulting regression equation?

e. Test to see whether the disturbances are positively autocorrelated. (What test should you use for this model?) State the hypotheses to be tested, the decision rule, the test statistic, and your decision. Use a 0.05 level of significance. On the basis of the test result, what action should be taken?

f. Use your choice of the "best" model to predict donations for 2003. Find a point prediction and a 95% prediction interval. (*Source*: Data courtesy of the *Fort Worth Star-Telegram*.)

17. **Estimating Residential Real Estate Values.** The Tarrant County Appraisal District must appraise properties for the entire county. The appraisal district uses data such as square footage of the individual houses as well as location, depreciation, and physical condition of an entire neighborhood to derive individual appraisal values on each house. This avoids labor-intensive reinspection each year.

Regression can be used to establish the weight assigned to various factors used in assessing values. For example, the file REALEST6 on the CD contains the value (VALUE), size in square feet (SIZE), a physical condition index (CONDITION), and a depreciation factor (DEPRECIATION) for a sample of 100 Tarrant County houses (in 1990). Using these data, develop an equation that might be useful to the appraisal district in assessing values.

If any of the assumptions have been violated, state which one or ones and suggest possible corrections. Try the new model to see if it is an improvement over the original one. Be sure to examine residual plots from the corrected model (or models) that you try.

Discuss how the equation developed here could be used to value houses. What would be the value assigned to a 1400-square-foot house with physical condition index 0.70 and depreciation factor 0.02?

18. Criminal Justice Expenditures. The file CRIMSPN6 on the CD contains the following data for each of the 50 states:

> total expenditures on a state's criminal justice system (in thousands of dollars) (EXPEND)
> total number of police employed in the state (POLICE)
> population of the state (in thousands) (POP)
> total number of incarcerated prisoners in the state (PRISONER)

State governments must try to project spending in many areas. Expenditure on the criminal justice system is one area of continually rising cost. Your job is to build a model that can be used to forecast spending on a state's criminal justice system. Any of the three possible explanatory variables can be used. Be sure to consider violations of any assumptions in building your model and correct for any violations. Once your model is complete, predict expenditures for a state that plans to hire 10,000 police personnel, has a population of 3 million, and expects 600 prisoners. Find a point prediction and a 95% prediction interval.

(*Source*: These data were obtained from the U.S. Department of Criminal Justice web site and are for the year 1999.)

19. Intersections. One factor related to the number of accidents at an intersection is the peak-hour volume of traffic. Data on both the volume of traffic during the peak hour (VOLUME) and the total number of accidents (ACCIDENT) are available for 62 intersections in Fort Worth, Texas. These data are available in a file named TRAFFIC6 on the CD.

(*Source*: City of Fort Worth)

The city wants to identify intersections that have an unusual number of accidents. Its definition of unusual involves first establishing a base-level forecast of the number of accidents, taking account of the peak-hour volume of traffic, and then judging which intersections still have an unusual number of accidents. As a consultant to the city, your job is to determine which, if any, of these 62 intersections appear to have an unusually high number of accidents given the volume of traffic.

20. Fuel Consumption. Table 5.4 showed the fuel consumption (FUELCON) in gallons per capita for each of the 50 states and Washington, D.C. The following variables are also shown: the population of the state (POP), the area of the state in square miles (AREA). From these data, create the variable population density (DENSITY) defined as population divided by area. The file FUELCON6 on the CD contains this information as well as data for the following variables:

> DRIVERS, the ratio of licensed drivers to private and commercial motor vehicles registered
> HWYMILES, the number of federally funded miles of highway
> GASTAX, the tax on a gallon of gas
> INCOME, the average personal income

The object is to develop a regression equation to predict fuel consumption. Any of the variables in the file can be used as explanatory variables. Find what you believe is the best equation to explain fuel consumption. Be sure to check for violations of any of the regression assumptions.

(*Source*: The population data and average personal income are from the U.S. Bureau of the Census web site. All other data are from the Federal Highway Administration: Office of Highway Information Management.)

USING THE COMPUTER

The Using the Computer section in each chapter describes how to perform the computer analyses in the chapter using Excel, MINITAB, and SAS. For further detail on Excel, MINITAB, and SAS, see Appendix C.

EXCEL

Storing Standardized Residuals and Fitted Values

In the Regression dialog box, check Standardized Residuals, as shown in Figure 6.67. Checking the box for the standardized residuals saves the standardized residuals and the fitted values (as well as the raw residuals).

FIGURE 6.67 Saving Fitted Values and Standardized Residuals with Excel.

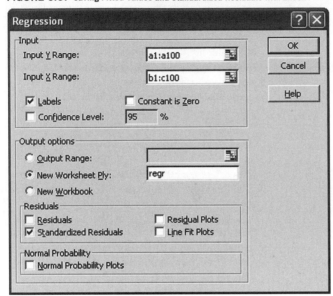

Requesting Other Influence Statistics and the Durbin–Watson Statistic

Excel does not provide options for automatically producing the other influence statistics discussed in the text or the Durbin–Watson statistic. Although these could be computed by creating formulas, this option is not discussed in this text.

A regression add-in is available that does compute a number of these additional statistics. See the Excel section in Appendix C for more information on the add-in.

Variable Transformations

See Variable Transformations in the Using the Computer section at the end of Chapter 5.

Testing for Normality

Although there is a normal probability plot option in the Excel Regression dialog box, it does not do you much good. It provides a normal probability plot of the dependent variable, when in fact, what you really want is a normal probability plot of the standardized residuals. Plus, I find the plot difficult to interpret given the way it is produced in Excel.

FIGURE 6.68 MINITAB Regression—Storage Dialog Box for Saving Fitted Values, Standardized Residuals, and Influence Statistics.

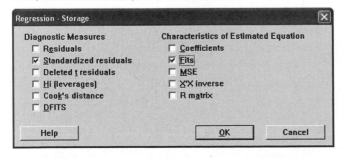

MINITAB

Storing Standardized Residuals and Fitted Values

STAT: REGRESSION: REGRESSION: STORAGE

Store standardized residuals and fitted values by using the Storage button in the Regression dialog box. Then click "Fits" and "Standardized residuals" as shown in Figure 6.68. The fitted values and standardized residuals are stored, respectively, in the next two free columns in the MINITAB worksheet.

Requesting Other Influence Statistics

Refer again to Figure 6.68. Requests for other influence statistics are made by checking the appropriate boxes in the Regression—Storage dialog box. Other influence statistics available in MINITAB include leverage values (Hi), Cook's Distance, and DFITS. These values are stored in a column in the worksheet by simply clicking the box next to the desired statistic.

FIGURE 6.69 MINITAB Regression—Options Dialog Box for Requesting Durbin–Watson Statistic.

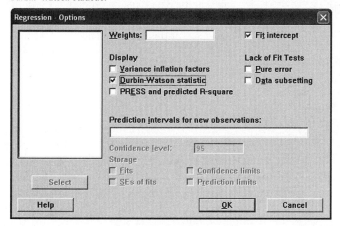

Requesting the Durbin–Watson Statistic

STAT: REGRESSION: REGRESSION: OPTIONS

To request the Durbin–Watson statistic, click the Durbin–Watson statistic box on the Regression—Options screen, as shown in Figure 6.69.

Variable Transformations

See Variable Transformations in the Using the Computer section at the end of Chapter 5.

FIGURE 6.70 MINITAB Normality Test for Disturbances.

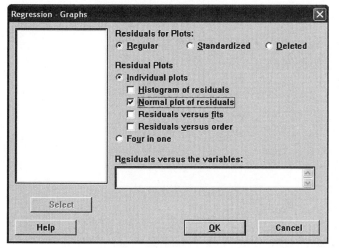

Testing for Normality of Disturbances

STAT: REGRESSION: REGRESSION: GRAPHS

To request a test for the normality of the regression disturbances, click the Graphs button in the Regression Dialog Box. Then check the "normal plot of residuals" option, as shown in Figure 6.70. Note that "Residuals for Plots" allows a choice of which residuals to use in the test. The Anderson–Darling test statistic is reported along with a p value.

STAT: BASIC STATISTICS: NORMALITY TEST

To request a test for normality (for the disturbances or any other variable), use the Normality Test option on the Basic Statistics menu. Fill in the variable to be tested (the standardized residuals from a regression are used in the example) and the test desired, as shown in Figure 6.71. You have a choice of the Anderson–Darling test, the Ryan–Joiner test, or the Kolmogorov–Smirnov test. Regardless of which test is chosen, a p value will be reported, which makes conducting any of the tests very easy.

Szroeter's Test for Nonconstant Variance

Although this test cannot be requested with a single command, it can be computed through a sequence of commands. This will be demonstrated for MINITAB. (I used commands here rather than menu items because I find it easier when there is a sequence

FIGURE 6.71 MINITAB Normality Test Dialog Box.

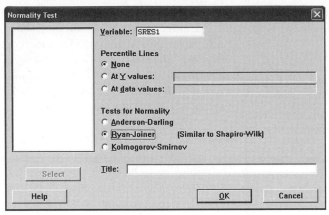

of operations to be performed. Obviously, these same procedures could be accomplished with menus.) Assume that the dependent (y) variable is in C1 and the (one) explanatory variable (x) is in C2. First, recall that the data must be ordered according to increasing values of x. The Sort command accomplishes this. The standardized residuals from the regression of y on x are then stored in C10.

```
SORT C2, CARRY C1, PUT IN C20, C21
REGR Y IN C21 ON 1 PRED IN C20, STORE IN
   C10
```

A patterned Set command is used to put integers 1 through n (the sample size) in C25.

```
SET IN C25
1:n
END
```

Now compute the quantities needed to compute h in Szroeter's test statistic:

```
LET C16 = C10*C10
LET C17 = C16*C25
SUM C17 K2
SUM C16 K1
LET K3 = K2/K1
```

The value of K3 printed will be h.

```
PRINT K3
```

Szroeter's Q statistic can then be computed using the formula

$$Q = \left(\frac{6n}{n^2 - 1} \right)^{1/2} \left(h - \frac{n + 1}{2} \right)$$

SAS

Storing Standardized Residuals and Fitted Values

The following command sequence produces a regression of SALES on two independent variables, ADV and BONUS, saves the fitted values and standardized residuals, and then plots the standardized residuals versus the fitted values and each of the independent variables.

```
PROC REG;
MODEL SALES = ADV BONUS;
OUTPUT PREDICTED = FITS STUDENT = STRES;
PROC GPLOT;
PLOT STRES*FITS STRES*ADV STRES*BONUS;
```

In the model statement, the dependent variable (SALES) is listed first; the independent variables (ADV and BONUS) follow the equal sign. In the output statement, the fitted (predicted) values are requested and labeled FITS. The standardized residuals (STUDENT) are also requested and labeled STRES. PROC GPLOT is used to produce residual plots. As shown, residual plots of the standardized residuals versus the fitted values and each of the two explanatory variables are generated.

Requesting Other Influence Statistics

Influence statistics other than standardized residuals and fitted values can be requested in SAS as follows:

```
PROC REG;
MODEL Y = X/INFLUENCE;
OUTPUT COOK = COOKD;
```

The INFLUENCE option requests that SAS print out the leverage values, the studentized deleted residuals, and the DFITS statistics (among others not discussed in this text). The OUTPUT command requests that Cook's D statistic be computed, stored, and labeled COOKD.

Requesting the Durbin–Watson Statistic

```
PROC REG;
MODEL SALES = ADV/DW;
```

requests that the Durbin–Watson statistic be printed. SAS also prints out an estimate of the first-order autocorrelation statistic.

Variable Transformations

See Variable Transformations in the Using the Computer section in Chapter 5.

Testing for Normality

In SAS, a normal probability plot for the variable STRES can be produced using the commands:

```
PROC UNIVARIATE PLOT NORMAL;
VAR STRES;
```

A variety of descriptive statistics on the variable STRES are printed along with a stem-and-leaf plot, a box plot, and a normal probability plot. Four tests for normality are performed: Shapiro–Wilk (essentially the same test as the Ryan–Joiner test produced in MINITAB), Kolmogorov–Smirnov, Cramer–von Mises, and Anderson–Darling. In all cases, a p value corresponding to the observed statistic is computed and printed.

Prais–Winsten Transformation

The Prais–Winsten transformation to correct for first-order autocorrelation was discussed earlier in this chapter.

```
PROC AUTOREG;
MODEL SALES = ADV/NLAG = 1 ITER;
```

These commands request the Prais–Winsten transformation through the PROC AUTOREG command. The dependent variable here is SALES and the explanatory variable is ADV (more than one explanatory variable can be included if necessary). The NLAG = 1 option tells the procedure that first-order autocorrelation is to be corrected (AUTOREG is a general procedure that can correct for higher-order autocorrelation not discussed in this text). The ITER option requests an iterative procedure, which has in general been shown superior in various studies.

7

Using Indicator and Interaction Variables

7.1 USING AND INTERPRETING INDICATOR VARIABLES

Indicator variables or *dummy variables* are a special type of variable used in a variety of ways in regression analysis. Indicator variables take on only two values, either 0 or 1. They can be used to indicate whether a sample unit either does (1) or does not (0) belong in a certain category. For example, a dummy variable could be used to indicate when an individual in the sample was employed by constructing the variable as

D_{1i} = 1 if individual i is employed

   = 0 if individual i is not employed

To indicate when an individual is unemployed, the variable could be constructed as

D_{2i} = 1 if individual i is unemployed

   = 0 if individual i is not unemployed

Obviously, any type of split into two groups can be easily represented by indicator variables.

If there are more than two groups into which individuals may be classified, this simply requires the use of additional indicator variables. Suppose firms in a sample are to be categorized according to the exchange on which they are listed: NYSE,

AMEX, or NASDAQ. This could be accomplished by constructing the following variables:

$$D_{1i} = 1 \text{ if firm } i \text{ is listed on the NYSE}$$
$$= 0 \text{ if firm } i \text{ is not listed on the NYSE}$$
$$D_{2i} = 1 \text{ if firm } i \text{ is listed on the AMEX}$$
$$= 0 \text{ if firm } i \text{ is not listed on the AMEX}$$
$$D_{3i} = 1 \text{ if firm } i \text{ is listed on the NASDAQ}$$
$$= 0 \text{ if firm } i \text{ is not listed on the NASDAQ}$$

Thus far, when sample individuals could belong to one of m different groups, m indicator variables were constructed, one for each group. When indicator variables are used in a regression analysis, however, only $m - 1$ of the indicator variables are included in the regression because only $m - 1$ indicator variables are needed to indicate m groups. The one group whose indicator is omitted serves as what might be called a *base-level* group. Consider the following example to clarify this point.

EXAMPLE 7.1   **Employment Discrimination**   Regression analysis has been used extensively in employment discrimination cases. The desire in such cases is typically to compare mean salaries of two groups of employees (say, male and female employees) to determine whether one group has significantly lower salaries than the other group. Evidence of lower average salaries can provide some support for a discrimination suit against the employer. It is recognized that a simple two-sample comparison of mean salaries is not sufficient to conclude that one group has been discriminated against. Obviously, there are many factors other than discrimination that affect salary to which differences in average salary might be attributed. Regression is used to adjust for the effects of these other factors before the two groups are compared. An indicator variable is added to the regression to separate the employees into two groups: male and female.

Table 7.1 presents data from the case of *United States Department of the Treasury* v. *Harris Trust and Savings Bank* (1981). The data, contained in the file HARRIS7 on the CD, include the salary of 93 employees of the bank (SALARY), their educational level (EDUCAT), and an indicator variable (MALES) signifying whether the employee is male (1) or female (0). Figure 7.1 shows a scatterplot of salary versus education for all 93 employees. Figure 7.2 shows the same plot but with the two groups indicated by different symbols (male = ■ and female = ●). There is some indication from the plot in Figure 7.2 that male salaries are higher than female salaries, even when differing education levels have been taken into account. To obtain a better sense of the magnitude of these differences and to provide a test for whether the differences are significant or whether they are small enough that they could have occurred by chance, the regression of SALARY on the two explanatory variables EDUCAT and MALES is shown in Figure 7.3. The resulting equation is:

$$\text{SALARY} = 4173 + 80.7\text{EDUCAT} + 692\text{MALES}$$

The next question is: How do we interpret this equation?

TABLE 7.1 Data for Employment Discrimination Example

SALARY	EDUCAT	MALES	SALARY	EDUCAT	MALES	SALARY	EDUCAT	MALES
3900	12	0	5220	12	0	5040	15	1
4020	10	0	5280	8	0	5100	12	1
4290	12	0	5280	8	0	5100	12	1
4380	8	0	5280	12	0	5220	12	1
4380	8	0	5400	8	0	5400	12	1
4380	12	0	5400	8	0	5400	12	1
4380	12	0	5400	12	0	5400	12	1
4380	12	0	5400	12	0	5400	15	1
4440	15	0	5400	12	0	5400	15	1
4500	8	0	5400	12	0	5700	15	1
4500	12	0	5400	12	0	6000	8	1
4620	12	0	5400	12	0	6000	12	1
4800	8	0	5400	15	0	6000	12	1
4800	12	0	5400	15	0	6000	12	1
4800	12	0	5400	15	0	6000	12	1
4800	12	0	5400	15	0	6000	12	1
4800	12	0	5520	12	0	6000	12	1
4800	12	0	5520	12	0	6000	15	1
4800	12	0	5580	12	0	6000	15	1
4800	12	0	5640	12	0	6000	15	1
4800	12	0	5700	12	0	6000	15	1
4800	16	0	5700	12	0	6000	15	1
4980	8	0	5700	15	0	6000	16	1
5100	8	0	5700	15	0	6300	15	1
5100	12	0	5700	15	0	6600	15	1
5100	12	0	6000	12	0	6600	15	1
5100	15	0	6000	15	0	6600	15	1
5100	15	0	6120	12	0	6840	15	1
5100	16	0	6300	12	0	6900	12	1
5160	12	0	6300	15	0	6900	15	1
5220	8	0	4620	12	1	8100	16	1

Consider a regression such as the one in Example 7.1 with one quantitative explanatory variable (x_1) and one indicator variable (D)

$$y = \beta_0 + \beta_1 x_1 + \beta_2 D$$

The indicator variable D is coded as 1 if an item in the sample belongs to a certain group and as 0 if the item does not belong to the group. The equation can be separated into two parts as follows. If the sample item is in the indicated group $(D = 1)$:

$$y = \beta_0 + \beta_1 x_1 + \beta_2(1) = (\beta_0 + \beta_2) + \beta_1 x_1$$

If the sample item is not in the indicated group $(D = 0)$:

$$y = \beta_0 + \beta_1 x_1 + \beta_2(0) = \beta_0 + \beta_1 x_1$$

FIGURE 7.1
Scatterplot of Salary
versus Education for
Employment
Discrimination
Example.

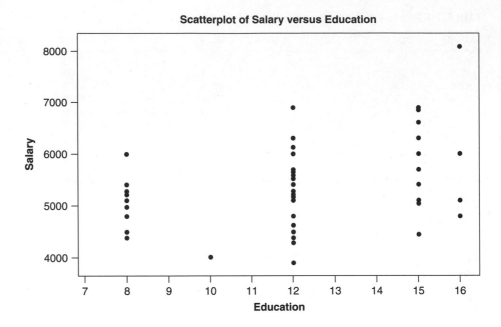

Scatterplot of Salary versus Education

FIGURE 7.2
Scatterplot of Salary
versus Education with
Males (■) and
Females (●)
Indicated.

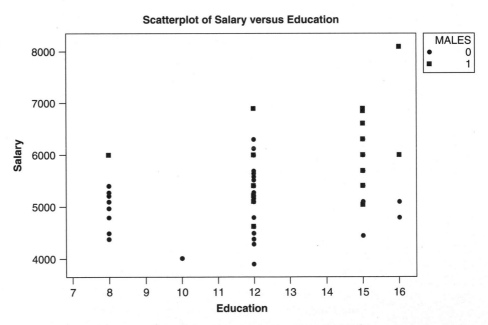

Scatterplot of Salary versus Education

Using the indicator variable results in one equation for the indicated group and another for the other group. Note that the equation for the indicated group has been rewritten with two components making up the intercept term: the original intercept, β_0, and the coefficient of the indicator variable, β_2. Two lines have been fit with the same slope but different intercepts, even though only one

FIGURE 7.3
Regression of SALARY on EDUCAT and MALES for Employment Discrimination Example.

Variable	Coefficient	Std Dev	T Stat	P Value
Intercept	4173.1	339.2	12.30	0.000
EDUCAT	80.7	27.7	2.92	0.004
MALES	691.8	132.2	5.23	0.000

Standard Error = 572.437   R-Sq = 36.3%   R-Sq(adj) = 34.9%

Analysis of Variance

Source	DF	Sum of Squares	Mean Square	F Stat	P Value
Regression	2	16831744	8415872	25.68	0.000
Error	90	29491546	327684		
Total	92	46323290			

FIGURE 7.4 Graph Showing Relative Placement of Regression Lines If β_2 Is Assumed to Be Positive.

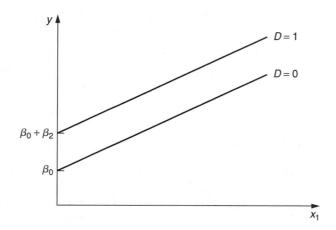

regression has been run. Figure 7.4 shows an example of how we might draw the two estimated lines. The difference in the intercepts is given by β_2 (which is assumed to be positive in the graph). This process allows us to answer the question of whether there is a difference in the average value of the y variable for the two groups after adjusting for the effect of the quantitative variable (or variables if there is more than one) and how much the average difference is. The adjusted difference in the averages is represented by the difference in the intercepts (that is, by the coefficient of the indicator variable). A t test on this coefficient will help decide whether the difference is large enough to be considered statistically significant.

EXAMPLE 7.2   **Employment Discrimination (continued)**   The equation in the employment discrimination example can be interpreted as follows: Salary increases by $80.70 for each year of education. Males earn, on average, $692 more per year than females (this was in 1977, so $692 a year was more than it sounds like now).

The next question is whether \$692 is large enough to be considered statistically significant. Or could this difference have occurred purely by chance? To determine this, the following hypotheses will be tested.

$$H_0: \beta_2 \leq 0$$
$$H_a: \beta_2 > 0$$

If the null hypothesis is not rejected, then there is not enough evidence to conclude that the average salary for males is higher (after taking into account education). If the null hypothesis is rejected, we conclude that the difference observed is too large to have occurred by chance. There is evidence that the average salary for males is higher, even after taking into account the effect of education. The test performed here is a one-tailed test because of the question we are asking: Is there evidence that the average salary for males is higher?

Decision rule: Reject H_0 if $t > 1.645$

Do not reject H_0 if $t \leq 1.645$

Because there are a large number of degrees of freedom (90) we use the z value, 1.645, as the critical value.

Test Statistic: t value $= 5.23$

Decision: Reject H_0

Our conclusion is that males, on average, appear to be earning significantly more than females, even after we take into account education. There is evidence of employment discrimination. Legal counsel for Harris Trust and Savings Bank in this case would point out that education is not the only factor that could cause a difference in salary. Factors like previous experience and time on the job might also have an effect. These factors, if measurable, could be included in the regression as well (and typically are in such applications).[1]

One concern that arises is whether the way the two groups were coded matters. Would the results be the same in our example if males were coded 0 and females were coded 1? Yes, they would. Figure 7.5 shows the regression of SALARY on EDUCAT and a new variable called FEMALES. This new variable is coded 0 for males and 1 for females. Note that the coefficient of EDUCAT is exactly the same as it was for the regression in Figure 7.3. The coefficient of the indicator variable is the same except for the sign. In the original regression, it was positive and now it is negative. The intercept of the new equation is larger by \$692. However, the interpretation of the results is exactly the same. When the variable MALES was used, our interpretation was that males earned, on average, \$692 per year *more* than females. Using the FEMALES variable, our interpretation is that females earned, on average, \$692 per year less than males. Also, the t test again tells us that the difference is statistically significant.

---

[1]For a more detailed examination of the use of regression in employment discrimination cases, see D. A. Conway and H. V. Roberts, "Regression Analyses in Employment Discrimination Cases," in *Statistics and the Law*, New York: Wiley, 1986.

FIGURE 7.5
Regression of SALARY on
EDUCAT and FEMALES
for Employment
Discrimination Example.

Variable	Coefficient	Std Dev	T Stat	P Value
Intercept	4864.9	387.9	12.54	0.000
EDUCAT	80.7	27.7	2.92	0.004
FEMALES	-691.8	132.2	-5.23	0.000

Standard Error = 572.437   R-Sq = 36.3%   R-Sq(adj) = 34.9%

Analysis of Variance

Source	DF	Sum of Squares	Mean Square	F Stat	P Value
Regression	2	16831744	8415872	25.68	0.000
Error	90	29491546	327684		
Total	92	46323290			

In the employment discrimination example, an indicator variable was used to indicate when sample observations fell into one of two groups. The groups represented qualitative variables (that is, variables that could not be quantified in a meaningful way). For example, there is no meaningful way to quantify the difference between males and females. An indicator variable was used to separate the two groups for separate analyses, but no numeric values were assigned to express the difference between being male or female. Indicator variables can be used in this manner to incorporate qualitative information into the regression equation. The next example illustrates the use of indicator variables to represent qualitative information when there are more than two categories.

EXAMPLE 7.3 **Meddicorp**   In the Meddicorp example in Chapter 4, the relationship between the dependent variable sales (SALES) and two explanatory variables, advertising (ADV) and bonus (BONUS), was examined. It also was noted in the example that Meddicorp sold in three different regions of the country: South, West, and Midwest. A variable has been added to the data set to indicate each of these three regions. The variable is called REGION and is coded as follows:

REGION = 1 if South

= 2 if West

= 3 if Midwest

Data for the new variable and the original variables are shown in Table 7.2 (see the MEDDICORP7 file on the CD). The new variable is included in a regression along with the ADV and BONUS variables. The results are shown in Figure 7.6.

The estimated regression equation is

$$\hat{y} = -84 + 1.55\text{ADV} + 1.11\text{BONUS} + 119\text{REGION}$$

The coefficient of REGION indicates that as the value assigned to the region increases, so does the amount of sales. On average, there is a difference of 119 units between sales in the territories of the three regions. Each unit is $1000. Sales in

TABLE 7.2 Meddicorp Data Including the Variable REGION and Three Resulting Indicator Variables SOUTH, WEST, and MIDWEST

SALES	ADV	BONUS	REGION	SOUTH	WEST	MIDWEST
963.50	374.27	230.98	1	1	0	0
893.00	408.50	236.28	1	1	0	0
1057.25	414.31	271.57	1	1	0	0
1183.25	448.42	291.20	2	0	1	0
1419.50	517.88	282.17	3	0	0	1
1547.75	637.60	321.16	3	0	0	1
1580.00	635.72	294.32	3	0	0	1
1071.50	446.86	305.69	1	1	0	0
1078.25	489.59	238.41	1	1	0	0
1122.50	500.56	271.38	2	0	1	0
1304.75	484.18	332.64	3	0	0	1
1552.25	618.07	261.80	3	0	0	1
1040.00	453.39	235.63	1	1	0	0
1045.25	440.86	249.68	2	0	1	0
1102.25	487.79	232.99	2	0	1	0
1225.25	537.67	272.20	2	0	1	0
1508.00	612.21	266.64	3	0	0	1
1564.25	601.46	277.44	3	0	0	1
1634.75	585.10	312.35	3	0	0	1
1159.25	524.56	292.87	1	1	0	0
1202.75	535.17	268.27	2	0	1	0
1294.25	486.03	309.85	2	0	1	0
1467.50	540.17	291.03	3	0	0	1
1583.75	583.85	289.29	3	0	0	1
1124.75	499.15	272.55	2	0	1	0

FIGURE 7.6
Regression of SALES on ADV, BONUS, and REGION.

Variable	Coefficient	Std Dev	T Stat	P Value
Intercept	-84.219	177.907	-0.47	0.641
ADV	1.546	0.306	5.05	0.000
BONUS	1.106	0.573	1.93	0.067
REGION	118.899	28.687	4.14	0.000

Standard Error = 68.8881   R-Sq = 92.0%   R-Sq(adj) = 90.9%

Analysis of Variance

Source	DF	Sum of Squares	Mean Square	F Stat	P Value
Regression	3	1149317	383106	80.73	0.000
Error	21	99657	4746		
Total	24	1248974			

territories of Region 2 (West) would be $119,000 more (on average) than sales in territories of Region 1 (South), and sales in territories of Region 3 (Midwest) would be $119,000 more than sales in Region 2. Note that the difference in sales between regions is forced to be $119,000. Although this situation may not be realistic, it is required by the direct use of the variable REGION in the regression equation. A more flexible representation of the changes in average sales between regions is allowed by using indicator variables.

Note also that the coding of the variable REGION is arbitrary. The coding could be done as follows: Midwest = 1, West = 2, and South = 3. The numbers assigned to the regions are merely naming devices. They represent qualitative categories rather than quantitative data. Changing the order of the number of the categories does change the results, however, when using the REGION variable. This does not happen if indicator variables are used.

The REGION variable can be transformed into three indicator variables. An indicator variable SOUTH can be developed that indicates whether or not a territory is in the South. SOUTH is made to take on the value 1 whenever the REGION variable is 1, and SOUTH is given the value 0 if REGION is 2 or 3. An indicator variable WEST can be developed that indicates whether or not a territory is in the West. WEST takes on the value 1 whenever the REGION variable is 2 (that is, when the territory is in the West) and is given the value 0 if REGION is 1 or 3. Similarly, an indicator variable for the Midwest can be developed. The three indicator variables are defined as follows:

SOUTH	= 1 if the territory is in the South
	= 0 otherwise
WEST	= 1 if the territory is in the West
	= 0 otherwise
MIDWEST	= 1 if the territory is in the Midwest
	= 0 otherwise

These three indicator variables—SOUTH, WEST, and MIDWEST—indicate into which of the three mutually exclusive regions each territory falls. Table 7.2 shows the SALES, ADV, BONUS, and REGION variables and the resulting indicators. As stated previously, only two of the three indicators need to be used in the regression. The territories indicated by the third indicator serve as a base-level group.

If the MIDWEST is used as the base-level group, the regression results shown in Figure 7.7 are obtained. The regression equation is

$$\hat{y} = 435 + 1.37\text{ADV} + 0.975\text{BONUS} - 258\text{SOUTH} - 210\text{WEST}$$

The interpretation of this equation is similar to that of the equation developed in the employment discrimination example. Here, however, the territories have been separated into three groups through the use of the indicator variables. The coefficient of each indicator variable represents the difference in the intercept between the indicated group and the base-level (MIDWEST) group. This can be expressed through the use of three separate equations:

SOUTH	$\hat{y} = 435 + 1.37\text{ADV} + 0.975\text{BONUS} - 258$
	$= 177 + 1.37\text{ADV} + 0.975\text{BONUS}$
WEST	$\hat{y} = 435 + 1.37\text{ADV} + 0.975\text{BONUS} - 210$
	$= 225 + 1.37\text{ADV} + 0.975\text{BONUS}$
MIDWEST	$\hat{y} = 435 + 1.37\text{ADV} + 0.975\text{BONUS}$

FIGURE 7.7
Regression of SALES on ADV, BONUS, and the Indicator Variables SOUTH and WEST.

Variable	Coefficient	Std Dev	T Stat	P Value
Intercept	435.099	206.234	2.11	0.048
ADV	1.368	0.262	5.22	0.000
BONUS	0.975	0.481	2.03	0.056
SOUTH	-257.892	48.413	-5.33	0.000
WEST	-209.746	37.420	-5.61	0.000

Standard Error = 57.6254   R-Sq = 94.7%   R-Sq(adj) = 93.6%

Analysis of Variance

Source	DF	Sum of Squares	Mean Square	F Stat	P Value
Regression	4	1182560	295640	89.03	0.000
Error	20	66414	3321		
Total	24	1248974			

The slopes of the equations are constrained to be the same, but the intercepts are allowed to differ. As a further interpretation, consider the values of $\hat{y}$ for a given level of ADV and BONUS—say, ADV = 500 and BONUS = 250. These values are

SOUTH $\hat{y} = 177 + 1.37(500) + 0.975(250) = 1105.8$

WEST $\hat{y} = 225 + 1.37(500) + 0.975(250) = 1153.8$

MIDWEST $\hat{y} = 435 + 1.37(500) + 0.975(250) = 1363.8$

The conditional mean sales for advertising equal to 500 and bonus payment equal to 250 are shown by these computations. The sales figures differ according to the coefficients of the indicator variables: \$1,105,800 for SOUTH; \$1,153,800 for WEST; \$1,363,800 for MIDWEST.

To determine whether there is a significant difference in sales for territories in different regions, the following hypotheses should be tested:

$H_0: \beta_3 = \beta_4 = 0$

H_a: At least one of β_3 and β_4 is not equal to zero

The model hypothesized is

$$y = \beta_0 + \beta_1 ADV + \beta_2 BONUS + \beta_3 SOUTH + \beta_4 WEST$$

so the null hypothesis states that the coefficients of the indicator variables are both zero. If this hypothesis is not rejected, then no difference between the various regions exists, and the indicator variables can be dropped from the model. The simpler model

$$y = \beta_0 + \beta_1 ADV + \beta_2 BONUS$$

explains just as much variation in sales. This hypothesis can be tested using the partial F test discussed in Chapter 4. The full model contains the indicator variables; the reduced model does not. The test statistic to be used is computed exactly as discussed in Chapter 4:

$$F = \frac{(SSE_R - SSE_F)/(K - L)}{MSE_F} = \frac{(181,176 - 66,414)/2}{3321} = 17.3$$

using the reduced model regression results in Figure 7.8 to get SSE_R and the full model results in Figure 7.7 to get SSE_F.

The decision rule is

Reject H_0 if $F > 3.49$

Do not reject H_0 if $F \le 3.49$

where 3.49 is the 5% F critical value with 2 numerator and 20 denominator degrees of freedom.

The null hypothesis is rejected because $17.3 > 3.49$. Thus, at least one of the coefficients of the indicator variables is not zero. This means that there are differences in average sales levels between the three regions in which Meddicorp does business.

When using indicator variables, the partial F statistic is used to test whether the variables are important as a group. The t test on individual coefficients should not be used to decide whether individual indicator variables should be retained or dropped from the equation (except when there are two groups represented and therefore only one indicator variable). The indicator variables are designed to have a particular meaning as a group. They are either all retained in the equation or all dropped from the equation as a group. Dropping individual indicators changes the meaning of the coefficients of the remaining indicators. In the Meddicorp example, each indicator

FIGURE 7.8
Regression of SALES on ADV and BONUS.

Variable	Coefficient	Std Dev	T Stat	P Value
Intercept	-516.444	189.876	-2.72	0.013
ADV	2.473	0.275	8.98	0.000
BONUS	1.856	0.716	2.59	0.017

Standard Error = 90.7485   R-Sq = 85.5%   R-Sq(adj) = 84.2%

Analysis of Variance

Source	DF	Sum of Squares	Mean Square	F Stat	P Value
Regression	2	1067797	533899	64.83	0.000
Error	22	181176	8235		
Total	24	1248974			

coefficient represents the difference in sales between the indicated group and the base level group (MIDWEST). If one of the other indicators is dropped, say, WEST, the remaining coefficients then represent the difference in sales between the indicated group and the new base-level group, which now becomes the MIDWEST and WEST regions combined. The interpretation of the coefficients is totally different due to the change in the base-level group. To answer the question of whether there is a difference in the intercepts for the groups involved, the indicators must be retained and tested as a group (although there is not universal acceptance of this point of view).

Recall that in the discrimination example, a t test was used to determine whether the intercepts for the males and females differed. Because there were only two groups and therefore only one indicator, the partial F test and the t test were equivalent. If more than two indicators are used, the partial F test is required.

Figure 7.9 shows the regression of SALES on ADV, BONUS, and the indicators WEST and MIDWEST. The base-level group is now the SOUTH region. The coefficients of the indicator variables measure the difference in sales between the base-level group and the indicated group. Although the regression coefficient estimates have different values from the previous regression (see Figure 7.7), the results are the same. For example, consider the value of $\hat{y}$ for ADV = 500 and BONUS = 250 for the SOUTH region:

$$\hat{y} = 177 + 1.37(500) + 0.975(250) = 1105.8$$

This is the same value determined from the regression using the WEST and SOUTH indicator variables. Comparisons for the other regions also show the same values regardless of which set of indicators is used. Any one of the three indicators can be omitted, and the omitted group simply serves as the base-level group. The values of the remaining indicator coefficients equal the difference between the indicated group and the chosen base-level group.

FIGURE 7.9
Regression of SALES on ADV, BONUS, and the Indicator Variables WEST and MIDWEST.

Variable	Coefficient	Std Dev	T Stat	P Value
Intercept	177.207	170.116	1.04	0.310
ADV	1.368	0.262	5.22	0.000
BONUS	0.975	0.481	2.03	0.056
WEST	48.146	32.801	1.47	0.158
MIDWEST	257.892	48.413	5.33	0.000

Standard Error = 57.6254   R-Sq = 94.7%   R-Sq(adj) = 93.6%

Analysis of Variance

Source	DF	Sum of Squares	Mean Square	F Stat	P Value
Regression	4	1182560	295640	89.03	0.000
Error	20	66414	3321		
Total	24	1248974			

EXERCISES

1. **Discrimination.** Data for the following variables for 93 employees of Harris Bank Chicago in 1977 are available in a file named HARRIS7 on the CD:

y = beginning salaries in dollars (SALARY)
x_1 = years of schooling at the time of hire (EDUCAT)
x_2 = number of months of previous work experience (EXPER)
x_3 = number of months after January 1, 1969, that the individual was hired (MONTHS)
x_4 = indicator variable coded 1 for males and 0 for females (MALES)

(*Source*: These data were obtained from D. Schafer, "Measurement-Error Diagnostics and the Sex Discrimination Problem," *Journal of Business and Economic Statistics.* Copyright 1987 by the American Statistical Association. Used with permission. All rights reserved).

The results for the regression of y on all four explanatory variables are shown in Figure 7.10. In this example, we are still concerned with whether there is evidence of discrimination, but we are now taking into account two other potentially important variables besides education. Use the results to answer the following questions:

a. Conduct the F test for the overall fit of the regression. State the hypotheses to be tested,
the decision rule, the test statistic, and your decision. Use a 5% level of significance.

b. What conclusion can be drawn from the test result in part a?

c. Is there a difference in salaries, on average, for male and female workers after accounting for the effects of the three other explanatory variables? Use a 5% level of significance to answer this question. State the hypotheses to be tested, the decision rule, the test statistic, and your decision.

d. Is there evidence that Harris Bank discriminated against female employees?

e. What salary would you forecast, on average, for males with 12 years education, 10 years of experience, and with time hired equal to 15? A point forecast is sufficient. What salary would you forecast, on average, for females if all other factors are equal?

2. **Graduation Rate.** The following data for 195 colleges were obtained from *Kiplinger's Personal Finance* and are available in a file named COLLEGE7 on the CD:

y = the percentage of students who earned a bachelor's degree in four years (GRADRATE4)

FIGURE 7.10
Regression Results for Discrimination Exercise.

Variable	Coefficient	Std Dev	T Stat	P Value
Intercept	3526.422	327.725	10.76	0.000
EDUCAT	90.020	24.694	3.65	0.000
EXPER	1.269	0.588	2.16	0.034
MONTHS	23.406	5.201	4.50	0.000
MALES	722.461	117.822	6.13	0.000

Standard Error = 507.422   R-Sq = 51.1%   R-Sq(adj) = 48.9%

Analysis of Variance

Source	DF	Sum of Squares	Mean Square	F Stat	P Value
Regression	4	23665351	5916338	22.98	0.000
Error	88	22657939	257477		
Total	92	46323290			

FIGURE 7.11
Regression Results for Graduation Rate Exercise.

Variable	Coefficient	Std Dev	T Stat	P Value
Intercept	0.58944	0.04034	14.61	0.000
ADMISRATE	-0.35044	0.05759	-6.09	0.000
PRIVATE	0.28196	0.02399	11.75	0.000

Standard Error = 0.139754   R-Sq = 65.5%   R-Sq(adj) = 65.1%

Analysis of Variance

Source	DF	Sum of Squares	Mean Square	F Stat	P Value
Regression	2	7.1215	3.5608	182.31	0.000
Error	192	3.7500	0.0195		
Total	194	10.8715			

x_1 = admission rate expressed as a percentage (ADMISRATE)

x_2 = an indicator variable coded as 1 for private schools and 0 for public schools (PRIVATE)

(*Source*: Used by permission from the November and December 2003 issues of *Kiplinger's Personal Finance*. © 2003 The Kiplinger Washington Editors, Inc. Visit our website at www.kiplingers.com for further information.)

Results for the regression of y on x_1 and x_2 are shown in Figure 7.11. Use the results to answer the following questions:

a. Conduct the F test for the overall fit of the regression. State the hypotheses to be tested, the decision rule, the test statistic, and your decision. Use a 5% level of significance.

b. What conclusion can be drawn from the test result in part a?

c. Is there a difference in graduation rate, on average, for public and private schools after the effect of admission rate is taken into account? State the hypotheses to be tested, the decision rule, the test statistic, and your decision. Use a 5% level of significance to answer this question.

d. What graduation rate would you forecast, on average, for a private school with a 15% admissions rate? A point forecast is sufficient. What graduation rate would you forecast, on average, for a public school with a 15% admissions rate?

7.2 INTERACTION VARIABLES

Another type of variable that is used in regression is called an *interaction* variable. An interaction variable is formed as the product of two (or more) variables. To illustrate the effect of using an interaction variable, consider a regression equation with dependent variable y and independent variables x_1 and x_2. Construct the interaction variable x_1x_2, which is the product of the two explanatory variables. Now consider two possible regression models, one with the interaction term and one without it:

$$y = \beta_0 + \beta_1 x_1 + \beta_2 x_2 + e$$

and

$$y = \beta_0 + \beta_1 x_1 + \beta_2 x_2 + \beta_3 x_1 x_2 + e$$

Now determine the change in y given a one-unit change in x_1 with each of these models. For the model without the interaction term, a one-unit change in x_1 produces a change in y of β_1 units. For the model with the interaction term, rewrite the equation as

$$y = \beta_0 + (\beta_1 + \beta_3 x_2) x_1 + \beta_2 x_2 + e$$

Then a one-unit change in x_1 produces a change of $\beta_1 + \beta_3 x_2$ units in y. As shown, the change in y resulting from a one-unit change in x_1 also depends on the value of the variable x_2. If x_2 is small, smaller changes result; if x_2 is large, larger changes result. Thus, the effect of movements in x_1 cannot be judged independently of the value of x_2 (and the effect of movements in x_2 cannot be judged independently of movements in x_1).

An important application of interaction variables is in testing for differences in the slopes of two regression lines. This is done in a manner similar to the procedure to test for differences in the intercepts. Consider a regression with one quantitative explanatory variable (x_1) and one indicator variable (D):

$$y = \beta_0 + \beta_1 x_1 + \beta_2 D$$

The indicator variable D is coded as 1 if an item in the sample belongs to a certain group and as 0 if the item does not belong to the group. Now add the variable representing the interaction between x_1 and D, $x_1 D$:

$$y = \beta_0 + \beta_1 x_1 + \beta_2 D + \beta_3 x_1 D$$

This equation can be separated into two parts as follows. If the sample item is in the indicated group ($D = 1$):

$$y = \beta_0 + \beta_1 x_1 + \beta_2(1) + \beta_3 x_1(1) = (\beta_0 + \beta_2) + (\beta_1 + \beta_3) x_1$$

If the sample item is not in the indicated group ($D = 0$):

$$y = \beta_0 + \beta_1 x_1 + \beta_2(0) + \beta_3 x_1(0) = \beta_0 + \beta_1 x_1$$

Using the indicator variable allows us to examine differences in the intercepts for the two groups. Using the interaction variable allows us to examine differences in the slopes. Note that the equation for the indicated group has been rewritten with two components making up the intercept term (the original intercept, β_0, and the coefficient of the indicator variable, β_2), and two components making up the slope term (the original slope, β_1, and the coefficient of the interaction term, β_3). Two lines have been fit with different slopes and different intercepts, even though only one regression has been run. The difference in the intercepts is given by β_2; the difference in the slopes by β_3.

Not only can we determine whether there is a difference in the average value of the y variable for the two groups after adjusting for the effect of the quantitative variable, but we can also tell whether there is a difference in the slopes for the two groups. A t test for whether β_3 is equal to zero helps to determine whether the slopes differ. Also, a partial F test of the hypotheses

H_0: $\beta_2 = \beta_3 = 0$

H_a: At least one of β_2 and β_3 is different from zero

can be used to tell us whether there is any difference in the regression lines for the two groups (intercept or slope). The following example illustrates.

EXAMPLE 7.4 **Employment Discrimination (again)** In Examples 7.1 and 7.2, we concluded that there is evidence of employment discrimination at Harris Bank even after taking into account education. Now suppose the following question is considered: Does the difference in average salaries increase between the two groups as education increases? This is one question an interaction term allows us to investigate. The equation can be written as

$$SALARY = \beta_0 + \beta_1 \, EDUCAT + \beta_2 \, MALES + \beta_3 \, MSLOPE$$

where MSLOPE represents the interaction between EDUCAT and MALES. Thus,

$$MSLOPE = EDUCAT * MALES$$

The regression results for this equation are shown in Figure 7.12.

We ask the question: Is there *any* difference between the two groups (males and females)? To answer this, the following hypotheses can be tested:

$H_0: \beta_2 = \beta_3 = 0$

H_a: At least one of β_2 and β_3 is different from zero

To test these hypotheses, a partial F test should be used. The reduced model has only the variable EDUCAT. This regression is shown in Figure 7.13. The F statistic is

$$F = \frac{(38{,}460{,}756 - 29{,}054{,}426)/2}{326{,}454} = 14.41$$

Using a 5% level of significance, the decision rule is

Reject H_0 if $F > 3.15$

Do not reject H_0 if $F \le 3.15$

FIGURE 7.12
Regression Results for Discrimination Example with Interaction Variable to Represent Different Slopes.

Variable	Coefficient	Std Dev	T Stat	P Value
Intercept	4395.32	389.21	11.29	0.000
EDUCAT	62.13	31.94	1.95	0.055
MALES	-274.86	845.75	-0.32	0.746
MSLOPE	73.59	63.59	1.16	0.250

Standard Error = 571.362   R-Sq = 37.3%   R-Sq(adj) = 35.2%

Analysis of Variance

Source	DF	Sum of Squares	Mean Square	F Stat	P Value
Regression	3	17268865	5756288	17.63	0.000
Error	89	29054426	326454		
Total	92	46323290			

FIGURE 7.13
Regression Results for Reduced Model in Discrimination Example with Interaction Variable.

Variable	Coefficient	Std Dev	T Stat	P Value
Intercept	3818.56	377.44	10.12	0.000
EDUCAT	128.09	29.70	4.31	0.000

Standard Error = 650.112   R-Sq = 17.0%   R-Sq(adj) = 16.1%

Analysis of Variance

Source	DF	Sum of Squares	Mean Square	F Stat	P Value
Regression	1	7862534	7862534	18.60	0.000
Error	91	38460756	422646		
Total	92	46323290			

The critical value used is the $F(0.05; 2,60)$ value since the value for $F(0.05; 2,89)$ is not in the tables. The decision is to reject H_0. (We should already have guessed that this would be the decision since the test in Example 7.2 showed that the coefficient of the indicator variable was not zero).

The coefficients can be tested individually to see whether one or both are different from zero. Note the conflicting results when we do this. The p values (or t ratios) on each of the individual coefficients for MALES and MSLOPE suggest that the coefficients are equal to zero. But the F test just told us that at least one of the coefficients is different from zero. The reason for these conflicting results is the high correlation between MALES and MSLOPE (0.986). This is an example of the multicollinearity problem discussed in Chapter 4. Because we already have concluded that the indicator variable is important in the regression and the addition of the interaction variable does not add much (R^2 only increases by 1%), it might be best in this example to stick to the simpler model in Example 7.2.

Example 7.4 illustrates one caution in using interaction variables. If the correlation is high between interaction variables and the original variables in the regression, multicollinearity problems can result. In a regression with several variables, the number of interaction variables that could be created is very large and the likelihood of multicollinearity problems is high. Therefore, it is wise not to use interaction variables indiscriminately. There should be some good reason to suspect that two variables might be related or some specific question that can be answered by an interaction variable before this type of explanatory variable is used.

EXERCISES

3. **More on Possible Discrimination.** Suppose that legal counsel representing Harris Bank suggests that an interaction exists between education and experience and that the introduction of this term into the regression may account for the difference in average salaries. The interaction term

$$EDUCEXPR = EDUCAT * EXPER$$

is created and introduced into the regression. The regression results are in Figure 7.14. Use the results to help answer the following questions:

a. What is the adjusted R^2 for this regression? Compare this value to the adjusted R^2 for the regression without the interaction variable (see Figure 7.10). Which model appears to be the best choice based on the adjusted R^2?

b. Test to see whether the interaction term is important in this regression model. Use a 5% level of significance. State the hypotheses to be tested, the decision rule, the test statistic, and your decision.

c. Does the interaction variable seem to be important in explaining SALARY?

d. How would you respond to the suggestion that introduction of the interaction term into the regression may account for the difference in average salaries?

4. Graduation Rate (continued). Consider again the data from Exercise 7.2:

y = the percentage of students who earned a bachelor's degree in four years (GRADRATE4)

x_1 = admission rate expressed as a percentage (ADMISRATE)

x_2 = an indicator variable coded as 1 for private schools and 0 for public schools (PRIVATE)

We define the interaction variable:

$$ADSLOPE = ADMISRATE * PRIVATE$$

When ADSLOPE is included in the regression, it allows the slopes of the regression lines for public and private schools to differ. The regression model can be written

$$GRADRATE4 = \beta_0 + \beta_1 \text{ ADMISRATE} + \beta_2 \text{ PRIVATE} + \beta_3 \text{ ADSLOPE} + e$$

The regression results for this model are shown in Figure 7.15. Figure 7.16 contains the results for the following model:

$$GRADRATE4 = \beta_0 + \beta_1 \text{ ADMISRATE} + e$$

Use the outputs to help answer the following questions:

a. Is there *any* difference between the regression lines for public and private schools? State the hypotheses to be tested, the decision rule, the test statistic, and your decision. Use a 5% level of significance.

b. Is there a difference, on average, in graduation rates for public and private schools? State the hypotheses to be tested, the decision rule, the test statistic, and your decision. Use a 5% level of significance.

c. Is there a difference between the slopes of the regression lines representing the relationship

FIGURE 7.14
Regression Results for Discrimination Exercise Using Interaction Term Between Education and Experience.

Variable	Coefficient	Std Dev	T Stat	P Value
Intercept	3006.167	490.660	6.13	0.000
EDUCAT	134.470	39.814	3.38	0.001
EXPER	5.679	3.164	1.79	0.076
MONTHS	22.421	5.218	4.30	0.000
MALES	687.630	119.697	5.74	0.000
EDUCEXPR	-0.364	0.257	-1.42	0.160

Standard Error = 504.530   R-Sq = 52.2%   R-Sq(adj) = 49.4%

Analysis of Variance

Source	DF	Sum of Squares	Mean Square	F Stat	P Value
Regression	5	24177362	4835472	19.00	0.000
Error	87	22145928	254551		
Total	92	46323290			

FIGURE 7.15
Regression Results for Graduation Rate Interaction Variable Exercise.

Variable	Coefficient	Std Dev	T Stat	P Value
Intercept	0.63554	0.06200	10.25	0.000
ADMISRATE	-0.42084	0.09213	-4.57	0.000
PRIVATE	0.21666	0.07088	3.06	0.003
ADSLOPE	0.11560	0.11800	0.98	0.329

Standard Error = 0.139769   R-Sq = 65.7%   R-Sq(adj) = 65.1%

Analysis of Variance

Source	DF	Sum of Squares	Mean Square	F Stat	P Value
Regression	3	7.1403	2.3801	121.83	0.000
Error	191	3.7313	0.0195		
Total	194	10.8715			

FIGURE 7.16
Regression Results for Reduced Model for Graduation Rate Interaction Variable Exercise.

Variable	Coefficient	Std Dev	T Stat	P Value
Intercept	0.93429	0.03620	25.81	0.000
ADMISRATE	-0.72332	0.06285	-11.51	0.000

Standard Error = 0.182775   R-Sq = 40.7%   R-Sq(adj) = 40.4%

Analysis of Variance

Source	DF	Sum of Squares	Mean Square	F Stat	P Value
Regression	1	4.4241	4.4241	132.43	0.000
Error	193	6.4475	0.0334		
Total	194	10.8715			

between GRADRATE and ADMISRATE for public and private schools? State the hypotheses to be tested, the decision rule, the test statistic, and your decision. Use a 5% level of significance.

7.3 SEASONAL EFFECTS IN TIME-SERIES REGRESSION

Seasonal effects are fairly regular patterns of movement in a time series, repeating within a 1-year period. For example, sales of swimsuits are expected to be higher in spring and summer months and lower in fall and winter. Although the influence of seasonal effects is not expected to be exactly the same every year, the same general pattern is expected to persist.

Seasonal patterns can be modeled in a regression equation by using indicator variables, which can be created to indicate the time period to which each observation belongs. For example, if quarterly data are being analyzed, the following indicator variables could be created:

$Q1$ = 1 if the observation is from the first quarter of any year
      = 0 otherwise
$Q2$ = 1 if the observation is from the second quarter of any year
      = 0 otherwise
$Q3$ = 1 if the observation is from the third quarter of any year
      = 0 otherwise
$Q4$ = 1 if the observation is from the fourth quarter of any year
      = 0 otherwise

Four indicator variables have been created to denote the quarter to which each observation belongs, but only three of the variables should be included in the regression equation. The excluded quarter simply serves as a base-level quarter from which changes in the seasonal levels are measured. The interpretation of the coefficients of the indicator variables is the same as for any set of indicator variables. To illustrate, consider the following estimated regression equation with only quarterly indicator variables:

$$\hat{y} = b_0 + b_1 Q1 + b_2 Q2 + b_3 Q3$$

Note that the fourth quarter serves as the base-level quarter. Predictions of observations in the fourth quarter would be computed using the equation

$$\hat{y} = b_0$$

because $Q1 = Q2 = Q3 = 0$. This point has been located on the graph in Figure 7.17.

FIGURE 7.17 Graph Showing Placement of Seasonal Levels and Interpretation of Seasonal Indicator Coefficients.

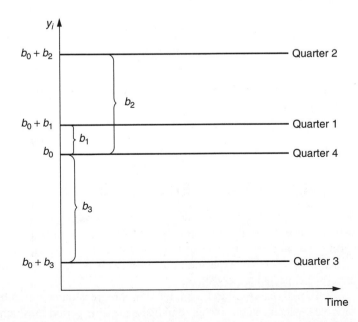

Predictions of observations in the first quarter would be computed using the equation

$$\hat{y} = b_0 + b_1$$

because $Q1 = 1$ and $Q2 = Q3 = 0$.

Similarly, predictions of observations in the second quarter would be computed using the equation

$$\hat{y} = b_0 + b_2$$

and in the third quarter by

$$\hat{y} = b_0 + b_3$$

The coefficients b_1, b_2, and b_3 represent the differences between the fitted values for the indicated quarter (first, second, and third, respectively) and the base-level quarter. The graph in Figure 7.17 has been constructed with the assumption that b_1 and b_2 are positive and b_3 is negative. The lines shown for each quarter have a zero slope because no term has been included in the regression except the indicator variables. Differences in the overall level of the series in different quarters are shown by the quarterly indicators. The regression coefficients estimate the differences in the mean levels of y in the indicated quarters. Obviously, variables could be used in the regression in addition to the seasonal indicator variables. This is illustrated in the following example.

EXAMPLE 7.5   **ABX Company Sales**   Consider again the ABX Company from Example 3.11 (ABXSALES7 on the CD). The ABX Company sells winter sports merchandise including skis, ice skates, sleds, and so on. Quarterly sales in thousands of dollars for the ABX Company are shown in Table 7.3. The time period represented starts in the first quarter of 1994 and ends in the fourth quarter of 2003.

A time-series plot of sales is shown in Figure 7.18. In Chapter 3, it was decided that a strong linear trend in sales appeared in this plot. The regression with the linear trend variable was estimated, and the results are shown again in Figure 7.19. The residual plot of the standardized residuals versus the fitted values is shown in Figure 7.20. From this plot, no obvious violations of assumptions can be observed.

A time-series plot of the standardized residuals is shown in Figure 7.21. Here a clear pattern emerges. The residuals tend to be higher in the first and fourth quarters and lower in the second and third quarters. This is not unexpected, because the ABX Company sells winter sports merchandise. This pattern is the result of seasonal variation in the data that has not been accounted for in the regression. Note that the pattern can also be seen in the time-series plot of the original data in Figure 7.18. Again, the residual plot emphasizes the pattern and points out some systematic variation in the data that we should try to model with our regression.

Figure 7.22 shows the regression of sales on the linear trend variable and indicator variables designed to indicate the first, second, and third quarters. The

TABLE 7.3 Data for ABX Company Sales Example Showing Seasonal Indicator

SALES	TREND	QUARTER	Q1	Q2	Q3	Q4
221.0	1	1	1	0	0	0
203.5	2	2	0	1	0	0
190.0	3	3	0	0	1	0
225.5	4	4	0	0	0	1
223.0	5	1	1	0	0	0
190.0	6	2	0	1	0	0
206.0	7	3	0	0	1	0
226.5	8	4	0	0	0	1
236.0	9	1	1	0	0	0
214.0	10	2	0	1	0	0
210.5	11	3	0	0	1	0
237.0	12	4	0	0	0	1
245.5	13	1	1	0	0	0
201.0	14	2	0	1	0	0
230.0	15	3	0	0	1	0
254.5	16	4	0	0	0	1
257.0	17	1	1	0	0	0
238.0	18	2	0	1	0	0
228.0	19	3	0	0	1	0
255.0	20	4	0	0	0	1
260.5	21	1	1	0	0	0
244.0	22	2	0	1	0	0
256.0	23	3	0	0	1	0
276.5	24	4	0	0	0	1
291.0	25	1	1	0	0	0
255.5	26	2	0	1	0	0
244.0	27	3	0	0	1	0
291.0	28	4	0	0	0	1
296.0	29	1	1	0	0	0
260.0	30	2	0	1	0	0
271.5	31	3	0	0	1	0
299.5	32	4	0	0	0	1
297.0	33	1	1	0	0	0
271.0	34	2	0	1	0	0
270.0	35	3	0	0	1	0
300.0	36	4	0	0	0	1
306.5	37	1	1	0	0	0
283.5	38	2	0	1	0	0
283.5	39	3	0	0	1	0
307.5	40	4	0	0	0	1

original sales data, the trend variable, a variable representing the number of the quarter (1 through 4) labeled QUARTER, and four indicator variables, labeled $Q1$, $Q2$, $Q3$, and $Q4$ are shown in Table 7.3. Remember that only three of the four indicator variables should be used in the regression. (Also, note that the variable QUARTER is not used in the regression; it is merely in the table to indicate the number of each quarter.)

FIGURE 7.18 Time-Series Plot of ABX Sales.

FIGURE 7.19 Regression of Sales on Linear Trend Variable for ABX Sales Example.

Variable	Coefficient	Std Dev	T Stat	P Value
Constant	199.017	5.128	38.81	0.000
TREND	2.556	0.218	11.73	0.000

Standard Error = 15.9126   R-Sq = 78.3%   R-Sq(adj) = 77.8%

Analysis of Variance

Source	DF	Sum of Squares	Mean Square	F Stat	P Value
Regression	1	34818	34818	137.50	0.000
Error	38	9622	253		
Total	39	44440			

FIGURE 7.20 Plot of Standardized Residuals versus Fitted Values for ABX Sales Example.

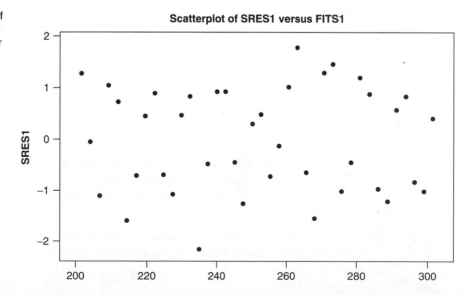

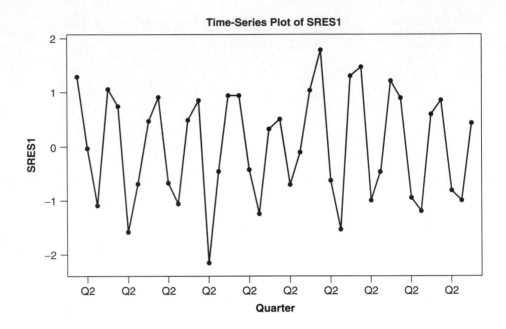

FIGURE 7.21
Time-Series Plot of Standardized Residuals for ABX Sales Example.

FIGURE 7.22
Regression of Sales on Linear Trend Variable and Quarterly Indicator Variables for ABX Sales Example.

Variable	Coefficient	Std Dev	T Stat	P Value
Intercept	210.846	3.148	66.98	0.000
TREND	2.566	0.099	25.93	0.000
Q1	3.748	3.229	1.16	0.254
Q2	-26.118	3.222	-8.11	0.000
Q3	-25.784	3.217	-8.01	0.000

Standard Error = 7.19028   R-Sq = 95.9%   R-Sq(adj) = 95.5%

Analysis of Variance

Source	DF	Sum of Squares	Mean Square	F Stat	P Value
Regression	4	42630	10658	206.14	0.000
Error	35	1810	52		
Total	39	44440			

The residual plot of the standardized residuals versus the fitted values for the regression with the quarterly indicators is shown in Figure 7.23. A time-series plot of the standardized residuals is shown in Figure 7.24. No further violations of any assumptions appear in either of these plots.

An important question to be asked in seasonal time-series models is, "Are there seasonal differences in the level of the dependent variable?" Another way of asking

FIGURE 7.23 Plot of Standardized Residuals versus Fitted Values for ABX Sales Example with Quarterly Indicator Variables.

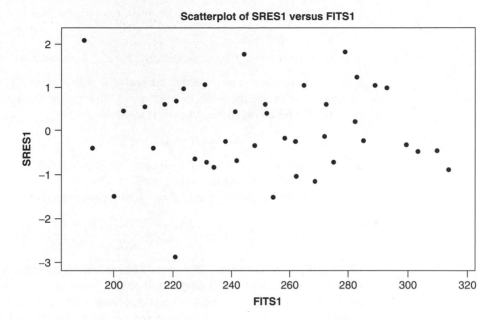

FIGURE 7.24 Time-Series Plot of Standardized Residuals for ABX Sales Example with Quarterly Indicator Variables.

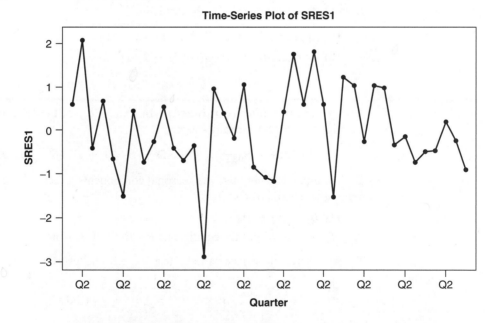

this question is, "Are the seasonal indicator variables necessary in the model?" The partial F test discussed in Chapter 4 can be used to answer the questions.

Consider the following model with explanatory variable x and quarterly seasonal indicators $Q1$, $Q2$, and $Q3$:

$$y = \beta_0 + \beta_1 x + \beta_2 Q1 + \beta_3 Q2 + \beta_4 Q3 + e$$

This is referred to as the full model. The hypotheses to be tested are

H_0: $\beta_2 = \beta_3 = \beta_4 = 0$

H_a: At least one of the coefficients β_2, β_3, and β_4 is not zero

If the null hypothesis is not rejected, the seasonal indicator variables add nothing to the model and can be removed. In other words, there are no seasonal differences. In this case, the following reduced model is adopted:

$$y = \beta_0 + \beta_1 x + e$$

If the null hypothesis is rejected, then there are seasonal differences, and the quarterly indicators should remain in the model.

To conduct the test, the partial F test statistic is computed as

$$F = \frac{(SSE_R - SSE_F)/(K - L)}{MSE_F}$$

where SSE_R is the error sum of squares from the reduced model, SSE_F is the error sum of squares from the full model, and MSE_F is the mean square error from the full model. Because the hypothesis test determines whether three coefficients are equal to zero, a divisor of 3 is used in the numerator in place of $K - L$ ($K = 4$, $L = 1$). The decision rule for the test is

Reject H_0 if $F > F(\alpha; K - L, n - K - 1)$

Do not reject H_0 if $F \leq F(\alpha; K - L, n - K - 1)$

EXAMPLE 7.6　**ABX Company Sales (continued)**　In Example 7.5, the following model was examined for ABX Company sales:

$$\text{SALES} = \beta_0 + \beta_1 \, \text{TREND} + \beta_2 \, Q1 + \beta_3 \, Q2 + \beta_4 \, Q3 + e$$

To determine whether there are seasonal components affecting sales, the following hypotheses should be tested:

H_0: $\beta_2 = \beta_3 = \beta_4 = 0$

H_a: At least one of the coefficients β_2, β_3, and β_4 is not zero

To test this hypothesis, the partial F test is used. The test statistic is

$$F = \frac{(9622 - 1810)/3}{52} = 50.1$$

Note that the reduced model regression is in Figure 7.19, and the full model regression is in Figure 7.22. Using a 5% level of significance, the decision rule for the test is

Reject H_0 if $F > F(0.05; 3.35) = 2.92$ (approximately)

Do not reject H_0 if $F \leq 2.92$

The null hypothesis is rejected. Thus, the conclusion is that seasonal components do affect sales and should be taken into account, as was done in Example 7.5.

When computing forecasts with seasonal models, the coefficients of the seasonal indicators are used to adjust the level of the forecast in the appropriate time periods. The quarterly forecasts for the year 2004 are as follows, using the seasonal model with trend:

Time Period	Point Forecast
2004 $Q1$	$211 + 2.57(41) + 3.75 = 320.12$
2004 $Q2$	$211 + 2.57(42) - 26.1 = 292.84$
2004 $Q3$	$211 + 2.57(43) - 25.8 = 295.71$
2004 $Q4$	$211 + 2.57(44) = 324.08$

Throughout this section, the use of quarterly indicator variables has been discussed. If monthly instead of quarterly data are used, the applications are similar. Instead of four quarterly indicators, twelve monthly indicators are created. Eleven of the twelve monthly indicators are used in the regression with the estimated coefficients interpreted in a manner similar to those of the quarterly coefficients. Tests for seasonal variation involve the set of eleven indicator variable coefficients. Since the use of monthly indicators is so similar to quarterly indicators, the demonstration of their use is reserved for the exercises.

EXERCISES

5. **Furniture Sales.** The file named FURNSALE7 on the CD contains data on monthly furniture sales (in millions of dollars) for the United States from January 1992 through December 2002. These data were obtained from the web site www.economagic.com. Figure 7.25 shows the time-series plot of the data. An extrapolative model to forecast furniture sales for each month in 2003 was developed. Figure 7.26 shows the regression results for the model. The model can be written as follows:

$$SALES = \beta_0 + \beta_1 TREND + \beta_2 LAGSALES + \beta_3 JAN + \beta_4 FEB + \beta_5 MARCH + \beta_6 APRIL + \beta_7 MAY + \beta_8 JUNE + \beta_9 JULY + \beta_{10} AUG + \beta_{11} SEPT + \beta_{12} OCT + \beta_{13} NOV + e$$

FIGURE 7.25
Time-Series Plot of Furniture Sales.

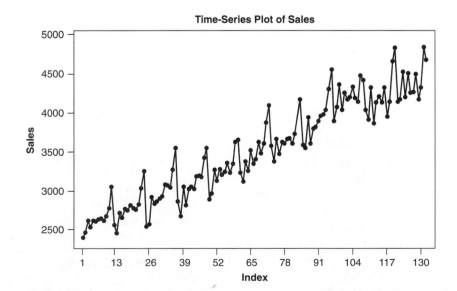

FIGURE 7.26
Regression Results with Seasonal Indicators for Furniture Sales Exercise.

Variable	Coefficient	Std Dev	T Stat	P Value
Intercept	1552.151	228.680	6.79	0.000
TREND	7.903	1.298	6.09	0.000
LAGSALES	0.484	0.083	5.84	0.000
JAN	-643.717	42.576	-15.12	0.000
FEB	-404.481	53.915	-7.50	0.000
MARCH	-80.369	56.921	-1.41	0.161
APRIL	-457.714	42.852	-10.68	0.000
MAY	-178.202	53.127	-3.35	0.001
JUNE	-316.796	45.155	-7.02	0.000
JULY	-278.410	47.612	-5.85	0.000
AUG	-183.319	47.143	-3.89	0.000
SEPT	-357.504	43.251	-8.27	0.000
OCT	-239.408	48.319	-4.95	0.000
NOV	6.826	45.791	0.15	0.882

Standard Error = 93.2329   R-Sq = 98.0%   R-Sq(adj) = 97.8%

Analysis of Variance

Source	DF	Sum of Squares	Mean Square	F Stat	P Value
Regression	13	49839825	3833833	441.06	0.000
Error	117	1017007	8692		
Total	130	50856832			

FIGURE 7.27
Regression Results with Seasonal Indicators for Furniture Sales Exercise.

Variable	Coefficient	Std Dev	T Stat	P Value
Intercept	1977.945	215.778	9.17	0.000
TREND	12.483	1.427	8.75	0.000
LAGSALES	0.201	0.087	2.32	0.022

Standard Error = 195.734   R-Sq = 90.4%   R-Sq(adj) = 90.2%

Analysis of Variance

Source	DF	Sum of Squares	Mean Square	F Stat	P Value
Regression	2	45952938	22976469	599.72	0.000
Error	128	4903894	38312		
Total	130	50856832			

TREND is a linear trend variable, LAGSALES is SALES lagged one period, and JAN through NOV are 11 monthly seasonal indicators. Note that December has been used as the base-level month.

Figure 7.27 shows the regression results for the model without the seasonal indicators.

Use the regression results to help answer the following questions.

a. Is there seasonal variation in furniture sales? State the hypotheses to be tested, your decision rule, the test statistic, decision, and conclusion. Use a 5% level of significance.

b. Are the TREND and LAGSALES variables important to the equation explaining furniture sales? Test the coefficient of each of these variables individually to answer this question. State the hypotheses to be tested, your decision rule, the test statistic, decision, and conclusion. Use a 5% level of significance.

c. Based on the test result in part a, use either the full or reduced model (with or without seasonal indicators) to develop a forecast of furniture sales for each month in 2003.

ADDITIONAL EXERCISES

6. Absenteeism. Data on 77 employees of the ABX
Company have been collected. The dependent
variable is absenteeism (ABSENT). The possible
explanatory variables are

COMPLX = measure of job complexity
SENIOR = seniority
SATIS = response to "How satisfied are
you with your foreman?"

These data are available in a file named
ABSENT7 on the CD. In this exercise, use
SENINV = 1/SENIOR, which is the reciprocal of
the seniority variable, and COMPLX as two of the
explanatory variables. The variable SATIS should
be transformed into indicator variables as follows:

FS1 = 1 if SATIS = 1 (very dissatisfied)
    = 0 otherwise
FS2 = 1 if SATIS = 2 (somewhat dissatisfied)
    = 0 otherwise
FS3 = 1 if SATIS = 3 (neither satisfied nor
    dissatisfied)
    = 0 otherwise
FS4 = 1 if SATIS = 4 (somewhat satisfied)
    = 0 otherwise
FS5 = 1 if SATIS = 5 (very satisfied)
    = 0 otherwise

Five indicator variables are created to represent all
five supervisor satisfaction categories. Recall that
only four need to be used in the regression. Run
the regression with the explanatory variables
described here. Answer the following questions:

a. Is there a difference in average absenteeism for
employees in different supervisor satisfaction
groups? Perform a hypothesis test to answer
this question. State the hypotheses to be tested,
the decision rule, the test statistic, and your
decision. Use a 5% level of significance.

b. Using the model chosen in part a (and keeping
the variables COMPLX and SENINV in the
model), what would be your estimate of the
average absenteeism rate for all employees
with COMPLX = 60 and SENIOR = 30
who were very dissatisfied with their supervisor? What if they were very satisfied with
their supervisor, but COMPLX and SENIOR
were the same values?

c. How do you account for the differences in the
estimates in part b?

d. How could this equation be used to help
identify employees who might be prone to
absenteeism?

7. Work-Order Closing. Management at the
Texas Christian University (TCU) Physical
Plant is interested in reducing the average time
to completion of routine work orders. The time
to completion is defined as the difference between the date of receipt of a work order and
the date closing information is entered. The
number of labor hours charged to each work
order and the cost of materials are two variables
believed to be related to the time to completion
of the work order. Management wants to know
if there is any difference in the time to completion of work orders, on average, for different
types of buildings. Buildings are classified into
four types on the TCU campus: residence halls,
athletic, academic, and administrative. In answering the question, take into account the
possible effect of labor hours charged and materials cost. The data for a random sample of 72
work orders (chosen from a population of
11,720) are available in a file named
WORKORD7 on the CD. The variables are as
follows:

y = DAYS   = number of days to complete each work order
x_1 = HOURS   = number of hours of labor
charged to each work
order
x_2 = MATERIAL = cost of materials charged
to each work order
x_3 = BUILDING = 1 for residence halls
2 for athletic buildings
3 for academic buildings
4 for administrative
buildings

8. Beer Production. The file named BEER7 on the
CD contains monthly U.S. beer production in
millions of barrels for January 1982 through
December 1991. Develop an extrapolative model
for these data and use it to examine whether there
is seasonal variation in beer production and
whether beer production seems to be increasing,
decreasing, or staying fairly constant over this
time period. Use the model you select as best for

beer production to forecast monthly production for each month in 1992.

(*Source*: Data were obtained from *Business Statistics 1963–91* and *Survey of Current Business*.)

9. **Monthly Temperatures.** Lone Star Gas recognizes that one of the simplest and most effective ways to forecast natural gas use is with average monthly temperatures. A model can be developed that relates gas usage to average temperature, and then forecasts can be made based on forecasts of average temperatures in the area of interest. Lone Star first needs a model to forecast average temperatures in the Dallas–Fort Worth area. The file named TEMPDFW7 on the CD contains the average monthly temperatures for January 1978 to December 2002 for the Dallas–Fort Worth area. Develop an extrapolative model for these data and use the model to forecast average monthly temperatures for each month in the years 2003 and 2004. Data were obtained from the web site www.srh.noaa.gov/fwd/clmdfw.html.

10. **BigTex Services.** BigTex Services is undergoing scrutiny for a possible wage discrimination suit. As consulting statistician hired by the corporate lawyer, you are to examine data on BigTex employees to further investigate the charges. The following data are available in a file named BIG-TEX7 on the CD:

> monthly salary for each employee (SALARY)
> years with the company (YEARS)
> position with the company (POSITION)
>   coded as
>
>   1 = manual labor
>   2 = secretary
>   3 = lab technician
>   4 = chemist
>   5 = management
> amount of education completed (EDUCAT)
>   coded as
>   1 = high school degree
>   2 = some college
>   3 = college degree
>   4 = graduate degree
> gender (GENDER) coded as
>   0 = female
>   1 = male

What would you conclude from the data? Should BigTex Services be worried about possible wage discrimination charges? Why or why not?

11. **FOC Sales.** Techcore is a high-tech company located in Fort Worth, Texas. The company produces a part called a fibre-optic connector (FOC) and wants to generate reasonably accurate but simple forecasts of the sales of FOC's over time. The company has weekly sales data for 265 weeks starting in January of 1995. The weekly sales data are in a file named FOC7 on the CD. You are to build an extrapolative regression model to forecast FOC sales. Do you detect any "seasonal" patterns where the use of indicator variables might be useful?

12. **Fort Worth Crime.** In the early 1990s, a concerted effort was put into place to revitalize downtown Fort Worth, Texas. This was a combined effort involving both the city and private investors. The city police and private security coordinated their communications networks. New entertainment and dining establishments were located in the area. Buildings were refurbished and remodeled. Prior to this time, the downtown area had experienced high crime rates. Because of police and private security efforts beginning in the summer of 1992, there is interest in determining whether these interventions had a significant effect in decreasing crime in downtown Fort Worth. You are to use an extrapolative regression model to examine the data and to produce a report for the city of Fort Worth outlining your findings regarding any changes in the downtown crime rate. The report should be written in two parts: an executive summary outlining your findings and a technical report to support those findings.

The data are in a file named FWCRIME7 on the CD. The variables included are YR/MONTH, which gives the year and month of the observation and the number of monthly occurrences of crimes in the categories shown: MURDER, RAPE, ROBBERY, ASSAULT, BURGLARY, LARCENY and AUTOTHEFT.

13. **Fund Returns.** An investment company has two different strategies that it follows resulting in hedge funds that will be called LONG/SHORT and EQUITY. The equity fund mirrors the S&P 500 whereas the LONG/SHORT strategy is intended to provide returns that are relatively uncorrelated with the market. Daily returns for these two strategies are available from January 3, 2002, through July 16, 2003. On November 1, 2002, a

change in the investment strategies is made for both of these funds. Management wants to know if the beta coefficients of these two funds changed as a result. The beta coefficient is the slope of the regression line relating the fund return to the return on the market, in this case, the S&P 500. The beta coefficient represents the systematic risk of the fund, which means that the beta indicates the amount of change in the fund return, on average, that will occur when the market return changes by 1%. If the beta coefficient is 2 then a 1% increase (decrease) in the market return will be mirrored, on average, by a 2% increase (decrease) in the fund return. The file FUND7 on the CD contains the following variables:

RET1, daily returns for the EQUITY strategy
RET2, daily returns for the LONG/SHORT
  strategy
SP500, daily returns for the S&P 500

Use the data to estimate the beta coefficient for the two funds and to determine whether the change in investment strategy resulted in a shift in the beta coefficient. (These data have been disguised at the request of the company to provide confidentiality.)

14. Productivity. A company that owns and operates 12 hospitals is interested in developing a measure of staff productivity. It would like to use a measure called EEOB, "equivalent employees per occupied bed." EEOB represents how many full-time employee equivalents (40 hours per week) are used for each patient. Corporate headquarters is set on using EEOB to measure productivity. It knows that average length of stay (ALOS) has an inverse impact on EEOB. In the last two years, the length of stay in hospitals has decreased drastically. As a result, headquarters believes that it can expect the EEOB to increase. Another factor that may influence productivity is outpatient factor (OUTFAC), which takes into account time spent on various outpatient processes.

You have been hired to investigate the relationships between EEOB and the two explanatory variables ALOS and OUTFAC. You have monthly data from January 2002 through May 2003 for 12 different hospitals. The variable labeled HOSP numbers each hospital 1 through 12. The relationships may differ between hospitals. You might want to use indicator variables in your regression to allow for these differences.

Examine the relationships using the data in the file HOSP7 on the CD. What conclusions can you make based on the results? Be sure to examine residual plots. Do the assumptions of the regression appear satisfied? Based on the residual plots, what questions might you have about the data? (These data have been disguised at the request of the company to provide confidentiality.)

15. Predicting Movie Grosses. The file named MOVIES7 on the CD contains data on movies released in the United States during the calendar year 1998. The variables in this file are

TDOMGROSS	the total domestic gross revenue (dependent variable)
BACTOR	the number of "best actors" in the movie; to qualify as a best actor, the person must be listed in *Entertainment Weekly*'s lists of the 25 Best Actors and 25 Best Actresses of the 1990s
TDACTOR	the number of actors or actresses appearing in the movie who were among the top 20 actors and top 20 actresses in average box office gross per movie in their careers ("top dollar actors"), according to The Movie Times web site at the beginning of the 1998 movie season
GENRE	classification of movie types coded as 1 = action, 2 = drama, 3 = children's, 4 = comedy, 5 = documentary, 6 = thriller, 7 = horror, 8 = science fiction
MPAA	MPAA rating coded as 1 = G, 2 = PG, 3 = PG13, 4 = R, 5 = NC17, 6 = unrated
COUNTRY	where the move was made coded as 1 = USA, 2 = English-speaking country other than USA, 3 = non-English speaking country
CHRISTMAS	coded as 1 if the movie was to be released during the

HOLIDAY — Christmas season, 0 otherwise

HOLIDAY — coded as 1 if the movie was released before any holiday weekend (President's Day, Memorial Day, Independence Day, Labor Day, Thanksgiving, or the Christmas season), 0 otherwise

SUMMER — coded as 1 if the movie was released during the summer (Memorial Day through Labor Day), 0 otherwise

SEQUEL — coded as 1 if the movie was a sequel, 0 otherwise

We would like to find an equation to predict the total domestic gross revenue of movies based on variables available prior to the release of the films. All of the variables in the data file are known or could be determined prior to release. Use regression to examine the relationships and determine what factors might be useful in helping to predict gross revenues. Use the log of TDOMGROSS as the dependent variable. Some of the explanatory variables are qualitative in nature and will need to be transformed into indicators. What relationships do you find that would be helpful in predicting revenues? What variables are not important? What recommendations would you make on the basis of your results?

These data are discussed in the article "Predicting Movie Grosses: Winners and Losers, Blockbusters and Sleepers," by Jeffrey S. Simonoff and Ilana R. Sparrow, *Chance*, Vol. 13, 2000, 15–24 and were obtained from Dr. Simonoff's web site.

16. **Rangers' Attendance.** The Texas Rangers major league baseball team needs your help. The team is interested in determining key identifiable variables that affect attendance at games played at the Ballpark in Arlington. It has provided you with an extensive data set, including the following variables:

DATE: — The date of the game played

WEEKDAY: — The day of the week, coded as 1 = Monday, 2 = Tuesday,..., 7 = Sunday

PROMOTION: — Equal to 1 if there was a special promotion (Dollar Decker Dog Night, Half Price Group Night, and so on); equal to 0 if there was no special promotion

WINS: — The cumulative number of wins for the Rangers prior to playing the game

AHEAD: — Equal to 1 if the Rangers were ahead in their division when the game was played; equal to zero otherwise

ATTENDANCE: — The attendance at all home games for the years 1994 through 1998; this provides a total of 375 observations

SCHOOL: — Equal to 1 if public schools were in session; equal to zero otherwise

OPPONENT: — Coded as
1 = New York
2 = Baltimore
3 = Cleveland
4 = Chicago
5 = Anaheim
6 = Minnesota
7 = Any National League team (coded this way since the teams played vary from year to year)
8 = Oakland
9 = Detroit
10 = Seattle
11 = Toronto
12 = Milwaukee
13 = Kansas City
14 = Boston

NIGHT: — Equal to 1 for a night game; 0 for a day game

These data are in a file named RANGERS7 on the CD.

Prepare a report summarizing the data and providing any information concerning what you see as factors that influence attendance. Recognize that the Rangers organization has little or no control over certain factors. It cannot, for example, schedule all the games with the Yankees, even if playing the Yankees produces higher attendance levels than any other team. Still, the team would be interested in finding out whether the opponent has an influence on attendance for planning purposes.

On the other hand, the Rangers do have control over some factors. If you find any controllable factors that might be of interest, be sure to single those out. Your report should consist of two parts: (1) an executive summary outlining your findings—any recommendations should be based on the results shown by the data; (2) a technical section including any analyses you need to support your position.

Data used courtesy of the Texas Rangers organization and Major League Baseball. Major League Baseball trademarks and copyrights are used with permission of Major League Baseball Properties, Inc.

17. **Book Cost.** A major publishing company would like to develop an equation that will help it in determining the cost of books that it publishes. It has a sample of 207 books that have been published recently. Of the 207 books in the sample, 83 are hardcover and 124 are softcover. Hardcover books obviously are priced at a premium, so some adjustment for this will need to be made. The variables in the data file named BOOKCOST7 on the CD are as follows:

COST	= the cost of producing the book
PAGES	= the number of pages in the book
SOFTCOVER	= 1 if the book is softcover and 0 if it is hardcover

Develop an equation that will help the publisher to predict cost for books to be published in the future.

USING THE COMPUTER

The Using the Computer section in each chapter describes how to perform the computer analyses in the chapter using Excel, MINITAB, and SAS. For further detail on Excel, MINITAB, and SAS, see Appendix C.

EXCEL

Creating and Using Indicator Variables

Indicator variables have values of either 0 or 1. Obviously, one way to create an indicator variable (or variables) is simply to type it into a data set. Logical if statements can be used to simplify the creation of indicators in some cases.

Consider the data in Table 7.2 for the Meddicorp example. The variable REGION was typed into the original data set. This variable numbers the three regions as 1, 2, or 3. It can be used to create the indicator variables for the three regions after they are defined as follows:

SOUTH	= 1 whenever REGION is 1 and 0 otherwise
WEST	= 1 whenever REGION is 2 and 0 otherwise
MIDWEST	= 1 whenever REGION is 3 and 0 otherwise

Suppose the REGION variable is in column D, as shown in Figure 7.28. The indicator SOUTH can be created as follows. In the first cell of the column where the SOUTH indicator is desired, type in the formula = if (d2 = 1,1,0). This formula puts 1 in the SOUTH column if the entry in cell d2 is 1; it puts 0 in the SOUTH column otherwise. Copy this formula down the SOUTH column. Now type the formula = if (d2 = 2,1,0) in the first cell of the WEST column. This formula puts 1 in the WEST column if the entry in cell d2 is 2; it puts 0 in the WEST column otherwise. Copy this formula down the WEST column. Now type the formula = if (d2 = 3,1,0) in the first cell of the MIDWEST column. This formula puts 1

FIGURE 7.28 Creating Indicator Variables in Excel.

	A	B	C	D	E	F	G
1	SALES	ADV	BONUS	REGION	SOUTH	WEST	MIDWEST
2	963.50	374.27	230.98	1	=IF(D2=1,1,0)		
3	893.00	408.50	236.28	1			
4	1057.25	414.31	271.57	1			
5	1183.25	448.42	291.20	2			
6	1419.50	517.88	282.17	3			
7	1547.75	637.60	321.16	3			
8	1580.00	635.72	294.32	3			
9	1071.50	446.86	305.69	1			
10	1078.25	489.59	238.41	1			
11	1122.50	500.56	271.38	2			
12	1304.75	484.18	332.64	3			
13	1552.25	618.07	261.80	3			
14	1040.00	453.39	235.63	1			
15	1045.25	440.86	249.68	2			

FIGURE 7.29 Creating Interaction Variables in Excel.

	A	B	C	D
1	SALES	ADV	BONUS	INTERACT
2	963.50	374.27	230.98	=B2*C2
3	893.00	408.50	236.28	
4	1057.25	414.31	271.57	
5	1183.25	448.42	291.20	
6	1419.50	517.88	282.17	
7	1547.75	637.60	321.16	
8	1580.00	635.72	294.32	
9	1071.50	446.86	305.69	
10	1078.25	489.59	238.41	
11	1122.50	500.56	271.38	
12	1304.75	484.18	332.64	
13	1552.25	618.07	261.80	
14	1040.00	453.39	235.63	
15	1045.25	440.86	249.68	

in the MIDWEST column if the entry in cell d2 is a 3; it puts 0 in the MIDWEST column otherwise. Copy this formula down the MIDWEST column. This creates the three indicator variables.

Creating and Using Interaction Variables

Interaction variables are created by multiplying one variable by another. This can be done in Excel by using formulas. Consider the example shown in Figure 7.29. To create an interaction variable between ADV and BONUS, type the formula= $b2*c2$ into the first cell of the interaction variable column. Then copy this formula down the column.

Creating and Using Seasonal Indicators for Time-Series Regression

To create quarterly seasonal indicators, follow this sequence: First, type in a variable that numbers the quarters 1 through 4 for each year. Then use the logical if statement to create the indicators. Consider the example shown in Figure 7.30. The column labeled "quarter" numbers the quarters of each year. In the column labeled Q1, type in the formula = if(c2 = 1,1,0). If cell c2 is 1, this results in 1 in the column for Q1. Otherwise, 0 is entered. Copy this formula down the column for Q1. In the column labeled Q2, type in the formula = if(c2 = 2,1,0). If cell c2 is 2, this results in 1 in the column for Q2. Otherwise, 0 is entered. Copy this formula down the column for Q2. Continue this process for each indicator variable. You can alter this process for monthly data by numbering the months from 1 to12 and then creating 12 monthly indicators.

FIGURE 7.30 Creating Quarterly Seasonal Indicator Variables in Excel.

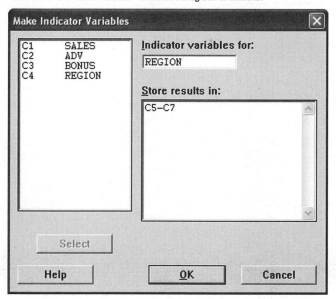

	A	B	C	D	E	F	G
1	SALES	TREND	QUARTER	Q1	Q2	Q3	Q4
2	221.00	1	1	=IF(C2=1,1,0)			
3	203.50	2	2	IF(**logical_test**, [value_if_true], [value_if_false])			
4	190.00	3	3				
5	225.50	4	4				
6	223.00	5	1				
7	190.00	6	2				
8	206.00	7	3				
9	226.50	8	4				
10	236.00	9	1				
11	214.00	10	2				
12	210.50	11	3				
13	237.00	12	4				
14	245.50	13	1				
15	201.00	14	2				

FIGURE 7.31 Make Indicator Variables Dialog Box in MINITAB.

MINITAB

Creating and Using Indicator Variables

CALC: MAKE INDICATOR VARIABLES

Indicator variables have values of either 0 or 1. Obviously, one way to create an indicator variable (or variables) is simply to type it into a data set. Make Indicator Variables on the CALC menu can be useful in creating indicator variables more quickly in some cases.

Consider the data in Table 7.2 for the Meddicorp example. The variable REGION was typed into the original data set. This variable numbers the three regions as 1, 2, or 3. It can be used to create the indicator variables for the three regions after they are defined as follows:

SOUTH = 1 whenever REGION is 1 and 0 otherwise

WEST = 1 whenever REGION is 2 and 0 otherwise

MIDWEST = 1 whenever REGION is 3 and 0 otherwise

To create the indicator variables SOUTH, WEST, and MIDWEST, click CALC and then Make Indicator Variables. The Make Indicator Variables dialog box is shown in Figure 7.31. Fill in the "Indicator variables for" box with the name of the variable to be transformed into indicators (REGION in this case). Then list the locations of the indicator variables in the "Store results in" box. Be sure to list locations for all m indicators (m being the number of categories) even though only $m - 1$ indicators are used in the regression.

Creating and Using Interaction Variables

CALC: CALCULATOR

Interaction variables are created by multiplying one variable by another. This can be done in MINITAB by using the CALCULATOR on the CALC menu. Put the

FIGURE 7.32 Using the Calculator to Create Interaction Variables in MINITAB.

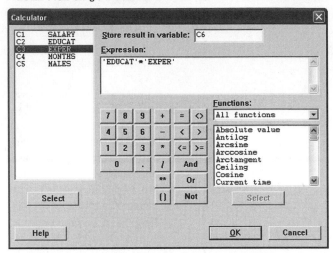

FIGURE 7.33 Step One in Creating Seasonal Indicators in MINITAB.

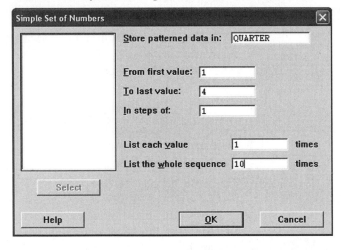

interaction variable column number in the "Store result in variable" box and put the action to be taken in the "Expression" box (see Figure 7.32). For example, put C6 in the "Store result in variable" box and 'EDUCAT'*'EXPER' in the "Expression" box and MINITAB multiplies C2 by C3 and stores the resulting interaction variable in C6. The interaction variable can then be used just like other variables.

Creating and Using Seasonal Indicators for Time-Series Regression

To create quarterly seasonal indicators, follow this sequence: First, create a variable that numbers each quarter 1 through 4: CALC: MAKE PATTERNED DATA: SIMPLE SET OF NUMBERS. In the Simple Set of Numbers dialog box, fill in a column number in the "Store patterned data in" box, put 1 in the "From first value" box, 4 in the "To last value" box, and 1s in the "In steps of" and "List each value" boxes. In the "List the whole sequence" box, put the number of years of data you have available. This creates a variable with the sequence of numbers 1,2,3,4 for each year of data, thus numbering the quarters in each year 1 through 4 (see Figure 7.33). Now click CALC: MAKE INDICATOR VARIABLES. Put the column number of the variable you just created in the "Indicator variables for" box and list four columns in the "Store results in" box. Click OK and the seasonal indicators are created. You can alter this process for monthly data by numbering the months from 1 to 12 and then creating 12 monthly indicators.

SAS

Creating and Using Indicator Variables

Indicator variables have values of either 0 or 1. Obviously, one way to create an indicator variable (or variables) is simply to type it into a data set. IF/THEN statements can be used to simplify the creation of indicator variables in some cases.

Consider the data in Table 7.2 for the Meddicorp example. The variable REGION was typed into the original data set. This variable numbers the three regions as 1, 2, or 3. It can be used to create the indicator variables for the three regions after they are defined as follows:

    SOUTH   = 1 whenever REGION is 1 and 0 otherwise
    WEST   = 1 whenever REGION is 2 and 0 otherwise
    MIDWEST = 1 whenever REGION is 3 and 0 otherwise

The three indicators can be created during the Input phase in SAS as follows:

```
INPUT SALES ADV BONUS REGION;
IF REGION = 1 THEN SOUTH = 1;
ELSE SOUTH = 0;
IF REGION = 2 THEN WEST = 1;
ELSE WEST = 0;
IF REGION = 3 THEN MIDWEST = 1;
ELSE MIDWEST = 0;
```

After creating the three indicator variables, any two of them can be used in PROC REG. For example,

```
PROC REG;
MODEL SALES = ADV BONUS SOUTH WEST;
```

runs a regression with SALES as the dependent variable and ADV, BONUS, SOUTH, and WEST as independent variables.

Creating and Using Interaction Variables

Interaction variables are created by multiplying one variable by another. In SAS, such variable transformations are performed during the data input phase. For example, the following commands create the interaction variable called EDUCEXPR, which is the product of EDUCAT and EXPER:

```
INPUT SALARY EDUCAT EXPER MONTHS MALES;
EDUCEXPR = EDUCAT*EXPER;
```

The interaction variable, EDUCEXPR, can then be used in PROC REG, PROC PLOT, and so on, just like other variables.

Creating and Using Seasonal Indicators for Time-Series Regression

Seasonal indicators are simply indicator variables, and their creation is like that of indicator variables discussed previously. The only difference is that, instead of the variable representing REGION in the example given, a variable numbering the quarters or the months in the seasonal cycle is needed.

For quarterly data, a variable numbering the quarters 1, 2, 3, or 4 is used with the following command sequence in the input phase. Call the variable that numbers the quarters QUARTER and proceed as follows:

```
IF QUARTER = 1 THEN Q1 = 1;
ELSE Q1 = 0;
IF QUARTER = 2 THEN Q2 = 1;
ELSE Q2 = 0;
IF QUARTER = 3 THEN Q3 = 1;
ELSE Q3 = 0;
IF QUARTER = 4 THEN Q4 = 1;
ELSE Q4 = 0;
```

Q1, Q2, Q3, and Q4 are the four quarterly indicator variables. Three of these four can then be used in PROC REG.

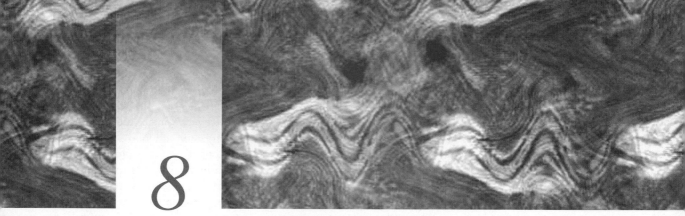

8

Variable Selection

8.1 INTRODUCTION

One of the primary tasks discussed in this text has been choosing which variables to include in the regression equation. Several hypothesis tests have been suggested to aid in this task. The F test for overall fit of the regression, the partial F test, and the t test are all designed to help decide whether certain variables should be included in the regression.

Several additional procedures, called *variable selection techniques*, also can be used to help choose which variables are important. These procedures are discussed in this chapter. The importance of choosing the correct variables is highlighted by examining what happens when either (a) important variables are omitted from the regression equation or (b) unimportant variables are included in the equation.

If an important variable is omitted from the regression, the effect of this variable is not taken into account. The estimates of the other regression coefficients become biased (systematically either too high or too low). Forecasts generated by the regression are also biased. If an unimportant variable is included in the regression equation, the standard errors of the coefficients and forecasts become inflated. Thus, forecasts and coefficient estimates are more variable than they would be with a proper choice of variables. The larger standard errors may also make results from inferences less dependable.

Whether identifying important variables to be included or unimportant variables to be deleted, variable choice is an important aspect of regression analysis. As a result, considerable work has been done in developing methods to help choose the "best" group of variables to include in the regression equation. A word of caution is in order before these techniques are introduced, however. None of these methods is guaranteed to automatically pick the best of all regression models. Otherwise, this text could have started with these procedures and stopped after one chapter.

Any regression analysis requires considerable input from the researcher performing the analysis. This person must define the problem to be solved, determine what variables might be useful in the regression equation, obtain data on these variables, and set up the data to be analyzed in a data file. During the analysis itself, the researcher must check the regression assumptions to ensure that none has been violated. The automatic methods do not do that. For example, if ln (y) rather than y is the appropriate dependent variable for analysis, the variable selection procedures cannot determine this. The functional form of the relationship must be determined separately from the use of these procedures. In fact, any correction for a violation of one of the assumptions of regression must be determined by the researcher. These methods also cannot suggest explanatory variables to add to the regression unless they have been initially determined by the researcher and included in the data set.

Finally, none of these automatic procedures has the ability to use the researcher's knowledge of the business or economic situation being analyzed. This knowledge should be used to help establish what form the regression model takes, what variables should be considered, and so on. This knowledge of the theory associated with the subject being analyzed is important. In many cases in business and economics, theoretical results are available that suggest what variables should be included in a relationship or what functional form the relationship should have. This theory will be provided throughout other courses in your business or economics major. Be sure to examine the theory in that field for suggestions that may help in building a model for your data.

The variable selection techniques discussed here are tools to aid the researcher in sifting through a number of explanatory variables to determine which ones should be included in the regression equation. With the knowledge that the techniques cannot be reliably applied without the judgment of the person researching the problem and the use of subject matter theory, the following variable selection procedures are discussed in this chapter:

1. all possible regressions (along with several criteria to choose which is the best regression)
2. backward elimination
3. forward selection
4. stepwise regression

8.2 ALL POSSIBLE REGRESSIONS

One of the variable selection techniques suggested to aid in choosing the best regression model is called *all possible regressions*. As the name suggests, the procedure is designed to run all possible regressions between the dependent variable and all possible subsets of explanatory variables. For example, if the three possible explanatory variables identified for consideration in the problem are denoted x_1, x_2, and x_3, then a total of eight possible regressions may be best. The possible regressions include the following subsets of the three explanatory variables:

1. no variables
2. x_1
3. x_2
4. x_3
5. x_1 and x_2
6. x_1 and x_3
7. x_2 and x_3
8. all three variables

The all possible regressions procedure evaluates each of these regressions and prints out summary statistics to aid in choosing which of the eight possibilities is best. The choice of the criterion to use and the final choice of a model are then up to the researcher. Commonly used criteria to help in choosing between the alternative regressions include:

1. R^2, adjusted or unadjusted, as the researcher prefers
2. C_p
3. s_e, the standard error of the regression

The R^2, R^2_{adj} and s_e have been discussed extensively in previous chapters, but the statistic C_p has not. C_p measures the total mean square error of the fitted values of the regression. The total mean square error involves two components: one resulting from random error and one resulting from bias. When there is no bias in the estimated regression model, the expected value of C_p is equal to p, which, in the notation of this text, is equal to $K + 1$, the number of coefficients to be estimated. When evaluating which regression is best, it is recommended that regressions with small C_p values and those with values near $K + 1$ be considered. If the value of C_p is large, then the mean square error of the fitted values is large, indicating either a poor fit, substantial bias in the fit, or both. If the value of C_p is much greater than $K + 1$, then there is a large bias component in the regression, usually indicating omission of an important variable.

The formula for computing C_p is

$$C_p = \frac{SSE_p}{MSE_F} - (n - 2p)$$

where SSE_p is the error sum of squares for the regression with p $(p = K + 1)$ coefficients to be estimated and MSE_F is the mean square error for the model with all possible explanatory variables included.

Although the C_p measure is highly recommended as a useful criterion in choosing between alternate regressions, keep in mind that the bias is measured with respect to the total group of variables provided by the researcher. This criterion cannot determine when the researcher has forgotten about some variable not included in the total group. In other words, the input of the researcher is still important. The all possible regressions procedure is illustrated in the following example.

EXAMPLE 8.1 **Meddicorp Revisited** Consider again Example 4.1, the study of sales in the Meddicorp Company (see the MEDDICORP8 file on the CD). The complete data set for this example is shown in Table 8.1 with descriptions of each variable as follows:

y, Meddicorp's sales (in thousands of dollars) in each territory for 2003 (SALES)

x_1, the amount (in hundreds of dollars) that Meddicorp spent on advertising in each territory in 2003 (ADV)

x_2, the total amount of bonuses paid (in hundreds of dollars) in each territory in 2003 (BONUS)

x_3, the market share currently held by Meddicorp in each territory (MKTSHR)

TABLE 8.1 Data for Meddicorp Example Territory

	SALES	ADV	BONUS	MKTSHR	COMPET	REGION
1	963.50	374.27	230.98	33.	202.22	1
2	893.00	408.50	236.28	29.	252.77	1
3	1057.25	414.31	271.57	34.	293.22	1
4	1183.25	448.42	291.20	24.	202.22	2
5	1419.50	517.88	282.17	32.	303.33	3
6	1547.75	637.60	321.16	29.	353.88	3
7	1580.00	635.72	294.32	28.	374.11	3
8	1071.50	446.86	305.69	31.	404.44	1
9	1078.25	489.59	238.41	20.	394.33	1
10	1122.50	500.56	271.38	30.	303.33	2
11	1304.75	484.18	332.64	25.	333.66	3
12	1552.25	618.07	261.80	34.	353.88	3
13	1040.00	453.39	235.63	42.	262.88	1
14	1045.25	440.86	249.68	28.	333.66	2
15	1102.25	487.79	232.99	28.	232.55	2
16	1225.25	537.67	272.20	30.	273.00	2
17	1508.00	612.21	266.64	29.	323.55	3
18	1564.25	601.46	277.44	32.	404.44	3
19	1634.75	585.10	312.35	36.	283.11	3
20	1159.25	524.56	292.87	34.	222.44	1
21	1202.75	535.17	268.27	31.	283.11	2
22	1294.25	486.03	309.85	32.	242.66	2
23	1467.50	540.17	291.03	28.	333.66	3
24	1583.75	583.85	289.29	27.	313.44	3
25	1124.75	499.15	272.55	26.	374.11	2

x_4, the largest competitor's sales (in thousands of dollars) in each territory (COMPET)

x_5, a variable coded to indicate the region in which each territory is located:

$1 = $ SOUTH, $2 = $ WEST, and $3 = $ MIDWEST (REGION)

The REGION variable was transformed to a set of three possible indicator variables—SOUTH, WEST, and MIDWEST—in Chapter 7. The interpretation of the coefficients of these variables was preferred to the single REGION variable. Recall from Chapter 7, however, that indicator variables should be treated as a group rather than individually. The all possible regressions technique combines each indicator variable with each other possible combination of variables. To keep the indicator variables grouped together, they are not included as possible explanatory variables in the all possible regressions procedure. Instead, they are examined later as a group (this also greatly reduces the amount of computation necessary and simplifies this example).

Figure 8.1 presents a summary of results for all possible regressions. Figure 8.2 shows the result as it would appear in SAS. In Figure 8.1 the summary measures included are: R^2, R^2_{adj}, C_p, and the standard error of the regression, s_e. These are descriptive measures that are often used to evaluate individual regression equations and to compare different equations. The R^2 does not compensate for the number of variables in the model, but the other three measures do. For this reason, the latter three measures are often considered more reliable for comparing equations with different numbers of variables.

Note that only 15 of the 16 ($2^4 = 16$) possible regressions are shown in the summary. The missing one is the "regression" with no variables included, which would be chosen as best only if none of the possible explanatory variables were linearly related to the dependent variable.

The researcher now must use the summary measures along with subject matter knowledge to choose from among the possible regressions. Recall that small values of C_p and values close to p are of interest in choosing good sets of explanatory variables. The smallest C_p value is for a two-variable regression. This is the regression with ADV and BONUS as explanatory variables; it has a C_p value of 1.61 and

	Variables in the Regression	R^2	R^2_{adj}	C_p	s_e
FIGURE 8.1 **Summary Results for All** **Possible Regressions.**	ADV	81.1%	80.2	5.90	101.42
	BONUS	32.3%	29.3	75.19	191.76
	COMPET	14.2%	10.5	100.85	215.83
	MKTSHR	0.0%	0.0	120.97	232.97
	ADV, BONUS	85.5%	84.2	1.61	90.75
	ADV, MKTSHR	81.2%	79.5	7.66	103.23
	ADV, COMPET	81.2%	79.5	7.74	103.38
	BONUS, COMPET	38.7%	33.2	68.03	186.51
	BONUS, MKTSHR	32.8%	26.7	76.46	195.33
	MKTSHR, COMPET	16.1%	8.5	100.18	218.20
	ADV, BONUS, MKTSHR	85.8%	83.8	3.11	91.75
	ADV, BONUS, COMPET	85.7%	83.6	3.33	92.26
	ADV, MKTSHR, COMPET	81.3%	78.6	9.59	105.52
	BONUS, MKTSHR, COMPET	40.9%	32.5	66.95	187.48
	ADV, BONUS, MKTSHR, COMPET	85.9%	83.1	5.00	93.77

FIGURE 8.2 SAS All Possible Regressions Output.

The REG Procedure

Model: MODEL1

Dependent Variable: SALES

R-Square Selection Method

Number in Model	R-Square	C(p)	Variables in Model
1	0.8106	5.9046	ADV
1	0.3228	75.1869	BONUS
1	0.1422	100.8498	COMPET
1	0.0005	120.9698	MKTSHR
2	0.8549	1.6052	ADV BONUS
2	0.8123	7.6612	ADV MKTSHR
2	0.8117	7.7404	ADV COMPET
2	0.3873	68.0320	BONUS COMPET
2	0.3280	76.4591	BONUS MKTSHR
2	0.1610	100.1763	MKTSHR COMPET
3	0.8585	3.1054	ADV BONUS MKTSHR
3	0.8569	3.3270	ADV BONUS COMPET
3	0.8128	9.5912	ADV MKTSHR COMPET
3	0.4090	66.9457	BONUS MKTSHR COMPET
4	0.8592	5.0000	ADV BONUS MKTSHR COMPET

explains 85.5% of the variation in sales ($R^2 = 85.5\%$). As competing models, for example, there are two three-variable models with relatively small C_p values:

Variables	R^2	R^2_{adj}	C_p	s_e
ADV,BONUS,COMPET	85.7%	83.6%	3.33	92.26
ADV,BONUS,MKTSHR	85.8%	83.8%	3.11	91.75

Note that only modest increases in R^2 are achieved in these models. The adjusted R^2 is highest for the two-variable model, again supporting this model as best. Other models with small C_p values could be examined, but there appears to be little bias in the two-variable model with ADV and BONUS (the small deviation of C_p from p probably results from random variation), and it has the smallest C_p value, largest adjusted R^2, and smallest standard error. Therefore, the all possible regressions procedure suggests using this model.

The number of computations involved in the all possible regressions technique is very large. (For example, with $K = 10$ explanatory variables, there are $2^{10} = 1024$ possible models). This is a limiting factor in using this procedure. With very large numbers of candidate explanatory variables, the number of possible models to investigate becomes unwieldy. As a result, several alternatives have been suggested. One is called best subsets regression.

Best subsets regression can be described as follows for subsets of size 2: Among all one-predictor regression models, the two models giving the largest R^2 are found and summary information is printed. Then the two models with the largest R^2 using two predictors are found and summary information is printed. This process continues until the model containing all predictors is examined. The number of subsets examined is chosen by the researcher. If all subsets are examined then the procedure is identical to all possible regressions.

EXAMPLE 8.2 **Meddicorp Revisited Again** The MINITAB best subsets procedure applied to the Meddicorp data produces the results shown in Figure 8.3. By default, MINITAB only prints out the regressions for the two subsets with the highest R^2 values for each number of explanatory variables. Requests for as many as five subsets are allowed, however. The result of the best subsets procedure asking for the five best subsets to be shown is in Figure 8.4. Note that five best subsets are not available for all combinations of explanatory variables. Note also that this procedure would be the same as all possible regressions if there were three or fewer explanatory variables.

FIGURE 8.3 MINITAB Best Subsets Results Using Default Number (2) of Regressions.

```
Best Subsets Regression: SALES versus ADV, BONUS, MKTSHR, COMPET
Response is SALES
                                                   M C
                                                 B K O
                                                 O T M
                                                 A N S P
                                    Mallows      D U H E
     Vars   R-Sq   R-Sq(adj)         C-p     S   V S R T
        1   81.1        80.2         5.9  101.42  X
        1   32.3        29.3        75.2  191.76    X
        2   85.5        84.2         1.6   90.749 X X
        2   81.2        79.5         7.7  103.23  X   X
        3   85.8        83.8         3.1   91.751 X X X
        3   85.7        83.6         3.3   92.255 X X   X
        4   85.9        83.1         5.0   93.770 X X X X
```

FIGURE 8.4 MINITAB Best Subsets Results Using Maximum Number (5) of Regressions.

```
Best Subsets Regression: SALES versus ADV, BONUS, MKTSHR, COMPET
Response is SALES
                                                   M C
                                                 B K O
                                                 O T M
                                                 A N S P
                                    Mallows      D U H E
     Vars   R-Sq   R-Sq(adj)         C-p     S   V S R T
        1   81.1        80.2         5.9  101.42  X
        1   32.3        29.3        75.2  191.76    X
        1   14.2        10.5       100.8  215.83        X
        1    0.1         0.0       121.0  232.97      X
        2   85.5        84.2         1.6   90.749 X X
        2   81.2        79.5         7.7  103.23  X   X
        2   81.2        79.5         7.7  103.38  X     X
        2   38.7        33.2        68.0  186.51    X X
        2   32.8        26.7        76.5  195.33    X X
        3   85.8        83.8         3.1   91.751 X X X
        3   85.7        83.6         3.3   92.255 X X   X
        3   81.3        78.6         9.6  105.52  X   X X
        3   40.9        32.5        66.9  187.48    X X X
        4   85.9        83.1         5.0   93.770 X X X X
```

SAS will also perform best subsets regressions. The output uses the same format as that shown for all possible regressions. See the Using the Computer section for more information.

8.3 OTHER VARIABLE SELECTION TECHNIQUES

As noted in the previous section, the all possible regressions technique becomes unwieldy when the number of possible explanatory variables is large. The techniques examined in this section attempt to cut down on the computational expense and the difficulty of examining a large number of potential models while still choosing variables that are important in explaining variation in the dependent variable.

Three procedures are discussed in this section:

1. backward elimination

2. forward selection

3. stepwise regression

Again, it should be stressed that none of these procedures is guaranteed to produce the best possible regression equation. The judgment of the researcher as well as careful examination of scatterplots, residual plots, and regression diagnostics is vital in choosing an appropriate model. The stepwise techniques are merely tools to help the researcher sort through a large number of possible explanatory variables. They help identify some important variables but by themselves are not sufficient to produce a good regression model.

8.3.1 BACKWARD ELIMINATION

The *backward elimination* procedure begins with a regression on all possible explanatory variables. After this regression is run, the explanatory variables are examined to determine which one has the smallest partial F statistic value. Calling this variable x_k, the following hypothesis test is performed:

$$H_0: \beta_k = 0$$
$$H_a: \beta_k \neq 0$$

The decision rule is

Reject H_0 if $F > F_c$

Do not reject H_0 if $F \leq F_c$

where F_c represents some critical value chosen as a cutoff for the test. If H_0 is rejected, the coefficient is judged to be nonzero, and the variable is considered important in the relationship. Because the partial F statistics for all other coefficients are known to exceed the partial F statistic for β_k, the null hypothesis is rejected for these coefficients also. The backward elimination procedure terminates at this point and produces summary statistics of the chosen regression.

On the other hand, if the null hypothesis is not rejected, then the variable is deleted from the equation, and a new regression is run with one less explanatory variable. The procedure is repeated until the null hypothesis is rejected, at which point the procedure terminates.

In this way, the backward elimination procedure sorts through the list of possible explanatory variables, eliminating those that are of little importance in explaining the variation in y and keeping those that are important. Importance is judged by the size of the partial F statistic for testing H_0: $\beta_k = 0$ relative to some critical value.

Two additional aspects of this procedure to note are

1. Although the test was described as a partial F test, it can just as easily be thought of as a t test. Rather than the partial F statistic, the t statistic is used and the decision rule is

   Reject H_0 if $t > t_c$ or $t < -t_c$
   Do not reject H_0 if $-t_c \leq t \leq t_c$

   where t_c is the chosen critical value. The t test and the partial F test for a *single* coefficient are equivalent, as was discussed in Chapter 4, as long as the same levels of significance are used for both tests.

2. Another way the test could be performed is by using p values. Whether the t or F statistic is computed, the p value could be calculated and used in the decision rule:

   Reject H_0 if p value $< \alpha$
   Do not reject H_0 if p value $\geq \alpha$

   where α is the chosen level of significance. When using p values, the critical value is α, while with the t or F tests it is a value chosen from the t or F tables. The p value form of the test is a better choice, if available, because the level of significance used to determine whether a variable stays or goes is the same regardless of the sample size or the number of variables in the equation. When a single t or F critical value is chosen, level of significance varies depending on the sample size and number of variables.

Example 8.3 demonstrates the backward elimination procedure as well as the other procedures. First, however, the remaining techniques are discussed.

8.3.2 FORWARD SELECTION

Forward selection starts by examining the list of possible explanatory variables and computing a simple regression for each one. The partial F statistic (or t statistic or the p value) is computed for the slope coefficient in each of these regressions, and the variable with the largest partial F statistic is noted. The hypothesis test

$$H_0: \beta_k = 0$$
$$H_a: \beta_k \neq 0$$

is conducted just as was done in the backward elimination procedure. If the null hypothesis is not rejected, then the conclusion is that x_k is of no importance to the regression, and none of the other variables are important because they have smaller partial F statistics. The forward selection procedure terminates at this point.

If the null hypothesis is rejected, then x_k is judged important and is retained in the regression. Next, each remaining variable is examined to determine which one will have the largest partial F statistic if added to the regression that already contains x_k. The hypothesis test is performed for the added variable, and the decision is made either to keep the variable in the regression or to discard it. When no more variables are judged to have nonzero coefficients, the procedure terminates and summary statistics are printed.

8.3.3 STEPWISE REGRESSION

The *stepwise regression* procedure combines elements of both backward elimination and forward selection. It begins like forward selection by examining the list of all possible explanatory variables in simple regressions and choosing the one with the largest partial F statistic. The hypothesis test for significance is performed, and if the variable is judged important, this variable is added to the model. Each of the remaining variables is then examined. The variable with the largest partial F statistic is chosen, and the hypothesis test for significance is performed on the coefficient of this variable to determine whether it should be added to the model. If the variable is judged important, it is added as in the forward selection procedure.

At this point, however, the stepwise procedure begins to act like the backward elimination procedure. After adding a new variable to the model, the stepwise procedure retests the coefficients of the previously added variables, deleting these variables if the test judges them to be unnecessary and retaining them otherwise. Because the addition of one variable can result in a change in the partial F statistic associated with another variable, it is possible for the stepwise procedure to allow a variable to enter the equation at one step, delete the variable at a later step, and allow the variable to reenter at an even later step.

Once none of the remaining out-of-equation variables test as significant and all of the variables in the equation are judged to be necessary, the stepwise regression procedure terminates.

8.4 WHICH VARIABLE SELECTION PROCEDURE IS BEST?

In an attempt to identify the best set of variables for a regression model, several techniques have been examined. This leaves the user to decide which technique is best for his or her purposes. As noted, none of the techniques examined is guaranteed to find the best possible regression model. The judgment of the researcher, including incorporation of relevant theory and subject matter knowledge, and careful examination of scatterplots, residual plots, and regression diagnostics are vital in choosing an appropriate model. With this caveat in mind, several trade-offs must be considered when choosing the variable selection technique to be used.

The all possible regressions technique is considered the best because it examines every possible model, given a certain list of variables. From the summary statistics such as R^2 and C_p, the researcher can decide which model is best. Note that even with the all possible regressions technique, a single best model might not be identifiable. There may be several competing models that have nearly identical summary statistics, leaving the researcher with the task of using judgment in choosing between these similar best models. This should not be looked upon as a drawback, however, but as a benefit. With a variety of models from which to choose, the researcher has more freedom to pick the one, say, with the most easily obtainable data or the simplest interpretation. The best subsets procedure is very similar to all possible regressions and would be a close second in its ability to identify the best possible models.

The remaining procedures are not as highly favored, but they do have one advantage over all possible regressions: computational cost. The all possible regressions procedure can be expensive in terms of the computer time needed to produce a solution and may result in too many models to consider if a large number of explanatory variables are used. If the variable list is a very large one, the researcher may be forced to avoid all possible regressions in favor of one of the computationally less expensive methods. As will be shown in Example 8.3, the forward selection, backward elimination, and stepwise regression procedures can be useful in identifying important variables. Research has shown, however, that these procedures can also choose unimportant variables for inclusion in the regressions by chance and may miss important variables. Thus, some caution must be exercised in their use. Among the three procedures, the stepwise regression and backward elimination procedures are very similar. The forward selection procedure is generally considered the least reliable of the techniques.

EXAMPLE 8.3 **Meddicorp Once Again** Figure 8.5 shows the MINITAB backward elimination output when applied to the Meddicorp data. The forward selection output is shown in Figure 8.6, and the stepwise output is in Figure 8.7. The equivalent outputs for SAS are shown in Figures 8.8, 8.9, and 8.10. The indicator variables have not been included in these analyses because these variables are treated as a group rather than individually.

The MINITAB outputs summarize the results of each step of these variable selection procedures in a column. For example, in the backward elimination output shown in Figure 8.5, the variables included in the model are in the left-hand column. In the column numbered Step 1, a summary of the regression equation at the first step is given. The estimated equation is

$$\text{SALES} = -593.5 + 2.51\text{ADV} + 1.91\text{BONUS} + 2.70\text{MKTSHR} - 0.12\text{COMPET}$$

Below the estimated coefficients in the columns are the t ratios and P values for testing $H_0: \beta_k = 0$. At the bottom of the column are the standard error of the regression, the R^2 (both unadjusted and adjusted) and the C_p. The variable with the smallest partial F statistic (or t ratio) is chosen: COMPET. The hypothesis test

$$H_0: \beta_k = 0$$
$$H_a: \beta_k \neq 0$$

FIGURE 8.5 MINITAB Output for Backward Elimination.

```
Stepwise Regression: SALES versus ADV, BONUS, MKTSHR, COMPET

Backward elimination. Alpha-to-Remove: 0.1

Response is SALES on 4 predictors, with N = 25
```

Step	1	2	3
Constant	−593.5	−620.6	−516.4
ADV	2.51	2.47	2.47
T-Value	8.00	8.87	8.98
P-Value	0.000	0.000	0.000
BONUS	1.91	1.90	1.86
T-Value	2.57	2.62	2.59
P-Value	0.018	0.016	0.017
MKTSHR	2.7	3.1	
T-Value	0.57	0.72	
P-Value	0.574	0.478	
COMPET	−0.12		
T-Value	−0.32		
P-Value	0.749		
S	93.8	91.8	90.7
R-Sq	85.92	85.85	85.49
R-Sq(adj)	83.10	83.82	84.18
Mallows C-p	5.0	3.1	1.6

FIGURE 8.6 MINITAB Output for Forward Selection.

```
Stepwise Regression: SALES versus ADV, BONUS, MKTSHR, COMPET

Forward selection. Alpha-to-Enter: 0.25

Response is SALES on 4 predictors, with N = 25
```

Step	1	2
Constant	−157.3	−516.4
ADV	2.77	2.47
T-Value	9.92	8.98
P-Value	0.000	0.000
BONUS		1.86
T-Value		2.59
P-Value		0.017
S	101	90.7
R-Sq	81.06	85.49
R-Sq(adj)	80.24	84.18
Mallows C-p	5.9	1.6

is performed, and the null hypothesis is not rejected. COMPET is removed from the regression, and a new equation is estimated. This regression is summarized in the Step 2 column. This process continues until the null hypothesis $H_0: \beta_k = 0$ is rejected for all remaining variables. The last column (Step 3 in this example) shows the result of the final regression. The summaries are similar for forward selection and stepwise regression.

FIGURE 8.7 MINITAB Output for Stepwise Regression.

Stepwise Regression: SALES versus ADV, BONUS, MKTSHR, COMPET

Alpha-to-Enter: 0.15   Alpha-to-Remove: 0.15

Response is SALES on 4 predictors, with N = 25

Step	1	2
Constant	−157.3	−516.4
ADV	2.77	2.47
T-Value	9.92	8.98
P-Value	0.000	0.000
BONUS		1.86
T-Value		2.59
P-Value		0.017
S	101	90.7
R-Sq	81.06	85.49
R-Sq(adj)	80.24	84.18
Mallows C-p	5.9	1.6

FIGURE 8.8 SAS Output for Backward Elimination.

The STEPWISE Procedure
Model: MODEL1
Dependent Variable: SALES

Backward Elimination: Step 0

All Variables Entered: R-Square = 0.8592 and C(p) = 5.0000

Analysis of Variance

Source	DF	Sum of Squares	Mean Square	F Value	Pr > F
Model	4	1073119	268280	30.51	<.0001
Error	20	175855	8792.75990		
Corrected Total	24	1248974			

Variable	Parameter Estimate	Standard Error	Type II SS	F Value	Pr > F
Intercept	−593.53745	259.19585	46107	5.24	0.0330
ADV	2.51314	0.31428	562260	63.95	<.0001
BONUS	1.90595	0.74239	57955	6.59	0.0184
MKTSHR	2.65101	4.63566	2875.57485	0.33	0.5738
COMPET	−0.12073	0.37181	927.06773	0.11	0.7488

Bounds on condition number: 1.4799, 20.838

(Continued).

FIGURE 8.8
(*Continued*).

Backward Elimination: Step 1

Variable COMPET Removed: R-Square = 0.8585 and C(p) = 3.1054

Analysis of Variance

Source	DF	Sum of Squares	Mean Square	F Value	Pr > F
Model	3	1072191	357397	42.46	<.0001
Error	21	176782	8418.20313		
Corrected Total	24	1248974			

Variable	Parameter Estimate	Standard Error	Type II SS	F Value	Pr > F
Intercept	-620.63774	240.10769	56245	6.68	0.0173
ADV	2.46979	0.27839	662567	78.71	<.0001
BONUS	1.90030	0.72620	57643	6.85	0.0161
MKTSHR	3.11646	4.31354	4394.15366	0.52	0.4780

Bounds on condition number: 1.2212, 10.325

Backward Elimination: Step 2

Variable MKTSHR Removed: R-Square = 0.8549 and C(p) = 1.6052

Analysis of Variance

Source	DF	Sum of Squares	Mean Square	F Value	Pr > F
Model	2	1067797	533899	64.83	<.0001
Error	22	181176	8235.29179		
Corrected Total	24	1248974			

Variable	Parameter Estimate	Standard Error	Type II SS	F Value	Pr > F
Intercept	-516.44428	189.87570	60924	7.40	0.0125
ADV	2.47318	0.27531	664572	80.70	<.0001
BONUS	1.85618	0.71573	55389	6.73	0.0166

Bounds on condition number: 1.2126, 4.8502

All variables left in the model are significant at the 0.1000 level.

Summary of Backward Elimination

Step	Variable Removed	Number Vars In	Partial R-Square	Model R-Square	C(p)	F Value	Pr > F
1	COMPET	3	0.0007	0.8585	3.1054	0.11	0.7488
2	MKTSHR	2	0.0035	0.8549	1.6052	0.52	0.4780

FIGURE 8.9
SAS Output for
Forward
Selection.

The STEPWISE Procedure
Model: MODEL1, Dependent Variable: SALES
Forward Selection: Step 1

Variable ADV Entered: R-Square = 0.8106 and C(p) = 5.9046

Analysis of Variance

Source	DF	Sum of Squares	Mean Square	F Value	P > F
Model	1	1012408	1012408	98.43	<.0001
Error	23	236566	10285		
Corrected Total	24	1248974			

Variable	Parameter Estimate	Standard Error	Type II SS	F Value	Pr > F
Intercept	−157.33011	145.19120	12077	1.17	0.2898
ADV	2.77212	0.27941	1012408	98.43	<.0001

Bounds on condition number: 1, 1

Forward Selection: Step 2

Variable BONUS Entered: R-Square = 0.8549 and C(p) = 1.6052

Analysis of Variance

Source	DF	Sum of Squares	Mean Square	F Value	Pr > F
Model	2	1067797	533899	64.83	<.0001
Error	22	181176	8235.29179		
Corrected Total	24	1248974			

Variable	Parameter Estimate	Standard Error	Type II SS	F Value	Pr > F
Intercept	−516.44428	189.87570	60924	7.40	0.0125
ADV	2.47318	0.27531	664572	80.70	<.0001
BONUS	1.85618	0.71573	55389	6.73	0.0166

Bounds on condition number: 1.2126, 4.8502

Forward Selection: Step 3

Variable MKTSHR Entered: R-Square = 0.8585 and C(p) = 3.1054

Analysis of Variance

Source	DF	Sum of Squares	Mean Square	F Value	Pr > F
Model	3	1072191	357397	42.46	<.0001
Error	21	176782	8418.20313		
Corrected Total	24	1248974			

(Continued).

FIGURE 8.9
(*Continued*).

Variable	Parameter Estimate	Standard Error	Type II SS	F Value	Pr > F
Intercept	−620.63774	240.10769	56245	6.68	0.0173
ADV	2.46979	0.27839	662567	78.71	<.0001
BONUS	1.90030	0.72620	57643	6.85	0.0161
MKTSHR	3.11646	4.31354	4394.15366	0.52	0.4780

Bounds on condition number: 1.2212, 10.325

No other variable met the 0.5000 significance level for entry into the model.

Summary of Forward Selection

Step	Variable Entered	Number Vars In	Partial R-Square	Model R-Square	C(p)	F Value	Pr > F
1	ADV	1	0.8106	0.8106	5.9046	98.43	<.0001
1	BONUS	2	0.0443	0.8549	1.6052	6.73	0.0166
2	MKTSHR	3	0.0035	0.8585	3.1054	0.52	0.4780

FIGURE 8.10
SAS Output for
Stepwise
Regression.

The STEPWISE Procedure
Model: MODEL1
Dependent Variable: SALES

Forward Selection: Step 1

Variable ADV Entered: R-Square = 0.8106 and C(p) = 5.9046

Analysis of Variance

Source	DF	Sum of Squares	Mean Square	F Value	P > F
Model	1	1012408	1012408	98.43	<.0001
Error	23	236566	10285		
Corrected Total	24	1248974			

Variable	Parameter Estimate	Standard Error	Type II SS	F Value	Pr > F
Intercept	−157.33011	145.19120	12077	1.17	0.2898
ADV	2.77212	0.27941	1012408	98.43	<.0001

Bounds on condition number: 1, 1

FIGURE 8.10
(*Continued*).

Stepwise Selection: Step 2

Variable BONUS Entered: R-Square = 0.8549 and C(p) = 1.6052

Analysis of Variance

Source	DF	Sum of Squares	Mean Square	F Value	Pr > F
Model	2	1067797	533899	64.83	<.0001
Error	22	181176	8235.29179		
Corrected Total	24	1248974			

Variable	Parameter Estimate	Standard Error	Type II SS	F Value	Pr > F
Intercept	-516.44428	189.87570	60924	7.40	0.0125
ADV	2.47318	0.27531	664572	80.70	<.0001
BONUS	1.85618	0.71573	55389	6.73	0.0166

Bounds on condition number: 1.2126, 4.8502

All variables left in the model are significant at the 0.1500 level.

No other variable met the 0.5000 significance level for entry into the model.

Summary of Stepwise Selection

Step	Variable Entered	Variable Removed	Number Vars In	Partial R-Square	Model R-Square	C(p)	F Value	Pr > F
1	ADV		1	0.8106	0.8106	5.9046	98.43	<.0001
1	BONUS		2	0.0443	0.8549	1.6052	6.73	0.0166

SAS also summarizes each step of the variable selection procedures, but prints out a more complete version of the regression results. Consider the backward elimination output in Figure 8.8. In the output labeled Step 0, the regression including all the potential explanatory variables is run. In Figures 8.9 (forward selection) and 8.10 (stepwise) the output labeled Step 1 is the regression using the variable with the largest partial F statistic. An ANOVA table is printed for the regression at each step. The table is the same as the standard ANOVA table for SAS. Below that, information about the regression coefficients is printed. This table contains the parameter estimate and the standard error for each variable in the regression as usual. However, an additional column is added: Type II SS. This column contains the sum of squares used to compute the partial F statistic to test if the coefficient of the variable in that row of the table is zero. The partial F statistic is given in the next column (rather than the usual t statistic) and the associated p value follows. The F statistic or the associated p value can be used as described earlier in the chapter to perform the tests needed for the variable selection procedures. SAS proceeds through the steps necessary to complete the various procedures and ends with a summary of the variables that were either omitted from or added to the regression equation at each step.

Regardless of the procedure or statistical package used, the result is the same in all but one instance. The equation chosen is SALES $= -516.4 + 2.47$ADV $+ 1.86$BONUS. The one exception is the SAS output for forward selection. In this case the equation chosen is SALES $= -620.64 + 2.47$ADV $+ 1.90$BONUS $+ 3.12$MKTSHR. The reason that MKTSHR is included in the equation developed by the SAS forward selection procedure is that different cutoffs for the hypothesis tests to choose the variables are used. SAS uses p values to determine what variables should and should not be included in the equations. For the forward selection procedure, the p value for inclusion of a variable is 0.5. MKTSHR is included in the final equation because the p value for its coefficient (0.4780) is less than 0.5. In contrast, MINITAB's forward selection procedure uses a p value of 0.25 as a cutoff and MKTSHR is not included in the resulting equation. For more on choosing cutoffs for these procedures, see the Using the Computer section at the end of this chapter.

Combining all information gathered from the all possible regressions procedure and the various stepwise procedures, the best model appears to be

$$\text{SALES} = -516.4 + 2.47\text{ADV} + 1.86\text{BONUS}$$

Before concluding, recall that the indicator variables for the region have not been included in any of this analysis. They could be added to the regression at this point to see whether they improve the model. The regression with the indicators included is shown in Figure 8.11. The partial F test to test the hypotheses

$$H_0: \beta_3 = \beta_4 = 0$$
$$H_a: \text{At least one of } \beta_3 \text{ and } \beta_4 \text{ is not equal to zero}$$

can be used to determine whether the indicator variables improve the model. The null hypothesis is rejected, and the conclusion is to retain the indicator variables as well.

FIGURE 8.11
Regression Results for Model Including Indicator Variables.

Variable	Coefficient	Std Dev	T Stat	P Value
Intercept	435.099	206.234	2.11	0.048
ADV	1.368	0.262	5.22	0.000
BONUS	0.975	0.481	2.03	0.056
SOUTH	-257.893	48.413	-5.33	0.000
WEST	-209.746	37.420	-5.61	0.000

Standard Error = 57.6254   R-Sq = 94.7%   R-Sq(adj) = 93.6%

Analysis of Variance

Source	DF	Sum of Squares	Mean Square	F Stat	P Value
Regression	4	1182560	295640	89.03	0.000
Error	20	66414	3321		
Total	24	1248974			

EXERCISES

1. Cost Control. Exercise 1 in Chapter 4 discussed data available for a firm that produces corrugated paper for use in making boxes and other packing materials. The variables discussed were

> y, total manufacturing cost per month in thousands of dollars (COST)
>
> x_1, total production of paper per month in tons (PAPER)
>
> x_2, total machine hours used per month (MACHINE)
>
> x_3, total variable overhead costs per month in thousands of dollars (OVERHEAD)
>
> x_4, total direct labor hours used each month (LABOR)

The data, available in a file named COST8 on the CD, are monthly and refer to the time period from January 2001 to March 2003. Use the backward elimination procedure to analyze these data, then answer the following questions:

a. What is the regression equation chosen by the backward elimination procedure?

b. What is the R^2 for the chosen equation?

c. What is the adjusted R^2 for the chosen equation?

d. What is the standard error of the chosen equation?

e. What variables were omitted? Do you feel these variables are unrelated to COST? Why or why not? Do you feel the omitted variables are necessary in the regression equation? Why or why not?

2. Sales Force Performance. Data on the following variables were obtained for a random sample of 25 sales territories for a company (the data have been transformed to preserve confidentiality).

> y, sales in units for the territory (SALES)
>
> x_1, length of time territory salesperson has been with the company (TIME)
>
> x_2, industry sales in units for the territory (POTENT)
>
> x_3, dollar expenditures on advertising (ADV)
>
> x_4, weighted average of past market share for 4 previous years (SHARE)
>
> x_5, change in market share over the 4 years before the time period analyzed (SHARECHG)

> x_6, total number of accounts assigned to salesperson (ACCTS)
>
> x_7, average workload per account using a weighted index based on annual purchases of accounts and concentration of accounts (WORKLOAD)
>
> x_8, an aggregate rating on a 1–7 scale on eight dimensions of performance by applicable field sales manager (RATING)

The data are available in a file named TERRITORY8 and were analyzed in D. W. Cravens, R. B. Woodruff, and J. C. Stamper, "An Analytical Approach for Evaluating Sales Territory Performance," *Journal of Marketing* 36 (1972): 31–37.

The goal of the study is to identify factors that influence territory sales (y). The equation to be developed will be used to assess whether salespersons in respective territories are performing up to standard.

Develop an appropriate model to explain sales territory performance. Use any of the techniques discussed to select appropriate variables. Be sure to examine scatterplots and residual plots for violations of assumptions and to correct for any such violations.

a. For the model you select, report the estimated regression equation. Be sure to define the variables used.

b. For the model you select, report any corrections for violations of assumptions.

c. For the model you select, report the R^2, R^2_{adj}, standard error, and C_p value.

d. Discuss how this equation could be used to set a performance standard for sales territories. How would average performance be determined? Below-average performance? What limitations would this approach have for setting performance standards?

3. 2003 Cars. Data on 147 cars were obtained from the October 2002 issue of *Road & Track: The New Cars*. The following data are available in a file named CARS8 on the CD:

> name of car
> weight, in pounds (WEIGHT)
> mileage in city driving (CITYMPG)
> mileage in highway driving (HWYMPG)

horsepower, @ 6300 rpm (HP)
number of cylinders (CYLIN)
displacement, in liters (LITER)

Using the available data, try to determine what factors involved in the construction of a car affect either mileage in city driving or mileage in highway driving. (Choose either CITYMPG or HWYMPG as your dependent variable. If you choose CITYMPG, do not use HWYMPG as a possible explanatory variable, and vice versa.) Use any of the techniques discussed to select appropriate variables. Be sure to examine scatterplots and residual plots for violations of assumptions and to correct for any such violations.

a. For the model you select, report the estimated regression. Be sure to define the variables used.

b. For the model you select, report any corrections for violations of assumptions. Explain why the correction was needed and justify the correction you used.

c. Justify your choice of variables from both a statistical and a practical standpoint.

d. State any limitations of the model.

(*Source*: From *Road & Track: The New Cars.* Copyright 2002 by Hachette Filipacchi Magazines, Inc. Used with permission.)

4. 1998 American League Pitchers. Data on 44 American League pitchers were obtained for the 1998 season. The pitchers included in the data set must have pitched in at least 40 innings to be listed here. Also, these pitchers were used only as starting pitchers (pitchers with even one appearance in relief were not included). The data are available in a file named ALPITCH8 on the CD as follows:

name of pitcher
team: coded as
    1 = Baltimore
    2 = Boston
    3 = Cleveland
    4 = Detroit
    5 = Anaheim
    6 = Chicago
    7 = Kansas City
    8 = Milwaukee
    9 = Minnesota
    10 = Seattle
    11 = New York
    12 = Oakland
    13 = Texas

    14 = Toronto
    15 = Tampa Bay
number of wins (W)
number of losses (L)
earned run average (ERA)
innings pitched (IP)
hits allowed (H)
home runs allowed (HR)
bases on balls (BB)
strikeouts (SO)

As a consultant to the Texas Rangers' coaching staff, you have been hired to determine what makes a starting pitcher successful. The Rangers are painfully aware of the need for good starting pitchers. They would like to improve their starting pitching before the next season. You have data available for American League starting pitchers in the 1998 season. Your goal is to determine what factors might be important in the success of a starting pitcher during a particular season. First, you must define success (the dependent variable). Then decide which of the variables available might make sense in evaluating a pitcher's success.

Write up a report to the Rangers with your recommendations. Your report should consist of two parts: (a) an executive summary with a brief nontechnical discussion of your recommendations and (b) a technical report to support your suggestions. The technical report should include a discussion of the analysis that led to your conclusions. This should include regression results, model validation, corrections for violations of assumptions, justification of your choice of variables, and an explanation of how this choice will help the Rangers in their decision making.

(*Source*: Used courtesy of the *Fort Worth Star-Telegram.*)

5. FOC Sales. Techcore is a high-tech company located in Fort Worth, Texas. The company produces a part called a fibre-optic connector (FOC) and wants to generate forecasts of the sales of FOCs over time. The company has weekly sales data for the past 265 weeks. The data are in the file FOC8 on the CD in the following columns: SALES, MONTH (numbers the month in which the observations were taken), FOV, COMPOSITE, INDUSTRIAL, TRANS, UTILITY, FINANCE, PROD, and HOUSE. (The SALES and FOV data have been disguised to provide confidentiality.) The variables are defined as follows:

FOV: Sales of a complementary product; sales of FOV are much easier to forecast than FOC sales

COMPOSITE: Friday close of the NYSE Composite Index

INDUSTRIAL: Friday close of the NYSE Industrial Stocks

TRANS: Friday close of the NYSE Transportation Stocks

UTILITY: Friday close of the NYSE Utility Stocks

FINANCE: Friday close of the NYSE Financials

PROD: Industrial Production—computers, communications equipment, and semiconductors, not seasonally adjusted

HOUSE: Monthly housing permits in thousands, seasonally adjusted rates

You have been hired as a consultant to Techcore to help build the forecasting model. Create a two-part report for Techcore that includes an executive summary with the essential nontechnical results of your study and a technical report that contains the details of your model building process.

USING THE COMPUTER

The Using the Computer section in each chapter describes how to perform the computer analyses in the chapter using Excel, MINITAB, and SAS. For further detail on Excel, MINITAB, and SAS, see Appendix C.

FIGURE 8.12 MINITAB Best Subsets Dialog Box.

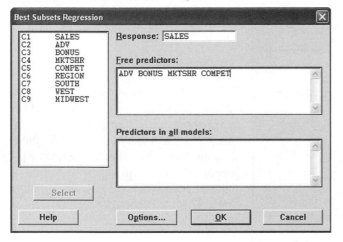

EXCEL

Excel does not come with any variable selection techniques built in. However, Appendix C describes an add-in that makes it possible to perform certain of the procedures. See the Appendix C section on Excel for more information.

MINITAB

Best Subsets Regression

STAT: REGRESSION: BEST SUBSETS

To use the best subsets procedure, click STAT, then REGRESSION from the STAT menu, and then BEST SUBSETS. The Best Subsets dialog box is shown in Figure 8.12. Fill in the "Response" variable (y) and the "Free predictors" (x variables) and click OK. Use "Predictors in all models" if you want certain x variables to be included in all regressions. The Options screen allows, among other things, the choice of from one to five subsets to be shown in the results.

STEPWISE REGRESSION, BACKWARD ELIMINATION, AND FORWARD SELECTION

Backward elimination, forward selection, and stepwise regression are all run from the STEPWISE Methods on the REGRESSION menu. Figure 8.13 shows the stepwise regression dialog box.

The MINITAB Stepwise—Options Dialog Box in Figure 8.14 allows the user to choose either F values or p values to be used in conducting the tests described in

FIGURE 8.13 MINITAB Stepwise Regression Dialog Box.

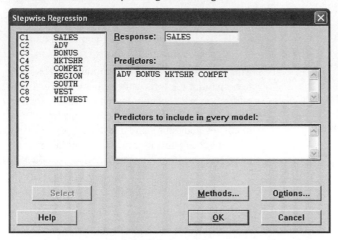

FIGURE 8.14 MINITAB Stepwise—Options Dialog Box.

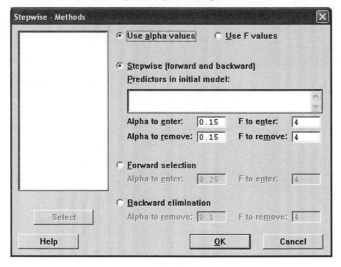

this chapter. If F values are chosen, the default critical value will be 4. The justification for the partial F test cutoff value of 4 is that this results in a test with a 5% level of significance (approximately) if a large number of observations are available. When using any of these three procedures, it is often recommended that a smaller cutoff value be used, say, $F_c = 2$ or $F_c = 1$. After examining the variables chosen by the procedure(s) at this critical value, the researcher is left to make the final choice about which variables should remain in the model. These critical values can be changed in the "F to enter" and/or "F to remove" boxes.

If p values are chosen, the default values are 0.15 both to enter and remove variables for stepwise regression, 0.25 to enter in forward selection, and 0.1 to remove in backward elimination. These default values can be changed at the discretion of the user.

Stepwise Regression

STAT: REGRESSION: STEPWISE

For stepwise regression, fill in the "Response" variable (y) and the names of the x variables in the "Predictors" box (Figure 8.13). Then click OK. Click Methods and the dialog box shown in Figure 8.14 opens. Be sure the stepwise button is clicked. Stepwise is the default method in MINITAB so it should be set unless it has been changed in a previous analysis.

Backward Elimination

STAT: REGRESSION: STEPWISE

For backward elimination, fill in the "Response" variable (y) and the names of the x variables in the "Predictors" box (Figure 8.13). Now click Methods, and the dialog box shown in Figure 8.14 opens. The backward elimination option is requested by clicking the button for that option.

Forward Selection

STAT: REGRESSION: STEPWISE

For forward selection, fill in the "Response" variable (y) and the names of the x variables in the "Predictors" box (Figure 8.13). Click Methods and the dialog box shown in Figure 8.14 opens. The forward selection option is requested by clicking the button for that option.

SAS

In SAS, the PROC REG command can be used to request all possible regressions, best subsets, backward elimination, forward selection, and stepwise. The choice of variable selection procedure is made using the SELECTION option in the MODEL statement. Each of the choices is demonstrated here.

All Possible Regressions and Best Subsets

```
PROC REG CP;
MODEL SALES=ADV BONUS MKTSHR COMPET/SELECTION=RSQUARE;
```

This command sequence produces all possible regressions output for the dependent variable SALES and independent variables ADV, BONUS, MKTSHR, and COMPET. The CP option requests that the C_p value associated with each regression be printed. In addition, the R^2 value (unadjusted) is printed. The MODEL statement works just like it does for PROC REG. The dependent variable is listed to the left of the equal sign, and the possible explanatory variables are listed to the right. The SELECTION = RSQUARE option requests all possible regressions. If BEST = k is added after SELECTION = RSQUARE, then the k best subsets are summarized rather than all possible regressions. Here, k is replaced by the number of desired subsets.

Stepwise Regression

```
PROC REG;
MODEL SALES=ADV BONUS MKTSHR COMPET/SELECTION=STEPWISE;
```

Control of the variables to enter or leave the regression is through the p values for the partial F test (rather than the F critical values). SLENTRY represents the maximum p value the coefficient of a variable can have and still have the variable enter the model. SLSTAY represents the maximum p value the coefficient of a variable can have for the variable to remain in the model. For stepwise regression, these values are set at SLENTRY = 0.15 and SLSTAY = 0.15. These default values can be changed. For example, to change SLENTRY and SLSTAY each to 0.2, the following sequence could be used:

```
PROC REG;
MODEL SALES=ADV BONUS MKTSHR COMPET/SELECTION=STEPWISE/SLENTRY=0.2 SLSTAY=0.2;
```

Backward Elimination

```
PROC REG;
MODEL SALES=ADV BONUS MKTSHR COMPET/SELECTION=BACKWARD;
```

Backward elimination uses SLSTAY = 0.1 in SAS.

Forward Selection

```
PROC REG;
MODEL SALES=ADV BONUS MKTSHR COMPET/SELECTION=FORWARD;
```

Forward selection uses SLENTRY = 0.5 in SAS.

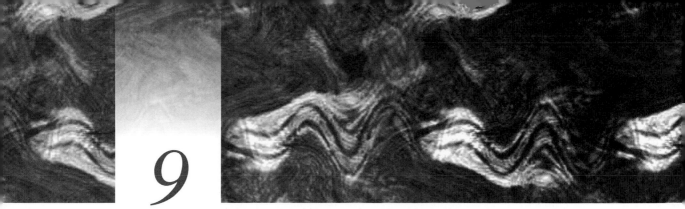

9

An Introduction to Analysis of Variance

9.1 ONE-WAY ANALYSIS OF VARIANCE

Analysis of variance (ANOVA) was a term used with regression to describe the decomposition of the total sum of squares into two parts: the regression ("explained") sum of squares and the error ("unexplained") sum of squares. ANOVA also describes a statistical technique used to test whether there is a difference between means of several populations. The procedure is very similar to regression analysis. In fact, for one-way analysis of variance, the procedure is the same as if a regression were performed using only indicator variables as explanatory variables. The actual analyses are usually performed with ANOVA routines rather than regression routines because the analysis of variance routines are more efficient computationally for this specific type of problem.

To describe the ANOVA procedure, consider a problem with K populations. One way that the ANOVA model can be written is

$$y_{ij} = \mu_i + e_{ij}$$

where

y_{ij} is the jth observation from population i

μ_i is the population mean for population i

e_{ij} is a random disturbance for the jth observation from population i

Because there are K populations, i ranges from 1 to K. Assuming that there are n_i observations from each population, j ranges from 1 to n_i. The use of the subscript on n_i implies that the number of sample observations from each population can differ. The total number of observations combining all samples is denoted

$$n = \sum_{i=1}^{K} n_i$$

The following assumptions are made concerning the disturbances e_{ij}:

1. The e_{ij} have mean zero.
2. The e_{ij} have constant variance, σ_e^2.
3. The e_{ij} are normally distributed.

ANOVA has its own special terminology. As in regression, y_{ij} is called the *dependent variable*. The explanatory variables are called *factors*. A *level* of the factor is a particular value of the explanatory variable. The μ_i are called *factor-level means*. The ANOVA model allows for a different mean for each factor level. Factor levels are also referred to as *treatments* in one-way ANOVA.

An alternative way of writing the ANOVA model is

$$y_{ij} = \mu + \gamma_i + e_{ij}$$

where μ is a constant component common to all observations and γ_i is the effect of the ith treatment (or factor level). Here the treatment means are

$$\mu_1 = \mu + \gamma_1$$
$$\mu_2 = \mu + \gamma_2$$
$$\vdots$$
$$\mu_K = \mu + \gamma_K$$

The question to be answered is, "Are the means of all K populations equal?" The hypotheses to be tested to answer this question can be stated as

$H_0: \mu_1 = \mu_2 = \cdots = \mu_K$
H_a: Not all means are equal

or as

$H_0: \gamma_1 = \gamma_2 = \cdots = \gamma_K$
H_a: Not all treatment effects are equal

depending on the form of the model used. The test procedure is the same regardless of the way the model is written.

The test statistic used to conduct the test is

$$F = \frac{MSTR}{MSE}$$

where *MSTR* is the *mean square due to treatments* and *MSE* is the *mean square error*. This statistic is similar to the F statistic used to test the overall fit of a regression.

The ANOVA F statistic has an F distribution with $K - 1$ numerator and $n - K$ denominator degrees of freedom. K is the number of populations and n is the total sample size. The decision rule for the test is

Reject H_0 if $F > F(\alpha; K - 1, n - K)$

Do not reject H_0 if $F \leq F(\alpha; K - 1, n - K)$

To examine the construction of the test statistic the following notation is needed:

y_{ij} is the jth sample observation from population i

$\bar{y}_{i.}$ is the mean of all the sample observations for the ith population:

$$\bar{y}_{i.} = \frac{\sum_{j=1}^{n_i} y_{ij}}{n_i}$$

$\bar{y}_{..}$ is the overall mean of all n observations:

$$\bar{y}_{..} = \frac{\sum_{i=1}^{K} \sum_{j=1}^{n_i} y_{ij}}{n}$$

The F statistic is composed of two components, $MSTR$ and MSE. The formula for $MSTR$ is

$$MSTR = \frac{SSTR}{K - 1}$$

The treatment sum of squares, $SSTR$, is

$$SSTR = \sum_{i=1}^{K} n_i (\bar{y}_{i.} - \bar{y}_{..})^2$$

and has $K - 1$ degrees of freedom. The treatment sum of squares measures the variability of the sample means for each group, $\bar{y}_{i.}$, around the overall mean $\bar{y}_{..}$. The bigger the variation, the more difference that can be attributed to the effect of the treatments and the less likely the null hypothesis is true. Another name used for $SSTR$ is the between-groups sum of squares.

The formula for MSE is

$$MSE = \frac{SSE}{n - K}$$

The error sum of squares, SSE, is

$$SSE = \sum_{i=1}^{K} \sum_{j=1}^{n_i} (y_{ij} - \bar{y}_{i.})^2$$

and has $n - K$ degrees of freedom. The error sum of squares measures the variability of the individual observations within each group. This variation is measured around the individual sample means, $\bar{y}_{i.}$. Another name for SSE is the within groups sum of squares.

To understand the use of the name *analysis of variance* note that the F statistic is the ratio of two mean squares, and each mean square is an estimate of the common population variance. One of the assumptions necessary for using ANOVA is that the variances of all populations are equal. This common variance is called σ_e^2. *MSE* provides an unbiased estimate of σ_e^2. Furthermore, if the means (μ_i) are all equal, *MSTR* also provides an unbiased estimate of σ_e^2. But when some of the means are not equal, *MSTR* is biased. Thus, if H_0 is false, *MSTR* tends to be bigger than *MSE*, and the null hypothesis is likely to be rejected. If H_0 is true, *MSTR* and *MSE* provide similar estimates, and the *F* statistic is close to 1, leading to acceptance of the null hypothesis.

As mentioned, the one-factor ANOVA model is equivalent to a regression model in which all the explanatory variables are indicator variables. The equivalent regression model can be written in two ways, just as the ANOVA model can be written in two ways. For example, the following regression equation is equivalent to the first form of the ANOVA model:

$$y_{ij} = \mu_1 x_{ij1} + \mu_2 x_{ij2} + \cdots + \mu_K x_{ijK} + e_{ij}$$

In this equation, the x_{ijk}'s are defined as

$X_{ij1} = 1$ if the observation corresponds to factor-level one

$\quad\quad = 0$ otherwise

$X_{ij2} = 1$ if the observation corresponds to factor-level two

$\quad\quad = 0$ otherwise

and so on.

Note that no constant term is included in the regression equation. The constant is unnecessary (and would result in a perfect multicollinearity problem if included). There is an alternate form of the regression model equivalent to the second ANOVA model, but it is not discussed here because regression is typically not used for ANOVA-type problems.

EXAMPLE 9.1   **Automobile Injuries**   Injury claims (INJURY) for 1984–1986 cars are listed in Table 9.1 (see the file INJURY9 on the CD) for cars in the following categories:

small two-door

midsized two-door

large two-door

small four-door

midsized four-door

large four-door

small station wagons and vans

midsized station wagons and vans

large station wagons and vans

The variable CARCLAS is used to indicate into which category each car falls. The categories are coded from 1 to 9 in the order shown.

TABLE 9.1 Data for Automobile Injuries Example

Car	CARCLAS	INJURY	Car	CARCLAS	INJURY	Car	CARCLAS	INJURY
Saab 900	1	70	Mazda 626	4	96	Dodge Diplomat	6	61
Honda Prelude	1	97	Volkswagen Jetta	4	102	Mercury Grand		
Mazda 626	1	103	Nissan Stanza	4	103	Marquis	6	63
Toyota Celica	1	109	Honda Civic	4	112	Plymouth Grand Fury	6	66
Subaru Hatchback	1	115	Toyota Corolla	4	115	Buick Electra	6	68
Volkswagen Scirocco	1	115	Chevrolet Nova	4	115	Pontiac Parisienne	6	70
Mitsubishi Starion	1	119	Plymouth Horizon	4	116	Oldsmobile		
Dodge Daytona	1	120	Toyota Tercel	4	121	Ninety-Eight	6	70
Plymouth Colt	1	124	Ford Escort	4	121	Toyota Van	7	75
Ford Escort	1	125	Renault Alliance	4	131	Volkswagen Vanagon	7	78
Mercury Lynx	1	137	Subaru Dl/Gl Sedan	4	132	Subaru Dl/Gl 4-Wh. Dr.	7	84
Mitsubishi Cordia	1	137	Mazda 323	4	137	Toyota Tercel 4-Wh. Dr.	7	90
Dodge Colt	1	139	Dodge Colt	4	140	Nissan Stanza	7	93
Renault Alliance	1	139	Nissan Sentra	4	142	Subaru Dl/Gl	7	98
Nissan Sentra	1	139	Mitsubishi Tredia	4	149	Honda Civic	7	100
Plymouth Turismo	1	140	Hyundai Excel	4	161	Volvo 240	8	58
Nissan Pulsar	1	151	Chevrolet Spectrum	4	177	Mercury Marquis	8	60
Chevrolet Sprint	1	152	Ford Taurus	5	73	Buick Century	8	73
Chevrolet Spectrum	1	153	Volvo 240	5	78	Chrysler Lebaron	8	74
Mitsubishi Mirage	1	158	Toyota Camry	5	80	Oldsmobile Cutlass		
Oldsmobile Cutlass			Pontiac Bonneville	5	81	Ciera	8	75
Ciera	2	80	Plymouth Caravelle	5	84	Nissan Maxima	8	81
Pontiac 6000	2	85	Pontiac 6000	5	84	Ford Celebrity	8	82
Chrysler Lebaron	2	91	Mercury Marquis	5	86	Pontiac Sunbird	8	89
Honda Accord	2	93	Honda Accord	5	87	Plymouth Reliant	8	91
Chevrolet Monte Carlo	2	95	Chrysler Lebaron Gts	5	87	Dodge Colt Vista	8	93
Buick Regal	2	97	Ford Ltd	5	90	Plymouth Colt Vista	8	97
Ford Thunderbird	2	102	Dodge Lancer	5	90	Chevrolet Cavalier	8	98
Mercury Cougar	2	111	Chevrolet Celebrity	5	91	Buick Skyhawk	8	103
Plymouth Reliant	2	117	Nissan Maxima	5	91	Pontiac Parisienne	9	52
Pontiac Grand Am	2	117	Dodge 600	5	94	Mercury Grand Marquis	9	53
Ford Tempo	2	123	Audi 4000	5	95	Oldsmobile Custom		
Buick Skyhawk	2	123	Buick Skyhawk	5	102	Cruiser	9	54
Chevrolet Cavalier	2	131	Mitsubishi Galant	5	106	Chevrolet Caprice	9	60
Mercury Grand Marquis	3	55	Plymouth Reliant	5	107	Plymouth Voyager	9	63
Ford Crown Victoria	3	70	Mercury Topaz	5	107	Dodge Caravan	9	65
Oldsmobile			Ford Tempo	5	110	Ford Aerostar	9	65
Ninety-Eight	3	73	Oldsmobile Firenza	5	113	Chevrolet Astro Van	9	68
Chevrolet Caprice	3	75	Chevrolet Cavalier	5	118			
Saab 900	4	66	Pontiac Sunbird	5	121			

Source: Copyright, September 28, 1987, *U.S. News and World Report*. Reprinted with permission.

In ANOVA terminology, the dependent variable is INJURY. Our concern is whether the average number of injuries differs for the nine different categories or populations. Our data comprise nine samples from each of these nine populations. The explanatory variable or factor is CARCLAS. Note that this is a qualitative variable. ANOVA is designed to investigate the relationship between a quantitative dependent variable and a qualitative independent variable or factor. There are nine levels of this factor because there are nine types of cars coded. The averages of the injuries for each type of car are the factor level means.

We want to test whether the factor level means are equal:

$H_0: \mu_1 = \mu_2 = \cdots = \mu_9$

H_a: Not all of the means are equal

Figure 9.1 shows the MINITAB ANOVA output for this problem. Figure 9.2 shows the Excel output and Figure 9.3 shows the SAS output. Figure 9.4 shows the general form of the ANOVA table for each of the software packages.

In Figure 9.4(a), the general form of the MINITAB output is shown. The sums of squares due to the treatment variable, error, and the total sum of squares are printed along with their degrees of freedom. *MSTR*, *MSE*, the *F* statistic, and a *p* value associated with the observed *F* statistic are computed. Below the ANOVA table, MINITAB provides the standard error, S (s_e in this text), which is the square root of MSE, the R^2 and R_{adj}^2. These values have interpretations similar to regression. This makes sense when you think of ANOVA as a special case of regression on indicator variables. MINITAB also lists summary information for the factor levels. The sample size, mean, and standard deviation for the observations in each factor-level are given. Individual confidence intervals for each factor-level (population) mean are shown graphically. (Note that these intervals should be used with caution for comparing two factor-level means. Better approaches are discussed later.)

FIGURE 9.1 MINITAB ANOVA Output for Automobile Injuries Example.

One-way ANOVA: INJURY versus CARCLAS

Source	DF	SS	MS	F	P
CARCLAS	8	54762	6845	22.80	0.000
Error	103	30917	300		
Total	111	85679			

S = 17.33   R-Sq = 63.92%   R-Sq(adj) = 61.11%

Individual 95% CIs For Mean Based on Pooled StDev

Level	N	Mean	StDev	
1	20	127.10	21.89	(--*--)
2	13	105.00	16.29	(---*---)
3	4	68.25	9.07	(------*------)
4	18	124.22	25.58	(---*--)
5	23	94.57	13.38	(--*--)
6	6	66.33	3.72	(-----*----)
7	7	88.29	9.64	(----*-----)
8	13	82.62	14.21	(---*---)
9	8	60.00	6.23	(----*----)

```
                  -+---------+--------- +---------+--------
                   50        75        100       125
```

Pooled StDev = 17.33

FIGURE 9.2 Excel ANOVA Output for Automobile Injuries Example.

Anova: Single Factor

SUMMARY

Groups	Count	Sum	Average	Variance
class1	20	2542	127.100	479.042
class2	13	1365	105.000	265.500
class3	4	273	68.250	82.250
class4	18	2236	124.222	654.418
class5	23	2175	94.565	178.893
class6	6	398	66.333	13.867
class7	7	618	88.286	92.905
class8	13	1074	82.615	201.923
class9	8	480	60.000	38.857

ANOVA

Source of Variation	SS	df	MS	F	P-value	F crit
Between Groups	54762.268	8	6845.283	22.805	0.000	2.030
Within Groups	30917.152	103	300.167			
Total	85679.420	111				

FIGURE 9.3 SAS ANOVA Output for Automobile Injuries Example.

The ANOVA Procedure

Class Level Information

Class	Levels	Values
CARCLAS	9	1 2 3 4 5 6 7 8 9

Number of observations   112

The ANOVA Procedure

Dependent Variable: INJURY

Source	DF	Sum of Squares	Mean Square	F Value	Pr > F
Model	8	54762.26753	6845.28344	22.80	<.0001
Error	103	30917.15211	300.16653		
Corrected Total	111	85679.41964			

R-Square	Coeff Var	Root MSE	INJURY Mean
0.639153	17.38585	17.32531	99.65179

Source	DF	Anova SS	Mean Square	F Value	Pr > F
CARCLAS	8	54762.26753	6845.28344	22.80	<.0001

FIGURE 9.4
Structure of the MINITAB, Excel, and SAS ANOVA Tables.

(a) MINITAB Analysis of Variance for (dependent variable name)

Source	DF	SS	MS	F	P
Treatment var	$K-1$	$SSTR$	$MSTR$	$\dfrac{MSTR}{MSE}$	p value
Error	$n-K$	SSE	MSE		
Total	$n-1$	SST			

(b) Excel ANOVA

Source of Variation	SS	df	MS	F	P-value	F crit
Between Groups	$SSTR$	$K-1$	$MSTR$	$\dfrac{MSTR}{MSE}$	p value	$F(\alpha; K-1, n-K)$
Within Groups	SSE	$n-K$	MSE			
Total	SST	$n-1$				

(c) SAS Dependent Variable: (dependent variable name)

Source	DF	Sum of Squares	Mean Square	F Value	Pr > F
Model	$K-1$	$SSTR$	$MSTR$	$\dfrac{MSTR}{MSE}$	p value
Error	$n-K$	SSE	MSE		
Corrected Total	$n-1$	SST			

The description of a general ANOVA table for Excel is given in Figure 9.4(b). Information similar to that shown in the MINITAB ANOVA table is represented in the Excel ANOVA table. One difference is that the Excel table also provides a critical value for the F statistic. Excel lists summary information for the factor levels above the ANOVA table. The sample size, the sum of the observations, the mean, and the variance for the observations in each factor level are given. Excel refers to the treatment sum of squares as the between-groups sum of squares and the error sum of squares as the within-groups sum of squares.

The description of a general ANOVA table for SAS is given in Figure 9.4(c). SAS names the sums of squares Model (in place of Treatment), Error, and Corrected Total and lists the same information as MINITAB and Excel. Below the ANOVA table SAS lists the R^2, the coefficient of variation (Coeff Var), which is the square root of the MSE divided by the mean of the dependent variable, the standard error (Root MSE), and the dependent variable mean (INJURY Mean). Below that SAS breaks out the sum of squares information for the treatment, in this case CARCLAS. (In some later sections the model will consist of more than one factor so this breakdown will be more useful.)

The decision rule to test the hypotheses using a 5% level of significance is

Reject H_0 if $F > 2.10$

Do not reject H_0 if $F \leq 2.10$

The critical value is chosen from the $\alpha = 0.05$ level F table with 8 numerator and 103 (approximately) denominator degrees of freedom. (Using Excel, the critical value provided is 2.03. Excel computes the exact critical value rather than approximating the value from a table.) The test statistic value, 22.80, exceeds the critical value, so the null hypothesis is rejected. There is a difference in the factor-level means. The average level of injuries differs depending on the type of car.   ◼

Analysis of variance is often used in an experimental design situation. The term *experiment* refers to the data collection process. The term *design* refers to the plan for conducting the experiment. In many situations, the researcher can assign objects upon which measurements are to be made (called *experimental units*) to the factor levels or treatments. This was not done in the previous example. Once a car was chosen for the example—say, a Ford Thunderbird—the researcher had no control over which factor level this case was assigned. A Ford Thunderbird is a midsized two-door, so it automatically is assigned to factor level two.

EXAMPLE 9.2   **Computer Sales**   Consider a situation where experimental design can be used. The effect of three different selling approaches on sales of computers is to be studied. The object is to determine whether there is a difference in the effectiveness of the selling approaches. We will judge differences in effectiveness by looking at the average sales of the three approaches (see the file COMPSAL9 on the CD). An approach with significantly higher average sales will be judged more effective. Fifteen salespeople are chosen to participate in the study. Five salespeople each are randomly assigned to use one of the three approaches for the next month. At the end of the month, sales figures will be computed for each salesperson. The data will be analyzed to determine whether the sales approaches produce the same or different average sales.

In this situation, the salespeople were randomly assigned to the factor levels or treatments. This random assignment is possible when an experiment is designed to help answer a particular question. The type of experimental design used in this case is called a *completely randomized design*. Analysis of variance can be used in this situation to analyze the data just as in the automobile-injuries example.

Suppose the following sales figures (in $1000) resulted from the experiment just described:

SELLING APPROACHES

Salesperson	A	B	C
1	15	19	28
2	17	17	25
3	21	17	22
4	13	25	31
5	12	30	34

The hypotheses to be tested are

$H_0: \mu_A = \mu_B = \mu_C$

$H_a:$ All three means are not equal

where μ_i is the population average sales for selling approach i. The ANOVA results for this example are shown in Figure 9.5. To test the hypotheses, either the F statistic

FIGURE 9.5

ANOVA Results for Computer Sales Example.

Source	DF	Sum of Squares	Mean Square	F Stat	P Value
Treatment	2	384.5	192.3	8.47	0.005
Error	12	272.4	22.7		
Total	14	656.9			

or its associated p value could be used. If the F statistic is used, the decision rule is (with a 5% level of significance):

Reject H_0 if $F > 3.89$

Do not reject H_0 if $F \leq 3.89$

where the critical value is chosen with 2 numerator and 12 denominator degrees of freedom. The test statistic is $F = 8.47$, so the decision is to reject the null hypothesis.

Rejection of the null hypotheses leads to the conclusion that the population average sales differ depending on what selling approach is used.

Note that rejection of the null hypothesis simply says that the population means are not all equal. It does not say they are all different or tell which ones are different from the others. In a case such as this, the researcher probably wants to know which population averages are different or whether two particular ones differ. A $(1 - \alpha)100\%$ confidence interval estimate of the difference between two means, μ_i and μ_j, can be constructed as follows:

$$(\bar{y}_{i.} - \bar{y}_{j.}) \pm t_{\alpha/2,\,df}\, s_e \sqrt{\left(\frac{1}{n_i}\right) + \left(\frac{1}{n_j}\right)}$$

where $t_{\alpha/2,\,df}$ is the value chosen to put $\alpha/2$ probability in the upper tail of the t distribution using df $= n - K$ degrees of freedom, s_e is the square root of MSE, and $\bar{y}_{i.}$ and $\bar{y}_{j.}$ are the sample means for samples i and j (or factors i and j).

EXAMPLE 9.3 **Computer Sales (continued)** In the computer sales example, the following is the 95% confidence interval estimate of the difference between selling approaches A and C:

$$(15.6 - 28.0) \pm (2.179)(4.76)\sqrt{\left(\frac{1}{5}\right) + \left(\frac{1}{5}\right)}$$

or $(-18.96, -5.84)$. Because the 95% confidence interval estimate of the difference between these two population means does not contain zero, this suggests that the population means for methods A and C are not equal.

This type of comparison works when only two means are compared. If this type of interval estimate is used for a series of comparisons, the level of significance will no longer be appropriate. Various procedures can be used when such multiple comparisons are desired. The Tukey method and the Bonferroni method will be discussed here.[1]

---

[1]See Neter, Wasserman, and Kutner, *Applied Linear Statistical Models*, p. 584, for a further discussion of multiple comparison methods. See References for complete publication information.

The *Bonferroni approach* is a method of comparing multiple quantities (means in this case) that assumes the user can specify in advance which quantities are to be compared. The Bonferroni intervals used to compare pairs of means are constructed as

$$(\bar{y}_{i.} - \bar{y}_{j.}) \pm Bs_e\sqrt{\left(\frac{1}{n_i}\right) + \left(\frac{1}{n_j}\right)}$$

where B is a t value chosen to put $\alpha/(2g)$ probability in the upper tail of the t distribution using $n - K$ degrees of freedom. Here g is the number of comparisons to be made.

EXAMPLE 9.4 **Computer Sales (continued)** Suppose all possible pairs of means are to be compared in the computer sales example with a 95% confidence level. Intervals are to be constructed to estimate $\mu_A - \mu_B$, $\mu_B - \mu_C$, and $\mu_A - \mu_C$. Because there are three comparisons to be made ($g = 3$), the Bonferroni confidence coefficient, B, is the t value that puts $0.05/[(2)(3)] = 0.008$ probability in the upper tail of the t distribution. In this example, the 0.005 t value is used in order to insure at least a 95% confidence level.

The interval estimates are

$$\mu_A - \mu_B: (15.6 - 21.6) \pm (3.055)(4.76)\sqrt{\left(\frac{1}{5}\right) + \left(\frac{1}{5}\right)} \qquad \text{or} \qquad (-15.20, 3.20)$$

$$\mu_B - \mu_C: (21.6 - 28.0) \pm (3.055)(4.76)\sqrt{\left(\frac{1}{5}\right) + \left(\frac{1}{5}\right)} \qquad \text{or} \qquad (-15.6, 2.80)$$

$$\mu_A - \mu_C: (15.6 - 28.0) \pm (3.055)(4.76)\sqrt{\left(\frac{1}{5}\right) + \left(\frac{1}{5}\right)} \qquad \text{or} \qquad (-21.60, -3.20)$$

The three estimates are said to have a *familywise* confidence level of 95%. This means that the estimates can be viewed simultaneously rather than one at a time and that the 95% confidence level applies.

The *Tukey approach* is used to examine all pairwise comparisons of the treatment means. The confidence intervals created by the Tukey method choose the confidence coefficient in a different way. If the sample sizes for all samples are equal, the Tukey confidence intervals for all pairwise comparisons of means are constructed as

$$(\bar{y}_{i.} - \bar{y}_{j.}) \pm q(p, v)\frac{s_e}{\sqrt{n_i}}$$

where $q(p,v)$ is the critical value of the studentized range, p is the number of means to be compared, and $n - K$ is the error degrees of freedom. The critical values for the studentized range are available in Appendix B.

If the sample sizes are not all equal, the following formula for the confidence intervals is used:

$$(\bar{y}_{i.} - \bar{y}_{j.}) \pm q(p, v)\frac{s_e}{\sqrt{2}}\sqrt{\frac{1}{n_i} + \frac{1}{n_j}}$$

Example 9.5 demonstrates the use of the Tukey intervals and shows how the results would appear when using MINITAB.

EXAMPLE 9.5   **Computer Sales (once more)**   Suppose all possible pairs of means are to be compared in the computer sales example as in Example 9.4. In this example we use the Tukey comparisons. For a 95% familywise confidence level, the studentized range confidence coefficient would be $q(3, 12) = 3.77$. The intervals would be as follows:

$$\mu_A - \mu_B: (15.6 - 21.6) \pm (3.77)\frac{4.76}{\sqrt{5}} \quad \text{or} \quad (-14.03, 2.03)$$

$$\mu_B - \mu_C: (21.6 - 28.0) \pm (3.77)\frac{4.76}{\sqrt{5}} \quad \text{or} \quad (-14.43, 1.63)$$

$$\mu_A - \mu_C: (15.6 - 28.0) \pm (3.77)\frac{4.76}{\sqrt{5}} \quad \text{or} \quad (-20.43, -4.37)$$

Figure 9.6 shows the MINITAB output when Tukey comparisons are requested (see Using the Computer section). Note that the subtraction to find the difference in each pair of sample means is done in the reverse order from the example worked out here. So the interval estimate of $\mu_A - \mu_B$ is given as $(-2.03, 14.03)$ on the MINITAB output rather than $(-14.03, 2.03)$ in the example. This just requires a little caution in interpreting the results, but any conclusions should be exactly the same regardless of the order of the subtraction.

EXERCISES

1. **Automobile Collisions.** The number of collision claims (COLLISION) reported for 1984–1986 cars are listed in the same categories as described for Example 9.1 and are available in a file named CRASH9 on the CD. For MINITAB, the file will have two columns, one with the data and one with a classification variable. The Excel spreadsheet will have nine columns containing the data for each type of car. (See Using the Computer for more on setting up data for analysis with ANOVA).

Using the classification variable (CARCLAS) described in that example, an ANOVA was run on the number of collisions. Figure 9.7 summarizes the results. Determine whether there is a difference in the average number of collisions for different types of cars. Use a 5% level of significance. State the hypotheses to be tested, the decision rule, the test statistic, and your decision. (*Source*: Copyright, September 28, 1987, *U.S. News and World Report*. Used with permission.)

FIGURE 9.6 Minitab ANOVA Ouput Showing Tukey 95% Simultaneous Confidence Intervals.

```
Source        DF      SS      MS      F       P
APPROACH       2    384.5   192.3   8.47    0.005
Error         12    272.4    22.7
Total         14    656.9


S = 4.764   R-Sq = 58.53% R-Sq(adj) = 51.62%


                            Individual 95% CIs For Mean Based on
                            Pooled StDev
Level   N     Mean   StDev   --+---------+---------+---------+-------
1       5    15.600  3.578   (-------*-------)
2       5    21.600  5.727              (-------*-------)
3       5    28.000  4.743                        (-------*------)
                            --+---------+---------+---------+-------
                            12.0      18.0      24.0      30.0
Pooled StDev = 4.764


Tukey 95% Simultaneous Confidence Intervals
All Pairwise Comparisons among Levels of APPROACH

Individual confidence level = 97.94%

APPROACH = 1 subtracted from:

APPROACH   Lower   Center   Upper   ----+---------+---------+---------+-----
2         -2.033    6.000  14.033                  (-------*-------)
3          4.367   12.400  20.433                        (-------*-------)
                                    ----+---------+---------+---------+-----
                                      -10         0        10        20


APPROACH = 2 subtracted from:
APPROACH   Lower   Center   Upper   ----+---------+---------+---------+-----
3         -1.633    6.400  14.433                  (-------*-------)
                                    ----+---------+---------+---------+-----
                                      -10         0        10        20
```

FIGURE 9.7 ANOVA Results for Automobile Collisions Example.

Source	DF	Sum of Squares	Mean Square	F Stat	P Value
Treatment	8	63790	7974	14.82	0.000
Error	103	55427	538		
Total	111	119218			

9.2 ANALYSIS OF VARIANCE USING A RANDOMIZED BLOCK DESIGN

In some situations, designs other than the completely randomized design discussed in the previous section can be beneficial. Another type of experimental design is called a *randomized block design*. Only randomized block designs with one value per treatment-block combination are considered in this text. (Randomized block designs with more than one value per treatment-block combination can be used, but there are adjustments that are needed such as different equations for various degrees of freedom.) Consider again the problem in Example 9.2 involving the study of the three different selling approaches. In the completely randomized design, 15 salespeople were used. Five people each were randomly assigned to each selling approach (treatment). One-way ANOVA was then used to determine whether the average sales for each group were significantly different. Two types of variation affect the sample averages computed in this case: (a) the variation that results from the differences in selling approaches and (b) the variation from the individuals involved in the study. Certain individuals simply may be better salespeople than others regardless of the sales approach used.

The goal of the study is to determine whether the difference in the sample means that results from differences in selling approaches is significant. The second type of variation is a hindrance to this goal because the additional variation may make it difficult to determine what is actually causing the differences in the means. Is it the selling approaches themselves or the individuals assigned to the three groups? To eliminate this second source of variation, a randomized block design can be used. The idea of the randomized block design is to compare the means of treatments within fairly homogeneous blocks of experimental units.

In the example given, instead of choosing 15 people and randomly assigning them to use a particular selling approach, it might be better to choose 5 people and let each person use all three of the selling approaches. The treatments in this study are still the three selling approaches. The blocks are the salespeople themselves. The experimental units within a block are instances when the different treatments can be "applied" to the salespeople. Table 9.2 illustrates this situation. The treatments (selling approaches) are denoted $T1$, $T2$, and $T3$. The three treatments have been applied to each block (salesperson) in a randomized order. Thus, during the first month, salespersons 1 and 5 use approach 1, salespersons 3 and 4 use approach 3, and salesperson 2 uses approach 2. The order of the treatments within each block is determined by using a random number table.

When the blocks are the subjects to be used in the experiment, the design is often referred to as a *repeated measures design*. One advantage of this type of randomized block design is that all sources of variability between subjects are removed. Another advantage is that fewer subjects are needed. In this example, 5 subjects were needed rather than the 15 needed when the completely randomized design was used.

TABLE 9.2 Assignment of Treatments to Experimental Units for Computer Sales Example

Blocks	Treatment Order		
1	$T1$	$T3$	$T2$
2	$T2$	$T1$	$T3$
3	$T3$	$T1$	$T2$
4	$T3$	$T2$	$T1$
5	$T1$	$T2$	$T3$

One serious potential disadvantage of the repeated measures design is the effect of time order in the application of the treatments. In this example, sales may vary somewhat by month—that is, those treatments applied in month 1 may produce higher sales than in month 2 simply because of some type of monthly seasonal variation. An attempt has been made to minimize this variation by randomizing the treatment order independently for each subject.

Another example of a randomized block design that is not a repeated measures design follows.

EXAMPLE 9.6   **Cereal Package Design**   A company wishes to evaluate the effect of package design on one of its products, a certain brand of cereal. The four package designs (treatments) are to be tested in different stores throughout a large city. There are 20 stores available for the study. The amount of cereal sold is known to vary depending on the size of the stores, so size is used as a blocking variable. The 20 stores are divided into five groups of four stores each by size. Table 9.3 illustrates the design. The treatments are randomized within each block. The randomization is performed independently between blocks (file CEREAL9 on the CD).

The model for a randomized block design can be written

$$y_{ij} = \mu + \gamma_i + B_j + e_{ij}$$

where

μ is an overall mean

γ_i is the ith treatment effect

B_j is the jth block effect

e_{ij} is the random disturbance for treatment i and block j

There are K treatments and b blocks, so $i = 1, \ldots, K$ and $j = 1, \ldots, b$.

An example of the type of ANOVA table used to analyze this model is shown in Figure 9.8. Three sources of variation are identified: blocks, treatments, and error. The sums of squares, the mean squares, and the degrees of freedom associated with each of these sources are shown, along with F statistics and p values. Note that the degrees of freedom from the three sources sum to the total degrees of freedom: $(b - 1) + (K - 1) + (b - 1)(K - 1) = bK - 1$. This is also true for the sums of squares: $SSBL + SSTR + SSE = SST$.

TABLE 9.3 Assignment of Treatments to Experimental Units for Cereal Package Design Example

Blocks	Treatment Order			
1	$T2$	$T1$	$T3$	$T4$
2	$T1$	$T2$	$T4$	$T3$
3	$T4$	$T3$	$T1$	$T2$
4	$T3$	$T1$	$T2$	$T4$
5	$T1$	$T4$	$T3$	$T2$

FIGURE 9.8 ANOVA Table for Randomized Block Design.

Source	DF	Sum of Squares	Mean Square	F Stat	p Value
Blocks	$b-1$	SSBL	MSBL	MSBL/MSE	p value
Treatments	$K-1$	SSTR	MSTR	MSTR/MSE	p value
Error	$(b-1)(K-1)$	SSE	MSE		
Total	$bK-1$	SST			

The goal of the analysis is to determine whether the treatment effects differ:

H_0: $\gamma_1 = \gamma_2 = \cdots = \gamma_K$

H_a: All treatment effects are not equal

The decision rule to perform the test is:

Reject H_0 if $F > F(\alpha; K-1,(b-1)(K-1))$

Do not reject H_0 if $F \leq F(\alpha; K-1,(b-1)(K-1))$

The test statistic is

$$F = \frac{MSTR}{MSE}$$

If the null hypothesis is rejected, then the conclusion is that the treatment effects differ.

A test can also be conducted to determine whether the block effects differ:

H_0: $B_1 = B_2 = \cdots = B_b$

H_a: All block effects are not equal

The decision rule to perform the test is

Reject H_0 if $F > F(\alpha; b-1,(b-1)(K-1))$

Do not reject H_0 if $F \leq F(\alpha; b-1,(b-1)(K-1))$

The test statistic is

$$F = \frac{MSBL}{MSE}$$

where MSBL is the mean square due to blocks.

If the null hypothesis is rejected, then the conclusion is that the block effects differ. In this case, the use of blocking has helped. If the null hypothesis is not rejected, then the conclusion is that the block effects are equal. The use of blocks appears unnecessary, and information may actually be lost by blocking. In many cases, however, blocking is used regardless of the outcome of this test if it is felt that blocking truly produces more homogeneous groups and therefore reduces variation. As a result, the test for block effects is often not used.

A $(1 - \alpha)100\%$ confidence interval estimate for the difference between two treatment means is given by

$$(\bar{y}_i - \bar{y}_{i'}) \pm t_{\alpha/2,(b-1)(K-1)} s_e \sqrt{\frac{2}{b}}$$

where $\bar{y}_i$ and $\bar{y}_{i'}$ are sample means for treatments i and i', the t value is chosen with $(b - 1)(K - 1)$ degrees of freedom, s_e is the square root of MSE, and b is the number of blocks. If more than one pair of treatment means is to be compared, a family-wise $(1 - \alpha)100\%$ level of confidence can be achieved by using the Bonferroni confidence coefficient described in the previous section.

EXAMPLE 9.7   **Cereal Package Design (continued)**   Consider again the design of cereal packages (file CEREAL9 on the CD). Four package designs (treatments) are to be tested in 20 stores. The 20 stores are divided into five groups (blocks) of four stores each by size. The treatments are randomized within each block and independently between blocks. The data obtained are shown in Table 9.4. This is the way the data would be entered in Excel. Table 9.5 shows the data in an alternate form. This is the way the data would be entered for analysis in MINITAB and SAS. See the Using the Computer section for more information.

The MINITAB output for this example is shown in Figure 9.9, the Excel output is in Figure 9.10 and the SAS output is in Figure 9.11. In the MINITAB and SAS output, the blocks are denoted SIZE and the treatments are denoted DESIGN. In the Excel output, the blocks are Rows and treatments are Columns. The MINITAB, Excel, and SAS ANOVA tables contain the information in the general ANOVA table of Figure 9.8. The general structure of the MINITAB, Excel, and SAS ANOVA tables is shown in Figure 9.12. The Excel output shows some additional information above the ANOVA table. Included are the number of observations, the sum, average, and variance of the observations in each block and at each treatment level. The ANOVA table for SAS is set up in two parts. In the first part the sources of variation are labeled Model, Error, and Corrected Total. Below that, the Model variation is partitioned into variation due to SIZE and DESIGN. Model variation combines the information from both the treatment and

TABLE 9.4 Data for Cereal Package Design Example as It Would Be Entered for Excel Analysis

	Sales (in $1000)			
Blocks	Treatments			
1	40	23	17	33
2	32	45	40	25
3	43	31	38	47
4	44	41	56	45
5	43	60	47	64

TABLE 9.5 Data for Cereal Package Design Example as It Would Be Entered in MINITAB or SAS

SALES	SIZE	DESIGN
40	1	1
23	1	2
17	1	3
33	1	4
32	2	1
45	2	2
40	2	3
25	2	4
43	3	1
31	3	2
38	3	3
47	3	4
44	4	1
41	4	2
56	4	3
45	4	4
43	5	1
60	5	2
47	5	3
64	5	4

FIGURE 9.9 MINITAB ANOVA Results for Cereal Package Design Example.

Source	DF	SS	MS	F	P
SIZE	4	1521.7	380.425	4.17	0.024
DESIGN	3	31.0	10.333	0.11	0.951
Error	12	1093.5	91.125		
Total	19	2646.2			

S = 9.546   R-Sq = 58.68%   R-Sq(adj) = 34.57%

FIGURE 9.10 Excel ANOVA Output for Cereal Package Design Example.

Anova: Two-Factor Without Replication

SUMMARY	Count	Sum	Average	Variance
SIZE1	4	113	28.25	104.92
SIZE2	4	142	35.50	77.67
SIZE3	4	159	39.75	47.58
SIZE4	4	186	46.50	43.00
SIZE5	4	214	53.50	101.67
DESIGN1	5	202	40.40	24.30
DESIGN2	5	200	40.00	199.00
DESIGN3	5	198	39.60	209.30
DESIGN4	5	214	42.80	221.20

ANOVA

Source of Variation	SS	df	MS	F	P-value	F crit
Rows	1521.700	4	380.425	4.175	0.024	3.259
Columns	31.000	3	10.333	0.113	0.951	3.490
Error	1093.500	12	91.125			
Total	2646.200	19				

FIGURE 9.11
SAS ANOVA Output
for Cereal Package
Design Example.

```
                          The ANOVA Procedure
                        Class Level Information

              Class          Levels    Values
              SIZE                5     1 2 3 4 5
              DESIGN              4     1 2 3 4

                  Number of observations    20

                          The ANOVA Procedure

Dependent Variable: SALES

                              Sum of
   Source            DF       Squares      Mean Square    F Value    Pr > F

   Model              7     1552.700000    221.814286       2.43     0.0842

   Error             12     1093.500000     91.125000

   Corrected Total   19     2646.200000


              R-Square     Coeff Var     Root MSE     SALES Mean
              0.586766     23.45440      9.545942      40.70000


   Source      DF      Anova SS      Mean Square     F Value     Pr > F

   SIZE         4    1521.700000     380.425000        4.17      0.0240
   DESIGN       3      31.000000      10.333333        0.11      0.9506
```

the block into a single measure of the quality of the model. The F statistic for Model tests the combined effect of the treatment and block in explaining variation in SALES. More typically, these effects are separated out into their individual components so that treatment effects can be examined separately from block effects.

To determine whether there is a difference in sales because of package design, the following hypotheses should be tested:

$H_0: \gamma_1 = \gamma_2 = \gamma_3 = \gamma_4$
H_a: All treatment effects are not equal

The decision rule for the test is

Reject H_0 if $F > F(0.05; 3,12) = 3.49$
Do not reject H_0 if $F \leq F(0.05; 3,12) = 3.49$

The test statistic is

$F = 0.11$

The null hypothesis cannot be rejected. There is no evidence of a difference in sales because of package design.

(a) MINITAB

Analysis of Variance for (dependent variable name)

Source	DF	SS	MS	F	P
Blocks	$b-1$	$SSBL$	$MSBL$	$MSBL/MSE$	p value
Treatments	$K-1$	$SSTR$	$MSTR$	$MSTR/MSE$	p value
Error	$(b-1)(K-1)$	SSE	MSE		
Total	$bK-1$	SST			

(b) Excel

ANOVA

Source of Variation	SS	df	MS	F	P-value	F crit
Rows	$SSBL$	$b-1$	$MSBL$	$MSBL/MSE$	p value	$F(\alpha; b-1,(b-1)(K-1))$
Columns	$SSTR$	$K-1$	$MSTR$	$MSTR/MSE$	p value	$F(\alpha; K-1,(b-1)(K-1))$
Error	SSE	$(b-1)(K-1)$	MSE			
Total	SST	$bK-1$				

(c) SAS

Dependent Variable (dependent variable name)

Source	DF	Sum of Squares	Mean Square	F Value	Pr > F
Model	$(b-1)+(K-1)$	$SSBL+SSTR$	SS Model/DF Model	$MSModel/MSE$	p value
Error	$(b-1)(K-1)$	SSE	MSE		
Corrected Total	$bK-1$	SST			
Blocks	$b-1$	$SSBL$	$MSBL$	$MSBL/MSE$	p value
Treatments	$K-1$	$SSTR$	$MSTR$	$MSTR/MSE$	p value

FIGURE 9.12 Structure of the MINITAB, Excel, and SAS ANOVA Tables for a Randomized Block Design.

EXERCISES

2. **Advertising.** A study was undertaken by a company that markets iced tea to determine the effectiveness of three types of advertising: (a) price discounts, (b) manufacturer's coupons, and (c) rebates. Three cities of nearly equal size are selected for the experiment. Each of the advertising strategies is to be used for a period of 2 months in each city, and the sales of the company's iced tea recorded for the 2-month period. It is known, however, that iced tea sales are highly seasonal. Thus, the months selected for a particular ad strategy could highly influence the level of sales. A randomized block design is used to eliminate the seasonal effects.

For each time period (block), the advertising strategies (treatments) are randomly assigned to the three cities. The results are shown in Table 9.6.

The variable ADV is coded as $1 =$ price discounts, $2 =$ manufacturer's coupons, and $3 =$ rebates. The variable BLOCK is coded as $1 =$ first 2-month period, $2 =$ second 2-month period, and $3 =$ third 2-month period.

Treating the experiment as a randomized block design, the ANOVA in Figure 9.13 was obtained. Use the results to determine whether the effects of the advertising strategies differ. Use a 5% level of significance. State any hypotheses to be tested, the decision rule, the test statistic, and your decision.

These data are available in a file named ADVERT9 on the CD in three columns: SALES, ADV, and BLOCK for MINITAB. The Excel Spreadsheet contains the data in the form needed for the Excel analysis as discussed earlier in this section.

TABLE 9.6 Data for Advertising Exercise

SALES	BLOCK	ADV
30	1	1
21	2	1
14	3	1
33	1	2
25	2	2
19	3	2
41	1	3
32	2	3
24	3	3

FIGURE 9.13 ANOVA Output for Advertising Exercise.

Source	DF	Sum of Squares	Mean Square	F Stat	P Value
BLOCK	2	369.556	184.778	302.36	0.000
ADV	2	174.222	87.111	142.55	0.000
Error	4	2.444	0.611		
Total	8	546.222			

9.3 TWO-WAY ANALYSIS OF VARIANCE

Two-way ANOVA refers to a situation where there are two factors or explanatory variables. For example, suppose a company wants to investigate the effect of selling price and type of advertising on sales of its top-of-the-line printer. The two possible prices investigated are $600 and $700. There are three types of advertising: television, radio, and newspaper.

In general, denote the two factors in the study as A and B and the number of levels of each factor as n_1 and n_2, respectively. Each combination of a factor level of A and a factor level of B is called a *treatment*. The total number of possible treatments in a two-factor study is n_1n_2. In the example given, there are $2 \times 3 = 6$ possible treatments:

1. $600 price and television advertising
2. $600 price and radio advertising
3. $600 price and newspaper advertising
4. $700 price and television advertising
5. $700 price and radio advertising
6. $700 price and newspaper advertising

This type of design is called a *factorial design*. When all treatment combinations (or factor-level combinations) are used, the experiment is termed a *complete factorial experiment*. This is the design considered in this section.

Equal sample sizes are assumed for each treatment. (Unequal sample sizes are not considered in this text.) The number of observations for each treatment is denoted r. The total number of observations is $n = n_1n_2r$. The number of observations for each treatment must be at least two ($r = 2$).

The two-way ANOVA model is written

$$y_{ijk} = \mu + \alpha_i + \beta_j + (\alpha\beta)_{ij} + e_{ijk}$$

where

y_{ijk} is the kth observation for factor level i of factor A and factor level j of factor B

μ is a common component for all treatments

α_i is the effect of factor A at level i

β_j is the effect of factor B at level j

$(\alpha\beta)_{ij}$ is called the interaction effect for factors A and B

e_{ijk} is a random disturbance

Here, $i = 1, 2, \ldots, n_1$; $j = 1, 2, \ldots, n_2$; and $k = 1, 2, \ldots, r$ with $r \geq 2$.

In the two-way ANOVA setting, hypotheses can be tested to determine whether the effects due to either factor A or factor B are equal or different. The factor effects due to A and B are called *main effects*. Hypotheses also can be tested to determine whether the interaction effects are equal. *Interaction effects* are used to represent situations when differences in mean levels of factor A may depend on the levels of factor B. The model has no interaction effects if the difference in mean levels of factor A is independent of the levels of B.

The first test that should be conducted is the test for interaction effects. The tests for main effects are relevant only if no interaction exists. If interaction exists, then the tests for factor-level means are inappropriate. In this case, individual treatment means must be examined, as is demonstrated in Example 9.8.

An example of the type of ANOVA table used in two-way analysis of variance problems is shown in Figure 9.14. Total variation in the dependent variable is attributed to four sources: the two factors A and B, the interaction between the two factors, and random error. The degrees of freedom, sum of squares, and mean square for each source are shown, along with F statistics and p values. The tests that can be performed using the ANOVA table are outlined next.

TEST FOR INTERACTION EFFECTS

H_0: No interaction between factors A and B exists—All $(\alpha\beta)_{ij} = 0$

H_a: Factors A and B interact—At least one $(\alpha\beta)_{ij}$ is not equal to zero

Source	DF	Sum of Squares	Mean Square	F Stat	P Value
Factor A	$n_1 - 1$	SSA	$MSA = SSA/(n_1 - 1)$	MSA/MSE	p value
Factor B	$n_2 - 1$	SSB	$MSB = SSB/(n_2 - 1)$	MSB/MSE	p value
Interaction	$(n_1 - 1)(n_2 - 1)$	$SSINT$	$MSINT = SSINT/[(n_1 - 1)(n_2 - 1)]$	$MSINT/MSE$	p value
Error	$n_1 n_2 (r - 1)$	SSE	$MSE = SSE/[n_1 n_2 (r - 1)]$		
Total	$n_1 n_2 r - 1$	SST			

FIGURE 9.14 Sample ANOVA Table for Two-Way Design.

The decision rule for the test is

Reject H_0 if $F > F(\alpha; (n_1 - 1)(n_2 - 1), n_1 n_2 (r - 1))$
Do not reject H_0 if $F \le F(\alpha; (n_1 - 1)(n_2 - 1), n_1 n_2 (r - 1))$

The test statistic is

$$F = \frac{MSINT}{MSE}$$

where *MSINT* is the mean square due to interaction. If the null hypothesis is accepted, then conclude that there is no interaction between the factors A and B. To determine whether the factor-level means are equal, the following tests can be performed.

TEST FOR FACTOR *A* EFFECTS

H_0: $\alpha_1 = \alpha_2 = \cdots = \alpha_{n_1}$
H_a: The α_i are not all equal

The decision rule for the test is

Reject H_0 if $F > F(\alpha; n_1 - 1, n_1 n_2 (r - 1))$
Do not reject H_0 if $F \le F(\alpha; n_1 - 1, n_1 n_2 (r - 1))$

The test statistic is

$$F = \frac{MSA}{MSE}$$

TEST FOR FACTOR *B* EFFECTS

H_0: $\beta_1 = \beta_2 = \ldots = \beta_{n_2}$
H_a: The β_j are not all equal

The decision rule for the test is

Reject H_0 if $F > F(\alpha; n_2 - 1, n_1 n_2 (r - 1))$
Do not reject H_0 if $F \le F(\alpha; n_2 - 1, n_1 n_2 (r - 1))$

The test statistic is

$$F = \frac{MSB}{MSE}$$

To construct a $(1 - \alpha)100\%$ confidence interval estimate of the difference between two treatment means, use the following formula

$$(\bar{y}_i - \bar{y}_{i'}) \pm t_{\alpha/2, n_1 n_2(r-1)} s_e \sqrt{\frac{2}{r}}$$

where s_e is the square root of *MSE*, the t value is chosen with $n_1 n_2 (r - 1)$ degrees of freedom, and $\bar{y}_i$ and $\bar{y}_{i'}$ are sample means using the r observations of the appropriate treatments (factor-level combinations).

If more than one pair of treatment means is compared, a familywise $(1 - \alpha)100\%$ level of confidence can be achieved by using the Bonferroni confidence coefficient described in the one-way ANOVA section.

EXAMPLE 9.8   **Printer Sales**   Consider again the company investigating sales of its top-of-the-line printer (file PRINTER9 on the CD). An experiment was conducted to determine the effects of type of advertising and selling price on sales. The two levels of selling price considered were $600 and $700. The three types of advertising were television, radio, and newspaper.

Table 9.7 shows the data obtained from the experiment. The first column of the table (SALES) shows total sales for a period of 1 month (in $1000). The second column (PRICE) indicates the level of price (1 = $600 and 2 = $700), and the third column (ADV) indicates the type of advertising (1 = television, 2 = radio, and 3 = newspaper).

The MINITAB output is shown in Figure 9.15, the Excel output is in Figure 9.16, and the SAS output in Figure 9.17. The MINITAB, Excel, and SAS ANOVA tables contain the information in the general ANOVA table in Figure 9.14. The general structure of the MINITAB, Excel, and SAS ANOVA tables is shown in Figure 9.18. Note that Excel has information above the ANOVA table on treatment means. The SAS output has two parts to the ANOVA table. First, variation is partitioned as Model, Error, and Corrected Total. Below that, the components that make up Model are listed separately. Model consists of the two factors and the interaction term. The F statistics associated with the individual components are used in testing whether there are interaction and main factor effects.

First, we test to see if interaction effects are present. The decision rule is

Reject H_0 if $F > F(0.05; 2,6) = 5.14$

Do not reject H_0 if $F \leq 5.14$

The test statistic is $F = 5.87$. The decision is to reject H_0.

TABLE 9.7 Data for Printer Sales Example

SALES (in $1000)	PRICE	ADV
18.0	1	1
16.8	1	1
12.0	1	2
13.2	1	2
7.8	1	3
9.0	1	3
14.0	2	1
14.7	2	1
10.8	2	2
9.6	2	2
9.8	2	3
8.4	2	3

FIGURE 9.15
MINITAB ANOVA Output for Printer Sales Example.

Source	DF	SS	MS	F	P
PRICE	1	7.521	7.5208	10.99	0.016
ADV	2	103.752	51.8758	75.82	0.000
Interaction	2	8.032	4.0158	5.87	0.039
Error	6	4.105	0.6842		
Total	11	123.409			

S = 0.8271   R-Sq = 96.67%   R-Sq(adj) = 93.90%

Anova: Two-Factor With Replication

SUMMARY	ADV1	ADV2	ADV3	Total
PRICE1				
Count	2	2	2	6
Sum	34.80	25.20	16.80	76.80
Average	17.40	12.60	8.40	12.80
Variance	0.72	0.72	0.72	16.66
PRICE2				
Count	2	2	2	6
Sum	28.70	20.40	18.20	67.30
Average	14.35	10.20	9.10	11.22
Variance	0.25	0.72	0.98	6.52
Total				
Count	4	4	4	
Sum	63.50	45.60	35.00	
Average	15.88	11.40	8.75	
Variance	3.42	2.40	0.73	

ANOVA

Source of Variation	SS	df	MS	F	P-value	F crit
Sample	7.521	1	7.521	10.993	0.016	5.987
Columns	103.752	2	51.876	75.823	0.000	5.143
Interaction	8.032	2	4.016	5.870	0.039	5.143
Within	4.105	6	0.684			
Total	123.409	11				

FIGURE 9.16 Excel ANOVA Output for Printer Sales Example.

We conclude that there are interaction effects. The F tests for main effects discussed in this section cannot be used to determine whether price effects differ or advertising effects differ because the two factors interact. In this case, it may be helpful to examine the individual treatment means. The treatment means are computed by averaging the r observations for each treatment (or factor-level combination). For example,

```
                          The ANOVA Procedure
                        Class Level Information

                    Class      Levels     Values
                    PRICE           2     1 2
                    ADV             3     1 2 3


                  Number of observations    12


                         The ANOVA Procedure
```

Dependent Variable: SALES

Source	DF	Sum of Squares	Mean Square	F Value	Pr > F
Model	5	119.3041667	23.8608333	34.88	0.0002
Error	6	4.1050000	0.6841667		
Corrected Total	11	123.4091667			

R-Square	Coeff Var	Root MSE	SALES Mean
0.966737	6.888081	0.827144	12.00833

Source	DF	Anova SS	Mean Square	F Value	Pr > F
PRICE	1	7.5208333	7.5208333	10.99	0.0161
ADV	2	103.7516667	51.8758333	75.82	<.0001
PRICE*ADV	2	8.0316667	4.0158333	5.87	0.0387

FIGURE 9.17 SAS ANOVA Output for Printer Sales Example.

for the treatment when price was \$600 and television advertising was used, the treatment mean is $(18.0 + 16.8)/2 = 17.4$. These means can be computed by hand, but some computer routines allow simple procedures that compute them. Figure 9.19 shows the output from MINITAB using the TABLE procedure (see Using the Computer section). Each of the treatment means is shown in this table.

Excel provides the treatment means as part of the ANOVA output. Above the ANOVA table are the number of observations, sum, average, and variance of the observations for each factor-level combination. The treatment means are found in the cell labeled Average.

Figure 9.20 shows a plot of the treatment means. It appears from the plot that the highest level of sales is associated with television advertising in combination with a \$600 price. Sales are higher for the \$600 price when either television or radio advertising is used. When newspaper advertising is used, however, the \$700 price is

(a) MINITAB

Analysis of Variance on (dependent variable name)

Source	DF	SS	MS	F	P
Factor A	$n_1 - 1$	SSA	MSA	MSA/MSE	p value
Factor B	$n_2 - 1$	SSB	MSB	MSB/MSE	p value
Interaction	$(n_1 - 1)(n_2 - 1)$	$SSINT$	$MSINT$	$MSINT/MSE$	p value
Error	$n_1 n_2 (r - 1)$	SSE	MSE		
Total	$n_1 n_2 r - 1$	SST			

(b) Excel

ANOVA

Source of Variation	SS	df	MS	F	P-value	F crit
Sample	SSA	$n_1 - 1$	MSA	MSA/MSE	p value	$F(\alpha; n_1 - 1, n_1 n_2 (r - 1))$
Columns	SSB	$n_2 - 1$	MSB	MSB/MSE	p value	$F(\alpha; n_2 - 1, n_1 n_2 (r - 1))$
Interaction	$SSINT$	$(n_1 - 1)(n_2 - 1)$	$MSINT$	$MSINT/MSE$	p value	$F(\alpha; (n_1 - 1)(n_2 - 1), n_1 n_2 (r - 1))$
Within	SSE	$n_1 n_2 (r - 1)$	MSE			
Total	SST	$n_1 n_2 r - 1$				

(c) SAS

Analysis of Variance on (dependent variable name)

Source	DF	Sum of Squares	Mean Square	F value	Pr > F
Model	$n_1 n_2 - 1$	$SSA + SSB + SSINT$	$SSModel/DFModel$	$MSModel/MSE$	p value
Error	$n_1 n_2 (r - 1)$	SSE	MSE		
Corrected Total	$n_1 n_2 r - 1$	SST			
Factor A	$n_1 - 1$	SSA	MSA	MSA/MSE	p value
Factor B	$n_2 - 1$	SSB	MSB	MSB/MSE	p value
Interaction	$(n_1 - 1)(n_2 - 1)$	$SSINT$	$MSINT$	$MSINT/MSE$	p value

FIGURE 9.18 Structure of the MINITAB, Excel, and SAS ANOVA Tables for a Two-Way Design.

FIGURE 9.19
MINITAB Output Showing
Treatment Means for
Printer Sales Example.

```
Tabulated statistics: PRICE, ADV

Rows:   PRICE       Columns:  ADV

              1          2         3        All

1          17.40      12.60     8.40      12.80

2          14.35      10.20     9.10      11.22

All        15.88      11.40     8.75      12.01


Cell Contents:    SALES  :   Mean
```

FIGURE 9.20
Plot of Treatment
Means for Printer
Sales Example.

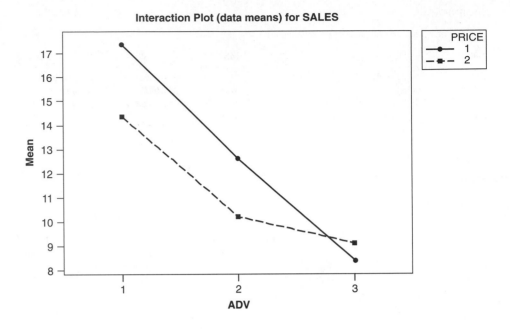

associated with slightly higher sales. This crossover of the lines illustrates the inter-action effect. If there were no interaction, the effects of factors A and B might ap-pear as illustrated in Figure 9.21. The two lines are almost parallel, which indicates a complete absence of interaction. Figure 9.22 illustrates another possible situation. Here the two lines are not parallel, but they do not cross. In this case, the interaction does not appear to be serious, and the model could be treated as one without interac-tion. Thus, the F tests for main factor effects are appropriate.

FIGURE 9.21
Example of Treatment
Mean Plot Showing No
Interaction Effects.

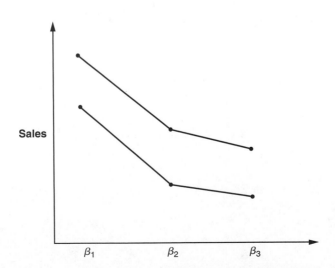

FIGURE 9.22
Example of Treatment Mean Plot Showing Weak Interaction Effects.

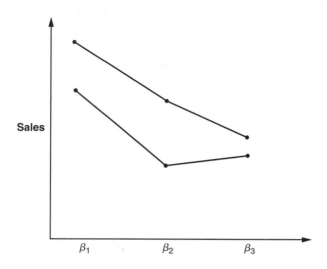

If there were no differences in the factor A effects, the two lines drawn would nearly coincide (see Figure 9.23, for example). If there were no differences in the factor B effects, the two lines drawn would be nearly horizontal (see Figure 9.24, for example).

A 95% confidence interval estimate for the difference between the treatment means for $700 price and television advertising and $700 price and radio advertising is given by

$$(14.35 - 10.2) \pm 2.447(0.827)\sqrt{\frac{2}{2}} \text{ or } (2.13, 6.17)$$

Note that the interval does not contain zero. If we were interested just in a comparison of these two means, we would conclude that the two means differ.

FIGURE 9.23
Example of Treatment Mean Plot Showing No Interaction Effects and No Factor A Main Effects.

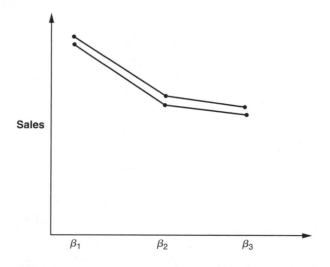

FIGURE 9.24
Example of Treatment
Mean Plot Showing No
Interaction Effects and
No Factor *B* Main
Effects.

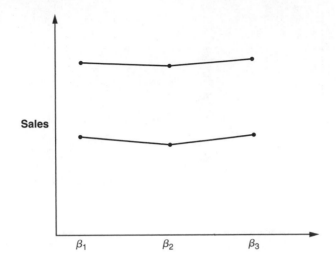

EXERCISES

3. Satisfaction. A company designs applications
software for a large number of firms. A study is
conducted to assess the level of satisfaction of
these firms with the software supplied. The effects
of two factors on the level of satisfaction are to be
investigated:

1. the industry of the firm that receives the software (INDUSTRY)

2. the contact person from whom the software
was purchased (CONTACT)

Four industries are served and there are three contact
people. For each combination of industries and
contacts, two firms are randomly selected and surveyed. A satisfaction score (SATIS) is obtained from
the questionnaire administered. The data are available
in a file named SATIS9 on the CD. The ANOVA for
this two-factor study is shown in Figure 9.25. Use the
results to determine whether industry or contact
person affects satisfaction. Use a 5% level of significance. State any hypotheses to be tested, the decision rule, the test statistic, and your decision.

9.4 ANALYSIS OF COVARIANCE

Analysis of covariance (ANCOVA) is a procedure sometimes used with models containing some quantitative and some qualitative independent variables. In ANCOVA,
however, the main interest is in the qualitative variables. In this respect, ANCOVA
might be viewed as a modification of ANOVA procedures rather than as a special
case of regression analysis (it is, in fact, both). The term ANCOVA is used when the

FIGURE 9.25 ANOVA
Output for Satisfaction
Exercise.

Source	DF	Sum of Squares	Mean Square	F Stat	P Value
INDUSTRY	3	716.125	238.708	49.82	0.000
CONTACT	2	0.250	0.125	0.03	0.974
Interaction	6	5.750	0.958	0.20	0.970
Error	12	57.500	4.792		
Total	23	779.625			

quantitative independent variables are added to the ANOVA model to reduce the variance of the error terms and thus provide more precise measurement of the treatment effects. These quantitative variables should be constructed in such a way that they are not influenced by the treatments.

Although some statistical programs do provide specific routines for performing ANCOVA, these are not discussed in this text. ANCOVA can be performed using regression routines with appropriately constructed indicator variables.

ADDITIONAL EXERCISES

4. ANOVA Tables

a. Assume that one-way analysis of variance is to be performed. Complete the ANOVA table for this analysis:

Source	DF	Sum of Squares	Mean Square	F Stat
Treatment	3			
Error		20		
Total	23	170		

b. Assume that a randomized block experiment is to be performed. Complete the ANOVA table for this analysis:

Source	DF	Sum of Squares	Mean Square	F Stat
Blocks	4			
Treatment		80	40	
Error	39		2	
Total	45	198		

5. Assembly Line.
Three assembly lines are used to produce a certain component for a computer. To examine the production rate of the assembly lines, a random sample of six hourly periods is chosen for each assembly line, and the number of components produced during these periods for each line is recorded. These data are available in a file named ASSEMBL9 on the CD. In the MINITAB file, the number of components produced will be in one column. A second column will indicate the production line (coded as 1, 2, or 3). In the Excel spreadsheet, the production numbers will be in a separate column for each production line.

Is there a difference in the average production rates for the three assembly lines? Use a 5% level of significance in answering this question. State any hypotheses to be tested, the decision rule, the test statistic, and your decision.

6. Salaries.
A large state university is interested in comparing salaries of its graduates (BA or BS) in the following areas: business, education, engineering, and liberal arts. Five graduates in each major are randomly selected and their starting salaries recorded. These data are available in a file named SALMAJ9 on the CD. In the MINITAB file, the variable SALARY will be one column. The variable MAJOR is a second column coded as $1 =$ business, $2 =$ education, $3 =$ engineering, and $4 =$ liberal arts. The salary data are in four separate columns in the Excel spreadsheet.

a. The university wants to know if there is a difference in the population average salaries for the four majors. Use a 5% level of significance in making the decision. State any hypotheses to be tested, the decision rule, the test statistic, and your decision.

b. Find a 95% interval estimate for the difference between the business and engineering mean salaries.

c. Use the Tukey or Bonferroni method to compare all possible pairs of means with a familywise confidence level of 90%. Use the familywise comparisons to determine if there is a significant difference in the population mean salaries for

(1) business and education majors

(2) business and engineering majors

(3) education and liberal arts majors

If so, which major has a higher mean salary in each comparison?

7. Test Scores.
A large financial planning firm wants to compare the results of three training programs

for its staff. Twenty-seven employees are selected for the study. These employees are grouped into three blocks of nine for the comparison. The blocks are based on number of years since college graduation, with block 1 being the most recent graduates and block 3 the most distant graduates. After the training programs are completed, the employees are tested and the test scores recorded. The data are in a file named TESTSCR9 on the CD. In the MINITAB file there are three columns: SCORE contains the test scores, PROGRAM coded as 1, 2, or 3, and BLOCK coded as 1, 2, or 3. The Excel spreadsheet is set up as it should be for an Excel randomized block analysis. See Using the Computer for more information.

a. Determine whether there is a difference in average test scores because of the training programs. Use a 5% level of significance. State any hypotheses to be tested, the decision rule, the test statistic, and your decision.

b. Construct a 95% confidence interval estimate of the difference between the means for program 1 and program 2.

8. **Employee Productivity.** A study of employee productivity is to be conducted with employees who enter data at computers. The amount of data entered is the dependent variable. Two factors that may influence the dependent variable are examined. One factor is the type of keyboard used (three types are available in the company). The second factor is the time of day (morning or afternoon). Four employees are randomly assigned to each type of keyboard. Two employees' production levels are recorded for a period of 1 hour in the morning. The other two are recorded in the afternoon. The production level is the number of forms completely entered by each employee. These data are available in a file named PRODRAT9 on the CD. In the MINITAB file, there are three columns: NUMBER is the number of forms processed by each employee, KEYBOARD is the keyboard type (coded as 1, 2, or 3), and TIME is coded as 1 for morning and 2 for afternoon. In the Excel spreadsheet, the data will be set up for a two-way analysis of variance. See Using the Computer for more information.

Do the two factors, KEYBOARD and TIME, appear to influence production rate (NUMBER)? Use a 5% level of significance for any hypothesis

tests used. State any hypotheses to be tested, the decision rule, the test statistic, and your decision. How would you describe the influence you observed in words?

9. **Bill's Sales.** Bill's is a popular restaurant/bar in southwest Fort Worth, Texas. Daily sales data for Bill's for the period from October 14 through December 8 are available in a file named BILLS9 on the CD. In the MINITAB file, SALES will be in a single column. The column labeled DAY represents the day of the week, coded as 1 = Monday, 2 = Tuesday, 3 = Wednesday, 4 = Thursday, 5 = Friday, 6 = Saturday, and 7 = Sunday. In the Excel spreadsheet sales for each day will be in a separate column. Determine whether there is a difference in Bill's average sales on different days of the week. Use a 5% level of significance for any hypothesis tests used. State any hypotheses to be tested, the decision rule, the test statistic, and your decision. If there is a difference, on which day (or days) do sales appear to be highest? What does this suggest to the owner of Bill's about staffing?

10. **Executives' Salaries.** The file EXECSAL9 on the CD contains the company name, name of the CEO, the 2002 salary, and the 2002 salary plus bonus for samples of executives from 11 different industries. The data are from the *Wall Street Journal* 2002 CEO Compensation Survey (Monday April 14, 2003). The industries are coded as follows:

    1 = Basic Materials
    2 = Energy
    3 = Industrial
    4 = Cyclical
    5 = Noncyclical
    6 = Technology
    7 = Financial
    8 = Utilities
    9 = Health Care
    10 = Telecommunications

SALARY and SALBONUS will each be in a single column. The column labeled INDUSTRY will contain the industry classification, coded as indicated.

a. Determine whether there is a difference in average salaries for CEOs in the different industry classifications. Use a 5% level of significance for any hypothesis tests used.

State any hypotheses to be tested, the decision rule, the test statistic, and your decision.

b. Determine whether there is a difference in the combined average salary and bonus for CEOs in the different industry classifications. Use a 5% level of significance for any hypothesis tests used. State any hypotheses to be tested, the decision rule, the test statistic, and your decision.

11. **Comparing Suppliers.** A company purchases diodes from four different suppliers. The company's engineers would like to determine if the average lifetimes of the diodes are essentially the same. They randomly select three diodes from each supplier's shipment and test the lifetimes. The resulting data are available in a file named SUPPLY9 on the CD. In the MINITAB file, the variable LIFETIME will be one column. The variable SUPPLIER is a second column coded as 1, 2, 3, or 4 to indicate the supplier. The lifetime data are in four separate columns in the Excel spreadsheet.

a. The engineers want to know if there is a difference in the population average lifetimes of the diodes for the four suppliers. Use a 5% level of significance in making the decision. State any hypotheses to be tested, the decision rule, the test statistic, and your decision.

b. Find a 95% interval estimate for the difference between the mean lifetimes for suppliers 1 and 2.

c. Use the Tukey or Bonferroni method to compare all possible pairs of means with a family-wise confidence level of 90%.

12. **Comparing Inspectors.** A manufacturer employs five persons who visually inspect circuit boards for flaws in the printed circuitry. A circuit board that is rejected at visual inspection but does not have the flaw claimed by the inspector is referred to as a "false reject." Since false rejects add to manufacturing costs, the manufacturer wants to determine whether the false reject averages are the same for the five visual inspectors. To do so, the boards rejected by each inspector are checked for false rejects over six 1-week periods and the numbers of false rejects (FALSE) recorded. The resulting data are available in a file named INSPECT9 on the CD. In the MINITAB file, the variable FALSE will be one column. The variable INSPECTOR is a second column coded as 1, 2, 3, 4, or 5 to indicate the inspector. The data are in five separate columns in the Excel spreadsheet. The manufacturer wants to know if there is a difference in the population average number of false rejects for the five inspectors. Use a 5% level of significance in making the decision. State any hypotheses to be tested, the decision rule, the test statistic, and your decision.

(*Source*: This example is from *Engineer Statistics: The Industrial Experience*, pages 476–477.[2])

USING THE COMPUTER

The Using the Computer section in each chapter describes how to perform the computer analyses in the chapter using Excel, MINITAB, and SAS. For further detail on Excel, MINITAB, and SAS, see Appendix C.

EXCEL

One-Way Analysis of Variance

Figure 9.26 shows the data arrangement for a one-way ANOVA in Excel. The data for each factor level must be in a separate column. In the example shown, there are nine factor levels, so there are nine columns of data. The columns do not have to be of equal length. For a one-way ANOVA, choose Anova: Single Factor from the Data

---

[2]See References for complete publication information.

FIGURE 9.26 Data for One-Way ANOVA in Excel.

	A	B	C	D	E	F	G	H	I
1	CLASS1	CLASS2	CLASS3	CLASS4	CLASS5	CLASS6	CLASS7	CLASS8	CLASS9
2	70	80	55	66	73	61	75	58	52
3	97	85	70	96	78	63	78	60	53
4	103	91	73	102	80	66	84	73	54
5	109	93	75	103	81	68	90	74	60
6	115	95		112	84	70	93	75	63
7	115	97		115	84	70	98	81	65
8	119	102		115	86		100	82	65
9	120	111		116	87			89	68
10	124	117		121	87			91	
11	125	117		121	90			93	
12	137	123		131	90			97	
13	137	123		132	91			98	
14	139	131		137	91			103	
15	139			140	94				
16	139			142	95				
17	140			149	102				
18	151			161	106				
19	152			177	107				
20	153				107				
21	158				110				
22					113				
23					118				
24					121				

FIGURE 9.27 Excel Dialog Box for One-Way ANOVA.

FIGURE 9.28 Data for Randomized Block ANOVA in Excel.

	A	B	C	D	E
1		DESIGN1	DESIGN2	DESIGN3	DESIGN4
2	SIZE1	40	23	17	33
3	SIZE2	32	45	40	25
4	SIZE3	43	31	38	47
5	SIZE4	44	41	56	45
6	SIZE5	43	60	47	64

Analysis menu. The dialog box is shown in Figure 9.27. Fill in the Input Range. For the example shown, the factor level labeled CLASS5 has 23 observations, so the column ends at row 24. To select the input range, use A1:I24. The fact that some columns do not have 24 rows of data is not a problem. Click Labels in First Row because the columns are labeled and specify the desired output option. (Note that data can be in columns or rows. Indicate which using the Grouped By option.) The default level of significance is 0.05, but this can be changed as desired. The level of significance is used to look up the critical value for the F test. Then click OK.

Analysis of Variance Using a Randomized Block Design

Figure 9.28 shows the data arrangement for a randomized block ANOVA in Excel. The names of the levels for the blocking variable are listed in the left-hand column. The names of the factor levels are listed in the first row. The columns of data represent the observations for each factor level. There is one observation for each level of the blocking variable. For a randomized block ANOVA, choose Anova: Two-Factor Without Replication from the Data Analysis menu. The dialog box is shown in Figure 9.29. Fill in the Input Range. For the example shown; use A1:E6 as the input range. Click Labels because the rows and columns are labeled and specify the desired output option. The default level of significance is 0.05, but this can be changed as desired. The level of significance is used to look up the critical values for the F tests. Then click OK.

Two-Way Analysis of Variance

Figure 9.30 shows the data arrangement for a two-way ANOVA in Excel. The names of the levels for factor A are listed

FIGURE 9.29 Excel Dialog Box for Randomized Block ANOVA.

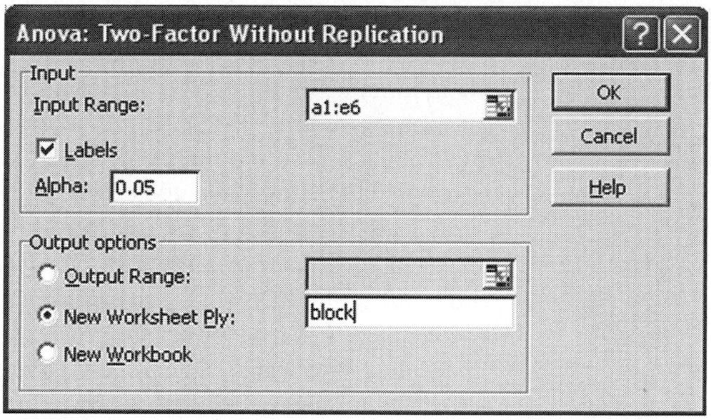

FIGURE 9.30 Data for Two-Way ANOVA in Excel.

	A	B	C	D	E
1		CONTACT1	CONTACT2	CONTACT3	
2	INDUST1	74	73	75	
3	INDUST1	72	71	69	
4	INDUST2	73	75	71	
5	INDUST2	71	70	72	
6	INDUST3	81	83	85	
7	INDUST3	84	82	82	
8	INDUST4	84	85	82	
9	INDUST4	81	83	85	

FIGURE 9.31 Excel Dialog Box for Two-Way ANOVA.

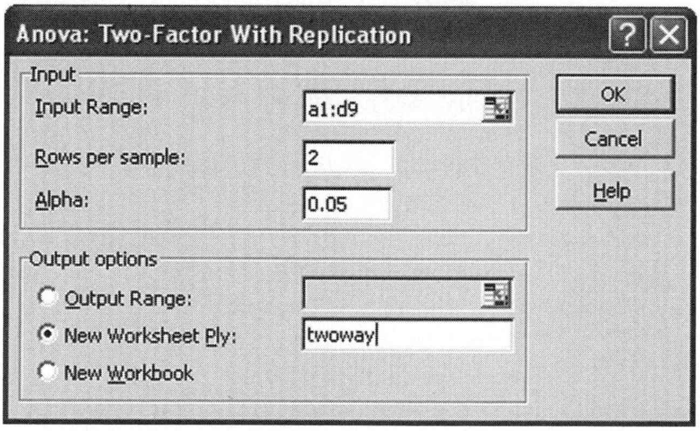

in the left-hand column. The names of the factor B levels are listed in the first row. The columns of data represent the observations for each level of factor B. There are at least two observations for each treatment. A treatment represents a factor A and factor B combination. For example, for the first level of factor A (INDUST1) and factor B (CONTACT1), the two observations are 74 and 72. These two values must be the first two entries in the CONTACT1 column. The next two values in that column are for the INDUST2/CONTACT1 treatment and so on. For a two-way ANOVA, choose Anova: Two-Factor With Replication from the Data Analysis menu. The dialog box is shown in Figure 9.31. Fill in the Input Range. For the example shown, use A1:D9 as the input range. Fill in "Rows per sample" with the number of replications at each treatment ($r = 2$, in this example). Specify the desired output option. The default level of significance is 0.05, but this can be changed as desired. The level of significance is used to look up the critical values for the F tests. Then click OK.

MINITAB

One-Way Analysis of Variance

STAT: ANOVA: ONE-WAY or STAT: ANOVA: ONE-WAY (UNSTACKED)

The ANOVA: ONE-WAY procedure is used when all data on the dependent variable are in a single column. A second column is used to specify the factor level assigned to each value of the dependent variable. Data would be arranged as in Table 9.1, for example, with one column containing the car classification and the other the number of injuries. The dialog box for this procedure is shown in Figure 9.32. Fill in the Response (dependent) variable and the Factor and click OK.

Another way of arranging the data is to separate (unstack) the values for each different factor level and put these values into separate columns. If this arrangement

FIGURE 9.32 MINITAB One-Way Analysis of Variance (Stacked Data) Dialog Box.

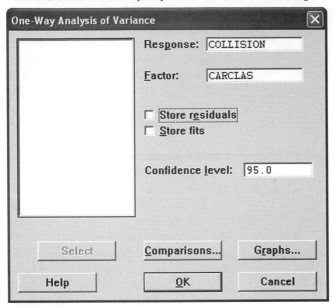

FIGURE 9.33 MINITAB One-Way Analysis of Variance (Unstacked Data) Dialog Box.

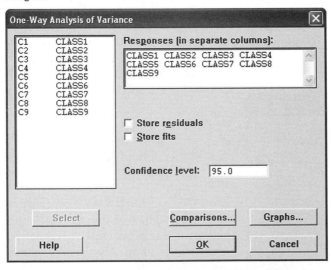

is used, the STAT: ANOVA: ONE-WAY (UNSTACKED) procedure is appropriate. Figure 9.33 shows the dialog box. Fill in the columns containing the observations for each factor level. Remember that these must be in separate columns to use this ANOVA procedure. Then click OK.

Click Comparisons to open another dialog box. This box allows the user a choice of methods for pairwise comparisons: Tukey's (family error rate), Fisher's (individual error rate), Dunnett's (family error rate against a control group), and Hsu's (family error rate against the best group). The Tukey comparison was discussed in the text.

Analysis of Variance Using a Randomized Block Design

STAT: ANOVA: TWO-WAY

To do a randomized block design in MINITAB, use the two-way analysis of variance option on the ANOVA menu. The dialog box is shown in Figure 9.34. Fill in the response (dependent) variable, the row factor (blocking variable), and the column factor (treatment). Click the "Fit additive model" box. This is important! Otherwise, MINITAB does a two-way analysis of variance that is not appropriate if you really have a randomized block design.

TWO-WAY ANALYSIS OF VARIANCE

STAT: ANOVA: TWO-WAY

The dialog box for a two-way ANOVA is shown in Figure 9.34. Fill in the response (dependent) variable, the Row factor (Factor A) and the Column factor (Factor B), and click OK.

To do an interactions plot such as the one in Figure 9.20, use STAT: ANOVA: INTERACTIONS PLOT. Fill in the Responses and the Factors and click OK. See the dialog box in Figure 9.35.

To produce a table of treatment means such as the one in Figure 9.19, use STAT: TABLES: DESCRIPTIVE STATISTICS. In the Table of Descriptive Statistics

FIGURE 9.34 MINITAB Dialog Box for Randomized Block Designs and Two-Way Analysis of Variance.

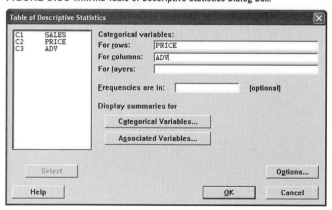

FIGURE 9.35 MINITAB Interactions Plot Dialog Box.

FIGURE 9.36 MINITAB Table of Descriptive Statistics Dialog Box.

dialog box (Figure 9.36), fill in "Categorical variables" as shown: The "For rows" box will contain one of the factors and the "For columns" box will contain the other. The categorical—variables are the two factors of the two-way ANOVA. Click "Associated Variables" and fill in the response variable in "Associated variables" in the Descriptive Statistics— Summaries for Associated Variables dialog box (Figure 9.37). Click the box for Means to display the treatment means.

SAS

One-Way Analysis of Variance

```
PROC ANOVA;
CLASS CARTYPE;
MODEL INJURY = CARTYPE;
```

CLASS indicates which variable is the classification variable—that is, which variable denotes the factor levels for each observation in the data set. The variable on the left-hand side of the equality in the MODEL statement is the dependent variable.

Analysis of Variance Using a Randomized Block Design

```
PROC ANOVA;
CLASS SIZE DESIGN;
MODEL SALES = SIZE DESIGN;
```

The PROC ANOVA statement is the same as the statement used for a one-way analysis of variance. The CLASS statement now specifies both the blocking and treatment variables. The MODEL statement lists the dependent variable on the left-hand side of the equality and the blocking and treatment variables on the right-hand side.

FIGURE 9.37 MINITAB Summaries for Associated Variables Dialog Box.

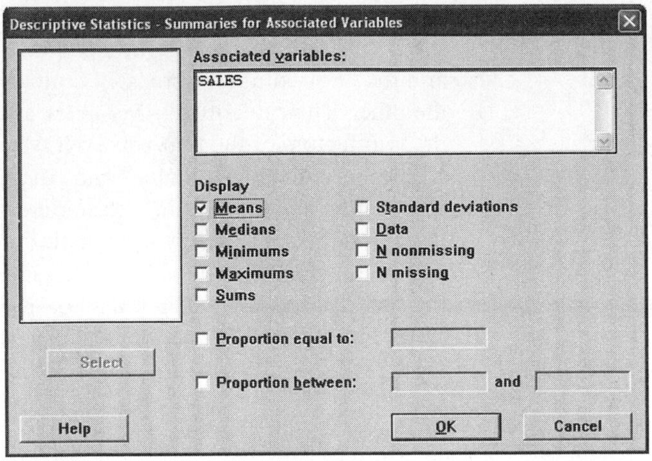

Two-Way Analysis of Variance

```
PROC ANOVA;
CLASS PRICE ADV;
MODEL SALES = PRICE ADV PRICE*ADV;
```

The SAS command sequence for two-way analysis of variance requires that the interaction term be explicitly included in the MODEL statement as shown: PRICE*ADV. Otherwise, the interaction term is omitted (as in the randomized block design). The MODEL statement lists the dependent variable on the left-hand side of the equality and the main effects and interaction variables on the right-hand side.

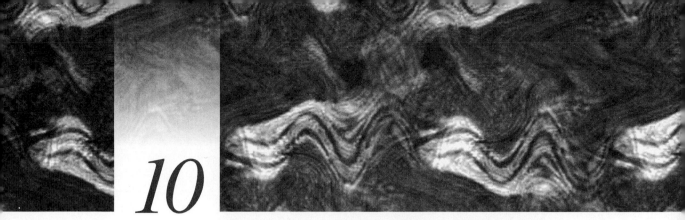

10

Qualitative Dependent Variables: An Introduction to Discriminant Analysis and Logistic Regression

10.1 INTRODUCTION

A bank is interested in determining whether certain borrowers are creditworthy. Should loans be made to these potential customers? The bank has a variety of quantitative information about each potential borrower and would like some way of classifying them into groups of qualifying and nonqualifying candidates.

An investment company is trying to determine the likelihood that certain firms will end up in bankruptcy. The investment company has quantitative information on the firms it is considering and wants to classify them into two groups: those that will go bankrupt and those that will not.

The director of personnel for a large corporation would like to know which of a group of new trainees will be successful in a certain position at the company. Using demographic information and the results of certain aptitude tests, the director wants to classify trainees into two groups: those who will succeed and those who will not.

The marketing department of a retail firm is interested in predicting whether people will buy a new product within the next year. The marketing department will have the results from surveys given to a sample of potential buyers and a quantitative score representing propensity to buy as a summary measure from these surveys. The department wants to use this score to divide the people into groups of potential buyers and nonbuyers so it can better target future advertising.

Each of these examples represents a case in which certain observations (people, firms, and so on) are to be divided into two groups: firms that do and do not qualify for credit; firms that will and will not go bankrupt; people that will and will not be successful in new positions; people who will and will not purchase a new product. The goal is to try to pick which of the two groups each observation will fall into based on available information. This type of problem can be placed in a regression setting. Write the regression model as

$$y = \beta_0 + \beta_1 x_1 + \cdots + \beta_K x_K + e$$

where

$y = 1$ if the observation falls into the group of interest

$y = 0$ if the observation does not fall into the group of interest

In Chapter 7, methods were discussed for dealing with explanatory variables representing qualitative information. Examples used were employed/unemployed, male/female, and so on. The way this information was incorporated into the regression was with indicator variables. These variables took on values of either 0 or 1 to indicate whether an item was or was not in a certain group (male $= 1$, female $= 0$, for example). In this chapter, we adopt a similar approach for situations where the dependent variable of interest is qualitative in nature. Situations are examined where the task is to try and determine whether items do or do not fall into a certain group. The dependent variable used to represent this situation will be a variable that has either the value 1 if the item is in the group or the value 0 if the item is not in the group.

Standard linear regression analysis is not designed for directly analyzing this situation. In the regression applications examined so far in this text, the dependent variable was a variable that could be effectively modeled as a continuous variable. The assumptions of the regression model are designed for such a situation. When a 0/1 dependent variable is used, several problems occur. To discuss the first problem, it is important to note that the conditional mean of y given x has a different interpretation when the dependent variable is a 0/1 variable than it did in the previous case when the dependent variable was modeled as continuous. It can be shown that $\mu_{y|x}$ is equal to the probability that the observation belongs to the indicated group: $\mu_{y|x} = p$, where $p = P(Y = 1)$. Probabilities must be between 0 and 1, of course, but when the regression is estimated, there is nothing to guarantee that the predictions from the estimated regression equation fall between 0 and 1. The actual predictions can vary considerably, with values above 1 or even negative values occurring.

Second, certain assumptions of the regression model will be violated. For example, the disturbances are not normally distributed and the variance around the regression line is not constant.

As a result of the problems that occur with applying standard regression techniques to situations with 0/1 dependent variables, alternative methods of analysis have been proposed. Two methods are discussed in this chapter: discriminant analysis and logistic regression. Both techniques have characteristics similar to the regression models that have been presented thus far in this text. Even though linear regression, discriminant analysis, and logistic regression differ, their similarities will help the reader understand the use of these procedures.

10.2 DISCRIMINANT ANALYSIS

EXAMPLE 10.1   **Employee Classification**   The personnel director for a firm that manufactures computers has classified the performance of each of the employees in a certain position as either satisfactory or unsatisfactory. The director has two tests that she would like to use to help determine which future employees will perform in a satisfactory manner and which will not before they are assigned to the position. With this knowledge, she will be better able to suggest jobs within the firm at which each employee will have a greater chance of success. To help determine whether the tests will be useful, she administers them to the current employees and records their scores. The resulting data are shown in Table 10.1 and in the file TRAIN10 on the CD. The first column notes whether the employee is currently classified as satisfactory (1) or unsatisfactory (0). The next two columns show the test score results on each of the two tests. The director's task is now to determine how to use the test scores to predict the correct classification of each employee.

Figure 10.1 shows a scatterplot of each employee's scores on the two tests. The employees classified as successful are denoted with ■ and those classified as unsuccessful as ●. From the plot, it does appear that knowledge of the test score may provide information useful in classifying the employees. There is no exact cutoff to separate the two groups precisely, but there is enough separation of the members to indicate that the information provided on the tests may be useful.   ▨

Discriminant analysis can be used to help solve the type of problem posed in Example 10.1. Discriminant analysis can be described as follows: Let *y* represent the dependent variable, defined as

y = 1 if the observation falls into the group of interest

y = 0 if the observation does not fall into the group of interest

Let $x_1, \ldots, x_K$ be explanatory variables that are used to help predict into which of the two groups each of the observations in the sample should be classified. The explanatory variables are assumed to be approximately normally distributed. Although discriminant analysis can be used when the explanatory variables are not normally distributed, it is not guaranteed to be optimal in such cases and may not provide good results. This is one of the more serious limitations of this technique because it limits the kinds of variables that can be used as explanatory variables in the equation. It should be noted, however, that a study by Amemiya and Powell[1] showed that discriminant analysis does well in prediction and estimation even when the explanatory variables are nonnormal if sample sizes are large. Discriminant analysis also assumes that the variation of the explanatory variables is the same for each group. Write the equation representing the relationship between *y* and the explanatory variables as

$$y = \beta_0 + \beta_1 x_1 + \cdots + \beta_K x_K + e$$

---

[1] See T. Amemiya and J. Powell, "A Comparison of the Logit Model and Normal Discriminant Analysis When Independent Variables Are Binary." See References for complete publication information.

TABLE 10.1 Data and Discriminant Analysis Results for Employee Classification Example

GROUP	TEST1	TEST2	Predicted Group	Discriminant Score
1	96	85	1	0.99694
1	96	88	1	1.04291
1	91	81	1	0.64276
1	95	78	1	0.83111
1	92	85	1	0.76262
1	93	87	1	0.85185
1	98	84	1	1.09878
1	92	82	1	0.71666
1	97	89	1	1.11681
1	95	96	1	1.10691
1	99	93	1	1.29526
1	89	90	1	0.66350
1	94	90	1	0.95639
1	92	94	1	0.90052
1	94	84	1	0.86446
1	90	92	1	0.75272
1	91	70	0*	0.47421
1	90	81	1	0.58418
1	86	81	0*	0.34986
1	90	76	0*	0.50757
1	91	79	1	0.61211
1	88	83	0*	0.49766
1	87	82	0*	0.42376
0	93	74	1*	0.65266
0	90	84	1*	0.63014
0	91	81	1*	0.64276
0	91	78	1*	0.59679
0	88	78	0	0.42105
0	86	86	0	0.42647
0	79	81	0	−0.06020
0	83	84	0	0.22009
0	79	77	0	−0.12149
0	88	75	0	0.37508
0	81	85	0	0.11825
0	85	83	0	0.32192
0	82	72	0	−0.02236
0	82	81	0	0.11554
0	81	77	0	−0.00433
0	86	76	0	0.27325
0	81	84	0	0.10293
0	85	78	0	0.24531
0	83	77	0	0.11283
0	81	71	0	−0.09626

* Indicates a misclassification

FIGURE 10.1
Scatterplot of TEST1 versus TEST2 for Satisfactory (■) and Unsatisfactory (●) Employees.

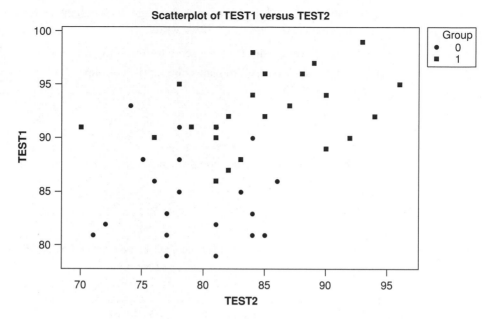

Apply linear regression to the data and estimate the previous equation. The estimated equation is written

$$d = b_0 + b_1 x_1 + \cdots + b_K x_K$$

where d is called the discriminant score. The *discriminant score* is just the predicted value from the estimated regression equation. These discriminant scores are used to classify each of the observations in the sample. A cutoff value is chosen, call it c, and the following classification rule is used:

If $d \le c$, assign the observation to group 0

If $d > c$, assign the observation to group 1

How is the cutoff value in the classification rule chosen? Our real goal in such a situation is not to classify the items in the sample correctly; we know what group these items are in. The goal is to classify future observations correctly. In the employee classification example, this means being able to correctly classify future employees as satisfactory or unsatisfactory in a particular position on the basis of their test scores. So ideally, the estimates $b_0, b_1, \ldots, b_K$ should be chosen so that the number of misclassified future observations is minimized.

In the problem considered here, this can be done as follows: Estimate the equation for the discriminant score with least-squares regression using the 0/1 dependent variable. Record the predicted or fitted values from this estimated equation. These are the discriminant scores, d. Choose a value, c, as a cutoff in such a way that the probability of misclassification for future observations is minimized. This is done by choosing c halfway between the average discriminant scores for the two groups if the sample sizes are equal. If we write

$\overline{d}_1$ = average discriminant score for the 1 group

$\overline{d}_0$ = average discriminant score for the 0 group

then

$$c = \frac{\overline{d}_0 + \overline{d}_1}{2}$$

If the sample sizes are not equal, then a weighted average of the two average discriminant scores can be used:

$$c = \frac{n_0\overline{d}_0 + n_1\overline{d}_1}{n_0 + n_1}$$

where n_0 is the sample size for the group labeled 0 and n_1 is the sample size for the group labeled 1.

EXAMPLE 10.2 **Employee Classification (continued)** Continuing with Example 10.1, the discriminant scores for the employees in the sample in Table 10.1 have been computed and are shown in the fifth column of the table. Column 4 shows the results of a classification rule applied to the discriminant scores. The employees who were incorrectly classified have been denoted with an *. Note that five employees who were in group 1 have been classified as being in group 0, and four employees who were in group 0 have been classified as being in group 1. These errors are termed misclassifications. The percentage of correct classifications in this example is 79.1%.

The average discriminant score computed from the linear regression for the satisfactory (1) group can be found from the data in column 5. This average is 0.7848. The average for the unsatisfactory (0) group is 0.2475. The weighted average of these two values is given by

$$\frac{20(0.2475) + 23(0.7848)}{20 + 23} = 0.5349$$

This number can be used to classify each of the observations in Table 10.1 and to check the classifications in column 4. Whenever the d score in column 5 is less than or equal to 0.5349, the observation should be classified as belonging to group 0. When the d score in column 5 is greater than 0.5349, the observation should be classified as belonging to group 1. This is how the classifications in the table were determined.

The regression used to determine the d scores is shown in Figure 10.2. For future employees, this regression can be used to determine the d scores for classification. ◼

Even though linear regression was used to perform the discriminant analysis, this is not typically the method of choice. There are specific computer routines available for discriminant analysis. Figure 10.3 shows the MINITAB output for the

FIGURE 10.2
Regression Used to Perform Discriminant Analysis for Employee Classification Example.

Variable	Coefficient	Std Dev	T Stat	P Value
Intercept	−5.9291	0.9633	−6.15	0.000
TEST1	0.0586	0.0112	5.23	0.000
TEST2	0.0153	0.0010	1.54	0.132

Standard Error = 0.351797   R-Sq = 53.7% R-Sq(adj) = 51.4%

Analysis of Variance

Source	DF	Sum of Squares	Mean Square	F Stat	P Value
Regression	2	5.7472	2.8736	23.22	0.000
Error	40	4.9505	0.1238		
Total	42	10.6977			

employee classification problem when a procedure designed specifically for discriminant analysis is used. Figure 10.4 shows the SAS discriminant analysis output. The results of the classification are identical to those from the regression approach, but the method used is somewhat different and the output differs considerably.

The primary reason that specific discriminant analysis routines are used rather than the regression approach is that discriminant analysis can be extended to more than two groups. When more than two groups are included in the analysis, the regression approach is not as straightforward. The case for more than two groups can be handled easily by discriminant analysis routines, however. Each group is given a different number. The numbers assigned do not have to include 0 and 1. (In fact, in the two-group case, the numbers that identify the two groups can be any integers, not necessarily 0 and 1. The values 0 and 1 were used here for convenience in extending the discussion to the topic of logistic regression.)

Discriminant analysis routines proceed by computing the means of each of the different groups for each explanatory variable. Then a measure of the distance from each observation to each set of means is computed. The observation is classified into the group whose set of means is closest. Example 10.3 illustrates a discriminant analysis routine applied to the employee classification data.

EXAMPLE 10.3   **Employee Classification (continued)**   Figure 10.3 shows the MINITAB discriminant analysis output applied to the employee classification data. In this example, the group counts are noted. There are 20 in the group denoted 0 and 23 in the group denoted 1. A Summary of Classification table is provided next. For example, for those employees who were actually unsatisfactory (True Group 0), 16 were classified as unsatisfactory (Put into Group 0) and 4 were classified as satisfactory (Put into Group 1) out of the total of 20. The 16 correct classifications for this group yield a proportion correct of 0.800 (80.0%). The same information is given for each group. Overall, there were 43 employees, 34 were correctly categorized, and the overall proportion correct was 0.791 (79.1%).

FIGURE 10.3
MINITAB Discriminant Analysis Output for Employee Classification Example.

```
Linear Method for Response: Group

Predictors: TEST1, TEST2

Group          0          1
Count         20         23

Summary of classification

                   True  Group
Put into Group       0      1
0                   16      5
1                    4     18
Total N             20     23
N correct           16     18
Proportion       0.800  0.783

N = 43            N Correct = 34        Proportion Correct = 0.791

Squared Distance Between Groups

         0          1
0   0.00000    4.44946
1   4.44946    0.00000

Linear Discriminant Function for Groups

              0          1
Constant  -298.27    -351.65
TEST1        5.20       5.68
TEST2        1.97       2.10

Summary of Misclassified Observations

                True   Pred          Squared
Observation    Group  Group  Group  Distance  Probability
     17**        1      0       0     6.657       0.586
                                1     7.352       0.414
     19**        1      0       0     0.1911      0.799
                                1     2.9454      0.201
     20**        1      0       0     2.561       0.518
                                1     2.703       0.482
     22**        1      0       0     1.034       0.538
                                1     1.340       0.462
     23**        1      0       0     0.5264      0.682
                                1     2.0567      0.318
     24**        0      1       0     6.438       0.244
                                1     4.177       0.756
     25**        0      1       0     2.2891      0.280
                                1     0.4008      0.720
     26**        0      1       0     2.6389      0.259
                                1     0.5416      0.741
     27**        0      1       0     2.900       0.339
                                1     1.564       0.661
```

The Squared Distance Between Groups table is not discussed in this text.

The next item in the output is the Linear Discriminant Function for Groups. When a discriminant analysis procedure is used, a separate equation is computed for each group. Note that these equations are not the same as the discriminant function computed using the regression approach. The two equations can be combined in a certain way to produce the overall discriminant function, however. The way this combination is achieved is not discussed here, but the use of the equations for the separate groups is demonstrated:

$$\text{Group 0: } -298.27 + 5.20\text{TEST1} + 1.97\text{TEST2}$$
$$\text{Group 1: } -351.65 + 5.68\text{TEST1} + 2.10\text{TEST2}$$

These equations are applied to each employee. The employee is then classified into the group for which his or her score is the highest. These equations can also be used to classify any future applicants. Administer the tests, record the test scores, and compute the values for each equation. The applicant is then assigned to the group for which the value from the equations is the highest. Although this method differs from the way the linear regression approach classified the employees, the results are the same, as can be seen from the Summary of Misclassified Observations. This is always true in the two-group case. For three or more groups, however, discriminant analysis procedures should be used.

In the Summary of Misclassified Observations, the number of each employee misclassified is shown along with the True Group and the Pred (predicted) Group. The Squared Distance column shows a measure of the distance computed from each observation to the means of each group. Each employee has been classified into the group with the smaller distance. In the last column, a Probability has been computed that can be thought of as the predicted probability that the employee belongs to a particular group. The employee has been classified into the group that has the highest probability.

The SAS output in Figure 10.4 provides information similar to MINITAB. Class Level Information provides the number in each group (and the proportion as well). Prior probability provides the researcher opportunity to weight the likelihood that individuals will fall into each category. The default probability is that the likelihood in each group is equal (0.5 when there are two groups). Under Linear Discriminant Function, the equations used to classify observations into groups are provided. The use of these equations was described earlier in this example. Number of Observations and Percent Classified into GROUPS provides a breakdown of the performance of the discriminant analysis in terms of correct and incorrect classifications. Error Count Estimates for Groups summarizes the incorrect classifications.

In linear regression, an equation was developed using a certain set of observations. It is typically not these observations for which predictions are desired, however. The quality of predictions is important for observations not included in the original sample. This situation is the same in discriminant analysis. What really matters in discriminant analysis is how well the discriminant equations classify future observations. The percentage of correctly classified observations given in the

FIGURE 10.4
SAS Discriminant
Analysis Output for
Employee Classification
Example.

The DISCRIM Procedure

Observations	43	DF Total	42
Variables	2	DF Within Classes	41
Classes	2	DF Between Classes	1

Class Level Information

GROUPS	Variable Name	Frequency	Weight	Proportion	Prior Probability
0	_0	20	20.0000	0.465116	0.500000
1	_1	23	23.0000	0.534884	0.500000

Pooled Covariance Matrix Information

Covariance Matrix Rank	Natural Log of the Determinant of the Covariance Matrix
2	6.05465

Generalized Squared Distance to GROUPS

From GROUPS	0	1
0	0	4.44946
1	4.44946	0

Linear Discriminant Function

Linear Discriminant Function for GROUPS

Variable	0	1
Constant	-298.26618	-351.64588
TEST1	5.19936	5.68452
TEST2	1.97076	2.09766

Number of Observations and Percent Classified into GROUPS

From GROUPS	0	1	Total
0	16	4	20
	80.00	20.00	100.00
1	5	18	23
	21.74	78.26	100.00
Total	21	22	43
	48.84	51.16	100.00
Priors	0.5	0.5	

Error Count Estimates for GROUPS

	0	1	Total
Rate	0.2000	0.2174	0.2087
Priors	0.5000	0.5000	

discriminant analysis output can be used as a guide for this, but this percentage will likely overstate the quality of future classifications. The same is true of the R^2 for a linear regression. The R^2 represents a measure of fit of the sample data, but may not reflect how well the equation will do in classifying future data.

There are other methods to assess the quality of future classifications when using discriminant analysis. One method is to split the original sample into two parts (if sample size is large enough) called an *estimation sample* and a *validation sample*. Use the estimation sample to determine the discriminant equations. Then use the discriminant equations to classify the items in the validation sample. The percentage of correct classifications in the validation sample should provide a better indication of how discriminant analysis will perform on future observations. After this has been done, the two samples can then be combined to compute the discriminant function for future use.[2]

10.3 LOGISTIC REGRESSION

Discriminant analysis is known to be statistically valid when we can assume that the independent variable in the regression equation is normally distributed. When the equation has more than one independent variable, the assumption is that the x variables have a multivariate normal distribution, a strong assumption that is not discussed in detail here. Suffice it to say that this normality assumption excludes many possible variables as explanatory variables in the discriminant function equation. *Logistic regression* is another procedure for modeling a 0/1 dependent variable that does not depend on the assumption that the independent variables are normally distributed. As a result, many other types of variables, including indicator variables, are in the possible set of explanatory variables.

The logistic regression approach does have its own set of assumptions, however. To briefly describe logistic regression, the notion presented earlier that the conditional mean of y given x has a different interpretation when the dependent variable is a 0/1 variable must be reconsidered. When the dependent variable is either 0 or 1, it can be shown that the conditional mean of y given x, $\mu_{y|x}$, is equal to the probability that the observation belongs to the indicated group:

$$\mu_{y|x} = p = P(Y = 1)$$

Probabilities must be between 0 and 1; thus, to model the conditional mean of y, a function that is restricted to lie between 0 and 1 must be used. The function considered in logistic regression is called the *logistic function* (what a coincidence!) and can be written as follows:

$$\mu_{y|x} = \frac{1}{1 + e^{-(\beta_0 + \sum_{j=1}^{K} \beta_j x_j)}}$$

---

[2] For more detail on discriminant analysis, including a discussion of discriminant analysis with more than two groups, see C. T. Ragsdale and A. Stam, "Introducing Discriminant Analysis to the Business Statistics Curriculum." See References for complete publication information.

FIGURE 10.5
The S-Shaped Curve of the Logistic Function.

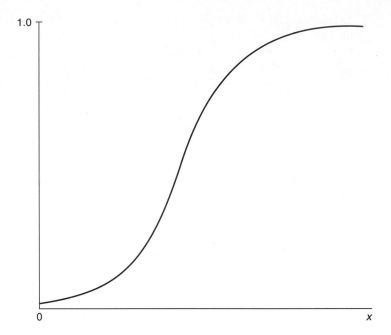

where the x_j are explanatory variables, K is the number of explanatory variables, and β_0 and the β_j are coefficients to be estimated.

This function works well for modeling probabilities because it is restricted to be between 0 and 1. The function forms an S-shaped curve such as the one in Figure 10.5. The logistic function is a nonlinear function of the regression coefficients and must be solved by a nonlinear regression routine. This makes the description of the solution process more complicated than that for linear least squares. However, logistic regression routines are available in certain statistical software packages and usually use a procedure called *maximum likelihood estimation* to estimate the regression coefficients in the logistic regression function. The following example illustrates.

EXAMPLE 10.4   **Logistic Regression and Employee Classification**   Consider again the employee classification problem discussed in Example 10.1. Consider trying to estimate the probability that each employee belongs to the satisfactory group (the group coded 1). A possible nonlinear model to estimate this probability using only the result from TEST1 can be written:

$$\mu_{y|x} = \frac{1}{1 + e^{-(\beta_0 + \beta_1 x_1)}}$$

where x_1 is the TEST1 result.

The MINITAB logistic regression output used to estimate this equation is shown in Figure 10.6. The Response Information table shows the number of observations

FIGURE 10.6
MINITAB Logistic Regression Output for Employee Classification Example Using Only TEST1 as an Independent Variable.

```
Link Function: Logit

Response Information

Variable   Value   Count
Group        1       23    (Event)
             0       20
           Total     43


Logistic Regression Table

                                                   Odds      95% CI
Predictor       Coef     SE Coef      Z       P    Ratio   Lower  Upper
Constant     -43.3684    12.9243    -3.36   0.001
TEST1          0.489722   0.144998    3.38   0.001   1.63    1.23   2.17


Log-Likelihood = -15.585
Test that all slopes are zero: G = 28.232, DF = 1, P-Value = 0.000


Goodness-of-Fit Tests

Method             Chi-Square   DF      P
Pearson              9.19454    17    0.934
Deviance             9.52955    17    0.922
Hosmer-Lemeshow      2.28161     8    0.971


Table of Observed and Expected Frequencies:
(See Hosmer-Lemeshow Test for the Pearson Chi-Square Statistic)

                                Group
Value    1    2    3    4    5    6    7    8    9    10   Total
1
   Obs   0    0    1    2    4    3    4    4    4    1     23
   Exp  0.1  0.2  1.0  1.6  3.2  3.8  4.3  3.8  3.9  1.0
0
   Obs   6    4    4    2    1    2    1    0    0    0     20
   Exp  5.9  3.8  4.0  2.4  1.8  1.2  0.7  0.2  0.1  0.0
Total    6    4    5    4    5    5    5    4    4    1     43


Measures of Association:
(Between the Response Variable and Predicted Probabilities)

Pairs        Number   Percent   Summary Measures
Concordant     411     89.3     Somers' D                 0.82
Discordant      35      7.6     Goodman-Kruskal Gamma     0.84
Ties            14      3.0     Kendall's Tau-a           0.42
Total          460    100.0
```

in each group (satisfactory = 1, unsatisfactory = 0). The Logistic Regression Table shows information pertinent to the estimation of β_0 (Constant row) and β_1 (TEST1 row). Estimates of the coefficients are in the Coef column, and standard errors of these estimates are in the SE Coef column. A Z statistic and the associated p value are shown. These are used just as in simple linear regression to test whether

β_0 or β_1 are equal to zero as follows, using β_1 to illustrate:

Hypotheses: H_0: $\beta_1 = 0$

$\qquad\qquad$ H_a: $\beta_1 \neq 0$

Decision rule: Reject H_0 if $Z > 1.96$ or $Z < -1.96$ (using a 5% level of significance)

$\qquad\qquad$ Do not reject H_0 if $-1.96 \leq Z \leq 1.96$

Test statistic: $Z = 3.38$

Decision: Reject H_0

Conclusion: TEST1 is useful in this model

The p value can also be used to conduct the test in the usual manner:

Decision rule: Reject H_0 if p value < 0.05

$\qquad\qquad$ Do not reject H_0 if p value ≥ 0.05

Test statistic: p value $= 0.001$

The same decision and conclusion are reached whether the Z statistic or its p value is used.

Information in the first row of the table can be used to test hypotheses about β_0, although interest is generally centered on the coefficients of the explanatory variables, as in linear regression.

Below this table is a statistic to test whether all slopes are zero. This is similar to the overall-fit F test in linear regression. The p value provided can be used in the usual way to conduct this test. This test is obviously of more interest when there is more than one explanatory variable. The remainder of the output is not discussed in this text.

The SAS logistic regression is in Figure 10.7. The Analysis of Maximum Likelihood Estimates table shows information pertinent to the estimation of β_0 (Intercept row) and β_1 (TEST1 row). Estimates of the coefficients are in the Estimate column, and standard errors of these estimates are in the Standard Error column. A chi-square statistic and the associated p value are shown. These are used to test whether β_0 or β_1 are equal to zero as follows, using β_1 to illustrate:

Hypotheses: H_0: $\beta_1 = 0$

$\qquad\qquad$ H_a: $\beta_1 \neq 0$

Decision rule: Reject H_0 if $\chi^2 > \chi^2 (0.05,1) = 3.841$

$\qquad\qquad$ Do not reject H_0 if $\chi^2 \leq \chi^2 (0.05,1) = 3.841$

where χ^2 is the test statistic and $\chi^2 (0.05,1)$ is a chi-square critical value chosen from the chi-square table in Appendix B. One degree of freedom is used in selecting the proper critical value for the α level of significance.

Test statistic: $\chi^2 = 11.407$

Decision: Reject H_0

Conclusion: TEST1 is useful in this model

FIGURE 10.7
SAS Logistic Regression
Output for Employee
Classification Example
Using Only TEST1 as an
Independent Variable.

The LOGISTIC Procedure

Model Information

Data Set	WORK.TRAIN
Response Variable	GROUP
Number of Response Levels	2
Number of Observations	43
Model	binary logit
Optimization Technique	Fisher's scoring

Response Profile

Ordered Value	GROUP	Total Frequency
1	1	23
2	0	20

Probability modeled is GROUP = 1.

Model Convergence Status

Convergence criterion (GCONV = 1E - 8) satisfied.

Model Fit Statistics

Criterion	Intercept Only	Intercept and Covariates
AIC	61.401	35.169
SC	63.162	38.692
-2 Log L	59.401	31.169

Testing Global Null Hypothesis: BETA = 0

Test	Chi-Square	DF	Pr > ChiSq
Likelihood Ratio	28.2321	1	<.0001
Score	21.9226	1	<.0001
Wald	11.4070	1	0.0007

The LOGISTIC Procedure

Analysis of Maximum Likelihood Estimates

Parameter	DF	Estimate	Standard Error	Wald Chi-Square	Pr > ChiSq
Intercept	1	-43.3684	12.9243	11.2599	0.0008
TEST1	1	0.4897	0.1450	11.4070	0.0007

Odds Ratio Estimates

Effect	Point Estimate	95% Wald Confidence Limits	
TEST1	1.632	1.228	2.168

Association of Predicted Probabilities and Observed Responses

Percent Concordant	89.3	Somers' D	0.817
Percent Discordant	7.6	Gamma	0.843
Percent Tied	3.0	Tau-a	0.416
Pairs	460	c	0.909

The p value can also be used to conduct the test in the usual manner:

Decision rule: Reject H_0 if p value < 0.05

Do not reject H_0 if p value ≥ 0.05

Test statistic: p value $= 0.0007$

The Z test in MINITAB and the χ^2 test in SAS are essentially equivalent, and serve the same purpose.

Above this table in the Testing Global Null Hypothesis: BETA $= 0$ table, information is provided for a test of whether all slopes are zero. This is similar to the overall-fit F test in linear regression. There are three different test statistics provided (Likelihood Ratio, Score and Wald) which can be used to conduct the test. For any of the three options, the p value provided can be used in the usual way to perform this test. The Likelihood Ratio appears to be one of the more dependable approaches for this test.

The remainder of the output is not discussed in this text.

When using logistic regression, the goal of the analysis may be to classify the observations into a particular group (as in discriminant analysis). In this case, some rule must be designed to help decide into which group each observation should be classified. Predicted values of the probability of group membership in the indicated group can be computed from the logistic regression. A rule using these predicted probabilities can then be designed. The form of such a rule is

Classify observations into the

$y = 0$ group if the predicted value is below the cutoff

$y = 1$ group otherwise

A cutoff value of 0.5 is reasonable when the 0 and 1 outcomes are equally likely and the costs of misclassification into each group are about equal. In other cases, a different cutoff may be considered superior.[3]

EXERCISES

1. **Employee Classification.** Figure 10.8 shows summary logistic regression output for the employee classification data from Example 10.4 using both test results as explanatory variables. Use the output shown to answer the following questions. The data are in a file named TRAIN10 on the CD.

   a. Which variable or variables appear to be useful in the logistic regression function?

Justify your answer. Use a 5% level of significance for any hypotheses tests.

   b. Using a cutoff probability of 0.5, classify potential employees whose test scores were as follows:

   1. TEST1 $= 94$ TEST2 $= 88$
   2. TEST1 $= 80$ TEST2 $= 87$
   3. TEST1 $= 82$ TEST2 $= 74$
   4. TEST1 $= 90$ TEST2 $= 80$

---

[3] For a more complete presentation of logistic regression, see C. E. Lunneborg, *Modeling Experimental and Observational Data*, Chapters 16, 17, and 18, or J. Neter, W. Wasserman, and M. Kutner, *Applied Linear Regression Models*, Chapter 16.

FIGURE 10.8
Logistic Regression Output for Employee Classification Example Using TEST1 and TEST2 as Independent Variables.

```
Response Information

Variable  Value   Count
Group       1       23   (Event)
            0       20
          Total     43

Logistic Regression Table

                                                   Odds      95% CI
Predictor   Coefficient   StdDev     Z      P      Ratio   Lower   Upper

Constant     -56.1704    17.4516   -3.22   0.001
TEST1          0.483314   0.157779   3.06   0.002   1.62    1.19    2.21
TEST2          0.165218   0.102070   1.62   0.106   1.18    0.97    1.44


Log-Likelihood = -13.959
Test that all slopes are zero: G = 31.483, DF = 2, P-Value = 0.000
```

2. **Harris Salaries.** In Exercise 1 in Chapter 7, data from Harris Bank were examined to test for possible discrimination. In that exercise, the dependent variable was salary and one of the explanatory variables was an indicator variable to separate the employees into male and female groups. The coefficient of the indicator variable served as a measure of whether males earned more (or less), on average, than females.

Another way of examining this problem might be to use the male/female indicator variable as the dependent variable and see if group membership can be predicted from knowledge of salary. This can be done with either discriminant analysis or logistic regression. Try discriminant analysis and/or logistic regression and see how well these methods do in predicting whether employees are male or female based only on knowledge of their salary. Does your result support the claim that Harris Bank discriminated by underpaying female employees?[4]

The data are defined as follows and are in the file named HARRIS10 on the CD:

x_1 = beginning salaries in dollars (SALARY)
x_2 = years of schooling at the time of hire (EDUCAT)
x_3 = number of months of previous work experience (EXPER)

x_4 = number of months after January 1, 1969, that the individual was hired (MONTHS)
y = variable coded 1 for males and 0 for females (MALES)

(*Source:* These data were obtained from D. Schafer, "Measurement-Error Diagnostics and the Sex Discrimination Problem," *Journal of Business and Economic Statistics.* Copyright 1987 by the American Statistical Association. Used with permission. All rights reserved.)

3. **Computer Purchase.** The Daleway Corporation owns stores throughout the United States that carry its brand of computer. Management would like some information on the purchasing practices of the American public. Specifically, it would like to know if certain factors affect the decision to upgrade a system with the purchase of a new computer. The company surveyed 40 recent customers who were considering upgrading and collected information on the following variables:

PURCHASE, coded as 1 if the customer purchased a computer, 0 if the customer did not
INCOME, the household income of the customer (in thousands of dollars)
AGE, the age of the customer's current computer

---

[4] Note: Using discriminant analysis or logistic regression in this manner would probably not be the preferred method of examining this question in a legal proceeding. The regression approach discussed in Chapter 7 would be preferred, but this makes an interesting exercise.

Are either of the variables INCOME or AGE of use in predicting whether the customer will purchase a new computer? Justify your answer. Use a 5% level of significance for any hypothesis tests. The file containing these data is named COMP-PURCH10 on the CD.

4. **March Madness (Men's Tournament).** Each March the games in the men's NCAA (National Collegiate Athletic Association) basketball tournament begin. Sixty-four teams are selected for the tournament. Each team is given a number from 1 to 16, called the *seed*. There are 4 teams assigned to each seed number (4 number 1s, 4 number 2s, etc). The lower the seed number, the higher the perceived quality of the team. In the first round of the tournament these 64 teams are paired according to their seed. Number 1 seeds play number 16 seeds; number 2 seeds play number 15 seeds; number 3 seeds play number 14 seeds and so on. The games are distributed in brackets with 16 teams (consisting of 1 through 16 seeds) in each bracket. The seed numbers are assigned by a committee formed by the NCAA. It is a commonly accepted fact that the seed numbers are of use in predicting the winner of the games. For example, a number 16 seed has never beaten a number 1 seed. This seems reasonable. Number 1 seeds, the teams of highest quality, should have a high probability of beating number 16 seeds. But do the seed numbers contain all the information useful in predicting the winners, or are there other team characteristics that might be of use? The file NCAAMEN10 on the CD contains data on the following variables:

WIN = 1 if the higher seed won the game
= 0 otherwise
DIFF = lower seed number–higher seed number; note that DIFF will always be negative (or zero if two teams of equal seed play)
PCTHIGH = winning percentage for the higher-seeded team
PCTLOW = winning percentage for the lower-seeded team

There is one additional column in the file labeled ROUND. This column indicates in which round of the tournament the game took place. Although it might be of interest as an explanatory variable, it was included primarily for information.

ROUND = 1, first-round game (32 total games)
= 2, second-round game (16 games pairing the 32 teams who survived the first round)
= 3, third-round games (8 games—these 16 teams are often referred to as the Sweet 16)
= 4, fourth-round games (4 games—these 8 teams are often referred to as the Elite 8 or the Great 8)
= 5, fifth-round games (2 games—Final 4 teams)
= 6, championship game

Use logistic regression to determine whether DIFF, PCTHIGH, or PCTLOW are useful in predicting the winners of games in the NCAA basketball tournament. Use a 5% level of significance in any hypothesis tests. Are there other variables that might be useful in addition to ones given in this problem?

5. **March Madness (Women's Tournament).** Each March the games in the women's NCAA (National Collegiate Athletic Association) basketball tournament begin. Sixty-four teams are selected for the tournament. Each team is given a number from 1 to 16, called the *seed*. There are 4 teams assigned to each seed number (4 number 1s, 4 number 2s etc). The lower the seed number, the higher the perceived quality of the team. In the first round of the tournament these 64 teams are paired according to their seed. Number 1 seeds play number 16 seeds; number 2 seeds play number 15 seeds; number 3 seeds play number 14 seeds, and so on. The games are distributed in brackets with 16 teams (consisting of 1 through 16 seeds) in each bracket. The seed numbers are assigned by a committee formed by the NCAA. It is a commonly accepted fact that the seed numbers are of use in predicting the winner of the games. But do the seed numbers contain all the information useful in predicting the winners, or are there other team characteristics that might be of use? The file NCAAWOMEN10 on the CD contains data on the following variables:

WIN = 1 if the higher seed won the game
= 0 otherwise

DIFF = lower seed number–higher seed number; note that DIFF will always be negative (or zero if two teams of equal seed play)

PCTHIGH = winning percentage for the higher-seeded team

PCTLOW = winning percentage for the lower-seeded team

There is one additional column in the file labeled ROUND. This column indicates in which round of the tournament the game took place. Although it might be of interest as an explanatory variable, it was included primarily for information.

ROUND = 1, first-round game (32 total games)

= 2, second-round game (16 games pairing the 32 teams who survived the first round)

= 3, third-round games (8 games— these 16 teams are often referred to as the Sweet 16)

= 4, fourth-round games (4 games—these 8 teams are often referred to as the Elite 8 or the Great 8)

= 5, fifth-round games (2 games— Final 4 teams)

= 6, championship game

Use logistic regression to determine whether DIFF, PCTHIGH, or PCTLOW are useful in predicting the winners of games in the NCAA basketball tournament. Use a 5% level of significance in any hypothesis tests. Are there other variables that might be useful in addition to ones given in this problem?

6. **Loan Performance.** The National Bank of Fort Worth, Texas, wants to examine methods for predicting sub-par payment performance on loans. It has data on unsecured consumer loans made over a 3-day period in October 1994 with a final maturity of 2 years. There are a total of 348 observations in the sample. The data, which have been transformed to provide confidentiality, include the following:

PASTDUE: Coded as 1 if the loan payment is past due and zero otherwise.

CBSCORE: Score generated by the CSC Credit Reporting Agency. Values range from 400 to 8390, with higher values indicating a better credit rating.

DEBT: This is a debt ratio calculated by taking required monthly payments on all debt and dividing it by the gross monthly income of the applicant and coapplicant. This ratio represents the amount of the applicant's income that will go toward repayment of debt.

GROSSINC: Gross monthly income of the applicant and coapplicant.

LOANAMT: Loan amount.

You have been asked to examine the feasibility of predicting past-due loan payment. Describe your results to the bank in a two-part report. The report should include an executive summary with a brief nontechnical description of your results and an accompanying technical report with the details of your analysis. The data are in a file named LOAN10 on the CD.

USING THE COMPUTER

The Using the Computer section in each chapter describes how to perform the computer analyses in the chapter using Excel, MINITAB, and SAS. For further detail on Excel, MINITAB, and SAS, see Appendix C.

EXCEL

Discriminant Analysis and Logistic Regression

There are currently no options for either discriminant analysis or logistic regression in Excel.

MINITAB

Discriminant Analysis

STAT: MULTIVARIATE: DISCRIMINANT ANALYSIS

Performs a discriminant analysis. The "Groups:" (dependent) variable, which designates the two (or more) groups into which the observations are classified, and the explanatory or predictor variables are requested in the dialog box. See Figure 10.9.

Logistic Regression

STAT: REGRESSION: BINARY LOGISTIC REGRESSION

FIGURE 10.9 MINITAB Dialog Box for Discriminant Analysis.

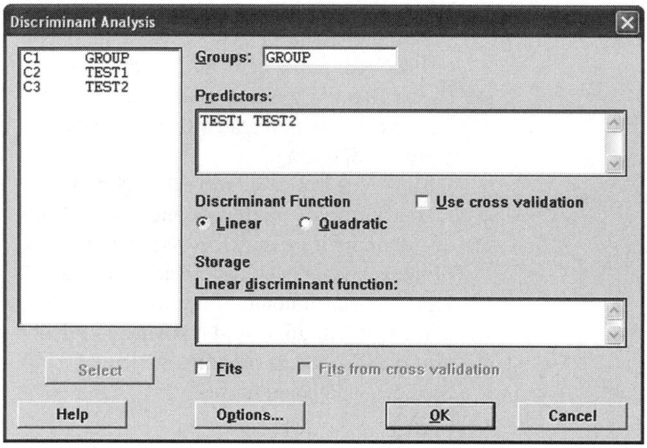

FIGURE 10.10 MINITAB Dialog Box for Binary Logistic Regression.

Performs a logistic regression. Binary logistic regression assumes the sample observations come from two groups. More than two groups can be considered by using ORDINAL LOGISTIC REGRESSION (categories are ordinal in nature) or NOMINAL LOGISTIC REGRESSION (categories have no natural ordering). Figure 10.10 shows the dialog box for a binary logistic regression.

SAS

Discriminant Analysis

PROC DISCRIM;
CLASS VAR1;
VAR VAR2;

Performs a discriminant analysis. The dependent variable, which designates the two (or more) groups into which the observations are classified, is denoted here as VAR1 and is listed on the CLASS statement. The explanatory or predictor variables are listed on the VAR statement. In this example, only one predictor variable, VAR2, is used.

Logistic Regression

PROC LOGISTIC DESCENDING;
MODEL VAR1 = VAR2;

Performs a logistic regression. VAR1 is assumed to be the dependent variable. This variable is a 0/1 variable (as discussed in this chapter). The DESCENDING option is used so that probabilities for the group labeled 1 (the highest value of the dependant variable) are predicted. If you do not use this option, the only difference will be that the signs of the coefficients will be reversed. The explanatory variables are listed on the right-hand side of the equality in the MODEL statement. In this example, the only variable listed is VAR2. The MODEL statement setup is similar to PROC REG.

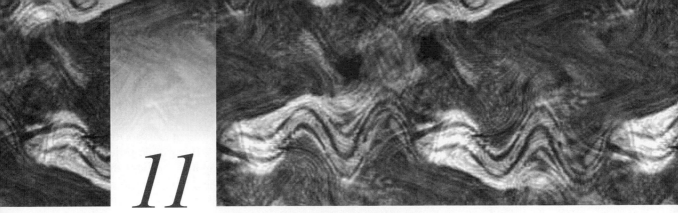

11

Forecasting Methods for Time-Series Data

11.1 INTRODUCTION

Time-series data are data gathered on a single individual (person, firm, and so on) over a sequence of time periods, which may be days, weeks, months, quarters, years, or virtually any other measure of time. In this text it is assumed that the data are gathered over only one interval of time (daily and weekly data are not combined, for example).

When dealing with time-series data, the primary goal often is to be able to produce forecasts of the dependent variable for future time periods. Two separate approaches to this problem have been discussed in this text. On the one hand, a researcher may identify variables that are related to the dependent variable in a causal manner and use these in developing a *causal regression model*. For example, when trying to forecast sales for a particular product, causal variables might include advertising expenditures and competitors' market share. Changes in these variables are believed to produce or cause changes in sales. Thus, the term *causal regression model* is used.

The researcher may, on the other hand, identify patterns of movement in past values of the dependent variable and extrapolate these patterns into the future using an *extrapolative regression model*. An extrapolative model uses explanatory variables, although they are not related to the dependent variable in a causal manner. They simply describe the past movements of the dependent variable so that these movements can be extended into future time periods. Variables that represent trend and seasonal components often are included in extrapolative models.

There are extrapolative methods other than regression that are often used in forecasting time-series data. These methods do not depend on knowledge of the values of any variable other than the one to be forecast.

As with extrapolative regression models, the success of extrapolative time-series techniques depends on the stability of the time series. If past time-series patterns are expected to continue into the future, then an extrapolative model should be relatively successful in making accurate forecasts. If these past patterns are altered for some reason, and future movements differ in general from past movements, then extrapolative models do not perform well. Thus, an assumption when using an extrapolative model for forecasting is that past patterns of data movement are reflective of future patterns.

The time-series forecasting techniques that will be discussed in this chapter include moving averages, exponential smoothing techniques, and decomposition methods. These methods are used extensively in business and economic applications. There are other, more complex extrapolative time-series forecasting methods, but these will not be discussed in this text.

11.2 NAÏVE FORECASTS

Baseline forecasting methods are methods that provide forecasts that are very simple to compute. For example, suppose we have a time series with T observations $y_1, y_2, \ldots, y_T$ and would like to compute a forecast for time period $T + 1$. One way we could do this is to use the most recent observation as the forecast for the next time period. The forecast for time period $T + 1$ is the actual value in time period T:

$$\hat{y}_{T+1} = y_T$$

This process is called the *naïve forecasting method* and the resulting forecasts are called *naïve forecasts*. Obviously, this is a very simple method of creating forecasts, and we might expect to be able to do better than this method. However, there are situations when the naïve forecasts are the "best" forecast that we can come up with. If the data follow what is called a random walk, as many financial and economic time series do, then the naïve forecast is best (among extrapolative methods, at least).

However, many time series do not follow a random walk and in many cases we can do better than using naïve forecasts. The naïve forecasts will be our baseline forecasts and can be used in the following way: Whenever a more complex forecasting method is considered, compare the results for that method to those from the baseline method. If the more complex method cannot do better than the naïve forecasts, then there is no reason to use the more complex method.

This is similar to what we did with regression when we built a regression model. We tested one model against a simpler model. If our conclusion was that the more complex model outperformed the simpler model, then we used the more complex model to generate forecasts. If not, then we would opt for the simpler model. In the case of extrapolative time-series methods considered in this chapter, we make a similar kind of comparison. We do it without the use of formal hypothesis tests, however, and simply use some method of comparing the accuracy of the forecasts generated from alternative forecasting methods.

11.3 MEASURING FORECAST ACCURACY

There are a variety of measures of forecast accuracy. In this chapter we use three commonly used measures: the mean square deviation (*MSD*), the mean absolute deviation (*MAD*), and the mean absolute percentage forecast error (*MAPE*).

Assume that we have *n* forecasts and the same number of actual values of the variable of interest. The forecast accuracy measures are computed as follows. The *mean square deviation (MSD)* is the average of the squared forecast errors:

$$MSD = \frac{\sum_{i=1}^{n}(y_i - \hat{y}_i)^2}{n}$$

The *mean absolute deviation (MSD)* is the average of the absolute values of the forecast errors:

$$MAD = \frac{\sum_{i=1}^{n}|y_i - \hat{y}_i|}{n}$$

The *mean absolute percentage error (MAPE)* is the average of the ratio of the absolute value of the forecast errors to the absolute value of the actual *y* value. This quantity is multiplied by 100 so that the forecast error is expressed as a percentage of the actual values:

$$MAPE = \frac{\sum_{i=1}^{n}\frac{|y_i - \hat{y}_i|}{|y_i|}}{n} \times 100\%$$

When choosing between two possible forecasting techniques we would choose the technique that had smaller values of *MSD*, *MAD*, or *MAPE*. Smaller values of the accuracy measures result from smaller forecast errors. Unfortunately, sometimes a method that has a smaller value on one of the measures will have a larger value on another. The forecaster will have to decide which measure is best in the situation and make a choice between the competing forecasting methods. Judgment of the forecaster is of importance in such a case.

11.4 MOVING AVERAGES

In subsequent sections of this chapter, we will consider forecasting methods that are somewhat more complex than our baseline method, the naïve forecast. In this section, the method discussed is called a *moving average*.

A moving average uses the average of the m most recent observations as the forecast for the next observation. Suppose we want to use an m-period moving average to forecast for time period $t + 1$. Then our forecast will be the average of the observations from time periods $t - m + 1$ through t.

$$\hat{y}_{t+1} = \frac{y_{t-m+1} + y_{t-m+2} + \cdots + y_t}{m}$$

This method is called a moving average because the forecast for the next time period will be computed by dropping the oldest observation and adding in the newest:

$$\hat{y}_{t+2} = \frac{y_{t-m+2} + y_{t-m+3} + \cdots + y_{t+1}}{m}$$

The average continues to "move" through the data in this manner. The technique is similar to a simple average but considers only the m most recent data values. If $m = n$, then the entire data set would be used in the average. On the other hand, if $m = 1$, we have the naïve forecast.

EXAMPLE 11.1   **XYZ Sales: Moving Average**   The data in Table 11.1 will be used to illustrate the moving average forecast. (See also the file XYZSALES11 on the CD.) There are 20 monthly observations for sales (in thousands of units) of computers assembled by the XYZ Corporation. XYZ Corporation needs short-term forecasts to help in production planning. Figure 11.1 shows a time-series plot of the data. In this example we will use a three-period moving average. The forecasts are also shown in Table 11.1 along with the forecast error. Figure 11.2 shows a plot of the time series and the moving average forecasts. For example, the forecast for time period 4 is

$$\hat{y}_4 = \frac{4 + 13 + 9}{3} = 8.667$$

The forecast error is $y_4 - \hat{y}_4 = 11 - 8.667 = 2.333$. The forecast errors can be used to compute the accuracy measures:

$$MSD = 20.608$$
$$MAD = 3.627$$
$$MAPE = 56.591\%$$

These measures provide an idea of the accuracy of the forecasts. However, without some basis for comparison, they do not really tell us how useful this forecasting technique is. This is where the baseline measure comes into play. We will compute the accuracy measures for our baseline measure and use these as a gauge for the accuracy of the moving average forecasts.

    Suppose we use the previous time period observation as our forecast for the next time period. To make things even for our comparison with the moving average,

TABLE 11.1 XYZ Corporation SALES with Three-Period Moving Average Forecast and Forecast Errors

Time Period	SALES	Three-Period Moving Average Forecasts	Forecast Error
1	4		
2	13		
3	9		
4	11	8.667	2.333
5	10	11.000	−1.000
6	3	10.000	−7.000
7	15	8.000	7.000
8	4	9.333	−5.333
9	4	7.333	−3.333
10	9	7.667	1.333
11	5	5.667	−0.667
12	7	6.000	1.000
13	8	7.000	1.000
14	8	6.667	1.333
15	18	7.667	10.333
16	8	11.333	−3.333
17	5	11.333	−6.333
18	15	10.333	4.667
19	5	9.333	−4.333
20	7	8.333	−1.333
21		9.000	

FIGURE 11.1
Time-Series Plot of Data for XYZ Corporation.

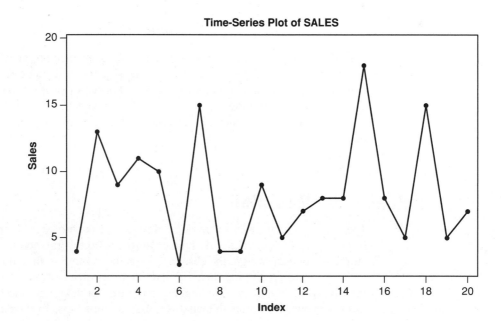

Time-Series Plot of SALES

FIGURE 11.2
Three-Period Moving
Average Plot of
SALES for XYZ
Corporation.

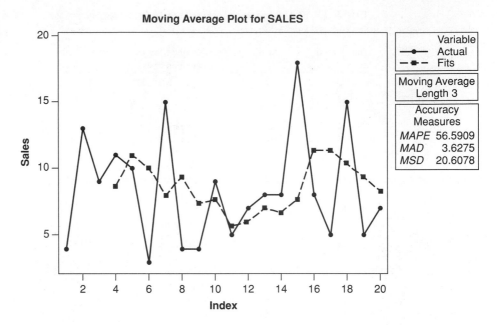

we will use the same 17 observations as the basis for our comparison. The computation of the forecast errors is shown in Table 11.2. The resulting values for the accuracy measures are:

$$MSD = 45.765$$
$$MAD = 5.294$$
$$MAPE = 78.173\%$$

For all three accuracy measures, using the moving average provides a smaller value than using the naïve forecast. In this case, the slightly more complicated moving average method does appear to improve forecast accuracy over our baseline measure, so we will choose the moving average method to generate forecasts. Also, periods other than three for the moving average could be tried to see if additional improvements are possible.

11.5 EXPONENTIAL SMOOTHING

The moving average uses the average of several of the most recent observations as a forecast for the next time period. One alternative to this method that may have occurred to you is to weight the observations in the average so that more recent observations are weighted more heavily than the older observations. This makes sense intuitively in many cases. We might expect sales for next month to be more similar to sales in the current month than it is to last month's sales. In such a case we would

TABLE 11.2 XYZ Corporation SALES with Naïve Forecast and Forecast Errors

Time Period	SALES	Naïve Forecasts	Forecast Error
1	4		
2	13		
3	9		
4	11	9	2
5	10	11	−1
6	3	10	−7
7	15	3	12
8	4	15	−11
9	4	4	0
10	9	4	5
11	5	9	−4
12	7	5	2
13	8	7	1
14	8	8	0
15	18	8	10
16	8	18	−10
17	5	8	−3
18	15	5	10
19	5	15	−10
20	7	5	2

like to weight the most recent observations more heavily and compute a weighted average. Exponential smoothing techniques compute a special type of weighted average. With exponential smoothing, all past observations are included in the average, but the most recent observations are weighted most heavily. Also, the user can adjust the weights so that recent observations can be weighted more or less heavily as desired. There are several types of exponential smoothing techniques. The three methods that will be discussed in this chapter are single exponential smoothing, double exponential smoothing, and Winters' exponential smoothing. Single exponential smoothing is appropriate when there is no trend or seasonal variation in the data. Double exponential smoothing is designed for cases when there is a trend in the data, but no seasonal variation. Winters' method takes into account both trend and seasonal variation.

11.5.1 SINGLE EXPONENTIAL SMOOTHING

As mentioned previously, exponential smoothing techniques compute a weighted average of all past observations, but the most recent observations are weighted most heavily. *Single exponential smoothing* is intended to perform best when the time series to be forecast has no trend or seasonal component. Suppose we have a time series with T observations $y_1, y_2, \ldots, y_T$ and want to compute a forecast for time period $T + 1$. The single exponential smoothing forecast can be written:

$$\hat{y}_{T+1} = \alpha y_T + \alpha(1 - \alpha)y_{T-1} + \alpha(1 - \alpha)^2 y_{T-2}$$
$$+ \alpha(1 - \alpha)^3 y_{T-3} + \cdots + \alpha(1 - \alpha)^{T-1} y_1 \quad (11.1)$$

The value α (alpha) is called the *smoothing constant* and is chosen either by the forecaster or by statistical software packages that are equipped to compute the "best" value for α. If α is constrained to be a value between 0 and 1, then Equation (11.1) can be interpreted as follows: The forecast for the next time period will be related to all past values of the data, but the most recent value will be weighted most heavily. The weights on successively older values will decline exponentially (thus the name, exponential smoothing) because $1 - \alpha$ will be between 0 and 1 as well. For example, suppose we choose $\alpha = 0.9$. Then Equation (11.1) becomes

$$\hat{y}_{T+1} = 0.9y_T + 0.09y_{T-1} + 0.009y_{T-2} + \cdots$$

If we choose $\alpha = 0.1$, then

$$\hat{y}_{T+1} = 0.1y_T + 0.09y_{T-1} + 0.081y_{T-2} + \cdots$$

When $\alpha = 0.9$, the most recent observation has about 90% of the weight and successive observations have much smaller weights attached. Choose a large value of the smoothing constant if successive observations in the time series tend to be similar to each other. On a graph, such a time series would appear as a fairly smooth curve.

When $\alpha = 0.1$, a smaller weight is assigned to the most recent observation and successive weights are more evenly distributed to past observations. Choose a small value of the smoothing constant if successive observations in the time series are likely to be very different from each other. On a graph, such a time series would appear as a very jagged curve with a lot of variation.

There is an alternative formula to Equation (11.1) that is typically used when computing single exponential smoothing forecasts. The forecast for time period $T + 1$ can be written

$$\hat{y}_{T+1} = \alpha y_T + (1 - \alpha)\hat{y}_T \tag{11.2}$$

The forecast for time period $T + 1$ is a weighted sum of the actual value of y in time period T and the forecast for that time period. Equations (11.1) and (11.2) are two different ways of expressing the single exponential smoothing forecast, but Equation (11.2) is much easier to work with if computing forecasts by hand (or programming software to do the computations). Note that once a value of α is chosen, only the most recent observation and the forecast for that time period are needed to find the forecast for the next time period (using Equation (11.2)).

EXAMPLE 11.2 **XYZ Sales: Exponential Smoothing** The 20 monthly observations for XYZ computer sales (file XYZSALES11 on the CD) will be used to illustrate the computations for single exponential smoothing. The results are shown in Table 11.3. To do the computations, a value for the smoothing constant must first be chosen. We will arbitrarily choose $\alpha = 0.3$. Also, Equation (11.2) will be used because it is easier to use than Equation (11.1). Starting with the first time period, we have

$$\hat{y}_1 = \alpha y_0 + (1 - \alpha)\hat{y}_0$$

TABLE 11.3 Data for Exponential Smoothing Example Including Forecasts Using $\alpha = 0.3$ and Forecast Errors

Time Period	SALES	Forecasts ($\alpha = 0.3$)	Forecast Error
1	4		
2	13	4.0000	9.0000
3	9	6.7000	2.3000
4	11	7.3900	3.6100
5	10	8.4730	1.5270
6	3	8.9311	−5.9311
7	15	7.1518	7.8482
8	4	9.5062	−5.5062
9	4	7.8544	−3.8544
10	9	6.6981	2.3019
11	5	7.3886	−2.3886
12	7	6.6720	0.3280
13	8	6.7704	1.2296
14	8	7.1393	0.8607
15	18	7.3975	10.6025
16	8	10.5783	−2.5783
17	5	9.8048	−4.8048
18	15	8.3633	6.6367
19	5	10.3543	−5.3543
20	7	8.7480	−1.7480

and immediately run into problems. There is no time period zero, so we will skip computing the forecast for time period one. To compute the forecast for time period two use

$$\hat{y}_2 = \alpha y_1 + (1 - \alpha)\hat{y}_1$$

Now we have another problem. Although y_1 is available, there is no forecast for time period one. These kinds of issues are called *initialization* problems. How do we start the forecasting process in these early time periods? There are a variety of different ways to do this but for our purposes a very simple solution is adopted: Use the value of y in the first time period as the forecast for the second time period $(\hat{y}_2 = 4)$. Then the forecast for the third time period can be computed as

$$\hat{y}_3 = \alpha y_2 + (1 - \alpha)\hat{y}_2 = 0.3(13) + (1.0 - 0.3)(4) = 6.7$$

Now that the process is started, the remaining forecasts are easy to compute. Plug the actual and forecasted values into Equation (11.2) and compute the resulting forecasts, which are shown in Table 11.3. The forecast errors (also shown) can be used to obtain the following accuracy measures.

$$MSD = 22.858$$
$$MAD = 3.948$$
$$MAPE = 58.813\%$$

Exponential smoothing with $\alpha = 0.3$ is an improvement over the naïve forecasts but the three-period moving average has slightly lower accuracy measures. But $\alpha = 0.3$ is just one possible value of the smoothing constant. Could we do better if we used $\alpha = 0.1$ or $\alpha = 0.4$? Some statistical software packages are programmed to find the "best" value of α. *Best* is usually defined as the value that minimizes the *MSD* over the sample data. For example, the MINITAB result is shown in Figure 11.3. MINITAB chooses $\alpha = 0.0654404$ as the optimal value of α. The resulting accuracy measures are

$$MSD = 18.254$$
$$MAD = 3.383$$
$$MAPE = 48.395\%$$

In this case, the accuracy measures for single exponential smoothing are lower than those for the naïve method and for the three-period moving average forecasts.

MINITAB uses a method called Box–Jenkins ARIMA models to find the optimal value of the smoothing constant. The details behind this method are complex and are not covered in this text. For more information on Box–Jenkins ARIMA models see *Forecasting and Time Series: An Applied Approach* by Bowerman and O'Connell.[1] One result of using the Box–Jenkins method to estimate the optimal smoothing constant is that the smoothing constant suggested by MINITAB will not always be

FIGURE 11.3
Single Exponential Smoothing Plot of XYZ SALES with Optimal Smoothing Constant.

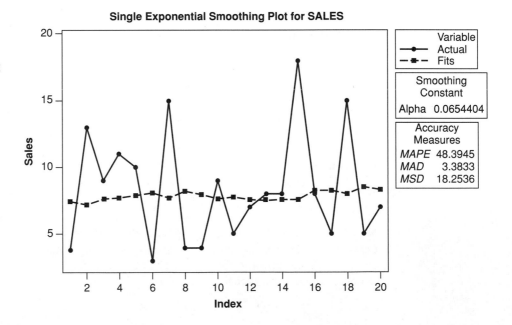

---

[1]See References section for complete citation.

between 0 and 1. The method does not restrict the smoothing constant to fall within that range. Often, if a value outside the 0 to 1 range occurs, it suggests that another technique such as double exponential smoothing might be preferred to single exponential smoothing (especially if the suggested smoothing constant value is greater than 1).

Recall that our solution to the initialization problems for single exponential smoothing was to skip the first time period and use the actual value for the first time period as the forecast for the second time period. There are a variety of other suggested ways to initialize the forecasting process. For example, if a value for the smoothing constant is specified by the user, MINITAB will use the average of the first six time period actual values as the forecast for the first time period and then apply the formula in Equation (11.2) after that. If an optimal value of the smoothing constant is requested, a method called *backcasting* is used to establish the initial values. This is a result of the fact that Box–Jenkins methods are used to find the optimal value. Again, these methods will not be discussed in this text. The result of all this is that computing these estimates by hand may result in somewhat different results than computer-generated forecasts. For example, Figure 11.4 shows the MINITAB result using a smoothing constant of 0.3. The results differ somewhat from the results obtained by hand computation in Table 11.3. Also, two different computer software packages might produce somewhat different forecasts. This does not mean that the results from one software package are right and the results from the other are wrong. It is likely just an artifact of the use of different initialization procedures or different optimization approaches. Typically, after cycling through several time periods, the differences in the forecasts using various techniques will be very similar so that initialization will not have a large effect on the final results.

FIGURE 11.4
Single Exponential
Smoothing Plot of XYZ
SALES using $\alpha = 0.3$.

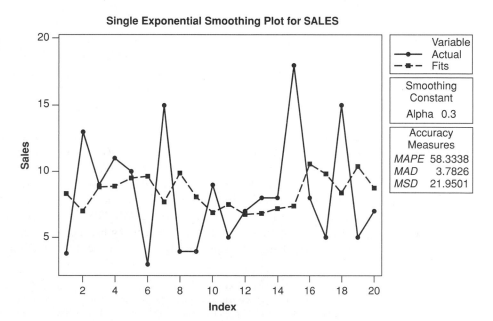

11.5.2 DOUBLE EXPONENTIAL SMOOTHING

Double exponential smoothing is a forecasting method intended to perform best when the time series to be forecast has a trend but no seasonal component. Single exponential smoothing computes a weighted average of all past observations, but the most recent observations are weighted most heavily. This process is sometimes referred to as *smoothing* the series. Applying single exponential smoothing to a series that has a trend produces forecasts that will consistently underestimate the true values. Double exponential smoothing uses the smoothed series produced by single exponential smoothing and essentially smoothes the series a second time. The difference between the single-smoothed values and the double-smoothed values provides a measure of the trend. This process is operationalized with the following formulas:

$$L_t = \alpha y_t + (1 - \alpha)(L_{t-1} + T_{t-1}) \tag{11.3}$$

$$T_t = \gamma(L_t - L_{t-1}) + (1 - \gamma)T_{t-1} \tag{11.4}$$

$$\hat{y}_{t+m} = L_t + mT_t \tag{11.5}$$

These formulas are also sometimes referred to as *Holt's two-parameter linear exponential smoothing*. The first thing to note about these formulas is that there are now two smoothing constants, α and γ (alpha and gamma). As with single exponential smoothing, the values of these smoothing constants is chosen either by the forecaster or by some statistical software packages that may compute the optimal values.[2]

Equation (11.3) represents the smoothed series or the estimate of the level (L) of the series. The equation is similar to that of single exponential smoothing in Equation (11.2) except that the term $L_{t-1} + T_{t-1}$ replaces the forecast. L_{t-1} represents the level in the previous time period and T_{t-1} adjusts that level for trend. Equation (11.4) is used to produce an estimate of the trend. It combines the old estimate of trend, T_{t-1}, and a new estimate computed as the difference between the two most recent levels of the series. These trend estimates are smoothed using a smoothing constant γ. Typically, α and γ are assumed to be between 0 and 1. A forecast for m periods into the future can be computed as in Equation (11.5). The m-period-ahead forecast is the level at time $t(L_t)$ plus m times the value of the trend component at time $t(mT_t)$.

EXAMPLE 11.3   **Double Exponential Smoothing: New Construction**   Our construction firm is interested in forecasting new construction in the United States for the years 2002 and 2003. We have data in billions of dollars for the years 1991 through 2001 from the Department of Commerce (on the CD, in the file NEWCON11). These data are shown in Table 11.4. Figure 11.5 shows a MINITAB single exponential smoothing plot for these data. Note that the forecasts lag behind (are typically less than) the actual values. This is a result of using single exponential smoothing on trended data.

---

[2] There is a procedure called *Brown's one-parameter linear exponential smoothing* that adjusts for trend but uses just one smoothing constant. Holt's method is covered in this text because the use of two smoothing constants provides greater flexibility by allowing the trend to be smoothed with a different parameter.

TABLE 11.4 Data for New Construction Example

Year	NEWCON
1991	432.6
1992	463.7
1993	491.0
1994	539.2
1995	557.8
1996	615.9
1997	653.4
1998	705.7
1999	765.9
2000	820.3
2001	842.5

FIGURE 11.5
Single Exponential Smoothing Plot of NEWCON with Optimal Smoothing Constant.

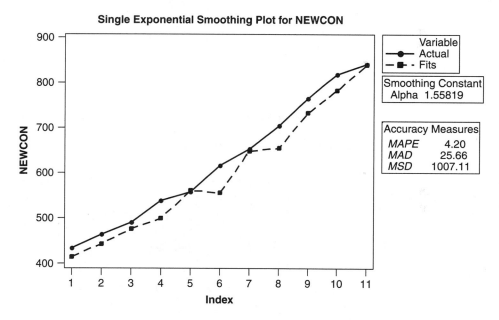

Also note that the optimal value of the smoothing constant chosen by MINITAB is greater than 1. This will often be the case when data are trended and single exponential smoothing is applied.

Figure 11.6 shows the double exponential smoothing plot. The accuracy measures for double exponential smoothing are all smaller than for single exponential smoothing. Double exponential smoothing is a better choice for the new construction data. The optimal smoothing parameters chosen are $\alpha = 0.782737$ and $\gamma = 0.225655$.

The hand computations for double exponential smoothing will not be shown.

FIGURE 11.6
Double
Exponential
Smoothing Plot
of NEWCON with
Optimal
Smoothing
Constants.

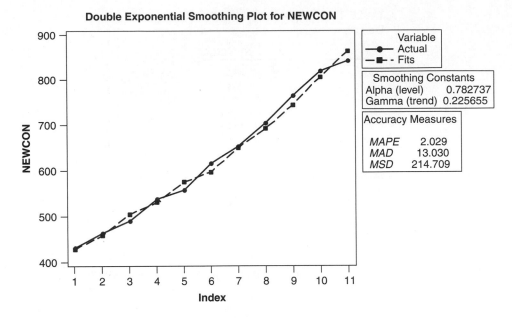

11.5.3 WINTERS' EXPONENTIAL SMOOTHING

Winters' exponential smoothing method is a forecasting method intended to take into account both trend and seasonal variation. It is an extension of Holt's two-parameter smoothing. There are two versions of Winters' method: additive and multiplicative.

The equations for the multiplicative version of Winters' method are

$$L_t = \alpha \frac{y_t}{S_{t-s}} + (1 - \alpha)(L_{t-1} + T_{t-1}) \tag{11.6}$$

$$T_t = \gamma(L_t - L_{t-1}) + (1 - \gamma)T_{t-1} \tag{11.7}$$

$$S_t = \delta \frac{y_t}{L_t} + (1 - \delta)S_{t-s} \tag{11.8}$$

$$\hat{y}_{t+m} = (L_t + mT_t)S_{t-s+p} \tag{11.9}$$

There are three smoothing constants, α, γ, and δ (alpha, gamma, and delta). The values of these smoothing constants are typically chosen by the forecaster.

Equation (11.6) represents the smoothed series or the estimate of the level (L) of the series after the seasonal effect has been removed. The seasonal effect is removed by dividing the data value, y_t, by the seasonal component S_{t-s} (because this is a multiplicative model, dividing by the seasonal component will remove the seasonal effect). Otherwise, the equation is the same as Equation (11.3) of double exponential smoothing. Equation (11.7) is the same as Equation (11.4) of double exponential smoothing and is used to produce an estimate of the trend. Equation (11.8) computes an estimate of the seasonal effect by smoothing the previous estimate, S_{t-s}, and the most recent observation (adjusted for level), (y_t/L_t). As with the other exponential smoothing approaches, the values of the smoothing parameters are usually restricted to be between 0 and 1. A forecast for m periods into the future can be computed as in Equation (11.9). The m-period-ahead forecast is the level at time $t(L_t)$ plus m times the value of the trend component at time $t(mT_t)$ multiplied by the seasonal component to adjust for seasonality.

The equations for the additive version of Winters' method are

$$L_t = \alpha(y_t - S_{t-s}) + (1 - \alpha)(L_{t-1} + T_{t-1}) \tag{11.10}$$

$$T_t = \gamma(L_t - L_{t-1}) + (1 - \gamma)T_{t-1} \tag{11.11}$$

$$S_t = \delta(y_t - L_t) + (1 - \delta)S_{t-s} \tag{11.12}$$

$$\hat{y}_{t+m} = L_t + mT_t + S_{t-s+p} \tag{11.13}$$

Equation (11.10) represents the smoothed series or the estimate of the level (L) of the series after the seasonal effect has been removed. This is done by subtracting the seasonal component, S_{t-s}, from the data value, y_t (because this is an additive model, subtracting the seasonal component will remove the seasonal effect) Equation (11.11) is the same as Equation (11.7) of the multiplicative method and is used to produce an estimate of the trend. Equation (11.12) computes an estimate of the seasonal effect by smoothing the previous estimate, S_{t-s}, and the most recent observation (adjusted for level), $y_t - L_t$. A forecast for m periods into the future can be computed as in Equation (11.13). The m-period-ahead forecast is the level at time $t(L_t)$ plus m times the value of the trend component at time $t(mT_t)$ plus the seasonal component to adjust for seasonality. Note that the additive version of Winters' method looks somewhat like a time-series regression with a constant, a trend variable, and seasonal indicators.

EXAMPLE 11.4   **ABX Company Sales: Winters' Additive Method**   The ABX Company sells winter sports merchandise including skis, ice skates, sleds, and so on. Quarterly sales (in thousands of dollars) for the ABX Company are shown in Table 11.5. The time period represented starts in the first quarter of 1994 and ends in the fourth quarter of 2003. (See the file ABXSALES11 on the CD.)

A time-series plot of the sales figures is shown in Figure 11.7. From the time-series plot it is clear that there is both a trend component and seasonal variation to this series. Winters' method would be appropriate for generating forecasts. But which of the Winters' approaches should be used, additive or multiplicative? Additive models are used when the seasonal component is constant; the amplitudes of the seasonal cycles are roughly the same. Multiplicative seasonal models are used when the seasonal variation is growing over time; the amplitudes of the seasonal cycles are proportional to the level of the time series. If there is any doubt about whether an additive or multiplicative model should be used, fit both and use the accuracy measures to help determine which method produces the most accurate forecasts. In this example, the additive model is used.

TABLE 11.5 Data for ABX Company Sales Example

Year.Qtr	SALES	Year.Qtr	SALES
1994.1	221.0	1999.2	244.0
1994.2	203.5	1999.3	256.0
1994.3	190.0	1999.4	276.5
1994.4	225.5	2000.1	291.0
1995.1	223.0	2000.2	255.5
1995.2	190.0	2000.3	244.0
1995.3	206.0	2000.4	291.0
1995.4	226.5	2001.1	296.0
1996.1	236.0	2001.2	260.0
1996.2	214.0	2001.3	271.5
1996.3	210.5	2001.4	299.5
1996.4	237.0	2002.1	297.0
1997.1	245.5	2002.2	271.0
1997.2	201.0	2002.3	270.0
1997.3	230.0	2002.4	300.0
1997.4	254.5	2003.1	306.5
1998.1	257.0	2003.2	283.5
1998.2	238.0	2003.3	283.5
1998.3	228.0	2003.4	307.5
1998.4	255.0		
1999.1	260.5		

FIGURE 11.7 Time-Series Plot of ABX Sales.

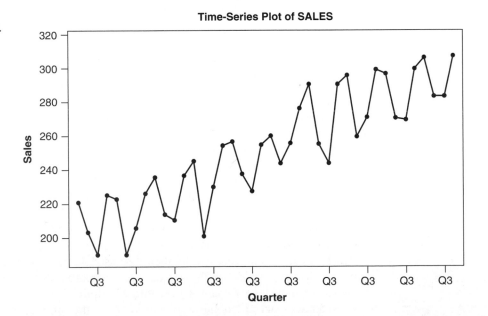

Figure 11.8 shows the Winters' exponential smoothing plot using a value of 0.2 for each of the three smoothing constants. These are the default values for MINITAB. There is nothing magical about these values, but they seem to provide a

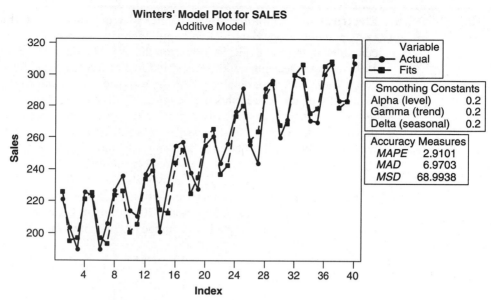

FIGURE 11.8
Winters' Exponential Smoothing Plot (Additive Model) of ABX Sales with Smoothing Parameters $\alpha = 0.2$, $\gamma = 0.2$, and $\delta = 0.2$.

good starting point for many time series. MINITAB does not provide an optimization option for Winters' method. Other values of the smoothing parameters could be tried and judged on the basis of the accuracy measures. Figure 11.9 shows the Winters' plot with $\alpha = 0.2$, $\gamma = 0.3$, and $\delta = 0.2$. The accuracy measures are slightly smaller for this combination of smoothing constants. There may be other combinations that improve on this result.

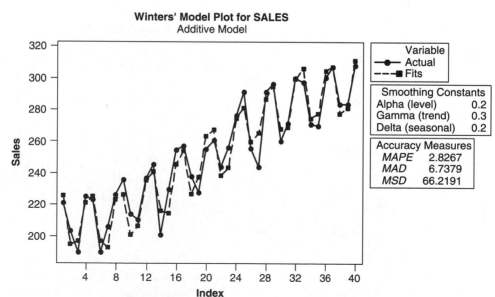

FIGURE 11.9
Winters' Exponential Smoothing Plot (Additive Model) of ABX Sales with Smoothing Parameters $\alpha = 0.2$, $\gamma = 0.3$, and $\delta = 0.2$.

EXAMPLE 11.5   **Electrical and Appliance Stores: Winters' Multiplicative Method**   Monthly sales amounts from electrical and appliance stores (in millions of dollars) are contained in the file ELECTAPPL11 on the CD. The time frame is from January 1992 through December 2002. We would like to forecast sales for each month in 2003. These data were obtained from the web site *www.economagic.com*. A time-series plot of the data is shown in Figure 11.10. Note that the amplitude of the seasonal cycles is increasing over time. As the level of the series increases, so does the magnitude of the seasonal cycle. This is a situation where a multiplicative model will typically perform better than an additive model. The Winters' exponential smoothing plot for a multiplicative model is shown in Figure 11.11. The additive model is shown in Figure 11.12. Both models use $\alpha = 0.2$, $\gamma = 0.2$, and $\delta = 0.2$. The accuracy measures show that the multiplicative model is preferred.

The following forecasts for January 2003 through December 2003 were computed using the resulting equations from the multiplicative model:

Period	Forecast
1/03	7372.5
2/03	6928.5
3/03	7435.4
4/03	6797.9
5/03	7157.6
6/03	7353.7
7/03	7403.4
8/03	7826.9
9/03	7390.4
10/03	7488.6
11/03	8688.2
12/03	12551.9

FIGURE 11.10 Time-Series Plot of Electrical and Appliance Store Sales.

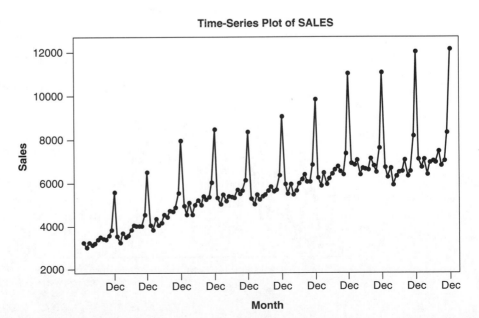

FIGURE 11.11
Winters'
Exponential
Smoothing Plot
(Multiplicative
Model) of Electrical
and Appliance
Store Sales.

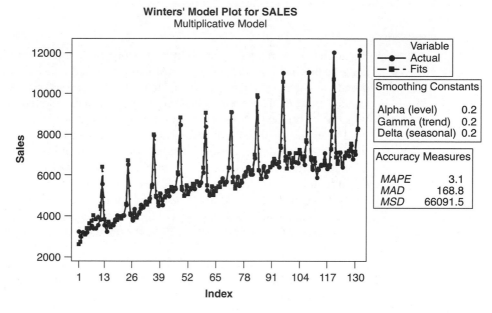

FIGURE 11.12
Winters'
Exponential
Smoothing Plot
(Additive Model) of
Electrical and
Appliance Store
Sales.

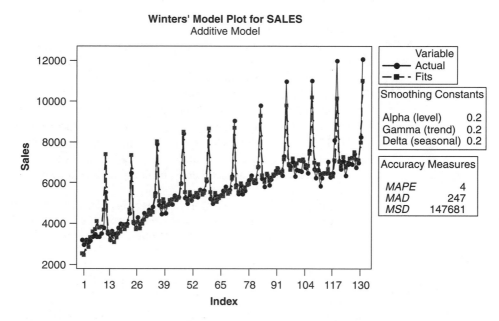

11.6 DECOMPOSITION

In previous sections the concepts of trend, seasonality, and random variation in a time series have been discussed. We use those concepts again in this section. For the purposes of the decomposition method, the time series will be written either as

$$y_t = T_t + S_t + E_t \tag{11.14}$$

or as

$$y_t = T_t \times S_t \times E_t \tag{11.15}$$

Equation (11.14) represents an additive version of the time series. Equation (11.15) is the multiplicative version. The term y_t represents the data. y_t has been decomposed into three parts: the trend component, T_t; the seasonal component, S_t; and the random error, E_t. These representations are similar to the additive and multiplicative versions of Winters' exponential smoothing. As in that case, additive models are used when the amplitudes of the seasonal cycles are roughly the same. Multiplicative seasonal models are used when the seasonal variation is growing over time. With multiplicative seasonality, the amplitudes of the seasonal cycles are proportional to the level of the time series. If there is any doubt about whether an additive or multiplicative model should be used, fit both and use the accuracy measures to help determine which method produces the most accurate forecasts.

The use of the additive model will be illustrated in the following example.

EXAMPLE 11.6   **ABX Sales: Additive Decomposition**   The ABX Company sells winter sports merchandise including skis, ice skates, sleds, and so on. Quarterly sales (in thousands of dollars) for the ABX Company are shown in Table 11.5 (file ABXSALES11 on the CD). The time period represented starts in the first quarter of 1994 and ends in the fourth quarter of 2003.

A time-series plot of the sales figures is shown in Figure 11.7. From the time-series plot it is clear that there is both a trend component and seasonal variation to this series. Also the magnitude of the seasonal cycles appears to be constant over time, so an additive decomposition will be used.

There is no single "correct" way to perform decomposition. In the following example we use a fairly standard approach (which is the same approach that is used by MINITAB Version 14). The goal is to estimate both the trend and seasonal components of the time series and then to use those estimates to produce forecasts. The steps in doing this are outlined here. The computations are shown in Table 11.6.

Step 1: Smooth out the seasonal variation in the time series using a centered moving average.   First compute a four-period moving average (four periods because these are quarterly data) with the results shown in the 4 MA column. For example, the first entry in that column is

$$\frac{221 + 203.5 + 190 + 225.5}{4} = 210$$

The second entry is

$$\frac{203.5 + 190 + 225.5 + 223}{4} = 210.5$$

In terms of the time sequence, the first entry, 210, can be thought of as occurring between the second and third data values. In other words, think of 210 as being between 203.5 and 190 in the time sequence. The second entry, 210.5, can be thought

TABLE 11.6 Computations for Additive Decomposition

SALES	4 MA	CMA	SALES-CMA	S HAT	DESEAS	T HAT	T HAT + S HAT
221				15.17188	205.8281	201.3753661	216.5472461
203.5				−12.7031	216.2031	203.9413722	191.2382722
190	210	210.25	−20.25	−14.7031	204.7031	206.5073783	191.8042783
225.5	210.5	208.8125	16.6875	12.23438	213.2656	209.0733844	221.3077644
223	207.125	209.125	13.875	15.17188	207.8281	211.6393905	226.8112705
190	211.125	211.25	−21.25	−12.7031	202.7031	214.2053966	201.5022966
206	211.375	213	−7	−14.7031	220.7031	216.7714027	202.0683027
226.5	214.625	217.625	8.875	12.23438	214.2656	219.3374088	231.5717888
236	220.625	221.1875	14.8125	15.17188	220.8281	221.9034149	237.0752949
214	221.75	223.0625	−9.0625	−12.7031	226.7031	224.469421	211.766321
210.5	224.375	225.5625	−15.0625	−14.7031	225.2031	227.0354271	212.3323271
237	226.75	225.125	11.875	12.23438	224.7656	229.6014332	241.8358132
245.5	223.5	225.9375	19.5625	15.17188	230.3281	232.1674393	247.3393193
201	228.375	230.5625	−29.5625	−12.7031	213.7031	234.7334454	222.0303454
230	232.75	234.1875	−4.1875	−14.7031	244.7031	237.2994515	222.5963515
254.5	235.625	240.25	14.25	12.23438	242.2656	239.8654576	252.0998376
257	244.875	244.625	12.375	15.17188	241.8281	242.4314637	257.6033437
238	244.375	244.4375	−6.4375	−12.7031	250.7031	244.9974698	232.2943698
228	244.5	244.9375	−16.9375	−14.7031	242.7031	247.5634759	232.8603759
255	245.375	246.125	8.875	12.23438	242.7656	250.129482	262.363862
260.5	246.875	250.375	10.125	15.17188	245.3281	252.695488	267.867368
244	253.875	256.5625	−12.5625	−12.7031	256.7031	255.2614941	242.5583941
256	259.25	263.0625	−7.0625	−14.7031	270.7031	257.8275002	243.1244002
276.5	266.875	268.3125	8.1875	12.23438	264.2656	260.3935063	272.6278863
291	269.75	268.25	22.75	15.17188	275.8281	262.9595124	278.1313924
255.5	266.75	268.5625	−13.0625	−12.7031	268.2031	265.5255185	252.8224185
244	270.375	271	−27	−14.7031	258.7031	268.0915246	253.3884246
291	271.625	272.1875	18.8125	12.23438	278.7656	270.6575307	282.8919107
296	272.75	276.1875	19.8125	15.17188	280.8281	273.2235368	288.3954168
260	279.625	280.6875	−20.6875	−12.7031	272.7031	275.7895429	263.0864429
271.5	281.75	281.875	−10.375	−14.7031	286.2031	278.355549	263.652449
299.5	282	283.375	16.125	12.23438	287.2656	280.9215551	293.1559351
297	284.75	284.5625	12.4375	15.17188	281.8281	283.4875612	298.6594412
271	284.375	284.4375	−13.4375	−12.7031	283.7031	286.0535673	273.3504673
270	284.5	285.6875	−15.6875	−14.7031	284.7031	288.6195734	273.9164734
300	286.875	288.4375	11.5625	12.23438	287.7656	291.1855795	303.4199595
306.5	290	291.6875	14.8125	15.17188	291.3281	293.7515856	308.9234656
283.5	293.375	294.3125	−10.8125	−12.7031	296.2031	296.3175917	283.6144917
283.5	295.25			−14.7031	298.2031	298.8835978	284.1804978
307.5				12.23438	295.2656	301.4496039	313.6839839

of as occurring between the third and fourth data values, that is, between 190 and 225.5. We want numbers that are centered on actual data values so we do a second moving average called a *centered* moving average. The results are in the column CMA. The first entry averages the first two four-period moving averages:

$$\frac{210 + 210.5}{2} = 210.25$$

In terms of the time sequence, this average is centered on the third time period value 190. It is the average of two other moving averages, one that is centered between 203.5 and 190 and one that is centered between 190 and 225.5. By proceeding in this manner, we end up with a moving average that smoothes out the seasonal variations in the data and has values centered on an actual data point in the time sequence. Because the CMA smoothes out the seasonal variation, what is left is the trend component and the random error.

Step 2: Remove the trend from the original data.   We can subtract the CMA from our data as in the column labeled SALES – CMA. This essentially removes the trend from the sales data, so each entry in this column consists only of the effect of the seasonal component and the random error.

Step 3: Find the seasonal indices.   We now collect these SALES – CMA values for like quarters and find the median. This median will represent the seasonal index for that quarter. Here are the individual values for the SALES – CMA column, grouped by quarter:

	1st Quarter	2nd Quarter	3rd Quarter	4th Quarter
			−20.25	16.6875
	13.875	−21.25	−7	8.875
	14.8125	−9.0625	−15.0625	11.875
	19.5625	−29.5625	−4.1875	14.25
	12.375	−6.4375	−16.9375	8.875
	10.125	−12.5625	−7.0625	8.1875
	22.75	−13.0625	−27	18.8125
	19.8125	−20.6875	−10.375	16.125
	12.4375	−13.4375	−15.6875	11.5625
	14.8125	−10.8125		
Medians	14.8125	−13.0625	−15.0625	11.875
Adj Medians	15.17188	−12.7031	−14.7031	12.23438

The medians are shown at the bottom of each column. An additional adjustment is made to the medians so that they sum to zero. The adjusted medians are the seasonal indices for each quarter and show the adjustment to any forecast that needs to be made to allow for seasonal effects. The seasonal indices are shown in the column labeled S HAT in Table 11.6.

Step 4: Deseasonalize the original data.   Subtract the seasonal indices from the original data as has been done in the column of Table 11.6 labeled DESEAS. Because we have deseasonalized the data, we can think of the numbers in this column as consisting of a trend component and random error.

Step 5: Fit a trend line to the deseasonalized data.   This can be done by applying a linear trend regression to the data in the DESEAS column in Table 11.6. The resulting fitted values for the regression are shown in the column labeled T HAT. The estimated trend-line regression equation is $\hat{y} = 198.8094 + 2.566006t$ where $t = 1, 2, 3, \ldots$ as is usual in a time-trend regression. (Note that a nonlinear trend could be used here if desired).

Step 6: Find the forecasted value for the in-sample data.   If we want to use the decomposition results to find forecasted value for the in-sample data, we use the trend-line regression to forecast the trend component and then use the seasonal indices to adjust this forecast. The resulting values are shown in the column labeled T HAT + S HAT in Table 11.6. For example, here are the forecasts for the first four values:

$$\hat{y}_1 = (198.8094 + 2.566006\,(1)) + 15.17188 = 216.5472$$
$$\hat{y}_2 = (198.8094 + 2.566006\,(2)) - 12.7031 = 191.2383$$
$$\hat{y}_3 = (198.8094 + 2.566006\,(3)) - 14.7031 = 191.8043$$
$$\hat{y}_4 = (198.8094 + 2.566006\,(4)) + 12.23438 = 221.3078$$

The trend-line regression is used to forecast the trend component, T_t, and the seasonal indices are used to adjust the trend forecast for seasonality. The seasonal indices are added to or subtracted from the trend forecast because this is an additive model. In a multiplicative model, the seasonal indices would be multiplied times the trend forecast to complete the forecast.

Step 7: Find the out-of-sample forecasts.   To forecast out-of-sample, this same process is used. The forecasts for each quarter of the next year would be

$$\hat{y}_{2004.1} = (198.8094 + 2.566006\,(41)) + 15.17188 = 319.1875$$
$$\hat{y}_{2004.2} = (198.8094 + 2.566006\,(42)) - 12.7031 = 293.8785$$
$$\hat{y}_{2004.3} = (198.8094 + 2.566006\,(43)) - 14.7031 = 294.4445$$
$$\hat{y}_{2004.4} = (198.8094 + 2.566006\,(44)) + 12.23438 = 323.948$$

The results using MINITAB to do the decomposition are shown in Figure 11.13. MINITAB shows a plot of the fitted values versus the actual values with the trend line included. Accuracy measures for the in-sample forecasts are displayed. Below the plot are the fitted trend line and the seasonal indices computed by MINITAB. The procedure demonstrated here is the procedure used by MINITAB so these are the same as the values computed in this example.   ◾

EXAMPLE 11.7   **Electrical and Appliance Stores: Multiplicative Decomposition**   Monthly sales amounts from electrical and appliance stores (in millions of dollars) are contained in the file ELECTAPPL11 on the CD. The time frame is from January 1992 through December 2002. We would like to forecast sales for each month in 2003. These data were obtained from the web site *www.economagic.com*. A time-series plot of the data is shown in Figure 11.10. Note that the amplitude of the seasonal cycles is increasing over time. As the level of the series increases, so does the magnitude of the seasonal cycle. This is typically a situation where a multiplicative model will perform better than an additive model. The multiplicative decomposition plot is shown in Figure 11.14. Below the plot is the trend line.

Also, the 12 monthly seasonal indices are given. The forecasts for the first four months in 2003 would be

$$\hat{y}_{2003.Jan} = (3694.79 + 32.1375\,(133))*0.95614 = 7619.6$$
$$\hat{y}_{2003.Feb} = (3694.79 + 32.1375\,(134))*0.89113 = 7130.1$$
$$\hat{y}_{2003.Mar} = (3694.79 + 32.1375\,(135))*0.96396 = 7743.8$$
$$\hat{y}_{2003.Apr} = (3694.79 + 32.1375\,(136))*0.89302 = 7202.6$$

The seasonal indices in a multiplicative model represent the percentage adjustment above or below the trend forecast due to seasonal effects. For example, the January index is 0.95614. This indicates that the forecast value for January should be 95.614% of the trend forecast. On average, actual January sales are 4.386% below the trend line.

The hand computations for the multiplicative decomposition will not be shown.

FIGURE 11.13
Decomposition Plot (Additive Model) of ABX Sales.

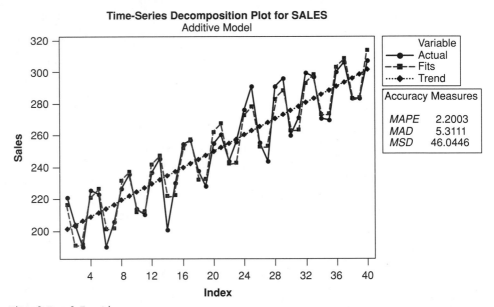

Time-Series Decomposition Plot for SALES
Additive Model

Variable
Actual
Fits
Trend

Accuracy Measures	
MAPE	2.2003
MAD	5.3111
MSD	46.0446

```
Fitted Trend Equation

Yt = 198.809 + 2.56601*t

Seasonal  Indices

Period     Index
    1     15.1719
    2    -12.7031
    3    -14.7031
    4     12.2344
```

FIGURE 11.14
Decomposition Plot (Multiplicative Model) of Electrical and Appliance Store Sales.

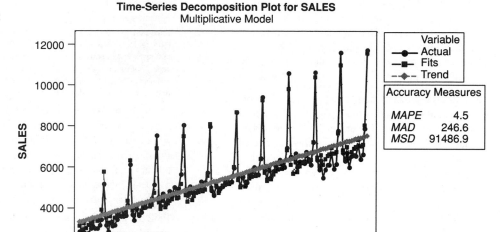

Time-Series Decomposition Plot for SALES
Multiplicative Model

Variable
Actual
Fits
Trend

Accuracy Measures	
MAPE	4.5
MAD	246.6
MSD	91486.9

Fitted Trend Equation

Yt = 3694.79 + 32.1375*t

Seasonal Indices

Period	Index
1	0.95614
2	0.89113
3	0.96396
4	0.89302
5	0.92695
6	0.94576
7	0.95183
8	0.98975
9	0.94734
10	0.95369
11	1.07095
12	1.50950

EXERCISES

1. **WestCo Envelopes.** WestCo manufactures various types of envelopes. The company has daily data on sales for the last 60 days. These data are in the file WESTCO11 on the CD. Use moving averages and single exponential smoothing to forecast sales. Compare your results to the naïve forecasting method. What is your choice for the "best" method to forecast WestCo sales?

2. **XYZ Purchases.** XYZ Corporation needs to track purchases of materials it uses in its production process for computers. It purchases shipments of hard drives weekly. There are 30 weekly observations for purchases of hard drives in the file

XYZPURCHASES11 on the CD. Use moving averages and single exponential smoothing to forecast the hard drive purchases. Compare your results to the naïve forecasting method. What is your choice for the "best" method to forecast hard drive purchases?

3. **Unemployment.** The file named UNEMP11 on the CD contains monthly unemployment rates from January 1983 until December 2002. (These data are found on the web site *www.economagic.com* and were obtained from the St. Louis Federal Reserve Bank.) The data have been seasonally adjusted, so you will need to consider only methods that do not allow for seasonality. Develop an extrapolative model to forecast the unemployment rate for each month in 2003.

Find the actual rates for each month in 2003 and compare them to your forecasts. How well did your model do? (How will you measure the accuracy of your forecasts?)

4. **Mortgage Rates.** The file named MRATES11 on the CD contains monthly 30-year conventional mortgage rates from January 1985 to December 2002. (These figures are found on the web site *www.economagic.com* and are obtained from the Federal Home Mortgage Corporation.) Develop an extrapolative model to forecast mortgage rates for each month in 2003.

Find the actual rates for each month in 2003 and compare them to your forecasts. How well did your model do? (How will you measure the accuracy of your forecasts?)

5. **Wheat Shipments.** The file named WHEATSHIP11 on the CD contains data on U.S. wheat export shipments. The data are observed monthly from January 1974 through March 1985. Develop an extrapolative model to forecast shipments for the remaining months in 1985.
(*Source*: Data are from D. A. Bessler and R. A. Babubla, "Forecasting Wheat Exports: Do Exchange Rates Really Matter?" *Journal of Business and Economic Statistics*, 5, 1987, pp. 397–406. Copyright 1987 by the American Statistical Association. Used with permission. All rights reserved.)

6. **Wheat Price.** The file named WHEATPRICE11 on the CD contains data on the per bushel real price of no. 1 red winter wheat. The data are

observed monthly from January 1974 through March 1985. Develop an extrapolative model to forecast price for the remaining months in 1985.
(*Source*: Data are from D. A. Bessler and R. A. Babubla, "Forecasting Wheat Exports: Do Exchange Rates Really Matter?" *Journal of Business and Economic Statistics*, 5, 1987, pp. 397–406. Copyright 1987 by the American Statistical Association. Used with permission. All rights reserved.)

7. **Prime Rate.** The file named PRIME11 on the CD contains monthly prime rates for the time period from January 1988 through December 2002. (These data are from the web site *www.economagic.com* and are obtained from the Federal Reserve Bank of St. Louis.) Develop an extrapolative model to forecast the prime rate for each month in 2003.

Find the actual rates for each month in 2003 and compare them to your forecasts. How well did your model do? (How will you measure the accuracy of your forecasts?)

8. **Beer Production.** The file named BEER11 on the CD contains monthly U.S. beer production in millions of barrels for January 1982 through December 1991.[3] Develop an extrapolative model for these data. Use the model you select as best for beer production to forecast monthly production for each month in 1992.

9. **Monthly Temperatures.** Lone Star Gas recognizes that one of the simplest and most effective ways to forecast natural gas use is with average monthly temperatures. A model can be developed that relates gas usage to average temperature, and then forecasts can be made based on forecasts of average temperatures in the area of interest. Lone Star first needs a model to forecast average temperatures in the Dallas–Fort Worth area. The file named TEMPDFW11 on the CD contains the average monthly temperatures for January 1978 to December 2002 for the Dallas–Fort Worth area. Develop an extrapolative model for these data and use the model to forecast average monthly temperatures for each month in the years 2003 and 2004. Data were obtained from the web site *www.srh.noaa.gov/fwd/clmdfw.html*.

10. **Furniture Sales.** The file FURNSALES11 on the CD contains monthly sales data (in millions of dollars) for retail furniture stores from January 1992 through December 2002. The data file

---

[3] Data were obtained from *Business Statistics 1963–91* and *Survey of Current Business*.

contains a column with the year, the month (coded 1 = Jan through 12 = Dec) and SALES. Develop an extrapolative model for these data and use the model to forecast monthly furniture sales for each month in the years 2003 and 2004.

USING THE COMPUTER

The Using the Computer section in each chapter describes how to perform the computer analyses in the chapter using Excel, MINITAB, and SAS. For further detail on Excel, MINITAB, and SAS, see Appendix C.

FIGURE 11.15 Excel Moving Average Dialog Box.

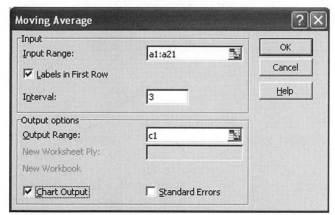

EXCEL

Moving Averages
Click Tools, Data Analysis, Moving Average. The dialog box is shown in Figure 11.15. Indicate the input range for the variable to be forecast and specify the moving average length (interval). Indicate whether you have a label in the first row and where you would like output placed. You can also choose to have a plot of the output. Click OK.

FIGURE 11.16 Excel Single Exponential Smoothing Dialog Box.

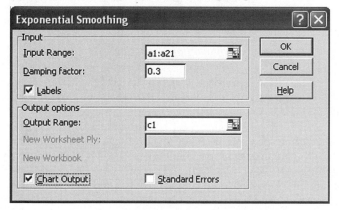

Exponential Smoothing
The only option available for exponential smoothing in Excel is for single exponential smoothing. Click Tools, Data Analysis, Exponential Smoothing. The dialog box is shown in Figure 11.16. Indicate the input range for the variable to be forecast and specify the smoothing constant (damping factor). Indicate whether you have a label in the first row and where you would like the output placed. You can also choose to have a plot of the output. Click OK.

Decomposition
There is not currently a decomposition method available in Excel. Creating your own using the standard Excel features is not a difficult task, however.

MINITAB

Figure 11.17 shows the MINITAB Time Series Menu. All time-series methods are accessed through this menu.

FIGURE 11.17 MINITAB Time Series Menu.

FIGURE 11.18 MINITAB Moving Average Dialog Box.

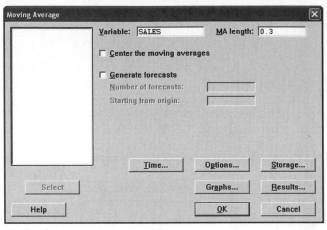

FIGURE 11.19 MINITAB Single Exponential Smoothing Dialog Box.

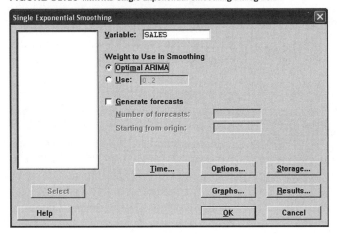

Moving Averages

STAT: TIME SERIES: MOVING AVERAGE

Click Stat, Time Series, then Moving Average. The dialog box is shown in Figure 11.18. Indicate the variable to be forecast, specify the moving average length, and click OK. There are a variety of options available regarding the amount of output to be shown, graphs produced etc.

Exponential Smoothing

STAT: TIME SERIES: SINGLE EXP SMOOTHING

Single Exponential Smoothing Click Stat, Time Series, then Single Exponential Smoothing. The dialog box is shown in Figure 11.19. Indicate the variable to be forecast and specify the weight to be used in smoothing. You can either indicate a specific value for the smoothing constant or you can request that MINITAB determine the best value (Optimal ARIMA). Click OK. There are a variety of options available regarding the amount of output to be shown, graphs produced, forecasts generated, etc.

STAT: TIME SERIES: DOUBLE EXP SMOOTHING

Double Exponential Smoothing Click Stat, Time Series, then Double Exponential Smoothing. The dialog box is shown in Figure 11.20. Indicate the variable to be forecast and specify the weights to be used in smoothing. There are two weights to be specified, one for level and one for trend. You can either indicate specific values for the smoothing constants or you can request that MINITAB determine the best values (Optimal ARIMA). Click OK. There are a variety of options available

FIGURE 11.20 MINITAB Double Exponential Smoothing Dialog Box.

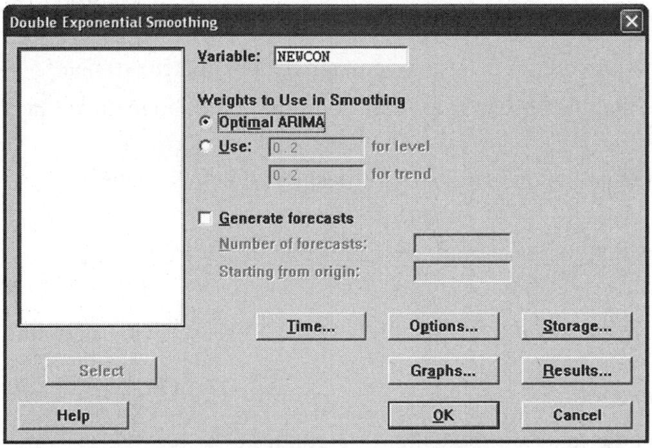

FIGURE 11.21 MINITAB Winters' Exponential Smoothing Dialog Box.

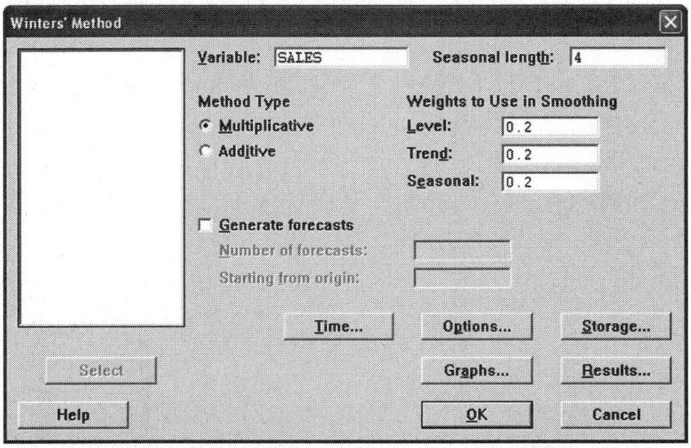

FIGURE 11.22 MINITAB Decomposition Dialog Box.

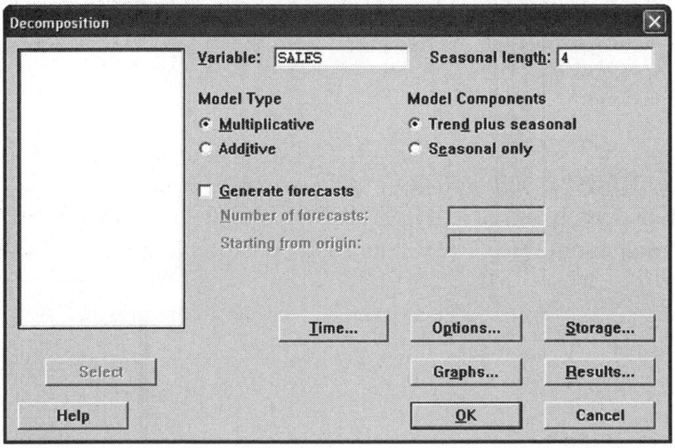

regarding the amount of output to be shown, graphs produced, forecasts generated, etc.

STAT: TIME SERIES: WINTERS' METHOD

Winters' Exponential Smoothing Click Stat, Time Series, then Winters' Method. The dialog box is shown in Figure 11.21. Indicate the variable to be forecast and the seasonal length (4 = quarterly, 12 = *monthly*, *etc.*), and specify the weights to be used in smoothing. There are three weights to be specified, one for level, one for trend, and one for the seasonal component. Note that MINITAB will not determine the optimal values when Winters' method is used. Specify whether you want an Additive or Multiplicative model. Click OK. There are a variety of options available regarding the amount of output to be shown, graphs produced, forecasts generated, etc.

Decomposition

STAT: TIME SERIES: DECOMPOSITION

Click Stat, Time Series, then Decomposition. The dialog box is shown in Figure 11.22. Indicate the variable to be forecast, the seasonal length (4 = quarterly, 12 = monthly, etc), the model type, either Multiplicative or Additive, and the model components desired. You have the option of including both trend and seasonal components or only seasonal components. Click OK. There are a variety of options available regarding the amount of output to be shown, graphs produced, forecasts generated, etc.

SAS

Exponential Smoothing

Single and double exponential smoothing and Winters' method are performed in SAS using PROC FORECAST. There are a wide variety of options for the PROC FORECAST command. The options used on the PROC FORECAST statements that follow include:

OUT = B names the output data set to contain the forecasts.

OUTEST = C names the output data set to contain the parameter estimates.

OUTDATA requests that the observations used to fit the model be included in the OUT = data set.

OUT1STEP requests that the one-step-ahead forecasts be output to the OUT = data set.

OUTLIMIT requests that forecast confidence limits be output to the OUT = data set.

LEAD = 12 specifies the number of periods ahead to forecast. The default is LEAD = 12.

METHOD = EXPO with

TREND = 1 specifies single exponential smoothing

TREND = 2 specifies double exponential smoothing

WEIGHT = 0.3 specifies the smoothing parameter(s) for single and double exponential smoothing.

METHOD = WINTERS with TREND = 2 specifies Winters' method

SEASONS = MONTH specifies the seasonality for the Winters' method. The interval can be QTR, MONTH, DAY, or HOUR.

INTERVAL = MONTH specifies the kind of time interval between observations.

There is much more detail on the options available for these methods in the SAS manuals.

Exponential Smoothing: Single Exponential Smoothing

PROC FORECAST with the combination of METHOD = EXPO and TREND = 1 requests single exponential smoothing. The following is an example command sequence:

```
PROC FORECAST OUT = B OUTEST = C METHOD = EXPO TREND = 1 OUTDATA OUTLIMIT
LEAD = 12 WEIGHT = 0.6;
VAR SALES;
PROC PRINT DATA = B;
PROC PRINT DATA = C;
```

Exponential Smoothing: Double Exponential Smoothing

PROC FORECAST with the combination of METHOD = EXPO and TREND = 2 requests double exponential smoothing. The following is an example command sequence:

```
PROC FORECAST OUT = B OUTEST = C METHOD = EXPO TREND = 2 OUTDATA OUTLIMIT
LEAD = 12;
VAR SALES;
PROC PRINT DATA = B;
PROC PRINT DATA = C;
```

Exponential Smoothing: Winters' Method

PROC FORECAST with METHOD = WINTERS and TREND = 2 requests WINTERS multiplicative method. An example command sequence along with an input data set is shown here:

```
DATA FURNSALES;
INPUT MONTH YEAR DATE :MONYY. SALES;
FORMAT DATE MONYY.;
CARDS;
1     1992     JAN92     2397
2     1992     FEB92     2465
3     1992     MAR92     2612
4     1992     APR92     2530
5     1992     MAY92     2609
6     1992     JUN92     2608
7     1992     JUL92     2632
8     1992     AUG92     2644
9     1992     SEP92     2616
10    1992     OCT92     2669
11    1992     NOV92     2773
12    1992     DEC92     3053
1     1993     JAN93     2557
2     1993     FEB93     2451
3     1993     MAR93     2718
4     1993     APR93     2650
5     1993     MAY93     2771
6     1993     JUN93     2748
7     1993     JUL93     2812
8     1993     AUG93     2780
9     1993     SEP93     2756
10    1993     OCT93     2827
11    1993     NOV93     3032
12    1993     DEC93     3254
1     1994     JAN94     2540
2     1994     FEB94     2562
3     1994     MAR94     2921
4     1994     APR94     2829
5     1994     MAY94     2865
6     1994     JUN94     2896
7     1994     JUL94     2930
8     1994     AUG94     3083
9     1994     SEP94     3069
10    1994     OCT94     3043
11    1994     NOV94     3277
12    1994     DEC94     3547
PROC FORECAST OUTEST = C METHOD = WINTERS SEASON = MONTH INTERVAL = MONTH
TREND = 2;
VAR SALES;
ID DATE;
PROC PRINT DATA = C;
```

Decomposition

The X11 procedure can be used to seasonally adjust data. Thus it is similar to the decomposition method discussed in this chapter. PROC X11 is fairly complex and is not discussed in detail here but a sample command sequence is given:

```
PROC X11;
VAR SALES;
MONTHLY START = JAN92;
```

This sequence will produce seasonally adjusted data for the SALES time series, a monthly series which starts in JAN92.

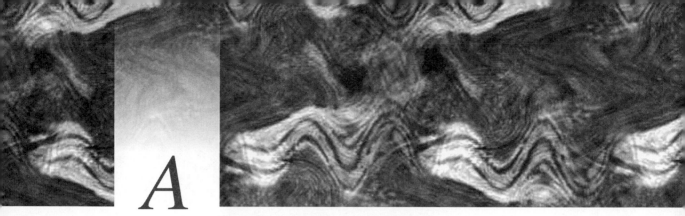

A

Summation Notation

A sample of n items chosen from a population can be denoted as $X_1, X_2, \ldots, X_n$. To represent the sum of these n items, the notation

$$\sum_{i=1}^{n} X_i$$

is used. This is simply shorthand notation for writing $X_1 + X_2 + \cdots + X_n$.

As an example, suppose a sample of four items is drawn, and the four sample values are 2, 3, 4, and 11. The sum of these four items can be represented by

$$\sum_{i=1}^{4} X_i = 2 + 3 + 4 + 11 = 20$$

Some other useful examples are

1. $\displaystyle\sum_{i=1}^{4} X_i^2 = 4 + 9 + 16 + 121 = 150$

2. $\displaystyle\sum_{i=1}^{4} (X_i - 5) = (2 - 5) + (3 - 5) + (4 - 5) + (11 - 5) = 0$

3. $\displaystyle\sum_{i=1}^{4} (X_i - 5)^2 = (2 - 5)^2 + (3 - 5)^2 + (4 - 5)^2 + (11 - 5)^2 = 50$

4. $\displaystyle\sum_{i=1}^{4} 4X_i = 8 + 12 + 16 + 44 = 80 = 4\sum_{i=1}^{4} X_i$

EXERCISES

Use the following information to complete each exercise. A sample of six items is chosen with the following values: 5, 8, 10, 11, 12, 20.

a. $\displaystyle\sum_{i=1}^{6} X_i$

b. $\displaystyle\sum_{i=1}^{6} X_i^2$

c. $\displaystyle\sum_{i=1}^{6} (X_i - 11) =$

d. $\displaystyle\sum_{i=1}^{6} (X_i - 11)^2 =$

e. $\displaystyle\sum_{i=1}^{6} 2X_i$

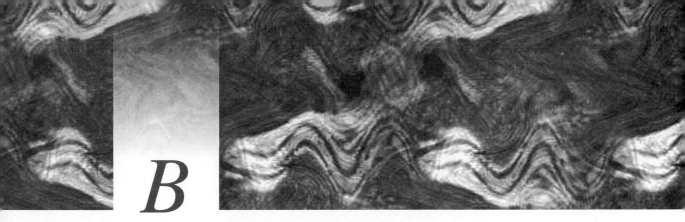

B

Statistical Tables

TABLE B.1 Standard Normal Distribution

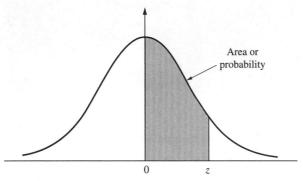

Entries in the table give the area under the curve between the mean and z standard deviations above the mean. For example, for $z =$ 1.25 the area under the curve between the mean and z is .3944.

z	.00	.01	.02	.03	.04	.05	.06	.07	.08	.09
.0	.0000	.0040	.0080	.0120	.0160	.0199	.0239	.0279	.0319	.0359
.1	.0398	.0438	.0478	.0517	.0557	.0596	.0636	.0675	.0714	.0753
.2	.0793	.0832	.0871	.0910	.0948	.0987	.1026	.1064	.1103	.1141
.3	.1179	.1217	.1255	.1293	.1331	.1368	.1406	.1443	.1480	.1517
.4	.1554	.1591	.1628	.1664	.1700	.1736	.1772	.1808	.1844	.1879
.5	.1915	.1950	.1985	.2019	.2054	.2088	.2123	.2157	.2190	.2224
.6	.2257	.2291	.2324	.2357	.2389	.2422	.2454	.2486	.2518	.2549
.7	.2580	.2612	.2642	.2673	.2704	.2734	.2764	.2794	.2823	.2852
.8	.2881	.2910	.2939	.2967	.2995	.3023	.3051	.3078	.3106	.3133
.9	.3159	.3186	.3212	.3238	.3264	.3289	.3315	.3340	.3365	.3389
1.0	.3413	.3438	.3461	.3485	.3508	.3531	.3554	.3577	.3599	.3621
1.1	.3643	.3665	.3686	.3708	.3729	.3749	.3770	.3790	.3810	.3830
1.2	.3849	.3869	.3888	.3907	.3925	.3944	.3962	.3980	.3997	.4015
1.3	.4032	.4049	.4066	.4082	.4099	.4115	.4131	.4147	.4162	.4177
1.4	.4192	.4207	.4222	.4236	.4251	.4265	.4279	.4292	.4306	.4319
1.5	.4332	.4345	.4357	.4370	.4382	.4394	.4406	.4418	.4429	.4441
1.6	.4452	.4463	.4474	.4484	.4495	.4505	.4515	.4525	.4535	.4545
1.7	.4554	.4564	.4573	.4582	.4591	.4599	.4608	.4616	.4625	.4633
1.8	.4641	.4649	.4656	.4664	.4671	.4678	.4686	.4693	.4699	.4706
1.9	.4713	.4719	.4726	.4732	.4738	.4744	.4750	.4756	.4761	.4767
2.0	.4772	.4778	.4783	.4788	.4793	.4798	.4803	.4808	.4812	.4817
2.1	.4821	.4826	.4830	.4834	.4838	.4842	.4846	.4850	.4854	.4857
2.2	.4861	.4864	.4868	.4871	.4875	.4878	.4881	.4884	.4887	.4890
2.3	.4893	.4896	.4898	.4901	.4904	.4906	.4909	.4911	.4913	.4916
2.4	.4918	.4920	.4922	.4925	.4927	.4929	.4931	.4932	.4934	.4936
2.5	.4938	.4940	.4941	.4943	.4945	.4946	.4948	.4949	.4951	.4952
2.6	.4953	.4955	.4956	.4957	.4959	.4960	.4961	.4962	.4963	.4964
2.7	.4965	.4966	.4967	.4968	.4969	.4970	.4971	.4972	.4973	.4974
2.8	.4974	.4975	.4976	.4977	.4977	.4978	.4979	.4979	.4980	.4981
2.9	.4981	.4982	.4982	.4983	.4984	.4984	.4985	.4985	.4986	.4986
3.0	.4986	.4987	.4987	.4988	.4988	.4989	.4989	.4989	.4990	.4990

Abridged from Table I of A. Hald, *Statistical Tables and Formulas* (New York: John Wiley and Sons, 1952). Reproduced by permission of the publisher.

TABLE B.2 *t* Distribution

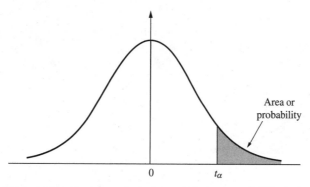

Entries in the table give t_α values, where α is the area or probability in the upper tail of the *t* distribution. For example, with 10 degrees of freedom and a .05 area in the upper tail, $t_{.05} = 1.812$

Degrees of Freedom	Area in Upper Tail				
	.10	.05	.025	.01	.005
1	3.078	6.314	12.706	31.821	63.657
2	1.886	2.920	4.303	6.965	9.925
3	1.638	2.353	3.182	4.541	5.841
4	1.533	2.132	2.776	3.747	4.604
5	1.476	2.015	2.571	3.365	4.032
6	1.440	1.943	2.447	3.143	3.707
7	1.415	1.895	2.365	2.998	3.499
8	1.397	1.860	2.306	2.896	3.355
9	1.383	1.833	2.262	2.821	3.250
10	1.372	1.812	2.228	2.764	3.169
11	1.363	1.796	2.201	2.718	3.106
12	1.356	1.782	2.179	2.681	3.055
13	1.350	1.771	2.160	2.650	3.012
14	1.345	1.761	2.145	2.624	2.977
15	1.341	1.753	2.131	2.602	2.947
16	1.337	1.746	2.120	2.583	2.921
17	1.333	1.740	2.110	2.567	2.898
18	1.330	1.734	2.101	2.552	2.878
19	1.328	1.729	2.093	2.539	2.861
20	1.325	1.725	2.086	2.528	2.845
21	1.323	1.721	2.080	2.518	2.831
22	1.321	1.717	2.074	2.508	2.819
23	1.319	1.714	2.069	2.500	2.807
24	1.318	1.711	2.064	2.492	2.797
25	1.316	1.708	2.060	2.485	2.787
26	1.315	1.706	2.056	2.479	2.779
27	1.314	1.703	2.052	2.473	2.771
28	1.313	1.701	2.048	2.467	2.763
29	1.311	1.699	2.045	2.462	2.756
30	1.310	1.697	2.042	2.457	2.750
40	1.303	1.684	2.021	2.423	2.704
60	1.296	1.671	2.000	2.390	2.660
120	1.289	1.658	1.980	2.358	2.617
∞	1.282	1.645	1.960	2.326	2.576

Reprinted by permission of Biometrika Trustees from Table 12. Percentage Points of the *t*-Distribution, in E. S. Pearson and H. O. Hartley, *Biometrika Tables for Statisticians, Vol. 1.*

TABLE B.3 Critical Values for the *F* Statistic ($\alpha = .10$)

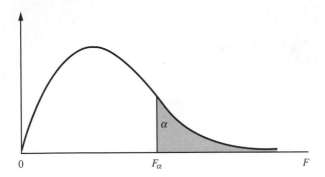

v2 \ v1	Numerator Degrees of Freedom								
	1	2	3	4	5	6	7	8	9
1	39.86	49.50	53.59	55.83	57.24	58.20	58.91	59.44	59.86
2	8.53	9.00	9.16	9.24	9.29	9.33	9.35	9.37	9.38
3	5.54	5.46	5.39	5.34	5.31	5.28	5.27	5.25	5.24
4	4.54	4.32	4.19	4.11	4.05	4.01	3.98	3.95	3.94
5	4.06	3.78	3.62	3.52	3.45	3.40	3.37	3.34	3.32
6	3.78	3.46	3.29	3.18	3.11	3.05	3.01	2.98	2.96
7	3.59	3.26	3.07	2.96	2.88	2.83	2.78	2.75	2.72
8	3.46	3.11	2.92	2.81	2.73	2.67	2.62	2.59	2.56
9	3.36	3.01	2.81	2.69	2.61	2.55	2.51	2.47	2.44
10	3.39	2.92	2.73	2.61	2.52	2.46	2.41	2.38	2.35
11	3.23	2.86	2.66	2.54	2.45	2.39	2.34	2.30	2.27
12	3.18	2.81	2.61	2.48	2.39	2.33	2.28	2.24	2.21
13	3.14	2.76	2.56	2.43	2.35	2.28	2.23	2.20	2.16
14	3.10	2.73	2.52	2.39	2.31	2.24	2.19	2.15	2.12
15	3.07	2.70	2.49	2.36	2.27	2.21	2.16	2.12	2.09
16	3.05	2.67	2.46	2.33	2.24	2.18	2.13	2.09	2.06
17	3.03	2.64	2.44	2.31	2.22	2.15	2.10	2.06	2.03
18	3.01	2.62	2.42	2.29	2.20	2.13	2.08	2.04	2.00
19	2.99	2.61	2.40	2.27	2.18	2.11	2.06	2.02	1.98
20	2.97	2.59	2.38	2.25	2.16	2.09	2.04	2.00	1.96
21	2.96	2.57	2.36	2.23	2.14	2.08	2.02	1.98	1.95
22	2.95	2.56	2.35	2.22	2.13	2.06	2.01	1.97	1.93
23	2.94	2.55	2.34	2.21	2.11	2.05	1.99	1.95	1.92
24	2.93	2.54	2.33	2.19	2.10	2.04	1.98	1.94	1.91
25	2.92	2.53	2.32	2.18	2.09	2.02	1.97	1.93	1.89
26	2.91	2.52	2.31	2.17	2.08	2.01	1.96	1.92	1.88
27	2.90	2.51	2.30	2.17	2.07	2.00	1.95	1.91	1.87
28	2.89	2.50	2.29	2.16	2.06	2.00	1.94	1.90	1.87
29	2.89	2.50	2.28	2.15	2.06	1.99	1.93	1.89	1.86
30	2.88	2.49	2.28	2.14	2.05	1.98	1.93	1.88	1.85
40	2.84	2.44	2.23	2.09	2.00	1.93	1.87	1.83	1.79
60	2.79	2.39	2.18	2.04	1.95	1.87	1.82	1.77	1.74
120	2.75	2.35	2.13	1.99	1.90	1.82	1.77	1.72	1.68
∞	2.71	2.30	2.08	1.94	1.85	1.77	1.72	1.67	1.63

Source: From M. Merrington and C. M. Thompson, "Tables of Percentage Points of the Inverted Beta (*F*)-Distribution," *Biometrika* 33 (1943): 73–88. Reproduced by permission of the *Biometrika* Trustees.

TABLE B.3 (*Continued*)

v2 \ v1	Numerator Degrees of Freedom									
	10	12	15	20	24	30	40	60	120	∞
1	60.19	60.71	61.22	61.74	62.00	62.26	62.53	62.79	63.06	63.33
2	9.39	9.41	9.42	9.44	9.45	9.46	9.47	9.47	9.48	9.49
3	5.23	5.22	5.20	5.18	5.18	5.17	5.16	5.15	5.14	5.13
4	3.92	3.90	3.87	3.84	3.83	3.82	3.80	3.79	3.78	3.76
5	3.30	3.27	3.24	3.21	3.19	3.17	3.16	3.14	3.12	3.10
6	2.94	2.90	2.87	2.84	2.82	2.80	2.78	2.76	2.74	2.72
7	2.70	2.67	2.63	2.59	2.58	2.56	2.54	2.51	2.49	2.47
8	2.54	2.50	2.46	2.42	2.40	2.38	2.36	2.34	2.32	2.29
9	2.42	2.38	2.34	2.30	2.28	2.25	2.23	2.21	2.18	2.16
10	2.32	2.28	2.24	2.20	2.18	2.16	2.13	2.11	2.08	2.06
11	2.25	2.21	2.17	2.12	2.10	2.08	2.05	2.03	2.00	1.97
12	2.19	2.15	2.10	2.06	2.04	2.01	1.99	1.96	1.93	1.90
13	2.14	2.10	2.05	2.01	1.98	1.96	1.93	1.90	1.88	1.85
14	2.10	2.05	2.01	1.96	1.94	1.91	1.89	1.86	1.83	1.80
15	2.06	2.02	1.97	1.92	1.90	1.87	1.85	1.82	1.79	1.76
16	2.03	1.99	1.94	1.89	1.87	1.84	1.81	1.78	1.75	1.72
17	2.00	1.96	1.91	1.86	1.84	1.81	1.78	1.75	1.72	1.69
18	1.98	1.93	1.89	1.84	1.81	1.78	1.75	1.72	1.69	1.66
19	1.96	1.91	1.86	1.81	1.79	1.76	1.73	1.70	1.67	1.63
20	1.94	1.89	1.84	1.79	1.77	1.74	1.71	1.68	1.64	1.61
21	1.92	1.87	1.83	1.78	1.75	1.72	1.69	1.66	1.62	1.59
22	1.90	1.86	1.81	1.76	1.73	1.70	1.67	1.64	1.60	1.57
23	1.89	1.84	1.80	1.74	1.72	1.69	1.66	1.62	1.59	1.55
24	1.88	1.83	1.78	1.73	1.70	1.67	1.64	1.61	1.57	1.53
25	1.87	1.82	1.77	1.72	1.69	1.66	1.63	1.59	1.56	1.52
26	1.86	1.81	1.76	1.71	1.68	1.65	1.61	1.58	1.54	1.50
27	1.85	1.80	1.75	1.70	1.67	1.64	1.60	1.57	1.53	1.49
28	1.84	1.79	1.74	1.69	1.66	1.63	1.59	1.56	1.52	1.48
29	1.83	1.78	1.73	1.68	1.65	1.62	1.58	1.55	1.51	1.47
30	1.82	1.77	1.72	1.67	1.64	1.61	1.57	1.54	1.50	1.46
40	1.76	1.71	1.66	1.61	1.57	1.54	1.51	1.47	1.42	1.38
60	1.71	1.66	1.60	1.54	1.51	1.48	1.44	1.40	1.35	1.29
120	1.65	1.60	1.55	1.48	1.45	1.41	1.37	1.32	1.26	1.19
∞	1.60	1.55	1.49	1.42	1.38	1.34	1.30	1.24	1.17	1.00

Denominator Degrees of Freedom (v2, row labels)

TABLE B.4 Critical Values for the F Statistic ($\alpha = .05$)

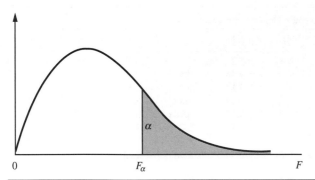

$v2$				Numerator Degrees of Freedom $v1$					
	1	2	3	4	5	6	7	8	9
1	161.40	199.50	215.70	224.60	230.20	234.00	236.80	238.90	240.50
2	18.51	19.00	19.16	19.25	19.30	19.33	19.35	19.37	19.38
3	10.13	9.55	9.28	9.12	9.01	8.94	8.89	8.85	8.81
4	7.71	6.94	6.59	6.39	6.26	6.16	6.09	6.04	6.00
5	6.61	5.79	5.41	5.19	5.05	4.95	4.88	4.82	4.77
6	5.99	5.14	4.76	4.53	4.39	4.28	4.21	4.15	4.10
7	5.59	4.74	4.35	4.12	3.97	3.87	3.79	3.73	3.68
8	5.32	4.46	4.07	3.84	3.69	3.58	3.50	3.44	3.39
9	5.12	4.26	3.86	3.63	3.48	3.37	3.29	3.23	3.18
10	4.96	4.10	3.71	3.48	3.33	3.22	3.14	3.07	3.02
11	4.84	3.98	3.59	3.36	3.20	3.09	3.01	2.95	2.90
12	4.75	3.89	3.49	3.26	3.11	3.00	2.91	2.85	2.80
13	4.67	3.81	3.41	3.18	3.03	2.92	2.83	2.77	2.71
14	4.60	3.74	3.34	3.11	2.96	2.85	2.76	2.70	2.65
15	4.54	3.68	3.29	3.06	2.90	2.79	2.71	2.64	2.59
16	4.49	3.63	3.24	3.01	2.85	2.74	2.66	2.59	2.54
17	4.45	3.59	3.20	2.96	2.81	2.70	2.61	2.55	2.49
18	4.41	3.55	3.16	2.93	2.77	2.66	2.58	2.51	2.46
19	4.38	3.52	3.13	2.90	2.74	2.63	2.54	2.48	2.42
20	4.35	3.49	3.10	2.87	2.71	2.60	2.51	2.45	2.39
21	4.32	3.47	3.07	2.84	2.68	2.57	2.49	2.42	2.37
22	4.30	3.44	3.05	2.82	2.66	2.55	2.46	2.40	2.34
23	4.28	3.42	3.03	2.80	2.64	2.53	2.44	2.37	2.32
24	4.26	3.40	3.01	2.78	2.62	2.51	2.42	2.36	2.30
25	4.24	3.39	2.99	2.76	2.60	2.49	2.40	2.34	2.28
26	4.23	3.37	2.98	2.74	2.59	2.47	2.39	2.32	2.27
27	4.21	3.35	2.96	2.73	2.57	2.46	2.37	2.31	2.25
28	4.20	3.34	2.95	2.71	2.56	2.45	2.36	2.29	2.24
29	4.18	3.33	2.93	2.70	2.55	2.43	2.35	2.28	2.22
30	4.17	3.32	2.92	2.69	2.53	2.42	2.33	2.27	2.21
40	4.08	3.23	2.84	2.61	2.45	2.34	2.25	2.18	2.12
60	4.00	3.15	2.76	2.53	2.37	2.25	2.17	2.10	2.04
120	3.92	3.07	2.68	2.45	2.29	2.17	2.09	2.02	1.96
∞	3.84	3.00	2.60	2.37	2.21	2.10	2.01	1.94	1.88

Denominator Degrees of Freedom

Source: From M. Merrington and C. M. Thompson, "Tables of Percentage Points of the Inverted Beta (F)-Distribution," *Biometrika* 33 (1943): 73–88. Reproduced by permission of the *Biometrika* Trustees.

TABLE B.4 (*Continued*)

$v2$ \ $v1$	10	12	15	20	24	30	40	60	120	∞
				Numerator Degrees of Freedom						
1	241.90	243.90	245.90	248.00	249.10	250.10	251.10	252.20	253.30	254.30
2	19.40	19.41	19.43	19.45	19.45	19.46	19.47	19.48	19.49	19.50
3	8.79	8.74	8.70	8.66	8.64	8.62	8.59	8.57	8.55	8.53
4	5.96	5.91	5.86	5.80	5.77	5.75	5.72	5.69	5.66	5.63
5	4.74	4.68	4.62	4.56	4.53	4.50	4.46	4.43	4.40	4.36
6	4.06	4.00	3.94	3.87	3.84	3.81	3.77	3.74	3.70	3.67
7	3.64	3.57	3.51	3.44	3.41	3.38	3.34	3.30	3.27	3.23
8	3.35	3.28	3.22	3.15	3.12	3.08	3.04	3.01	2.97	2.93
9	3.14	3.07	3.01	2.94	2.90	2.86	2.83	2.79	2.75	2.71
10	2.98	2.91	2.85	2.77	2.74	2.70	2.66	2.62	2.58	2.54
11	2.85	2.79	2.72	2.65	2.61	2.57	2.53	2.49	2.45	2.40
12	2.75	2.69	2.62	2.54	2.51	2.47	2.43	2.38	2.34	2.30
13	2.67	2.60	2.53	2.46	2.42	2.38	2.34	2.30	2.25	2.21
14	2.60	2.53	2.46	2.39	2.35	2.31	2.27	2.22	2.18	2.13
15	2.54	2.48	2.40	2.33	2.29	2.25	2.20	2.16	2.11	2.07
16	2.49	2.42	2.35	2.28	2.24	2.19	2.15	2.11	2.06	2.01
17	2.45	2.38	2.31	2.23	2.19	2.15	2.10	2.06	2.01	1.96
18	2.41	2.34	2.27	2.19	2.15	2.11	2.06	2.02	1.97	1.92
19	2.38	2.31	2.23	2.16	2.11	2.07	2.03	1.98	1.93	1.88
20	2.35	2.28	2.20	2.12	2.08	2.04	1.99	1.95	1.90	1.84
21	2.32	2.25	2.18	2.10	2.05	2.01	1.96	1.92	1.87	1.81
22	2.30	2.23	2.15	2.07	2.03	1.98	1.94	1.89	1.84	1.78
23	2.27	2.20	2.13	2.05	2.01	1.96	1.91	1.86	1.81	1.76
24	2.25	2.18	2.11	2.03	1.98	1.94	1.89	1.84	1.79	1.73
25	2.24	2.16	2.09	2.01	1.96	1.92	1.87	1.82	1.77	1.71
26	2.22	2.15	2.07	1.99	1.95	1.90	1.85	1.80	1.75	1.69
27	2.20	2.13	2.06	1.97	1.93	1.88	1.84	1.79	1.73	1.67
28	2.19	2.12	2.04	1.96	1.91	1.87	1.82	1.77	1.71	1.65
29	2.18	2.10	2.03	1.94	1.90	1.85	1.81	1.75	1.70	1.64
30	2.16	2.09	2.01	1.93	1.89	1.84	1.79	1.74	1.68	1.62
40	2.08	2.00	1.92	1.84	1.79	1.74	1.69	1.64	1.58	1.51
60	1.99	1.92	1.84	1.75	1.70	1.65	1.59	1.53	1.47	1.39
120	1.91	1.83	1.75	1.66	1.61	1.55	1.50	1.43	1.35	1.25
∞	1.83	1.75	1.67	1.57	1.52	1.46	1.39	1.32	1.22	1.00

Denominator Degrees of Freedom (left-side label, $v2$)

TABLE B.5 Critical Values for the F Statistic ($\alpha = .01$)

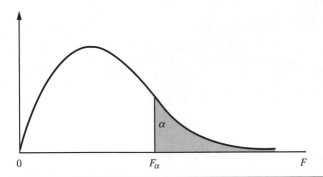

$v1$	Numerator Degrees of Freedom								
$v2$	1	2	3	4	5	6	7	8	9
1	4,052.00	4,999.50	5,403.00	5,625.00	5,764.00	5,859.00	5,928.00	5,982.00	6,022.00
2	98.50	99.00	99.17	99.25	99.30	99.33	99.36	99.37	99.39
3	34.12	30.82	29.46	28.71	28.24	27.91	27.67	27.49	27.35
4	21.20	18.00	16.69	15.98	15.52	15.21	14.98	14.80	14.66
5	16.26	13.27	12.06	11.39	10.97	10.67	10.46	10.29	10.16
6	13.75	10.92	9.78	9.15	8.75	8.47	8.26	8.10	7.98
7	12.25	9.55	8.45	7.85	7.46	7.19	6.99	6.84	6.72
8	11.26	8.65	7.59	7.01	6.63	6.37	6.18	6.03	5.91
9	10.56	8.02	6.99	6.42	6.06	5.80	5.61	5.47	5.35
10	10.04	7.56	6.55	5.99	5.64	5.39	5.20	5.06	4.94
11	9.65	7.21	6.22	5.67	5.32	5.07	4.89	4.74	4.63
12	9.33	6.93	5.95	5.41	5.06	4.82	4.64	4.50	4.39
13	9.07	6.70	5.74	5.21	4.86	4.62	4.44	4.30	4.19
14	8.86	6.51	5.56	5.04	4.69	4.46	4.28	4.14	4.03
15	8.68	6.36	5.42	4.89	4.56	4.32	4.14	4.00	3.89
16	8.53	6.23	5.29	4.77	4.44	4.20	4.03	3.89	3.78
17	8.40	6.11	5.18	4.67	4.34	4.10	3.93	3.79	3.68
18	8.29	6.01	5.09	4.58	4.25	4.01	3.84	3.71	3.60
19	8.18	5.93	5.01	4.50	4.17	3.94	3.77	3.63	3.52
20	8.10	5.85	4.94	4.43	4.10	3.87	3.70	3.56	3.46
21	8.02	5.78	4.87	4.37	4.04	3.81	3.64	3.51	3.40
22	7.95	5.72	4.82	4.31	3.99	3.76	3.59	3.45	3.35
23	7.88	5.66	4.76	4.26	3.94	3.71	3.54	3.41	3.30
24	7.82	5.61	4.72	4.22	3.90	3.67	3.50	3.36	3.26
25	7.77	5.57	4.68	4.18	3.85	3.63	3.46	3.32	3.22
26	7.72	5.53	4.64	4.14	3.82	3.59	3.42	3.29	3.18
27	7.68	5.49	4.60	4.11	3.78	3.56	3.39	3.26	3.15
28	7.64	5.45	4.57	4.07	3.75	3.53	3.36	3.23	3.12
29	7.60	5.42	4.54	4.04	3.73	3.50	3.33	3.20	3.09
30	7.56	5.39	4.51	4.02	3.70	3.47	3.30	3.17	3.07
40	7.31	5.18	4.31	3.83	3.51	3.29	3.12	2.99	2.89
60	7.08	4.98	4.13	3.65	3.34	3.12	2.95	2.82	2.72
120	6.85	4.79	3.95	3.48	3.17	2.96	2.79	2.66	2.56
∞	6.63	4.61	3.78	3.32	3.02	2.80	2.64	2.51	2.41

Denominator Degrees of Freedom

Source: From M. Merrington and C. M. Thompson, "Tables of Percentage Points of the Inverted Beta (F)-Distribution," *Biometrika* 33 (1943): 73–88. Reproduced by permission of the *Biometrika* Trustees.

TABLE B.5 (*Continued*)

v2	v1 10	12	15	20	24	30	40	60	120	∞
					Numerator Degrees of Freedom					
1	6,056.00	6,106.00	6,157.00	6,209.00	6,235.00	6,261.00	6,287.00	6,313.00	6,339.00	6,366.00
2	99.40	99.42	99.43	99.45	99.46	99.47	99.47	99.48	99.49	99.50
3	27.23	27.05	26.87	26.69	26.60	26.50	26.41	26.32	26.22	26.13
4	14.55	14.37	14.20	14.02	13.93	13.84	13.75	13.65	13.56	13.46
5	10.05	9.89	9.72	9.55	9.47	9.38	9.29	9.20	9.11	9.02
6	7.87	7.72	7.56	7.40	7.31	7.23	7.14	7.06	6.97	6.88
7	6.62	6.47	6.31	6.16	6.07	5.99	5.91	5.82	5.74	5.65
8	5.81	5.67	5.52	5.36	5.28	5.20	5.12	5.03	4.95	4.86
9	5.26	5.11	4.96	4.81	4.73	4.65	4.57	4.48	4.40	4.31
10	4.85	4.71	4.56	4.41	4.33	4.25	4.17	4.08	4.00	3.91
11	4.54	4.40	4.25	4.10	4.02	3.94	3.86	3.78	3.69	3.60
12	4.30	4.16	4.01	3.86	3.78	3.70	3.62	3.54	3.45	3.36
13	4.10	3.96	3.82	3.66	3.59	3.51	3.43	3.34	3.25	3.17
14	3.94	3.80	3.66	3.51	3.43	3.35	3.27	3.18	3.09	3.00
15	3.80	3.67	3.52	3.37	3.29	3.21	3.13	3.05	2.96	2.87
16	3.69	3.55	3.41	3.26	3.18	3.10	3.02	2.93	2.84	2.75
17	3.59	3.46	3.31	3.16	3.08	3.00	2.92	2.83	2.75	2.65
18	3.51	3.37	3.23	3.08	3.00	2.92	2.84	2.75	2.66	2.57
19	3.43	3.30	3.15	3.00	2.92	2.84	2.76	2.67	2.58	2.49
20	3.37	3.23	3.09	2.94	2.86	2.78	2.69	2.61	2.52	2.42
21	3.31	3.17	3.03	2.88	2.80	2.72	2.64	2.55	2.46	2.36
22	3.26	3.12	2.98	2.83	2.75	2.67	2.58	2.50	2.40	2.31
23	3.21	3.07	2.93	2.78	2.70	2.62	2.54	2.45	2.35	2.26
24	3.17	3.03	2.89	2.74	2.66	2.58	2.49	2.40	2.31	2.21
25	3.13	2.99	2.85	2.70	2.62	2.54	2.45	2.36	2.27	2.17
26	3.09	2.96	2.81	2.66	2.58	2.50	2.42	2.33	2.23	2.13
27	3.06	2.93	2.78	2.63	2.55	2.47	2.38	2.29	2.20	2.10
28	3.03	2.90	2.75	2.60	2.52	2.44	2.35	2.26	2.17	2.06
29	3.00	2.87	2.73	2.57	2.49	2.41	2.33	2.23	2.14	2.03
30	2.98	2.84	2.70	2.55	2.47	2.39	2.30	2.21	2.11	2.01
40	2.80	2.66	2.52	2.37	2.29	2.20	2.11	2.02	1.92	1.80
60	2.63	2.50	2.35	2.20	2.12	2.03	1.94	1.84	1.73	1.60
120	2.47	2.34	2.19	2.03	1.95	1.86	1.76	1.66	1.53	1.38
∞	2.32	2.18	2.04	1.88	1.79	1.70	1.59	1.47	1.32	1.00

Denominator Degrees of Freedom

TABLE B.6 Critical Values for the Ryan-Joiner Test for Normality

n	α		
	0.01	0.05	0.10
4	.8951	.8734	.8318
5	.9033	.8804	.8320
10	.9347	.9180	.8804
15	.9506	.9383	.9110
20	.9600	.9503	.9290
25	.9662	.9582	.9408
30	.9707	.9639	.9490
40	.9767	.9715	.9597
50	.9807	.9764	.9664
60	.9835	.9799	.9710
75	.9865	.9835	.9757

Source: *MINITAB Statistical Software Reference Manual—Release 6.1,* 1988, p. 63. Reproduced by permission of MINITAB, Inc.

TABLE B.7 Critical Values for the Durbin-Watson Statistic ($\alpha = .05$)

n	K=1		K=2		K=3		K=4		K=5	
	d_L	d_U	d_L	d_U	d_L	d_U	d_L	d_U	d_L	d_U
15	1.08	1.36	0.95	1.54	0.82	1.75	0.69	1.97	0.56	2.21
16	1.10	1.37	0.98	1.54	0.86	1.73	0.74	1.93	0.62	2.15
17	1.13	1.38	1.02	1.54	0.90	1.71	0.78	1.90	0.67	2.10
18	1.16	1.39	1.05	1.53	0.93	1.69	0.82	1.87	0.71	2.06
19	1.18	1.40	1.08	1.53	0.97	1.68	0.86	1.85	0.75	2.02
20	1.20	1.41	1.10	1.54	1.00	1.68	0.90	1.83	0.79	1.99
21	1.22	1.42	1.13	1.54	1.03	1.67	0.93	1.81	0.83	1.96
22	1.24	1.43	1.15	1.54	1.05	1.66	0.96	1.80	0.86	1.94
23	1.26	1.44	1.17	1.54	1.08	1.66	0.99	1.79	0.90	1.92
24	1.27	1.45	1.19	1.55	1.10	1.66	1.01	1.78	0.93	1.90
25	1.29	1.45	1.21	1.55	1.12	1.66	1.04	1.77	0.95	1.89
26	1.30	1.46	1.22	1.55	1.14	1.65	1.06	1.76	0.98	1.88
27	1.32	1.47	1.24	1.56	1.16	1.65	1.08	1.76	1.01	1.86
28	1.33	1.48	1.26	1.56	1.18	1.65	1.10	1.75	1.03	1.85
29	1.34	1.48	1.27	1.56	1.20	1.65	1.12	1.74	1.05	1.84
30	1.35	1.49	1.28	1.57	1.21	1.65	1.14	1.74	1.07	1.83
31	1.36	1.50	1.30	1.57	1.23	1.65	1.16	1.74	1.09	1.83
32	1.37	1.50	1.31	1.57	1.24	1.65	1.18	1.73	1.11	1.82
33	1.38	1.51	1.32	1.58	1.26	1.65	1.19	1.73	1.13	1.81
34	1.39	1.51	1.33	1.58	1.27	1.65	1.21	1.73	1.15	1.81
35	1.40	1.52	1.34	1.58	1.28	1.65	1.22	1.73	1.16	1.80
36	1.41	1.52	1.35	1.59	1.29	1.65	1.24	1.73	1.18	1.80
37	1.42	1.53	1.36	1.59	1.31	1.66	1.25	1.72	1.19	1.80
38	1.43	1.54	1.37	1.59	1.32	1.66	1.26	1.72	1.21	1.79
39	1.43	1.54	1.38	1.60	1.33	1.66	1.27	1.72	1.22	1.79
40	1.44	1.54	1.39	1.60	1.34	1.66	1.29	1.72	1.23	1.79
45	1.48	1.57	1.43	1.62	1.38	1.67	1.34	1.72	1.29	1.78
50	1.50	1.59	1.46	1.63	1.42	1.67	1.38	1.72	1.34	1.77
55	1.53	1.60	1.49	1.64	1.45	1.68	1.41	1.72	1.38	1.77
60	1.55	1.62	1.51	1.65	1.48	1.69	1.44	1.73	1.41	1.77
65	1.57	1.63	1.54	1.66	1.50	1.70	1.47	1.73	1.44	1.77
70	1.58	1.64	1.55	1.67	1.52	1.70	1.49	1.74	1.46	1.77
75	1.60	1.65	1.57	1.68	1.54	1.71	1.51	1.74	1.49	1.77
80	1.61	1.66	1.59	1.69	1.56	1.72	1.53	1.74	1.51	1.77
85	1.62	1.67	1.60	1.70	1.57	1.72	1.55	1.75	1.52	1.77
90	1.63	1.68	1.61	1.70	1.59	1.73	1.57	1.75	1.54	1.78
95	1.64	1.69	1.62	1.71	1.60	1.73	1.58	1.75	1.56	1.78
100	1.65	1.69	1.63	1.72	1.61	1.74	1.59	1.76	1.57	1.78

TABLE B.8 Critical Values for the Chi-Square Statistic

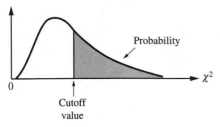

Note: Entries on this page are cutoff values to place a specified probability amount in the right tail. For example, to have probability = .10 in the right tail when $df = 4$, the table value is $\chi^2 =$ 7.779.

Degrees of Freedom, df	Probability in Right Tail				
	.10	.05	.025	.01	.005
1	2.706	3.841	5.024	6.635	7.879
2	4.605	5.991	7.378	9.210	10.597
3	6.251	7.815	9.348	11.345	12.838
4	7.779	9.488	11.143	13.277	14.860
5	9.236	11.070	12.833	15.086	16.750
6	10.645	12.592	14.449	16.812	18.548
7	12.017	14.067	16.013	18.475	20.278
8	13.362	15.507	17.535	20.090	21.955
9	14.684	16.919	19.023	21.666	23.589
10	15.987	18.307	20.483	23.209	25.188
11	17.275	19.675	21.920	24.725	26.757
12	18.549	21.026	23.337	26.217	28.300
13	19.812	22.362	24.736	27.688	29.819
14	21.064	23.685	26.119	29.141	31.319
15	22.307	24.996	27.488	30.578	32.801
16	23.542	26.296	28.845	32.000	34.267
17	24.769	27.587	30.191	33.409	35.718
18	25.989	28.869	31.526	34.805	37.156
19	27.204	30.144	32.852	36.191	38.582
20	28.412	31.410	34.170	37.566	39.997
21	29.615	32.671	35.479	38.932	41.401
22	30.813	33.924	36.781	40.289	42.796
23	32.007	35.172	38.076	41.638	44.181
24	33.196	36.415	39.364	42.980	45.558
25	34.382	37.652	40.647	44.314	46.928
26	35.563	38.885	41.923	45.642	48.290
27	36.741	40.113	43.194	46.963	49.645
28	37.916	41.337	44.461	48.278	50.993
29	39.087	42.557	45.722	49.588	52.336
30	40.256	43.773	46.979	50.892	53.672
50	63.167	67.505	71.420	76.154	79.490
60	74.397	79.082	83.298	88.379	91.952
80	96.578	101.879	106.629	112.329	116.321
100	118.498	124.342	129.561	135.807	140.169

Source: Computed by R. E. Shiffler and A. J. Adams.

TABLE B.9 Critical Values for the Studentized Range ($\alpha = .05$)

v = DF	\multicolumn{11}{c}{p = Number of Means}										
	2	3	4	5	6	7	8	9	10	11	12
2	6.08	8.33	9.80	10.88	11.73	12.43	13.03	13.54	13.99	14.39	14.75
3	4.50	5.91	6.82	7.50	8.04	8.48	8.85	9.18	9.46	9.72	9.95
4	3.93	5.04	5.76	6.29	6.71	7.05	7.35	7.60	7.83	8.03	8.21
5	3.64	4.60	5.22	5.67	6.03	6.33	6.58	6.80	6.99	7.17	7.32
6	3.46	4.34	4.90	5.30	5.63	5.90	6.12	6.32	6.49	6.65	6.79
7	3.34	4.16	4.68	5.06	5.36	5.61	5.82	6.00	6.16	6.30	6.43
8	3.26	4.04	4.53	4.89	5.17	5.40	5.60	5.77	5.92	6.05	6.18
9	3.20	3.95	4.41	4.76	5.02	5.24	5.43	5.59	5.74	5.87	5.98
10	3.15	3.88	4.33	4.65	4.91	5.12	5.30	5.46	5.60	5.72	5.83
11	3.11	3.82	4.26	4.57	4.82	5.03	5.20	5.35	5.49	5.61	5.71
12	3.08	3.77	4.20	4.51	4.75	4.95	5.12	5.26	5.39	5.51	5.61
13	3.06	3.73	4.15	4.45	4.69	4.88	5.05	5.19	5.32	5.43	5.53
14	3.03	3.70	4.11	4.41	4.64	4.83	4.99	5.13	5.25	5.36	5.46
15	3.01	3.67	4.08	4.37	4.59	4.78	4.94	5.08	5.20	5.31	5.40
16	3.00	3.65	4.05	4.33	4.56	4.74	4.90	5.03	5.15	5.26	5.35
17	2.98	3.63	4.02	4.30	4.52	4.70	4.86	4.99	5.11	5.21	5.31
18	2.97	3.61	4.00	4.28	4.49	4.67	4.82	4.96	5.07	5.17	5.27
19	2.96	3.59	3.98	4.25	4.47	4.65	4.79	4.92	5.04	5.14	5.23
20	2.95	3.58	3.96	4.23	4.45	4.62	4.77	4.90	5.01	5.11	5.20
30	2.89	3.49	3.85	4.10	4.30	4.46	4.60	4.72	4.82	4.92	5.00
40	2.86	3.44	3.79	4.04	4.23	4.39	4.52	4.63	4.73	4.82	4.90
60	2.83	3.40	3.74	3.98	4.16	4.31	4.44	4.55	4.65	4.73	4.81
120	2.80	3.36	3.68	3.92	4.10	4.24	4.36	4.47	4.56	4.64	4.71
Infinity	2.77	3.31	3.63	3.86	4.03	4.17	4.29	4.39	4.47	4.55	4.62

Source: From *Biometrika Tables for Statisticians*, Vol. I, by E. S. Pearson and H. O. Hartley, published by the Biometrika Trustees, Cambridge University Press, Cambridge, 1954. Reproduced with the permission of the authors and the publisher.

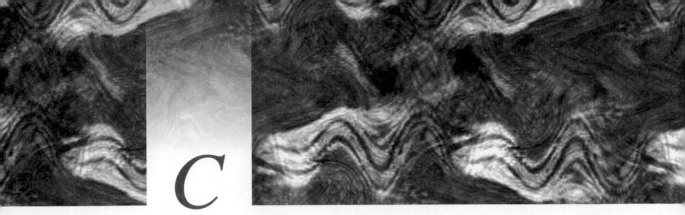

A Brief Introduction to Microsoft® Excel®, MINITAB®, and SAS®

C.1 INTRODUCTION

This appendix presents a brief summary of procedures in Excel, MINITAB, and SAS. Excel 2000 (used with Windows XP operating system), MINITAB Version 14, and Version 8.2 of SAS are used. Depending on the version of Excel, MINITAB, or SAS, the available procedures and output may differ slightly from what is presented in this text (but should be fairly similar). Only a general description of the use of these software packages is given here. Further details on statistical procedures are provided in the Using the Computer sections at the end of various chapters.

C.2 EXCEL

C.2.1 DATA INPUT

Data that have been stored in an Excel spreadsheet can be accessed through the File menu. Figure C.1 shows the contents of the File menu. Click Open and a dialog box opens requesting the name of the spreadsheet you want to open, as shown in Figure C.2. If you highlight cars8 and click Open, the cars8 spreadsheet opens, as shown in Figure C.3. Once you open the spreadsheet, you can begin to analyze the data.

If you have data of your own to analyze and you want to put the data into an Excel spreadsheet, you can just begin typing. For statistical analysis, columns typically serve as variables, so you can assign a name to each of the variables (columns) used (although the use of columns as variables is not required in Excel).

Certain Excel procedures that may be useful for data manipulation and analysis are discussed next.

FIGURE C.1
Excel File Menu.

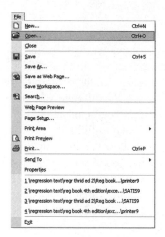

FIGURE C.2
Excel Request for
Spreadsheet to be
Opened.

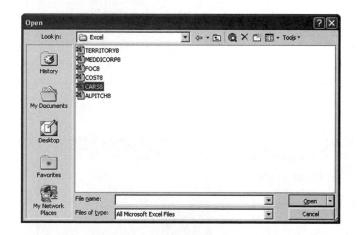

FIGURE C.3
Data File cars8 Opened
as Excel Spreadsheet.

	A	B	C	D	E	F	G
1	Name	WEIGHT	CITYMPG	HWYMPG	HP	CYLIN	LITER
2	Acura CL	3470	19	29	225	6	3.2
3	Acura NSX	3155	17	24	290	6	3.2
4	Acura RL	3920	18	24	225	6	3.5
5	Acura RSX	2750	25	31	160	4	2
6	Acura TL	3495	19	29	225	6	3.2
7	Aston Martin DB7 Vantage	4235	11	18	420	12	5.9
8	Aston Martin Vanquish	4110	12	19	460	12	5.9
9	Audi A4 Cabriolet	3250	22	31	170	4	1.8
10	Audi A6	3515	17	27	220	6	3
11	Audi A8	4070	17	24	310	8	4.2
12	Audi TT	2920	22	31	180	4	1.8
13	Bentley Arnage R	5700	10	15	400	8	6.8
14	Bentley Continental R	5400	11	16	420	8	6.8
15	BMW Z4 Roadster	3100	20	27	184	6	2.5

C.2.2 EXCEL'S CHART WIZARD

Scatterplots, time-series plots, pie charts, and bar charts are all constructed using the Chart Wizard. As an example, consider constructing a scatterplot. Click Excel's Chart Wizard button and the Chart Wizard appears, as in Figure C.4. You have a variety of chart types from which to choose. In Figure C.4, the scatterplot (or XY plot) option has been chosen. The chart subtypes show you the different variations on the basic scatter-plot. Choose the subtype that works best for your application. I like the plain old vanilla-flavored scatterplot, so that's the one I've highlighted. Click the Next button. Excel leads you through a series of screens that allow you to set up the scatterplot as you want it to appear. In Figure C.5, Excel is asking you to specify the data range and to indicate whether the data are arranged in rows or columns.

Note: To create a scatterplot in Excel, the variable you want to appear on the horizontal axis (the *x* axis) must be in a column directly to the left of the variable you want to appear on the vertical axis (the *y* axis).

In Figure C.6, the titles for the chart and the axis labels can be specified. (You can do this later if you want.) In Figure C.7, Excel allows you to place the chart on a separate sheet in the workbook or to put it on the current worksheet ply. Finally, Figure C.8 shows the completed scatterplot. You can size the chart and make changes to its appearance as you like.

To construct a time-series plot with Excel, click Excel's Chart Wizard button and highlight Line. Click the Next button. Excel leads you through a series of screens that allow you to set up the time-series plot as you want it to appear. The

FIGURE C.4
Excel's Chart Wizard.

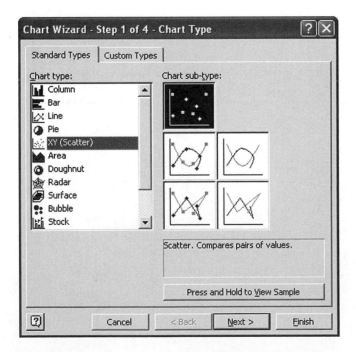

FIGURE C.5

Excel's Chart Wizard: Specifying the Data Range for a Scatterplot.

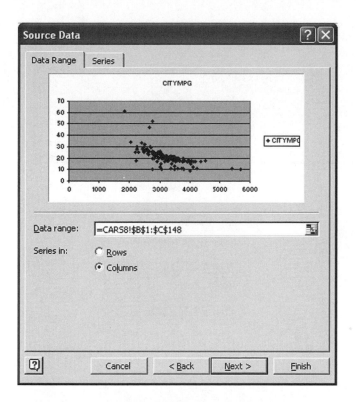

FIGURE C.6

Excel's Chart Wizard: Chart Title and Naming Axes on a Scatterplot.

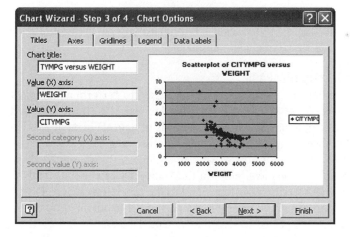

sequence of screens is similar to the sequence used in illustrating the creation of a scatterplot.

To construct a pie chart with Excel, click Excel's Chart Wizard button and highlight Pie. Click the Next button. Excel leads you through a series of screens that allow you to set up the pie chart as you want it to appear. The sequence of screens is

FIGURE C.7

Excel's Chart Wizard: Choosing a Location for the Chart.

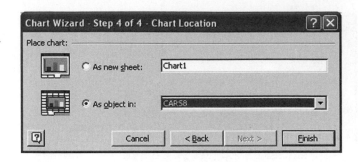

FIGURE C.8

Excel's Chart Wizard: The Completed Scatterplot.

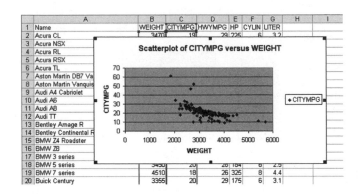

similar to the sequence used in illustrating the creation of a scatterplot. Bar charts are constructed in a similar manner.

C.2.3 DATA ANALYSIS TOOLPACK

The data analysis toolpack can be used to perform a variety of statistical analyses. Go to the Tools menu. Click Data Analysis,[1] as shown in Figure C.9. You then see a list of options for statistical analysis of data. This menu is used throughout this text because many of the procedures discussed in the text are performed in Excel by the use of the options provided. See Figure C.10 for a screen showing a listing of some of the options available on the Data Analysis menu. Both single and two-factor ANOVA options are explained in the text. Correlation, Descriptive Statistics, and Histogram are all used. On the part of the menu that you cannot see in Figure C.10, there are options for Regression and Two-sample t tests that are used as well. To use any of these procedures, click the desired option, click OK, and follow the instructions in the dialog box that appears.

---

[1] If Data Analysis does not appear on your Tools menu, click Add-Ins (also on the Tools menu) and make sure the square for Analysis Tool Box is checked on the Add-Ins menu. If Analysis Tool Box does not appear on your Add-Ins menu, you may need to reinstall Excel.

FIGURE C.9

Excel Data Analysis
Option on Tools Menu.

FIGURE C.10

Excel Screen Showing
Part of Data Analysis
Menu.

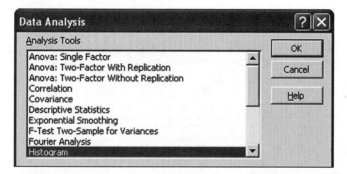

C.2.4 SMART REGRESSION ADD-IN

The SMARTReg add-in for Excel is included on the CD that comes with this text. It provides some features for regression that are not available with the standard Excel regression feature. To use the SMARTReg add-in, you must install it on your system. See the accompanying text file on the CD for installation instructions.

Once you have the SMARTReg add-in installed, you can use it from Excel. It will appear on the tools menu as shown in Figure C.11. Click Regression and the screen in Figure C.12 will open. Fill in the data range for the y variable, the x variables, click count, check labels if the variables have labels in the first row and indicate where the output should be placed (starting in cell k1 in this example). Under Regression Type, click Full to get the standard regression. In addition to the standard Excel regression output discussed in the text, you will get VIFs for each variable in the regression (discussed in Chapter 4) and Q (not discussed in this text). If

FIGURE C.11
Excel Screen Showing Tools Menu with SMARTReg Regression Feature Added.

FIGURE C.12
Excel Screen Showing SMARTReg Regression Feature.

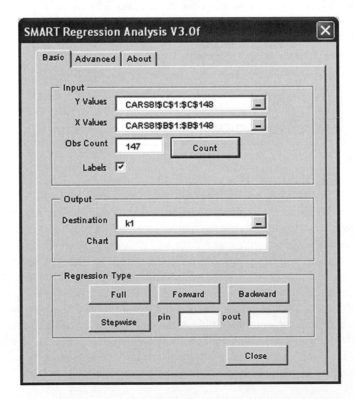

FIGURE C.13
Excel Screen Showing
SMARTReg Advanced
Tab.

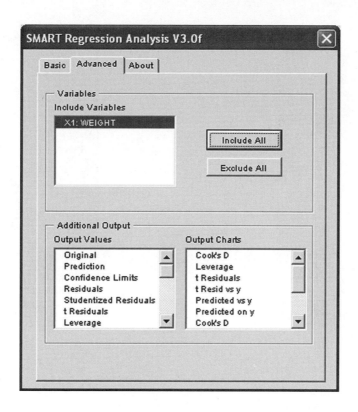

Forward, Backward, or Stepwise are chosen, you will obtain the variable selection results discussed in Chapter 8. You can specify the *p* value you want to allow variables to enter (pin) or leave (pout) the regression.

On the Advanced tab shown in Figure C.13 there are a number of additional options available. For example, leverage values, Cook's D, DFITS, the Durbin–Watson statistic, and the first-order autocorrelation coefficient were all discussed in Chapter 6 and are available using this feature.

For updates and more information about the add-in, go to www.gierus.ca/alex/smartreg. The SMARTReg feature was designed by J. D. Jobson of the University of Alberta. (See references for texts by Professor Jobson).

C.3 MINITAB

C.3.1 DATA INPUT

Storage of data during a MINITAB session is accomplished through the use of columns. (Note: There are two other storage modes in MINITAB: constants, which represent single numbers and are denoted K1, K2, . . . , and matrices, which represent arrays of numbers and are denoted M1, M2, ... These storage modes are not discussed.)

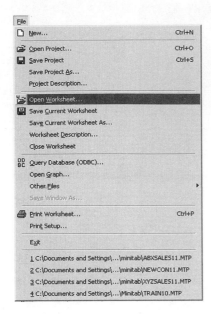

FIGURE C.15
MINITAB Open
Worksheet Dialog Box.

FIGURE C.14
MINITAB File Menu.

Data stored in a MINITAB file can be accessed through the File menu. Figure C.14 shows the contents of the File menu. Click Open Worksheet and a dialog box opens requesting the name of the worksheet you want to open, as shown in Figure C.15. If you highlight cars8 and click Open, the data from cars8 open into the current worksheet, as shown in Figure C.16. Once you have data in the worksheet columns, you can begin to analyze the data.

If you have data of your own to analyze and you want to put the data into a MINITAB worksheet, you can just begin typing. In MINITAB, it is best to start typing data in the first row of each column and to leave no empty rows. These columns serve as variables, so you can assign a name to each of the variables (columns) used.

Certain MINITAB procedures that may be useful for data manipulation and analysis are discussed next.

C.3.2 THE STAT MENU

Statistical procedures are accessed in MINITAB through the Stat menu. The Stat menu is shown in Figure C.17. In this text, the Basic Statistics, ANOVA, Regression, Multivariate, and Time Series categories from this menu were used.

FIGURE C.16
MINITAB Worksheet.

↓	C1-T Name	C2 WEIGHT	C3 CITYMPG	C4 HWYMPG	C5 HP	C6 CYLIN	C7 LITER
1	Acura CL	3470	19	29	225	6	3.2
2	Acura NSX	3155	17	24	290	6	3.2
3	Acura RL	3920	18	24	225	6	3.5
4	Acura RSX	2750	25	31	160	4	2.0
5	Acura TL	3495	19	29	225	6	3.2
6	Aston Martin DB7 Vantage	4235	11	18	420	12	5.9
7	Aston Martin Vanquish	4110	12	19	460	12	5.9
8	Audi A4 Cabriolet	3250	22	31	170	4	1.8
9	Audi A6	3515	17	27	220	6	3.0
10	Audi A8	4070	17	24	310	8	4.2
11	Audi TT	2920	22	31	180	4	1.8
12	Bentley Arnage R	5700	10	15	400	8	6.8
13	Bentley Continental R	5400	11	16	420	8	6.8
14	BMW Z4 Roadster	3100	20	27	184	6	2.5
15	BMW Z8	3495	13	21	394	8	4.9

FIGURE C.17
MINITAB Stat Menu.

Figure C.18 shows the Basic Statistics menu. Display Descriptive Statistics computes a number of descriptive statistics for specified columns. The statistics requested are the count of the number of observations in the column, the mean, median, the trimmed mean, the standard deviation, the standard error of the mean, the minimum and maximum, and the first and third quartiles. 1-Sample Z provides confidence intervals and hypothesis tests for a population mean when the population standard deviation is known. 1-Sample t provides confidence intervals and hypothesis tests for a population mean when the population standard deviation is unknown. 2-Sample t provides confidence intervals and hypothesis tests about the difference between two population means. Correlation provides pairwise correlations between two or more variables. Covariance provides pairwise covariances between two or

FIGURE C.18
MINITAB Basic Statistics Menu.

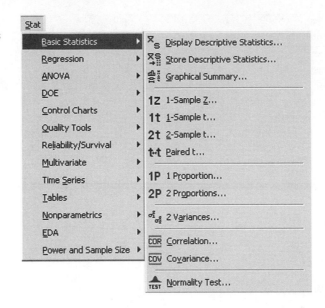

more variables. Normality Test provides a choice of three different tests for normality. The other items on this menu were not discussed in this text.

The Regression menu is shown in Figure C.19. Regression allows simple or multiple regressions and a variety of optional statistical measures to be computed. Stepwise provides for computation of stepwise regression with several options. Best Subsets provides for computation of a selection of the "best" regressions for a given group of variables along with summary statistics. The regressions computed as best subset regressions are those with the highest R^2 values for various subsets of

FIGURE C.19
MINITAB Regression Menu.

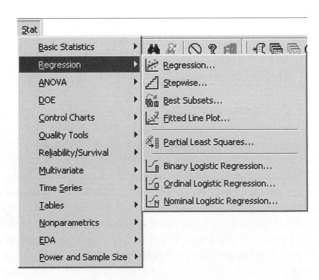

variables. Fitted Line Plot can be used to perform a simple regression and to plot the regression line on a scatterplot. Residual Plots provide a variety of residual plots to be used for diagnostic purposes for a regression. Binary Logistic Regression performs a logistic regression for a dependent variable with two possible values. Ordinal and Nominal Logistic Regression perform logistic regressions when there are more than two groups for classification.

The ANOVA menu is shown in Figure C.20. One-way performs a one-way analysis of variance (ANOVA). This form of the command assumes that the data are in one column, and a second column is used to designate which group or sample each observation belongs to. One-way (Unstacked) also performs a one-way ANOVA but assumes the items in each sample are in separate columns. Two-way performs a two-way ANOVA. Main Effects Plot and Interactions Plot provide useful graphs for two-way analysis of variance. The other items on this menu were not discussed in this text.

The Multivariate menu is shown in Figure C.21. Discriminant Analysis provides a way of examining the relationship between a qualitative dependent variable and one or more quantitative independent variables. The other items on this menu were not discussed in this text.

The Time Series menu is shown in Figure C.22. Time Series Plot produces a graph of a variable or variables over time (also available on the Graph menu). Trend Analysis provides for the fitting of a variety of trends to data. As shown in this text, the fitting of these trends to data can also be accomplished through the use of Regression. Lag allows for the creation of lagged variables. Decomposition, moving average, single exponential smoothing, double exponential smoothing, and Winters'

FIGURE C.20
MINITAB ANOVA Menu.

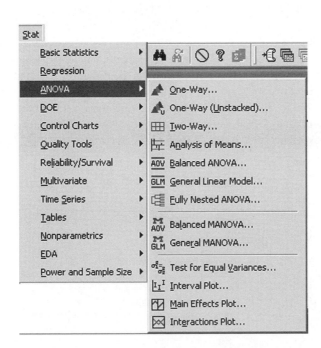

FIGURE C.21
MINITAB Multivariate
Menu.

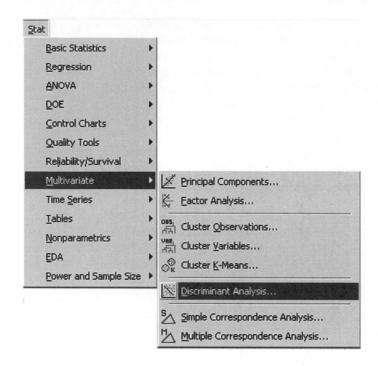

method are time-series forecasting methods discussed in Chapter 11. The other items on this menu were not discussed in this text.

C.3.3 THE CALC MENU

The CALC menu is shown in Figure C.23. Arithmetic operations may be performed on the columns. This is accomplished through the use of the Calculator (see Figure C.24). By placing the column for the result in "Store result in variable" and a mathematical expression in "Expression," columns can be added, subtracted, divided, and transformed in many other ways. Examples of various operations available in MINITAB and the symbols used to denote them are

Addition	+
Subtraction	−
Multiplication	*
Division	/
Exponentiation	**

A variety of functions are also available:

ABSOLUTE	Compute absolute value
SQRT	Compute square root
LOGE	Compute logarithm to base e
LOGT	Compute logarithm to base 10
SIN, COS, TAN	Compute sine, cosine, or tangent for an angle given in radians

FIGURE C.22
MINITAB Time Series
Menu.

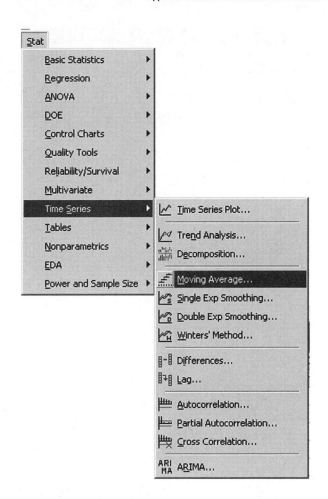

FIGURE C.23
MINITAB Calc Menu.

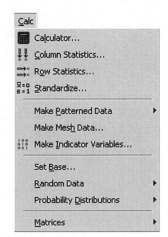

FIGURE C.24
MINITAB Calculator.

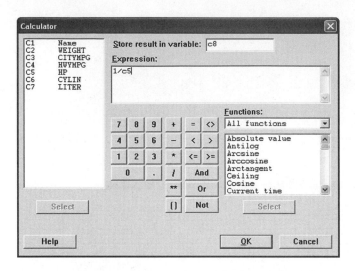

FIGURE C.25
MINITAB Make
Patterned Data Menu.

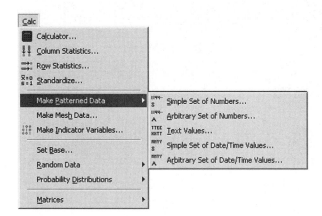

Figure C.24 shows the operation that finds the inverse of the variable HP and stores the result in c8.

Make Patterned Data (see Figure C.25) is useful when creating variables where the data take on certain regular patterns. The creation of a time-trend variable (see Figure C.26) and other patterned variables is explained more fully in Using the Computer sections throughout the text. Make Indicator Variables is another useful procedure that is explained more fully in the text. The remaining procedures on the Calc menu are not used in this text.

C.4 SAS

C.4.1 DATA INPUT

SAS analyses consist of at least two steps: (a) a DATA step to input the data set and create any new variables and (b) a PROC step to conduct any analyses.

FIGURE C.26
Creating a MINITAB
Time-Trend Variable.

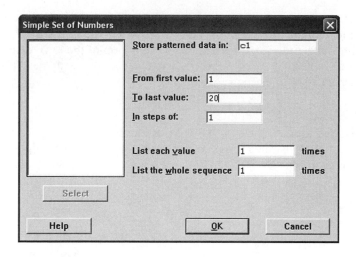

Consider the following measurements of height and weight made on five people:

Height (in inches)	Weight (in pounds)
60	140
66	195
70	185
71	190
73	210

To input these data in SAS, the following lines are used:

```
DATA SIZE;
INPUT HEIGHT WEIGHT;
CARDS;
60 140
66 195
70 185
71 190
73 210
```

The DATA command on the first line names the data set to be used. The name chosen for this data set was SIZE. The INPUT command on the second line specifies the names of the variables to be used. The variable names chosen for the two variables in this data set were HEIGHT and WEIGHT. The names can be at most eight characters long. The CARDS command indicates to SAS that the data are to follow. The data are then typed in with one entry for each variable per line with each entry on a line separated by at least one space. Note that each command in SAS is followed by a semicolon. The data values, however, are not.

When reading data from a file that has already been set up, the following commands might be used:

```
DATA DIV;
INFILE DIV3.DAT;
INPUT DIVYIELD EPS;
```

The DATA line indicates the name assigned to the data set in SAS. The INFILE command indicates the file name from which the data are to be read. The INPUT line names the variables to be read. This form of the DATA step assumes the data are arranged in the file DIV3.DAT in two columns with the entries in each column separated by at least one space.

C.4.2 ARITHMETIC WITH SAS VARIABLES

During the DATA step in SAS, other variables can also be created. The following example illustrates some of the possible variables that might be of use in a SAS analysis:

```
DATA EXAMPLE;
INFILE EXAMPLE.DAT;
INPUT Y X1 X2 X3 X4;
X1SQR=X1**2;
X2INV=1/X2;
X3X4=X3*X4;
```

In this example, data on five variables were read in from a file called EXAMPLE.DAT. The variables were named Y, X1, X2, X3, and X4. Several new variables were then created during the data input. The variable X1SQR is the square of X1, the variable X2INV is the inverse of X2, and the variable X3X4 is the product of the two variables X3 and X4.

Examples of various arithmetic operations available in SAS and the symbols used to denote them are

Addition	+
Subtraction	−
Multiplication	*
Division	/
Exponentiation	**

SAS can also be used to produce certain functions of variables. For example, to create a variable equal to the natural logarithm of X1, the following command could be used:

```
LOGX1=LOG(X1)
```

A variable named LOGX1 has been created by finding the natural logarithm of the values of the original variable.

Other functions available in SAS include

ABS	Compute absolute value
SQRT	Compute square root
LOG10	Compute logarithm to base 10
SIN, COS, TAN	Compute sine, cosine, or tangent for an angle given in radians

These are used in the same manner as the LOG function.

C.4.3 STATISTICAL PROCEDURES IN SAS

Statistical analyses of all types in SAS are performed in the PROC steps. PROC stands for "procedure," and each PROC statement in SAS specifies a certain procedure to be performed with the data. The following PROCS may be useful in data analysis.

```
PROC ANOVA;
CLASS X1;
MODEL Y=X1;
```

performs a one-way analysis of variance. The dependent variable is listed first on the MODEL statement, and the variable used to specify the factor levels is indicated in the CLASS statement and listed second on the MODEL statement.

```
PROC ANOVA;
CLASS X1 X2;
MODEL Y=X1 X2 X1*X2;
```

performs a two-way analysis of variance. The dependent variable is listed first on the MODEL statement, and the variables used to specify the factor levels are indicated in the CLASS statement and listed second on the MODEL statement. If an interaction term is desired, it is listed on the model statement as shown: X1*X2.

```
PROC CHART;
VBAR Y;
```

produces a bar chart for the variable named Y. PROC CHART should be used with discrete data.

```
PROC CORR;
VAR Y X1;
```

produces pairwise correlations for the variables Y and X1.

```
PROC FREQ;
TABLES Y;
```

produces a frequency distribution for the variable named Y. PROC FREQ should be used with discrete data.

```
PROC MEANS;
VAR Y;
```

produces the following summary statistics for the variable named Y: a count of the number of observations, the mean, standard deviation, minimum, maximum, and standard error of the mean.

```
PROC PLOT;
PLOT Y*X1 Y*X2 Y*X3 Y*X4;
```

produces a scatterplot of each indicated pair of variables with the first variable of each pair plotted on the vertical axis and the second variable on the horizontal axis.

```
PROC REG;
MODEL Y=X1 X2 X3 X4/OPTIONS;
OUTPUT PREDICTED=FITS STUDENT=STRES;
```

performs a regression analysis. The MODEL statement specifies the Y variable (listed first) and the explanatory variables (X1, X2, X3, and X4 in this example). In place of OPTIONS on the MODEL statement, there are a variety of choices. These include

DW to request the Durbin–Watson statistic.

VIF to request variance inflation factors.

INFLUENCE to request a variety of influence diagnostics.

P to request that predicted values be printed.

R to request that residuals be printed.

CLM to print 95% upper and lower confidence interval limits for the estimate of the point on the regression line (the estimate of the conditional mean) for each observation.

CLI to print 95% upper and lower prediction interval limits for the prediction of an individual point for each observation.

The OUTPUT statement can be used to create a variety of new variables. As shown, the predicted or fitted values are saved in a variable named FITS, and the standardized residuals are saved in a variable named STRES.

```
PROC STEPWISE;
MODEL Y=X1 X2 X3 X4/OPTIONS;
```

performs a stepwise regression. In place of OPTIONS, there are a variety of choices. These include

FORWARD if forward selection is to be used.

BACKWARD if backward elimination is to be used.

STEPWISE if stepwise regression is to be used.

MAXR if the maximum R-squared improvement technique is to be used.

SLE = desired level of significance for entering a variable when using STEPWISE or FORWARD.

SLS = desired level of significance for deleting a variable when using STEPWISE or BACKWARD.

INCLUDE = variables to be included in all models.

```
PROC TTEST;
CLASS X1;
VAR X2;
```

produces a test for whether the difference between two population means is zero or not. The data from the two samples are in the variable X2. The variable X1 is used to indicate from which population the sample value was chosen.

```
PROC UNIVARIATE PLOT NORMAL;
VAR Y;
```

produces a variety of descriptive statistics for the variable named Y: the count of the number of observations, the mean, standard deviation, variance, standard error of the mean, coefficient of variation, measures of skewness and kurtosis, a t value for testing whether the population mean is zero, the minimum, maximum, quartiles, median, range, interquartile range, mode, 1st, 5th, 10th, 90th, 95th, and 99th percentiles, and the five largest and smallest values.

PLOT and NORMAL are options. If PLOT is specified, a stem-and-leaf plot, box plot, and a normal probability plot are produced. If NORMAL is specified, a test for whether the data came from a normal distribution is conducted.

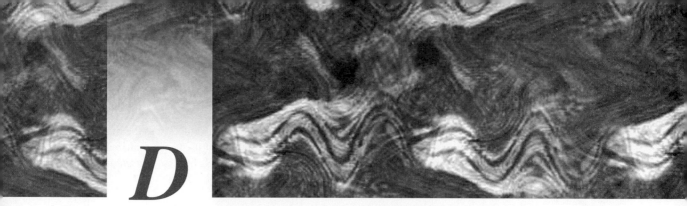

D

Matrices and Their Application to Regression Analysis

D.1 INTRODUCTION

In Chapter 3, the equations were provided to compute b_0 and b_1, the least-squares estimates of the simple regression coefficients. When multiple regression was discussed in Chapter 4, no equations were shown for the coefficient estimates because of the complexity involved. Instead, the computer was used to solve for the estimates of the multiple regression coefficients. There is, however, a very general way to represent the equations for the estimates of the regression coefficients (either simple or multiple). This involves the use of matrices. A *matrix* is a rectangular array of numbers. For example, the following are matrices:

$$\mathbf{A} = \begin{bmatrix} 5 & 7 & 3 \\ 4 & 2 & 6 \end{bmatrix} \qquad \mathbf{B} = \begin{bmatrix} 2 & 4 \\ 1 & 5 \end{bmatrix} \qquad \mathbf{C} = \begin{bmatrix} 7 \\ 4 \\ 6 \end{bmatrix}$$

The *dimensions* of a matrix are the number of its rows and columns. The matrix $\mathbf{A}$ has two rows and three columns, so $\mathbf{A}$ is referred to as a 2×3 matrix. Similarly, $\mathbf{B}$ is a 2×2 matrix and $\mathbf{C}$ is a 3×1 matrix. $\mathbf{B}$ is called a *square matrix* because it has equal numbers of rows and columns. The *diagonal elements* of a square matrix are the elements that are located on the diagonal that runs from the upper left-hand corner of the matrix to the lower right-hand corner. In the case of the matrix $\mathbf{B}$, the diagonal elements are 2 and 5. A matrix with only one column is usually referred to as a *vector*. In the previous example, the matrix $\mathbf{C}$ is a vector.

Certain arithmetic operations can be performed with matrices—addition, subtraction, and multiplication. Another operation of importance in working with matrices is determining the *transpose* of the matrix. Square matrices can also be inverted. In this appendix, each of these matrix operations is defined, and then the use of matrices to represent a regression equation and the computations necessary to produce the estimates of the regression coefficients are shown. This treatment is not intended to be complete by any means, but it may serve as a brief introduction to the matrix approach to regression analysis. For a more complete treatment using the matrix approach, see, for example, A. Hadi, *Matrix Algebra as a Tool*, or R. Myers, *Classical and Modern Regression with Applications*.[1]

D.2 MATRIX OPERATIONS

D.2.1 MATRIX ADDITION

Two matrices can be added together if their dimensions are the same. Consider the following matrices denoted **A** and **B**:

$$\mathbf{A} = \begin{bmatrix} 3 & 6 & 1 \\ 3 & 4 & 6 \end{bmatrix} \qquad \mathbf{B} = \begin{bmatrix} 2 & 4 & 2 \\ 4 & 5 & 4 \end{bmatrix}$$

The sum of these two matrices is **A** + **B**:

$$\mathbf{A} + \mathbf{B} = \begin{bmatrix} 3+2 & 6+4 & 1+2 \\ 3+4 & 4+5 & 6+4 \end{bmatrix} = \begin{bmatrix} 5 & 10 & 3 \\ 7 & 9 & 10 \end{bmatrix}$$

Note that the corresponding elements of the original matrices **A** and **B** have simply been added together to obtain the sum of the two matrices.

D.2.2 TRANSPOSE OF A MATRIX

The *transpose* of a matrix is formed by exchanging its rows and columns. For example, consider the matrix

$$\mathbf{A} = \begin{bmatrix} 3 & 6 & 1 \\ 3 & 4 & 6 \end{bmatrix}$$

The transpose of **A** (denoted $\mathbf{A}^T$) is

$$\mathbf{A}^T = \begin{bmatrix} 3 & 3 \\ 6 & 4 \\ 1 & 6 \end{bmatrix}$$

The rows of the matrix **A** have now become the columns of $\mathbf{A}^T$, and the columns of **A** have become the rows of $\mathbf{A}^T$.

---

[1] See references for complete publication information.

D.2.3 MATRIX MULTIPLICATION

Consider the following two matrices:

$$\mathbf{A} = \begin{bmatrix} 4 & 1 \\ 2 & 6 \end{bmatrix} \qquad \mathbf{B} = \begin{bmatrix} 3 & 2 & 4 \\ 1 & 4 & 5 \end{bmatrix}$$

The product, $\mathbf{AB}$, of the two matrices is found by multiplying the elements of each row of $\mathbf{A}$ by the corresponding elements of each column of $\mathbf{B}$ and then summing the individual products. In this example,

$$\mathbf{AB} = \begin{bmatrix} (4 \times 3) + (1 \times 1) & (4 \times 2) + (1 \times 4) & (4 \times 4) + (1 \times 5) \\ (2 \times 3) + (6 \times 1) & (2 \times 2) + (6 \times 4) & (2 \times 4) + (6 \times 5) \end{bmatrix}$$

$$= \begin{bmatrix} 13 & 12 & 21 \\ 12 & 28 & 38 \end{bmatrix}$$

To obtain the product matrix, $\mathbf{AB}$, the first element in row 1 of $\mathbf{A}$ is multiplied by the first element in column 1 of $\mathbf{B}$. Then the second element in row 1 of $\mathbf{A}$ is multiplied by the second element in column 1 of $\mathbf{B}$. These two products are then added together to obtain the element in the first row and first column of $\mathbf{AB}$. To obtain the element in the first row and second column of $\mathbf{AB}$, the first element in row 1 of $\mathbf{A}$ is multiplied by the first element in column 2 of $\mathbf{B}$. Then the second element in row 1 of $\mathbf{A}$ is multiplied by the second element in column 2 of $\mathbf{B}$. This row-by-column multiplication process continues until all elements of the product matrix $\mathbf{AB}$ have been computed. For this process to work, the number of columns in $\mathbf{A}$ must equal the number of rows in $\mathbf{B}$. In terms of the dimensions of the matrices, if $\mathbf{A}$ is an $m \times n$ matrix and $\mathbf{B}$ is a $p \times q$ matrix, n and p must be equal before the product matrix can be computed. The product matrix, $\mathbf{AB}$, will be an $m \times q$ matrix. In this example, $\mathbf{A}$ is a 2×2 matrix, $\mathbf{B}$ is a 2×3 matrix, and the product matrix, $\mathbf{AB}$, is a 2×3 matrix.

Consider another example. If

$$\mathbf{A} = \begin{bmatrix} 5 & 7 & 3 \\ 4 & 2 & 6 \end{bmatrix} \qquad \text{and} \qquad \mathbf{B} = \begin{bmatrix} 7 \\ 4 \\ 6 \end{bmatrix}$$

then

$$\mathbf{AB} = \begin{bmatrix} (5 \times 7) + (7 \times 4) + (3 \times 6) \\ (4 \times 7) + (2 \times 4) + (6 \times 6) \end{bmatrix} = \begin{bmatrix} 81 \\ 72 \end{bmatrix}$$

D.2.4 MATRIX INVERSION

A familiar property of multiplication is that any number multiplied by its multiplicative inverse results in the answer 1. For example, $2 \times \frac{1}{2} = 1$, $3 \times \frac{1}{3} = 1$, and so on. Thus, $\frac{1}{2}$ is the multiplicative inverse of the number 2 and $\frac{1}{3}$ is the multiplicative inverse of 3. The inverse of a matrix $\mathbf{A}$, denoted $\mathbf{A}^{-1}$ is defined as the matrix that produces an identity matrix when multiplied by the original matrix:

$$\mathbf{A}\mathbf{A}^{-1} = \mathbf{I}$$

where $\mathbf{I}$ is the identity matrix. An *identity matrix* is a matrix with 1s as diagonal elements and 0s everywhere else. For example, consider the matrix

$$\mathbf{A} = \begin{bmatrix} 4 & 2 \\ 5 & 3 \end{bmatrix}$$

The inverse of $\mathbf{A}$ is

$$\mathbf{A}^{-1} = \begin{bmatrix} 1.5 & -1 \\ -2.5 & 2 \end{bmatrix}$$

Multiplying $\mathbf{A}$ by $\mathbf{A}^{-1}$ verifies that these two matrices are inverses:

$$\mathbf{A}\mathbf{A}^{-1} = \begin{bmatrix} (4 \times 1.5) + (2 \times (-2.5)) & (4 \times (-1)) + (2 \times 2) \\ (5 \times 1.5) + (3 \times (-2.5)) & (5 \times (-1)) + (3 \times 2) \end{bmatrix}$$

$$= \begin{bmatrix} 1 & 0 \\ 0 & 1 \end{bmatrix}$$

The resulting product is an identity matrix, denoted $\mathbf{I}$. It has the property that any square matrix of appropriate dimension multiplied by an identity matrix results in the original matrix. For example, it is easy to verify that $\mathbf{A}\mathbf{I} = \mathbf{A}$. When working with numbers rather than matrices, the number 1 serves as the multiplicative identity. Thus, in matrix multiplication, the identity matrix serves the same purpose as the number 1.

Note that inverses can only be computed for square matrices. Any identity matrix $\mathbf{I}$ also must be a square matrix.

One method of finding the inverse of a matrix is demonstrated through the following example. Consider the matrix $\mathbf{A}$:

$$\mathbf{A} = \begin{bmatrix} 4 & 2 \\ 5 & 3 \end{bmatrix} \qquad \mathbf{I} = \begin{bmatrix} 1 & 0 \\ 0 & 1 \end{bmatrix}$$

The identity matrix of the same dimension has been written next to $\mathbf{A}$. A series of identical computations will now be performed on $\mathbf{A}$ and $\mathbf{I}$. The intent of these computations is to transform $\mathbf{A}$ to an identity matrix. The same operations applied to $\mathbf{I}$

transform this identity matrix to the inverse of $\mathbf{A}$, $\mathbf{A}^{-1}$. There are two types of computations that are allowed:

1. Any row of $\mathbf{A}$ can be multiplied or divided by a number.
2. One row of $\mathbf{A}$ can be added to or subtracted from another row of $\mathbf{A}$.

The computations to produce $\mathbf{A}^{-1}$ are as follows:

Step 1: Because the goal is to transform $\mathbf{A}$ to an identity matrix, start by dividing the first row of $\mathbf{A}$ by the value of the element in the first row and first column (4). This results in 1 as the first diagonal element. Perform the identical computation on $\mathbf{I}$; that is, divide the elements in the first row by 4.

$$\mathbf{A} = \begin{bmatrix} 1 & 0.5 \\ 5 & 3 \end{bmatrix} \qquad \mathbf{I} = \begin{bmatrix} 0.25 & 0 \\ 0 & 1 \end{bmatrix}$$

Step 2: Subtract 5 times row 1 from row 2; the result is 0 in the first element of row 2. As always, do the same to $\mathbf{I}$. Also note that row 1 is not actually changed in this computation. A multiple of row 1 is subtracted from row 2, producing a new row 2, but row 1 remains as it was originally:

$$\mathbf{A} = \begin{bmatrix} 1 & 0.5 \\ 0 & 0.5 \end{bmatrix} \qquad \mathbf{I} = \begin{bmatrix} 0.25 & 0 \\ -1.25 & 1 \end{bmatrix}$$

Step 3: Multiply row 2 by 2. The result is 1 for the diagonal element in that row.

$$\mathbf{A} = \begin{bmatrix} 1 & 0.5 \\ 0 & 1 \end{bmatrix} \qquad \mathbf{I} = \begin{bmatrix} 0.25 & 0 \\ -2.5 & 2 \end{bmatrix}$$

Step 4: Subtract 0.5 times row 2 from row 1 to eliminate the nondiagonal element.

$$\mathbf{A} = \begin{bmatrix} 1 & 0 \\ 0 & 1 \end{bmatrix} \qquad \mathbf{I} = \begin{bmatrix} 1.5 & -1 \\ -2.5 & 2 \end{bmatrix}$$

The matrix $\mathbf{A}$ has now been transformed into an identity matrix. The inverse of $\mathbf{A}$ is

$$\mathbf{A}^{-1} = \begin{bmatrix} 1.5 & -1 \\ -2.5 & 2 \end{bmatrix}$$

D.3 MATRICES AND REGRESSION ANALYSIS

In Chapter 4, the multiple regression model was written as follows:

$$y_i = \beta_0 + \beta_1 x_{1i} + \beta_2 x_{2i} + \cdots + \beta_K x_{Ki} + e_i \qquad (D.1)$$

In matrix notation, the multiple regression model can be written

$$Y = X\beta + e \tag{D.2}$$

where Y, X, β, and e are matrices defined as follows:

$$Y = \begin{bmatrix} y_1 \\ y_2 \\ \cdot \\ \cdot \\ \cdot \\ y_n \end{bmatrix} \qquad X = \begin{bmatrix} 1 & x_{11} & x_{12} & \cdot & \cdot & \cdot & x_{1K} \\ 1 & x_{21} & x_{22} & \cdot & \cdot & \cdot & x_{2K} \\ \cdot & \cdot & \cdot & & & & \cdot \\ \cdot & \cdot & \cdot & & & & \cdot \\ \cdot & \cdot & \cdot & & & & \cdot \\ 1 & x_{n1} & x_{n2} & \cdot & \cdot & \cdot & x_{nK} \end{bmatrix}$$

$$\beta = \begin{bmatrix} \beta_0 \\ \beta_1 \\ \cdot \\ \cdot \\ \cdot \\ \beta_K \end{bmatrix} \qquad e = \begin{bmatrix} e_1 \\ e_2 \\ \cdot \\ \cdot \\ \cdot \\ e_n \end{bmatrix}$$

The matrix Y is an $n \times 1$ matrix containing all n observations on the dependent variable. These observations are denoted y_1, y_2, and so on, as in Chapter 4.

The matrix X is an $n \times (K + 1)$ matrix. The first column of the matrix X is a column of 1s. The second column of X consists of all n values of the first explanatory variable $(k = 1)$, denoted $x_{11}, x_{21}, \ldots, x_{n1}$. The third column of X consists of all n values of the second explanatory variable $(k = 2)$ denoted, $x_{12}, x_{22}, \ldots, x_{n2}$, and so on. The initial column of 1s in the X matrix is necessary because there is a constant (or intercept) in the equation. These 1s can be thought of as multipliers of β_0 just as the appropriate x values can be thought of as multipliers of β_k (k not equal to zero).

The matrix β is a $(K + 1) \times 1$ matrix containing all of the population regression coefficients to be estimated. These are denoted $\beta_0, \beta_1, \ldots, \beta_K$.

The matrix e is an $n \times 1$ matrix containing the disturbances $e_1, e_2, \ldots, e_n$.

Equation (D.2) represents the same relationship as Equation (D.1) except in matrix form. This can be verified by actually performing the multiplication and addition of the matrices shown in Equation (D.2).

Equation (D.2) is the general representation of any multiple regression model. As a specific example, consider again the data from Example 3.1:

x	1	2	3	4	5	6
y	3	2	8	8	11	13

The **X** and **Y** matrices for this example are as follows:

$$\mathbf{Y} = \begin{bmatrix} 3 \\ 2 \\ 8 \\ 8 \\ 11 \\ 13 \end{bmatrix} \qquad \mathbf{X} = \begin{bmatrix} 1 & 1 \\ 1 & 2 \\ 1 & 3 \\ 1 & 4 \\ 1 & 5 \\ 1 & 6 \end{bmatrix}$$

The matrix $\boldsymbol{\beta}$ is

$$\boldsymbol{\beta} = \begin{bmatrix} \beta_0 \\ \beta_1 \end{bmatrix}$$

and the matrix **e** is

$$\mathbf{e} = \begin{bmatrix} e_1 \\ e_2 \\ e_3 \\ e_4 \\ e_5 \\ e_6 \end{bmatrix}$$

(There are no actual numbers in **e** because the disturbances are not observable.)

As least-squares estimates of β_0 and β_1 in Chapter 3, those values b_0 and b_1 that minimized the error sum of squares (*SSE*) were used. *SSE* was written as

$$\sum_{i=1}^{n} (y_i - \hat{y}_i)^2$$

where the $\hat{y}_i$ were the points on the regression line determined by b_0 and b_1. In matrix notation, the equation for the error sum of squares can be written as

$$(\mathbf{Y} - \mathbf{Xb})^{\mathrm{T}}(\mathbf{Y} - \mathbf{Xb})$$

where **b** is the matrix of estimated regression coefficients:

$$\mathbf{b} = \begin{bmatrix} b_0 \\ b_1 \end{bmatrix}$$

The equation representing the least-squares estimates of the regression coefficients is

$$\mathbf{b} = (\mathbf{X}^{\mathrm{T}}\mathbf{X})^{-1}\mathbf{X}^{\mathrm{T}}\mathbf{Y} \tag{D.3}$$

This equation represents the least-squares estimates of the regression coefficients in any simple or multiple regression. The matrices **b**, **X**, and **Y** just need to be defined appropriately. In the example used in this section, the least-squares estimates are found as follows:

$$\mathbf{X^T} = \begin{bmatrix} 1 & 1 & 1 & 1 & 1 & 1 \\ 1 & 2 & 3 & 4 & 5 & 6 \end{bmatrix}$$

$$\mathbf{X^TX} = \begin{bmatrix} 1 & 1 & 1 & 1 & 1 & 1 \\ 1 & 2 & 3 & 4 & 5 & 6 \end{bmatrix} \begin{bmatrix} 1 & 1 \\ 1 & 2 \\ 1 & 3 \\ 1 & 4 \\ 1 & 5 \\ 1 & 6 \end{bmatrix} = \begin{bmatrix} 6 & 21 \\ 21 & 91 \end{bmatrix}$$

$$\mathbf{(X^TX)^{-1}} = \begin{bmatrix} 91/105 & -21/105 \\ -21/105 & 6/105 \end{bmatrix}$$

$$\mathbf{X^TY} = \begin{bmatrix} 1 & 1 & 1 & 1 & 1 & 1 \\ 1 & 2 & 3 & 4 & 5 & 6 \end{bmatrix} \begin{bmatrix} 3 \\ 2 \\ 8 \\ 8 \\ 11 \\ 13 \end{bmatrix} = \begin{bmatrix} 45 \\ 196 \end{bmatrix}$$

$$\mathbf{(X^TX)^{-1}\,(X^TY)} = \begin{bmatrix} 91/105 & -21/105 \\ -21/105 & 6/105 \end{bmatrix} \begin{bmatrix} 45 \\ 196 \end{bmatrix} = \begin{bmatrix} 39 - 39.2 \\ -9 + 11.2 \end{bmatrix} = \begin{bmatrix} -0.2 \\ 2.2 \end{bmatrix}$$

The least-squares estimate of $\boldsymbol{\beta}$ is

$$\mathbf{b} = \begin{bmatrix} -0.2 \\ 2.2 \end{bmatrix}$$

The regression equation can be written

$$\hat{y} = -0.2 + 2.2x$$

as in Chapter 3.

Many of the other relationships discussed throughout this text also can be expressed in matrix form. Some of these are listed but not discussed in detail.

1. $\hat{\mathbf{Y}} = \mathbf{Xb}$ is the vector of predicted or fitted values.

2. $\mathbf{Y} - \hat{\mathbf{Y}}$ is the vector of residuals.

3. $(\mathbf{Y} - \hat{\mathbf{Y}})^T (\mathbf{Y} - \hat{\mathbf{Y}})$ is the error sum of squares (*SSE*). *SSE* can also be written as $\mathbf{Y^TY} - \mathbf{b^TX^TY}$.

4. $\mathbf{b^T X^T Y} - n\overline{Y}^2$ is the regression sum of squares.

5. The variances of the regression coefficients are the diagonal elements of the matrix $s_e^2 (\mathbf{X^T X})^{-1}$

6. The estimate of s_e^2 is $SSE/(n - K - 1)$ with SSE as given in item 3.

7. The variance of the estimate of a point on the regression line, denoted in Chapters 3 and 4 as s_m^2, is given by $s_m^2 = s_e^2(\mathbf{x(X^T X)}^{-1}\mathbf{x^T})$, and the variance of the prediction for a single individual is $s_p^2 = s_e^2(1 + \mathbf{x(X^T X)}^{-1}\mathbf{x^T})$. The $\mathbf{x}$'s in these formulas represent the vectors of the values of the explanatory variables used to generate estimates or predictions.

EXERCISES

In Exercises 1 through 3, find the sum of the following matrices:

1. $\mathbf{A} = \begin{bmatrix} 1 & 3 & 4 \\ 2 & 1 & 2 \\ 3 & 1 & 5 \end{bmatrix}$ $\mathbf{B} = \begin{bmatrix} 4 & 1 & 6 \\ 2 & 1 & 5 \\ 1 & 4 & 3 \end{bmatrix}$

2. $\mathbf{A} = \begin{bmatrix} 1 & 2 \\ 7 & 6 \end{bmatrix}$ $\mathbf{B} = \begin{bmatrix} 4 & 5 \\ 1 & 9 \end{bmatrix}$

3. $\mathbf{A} = \begin{bmatrix} 1 \\ 3 \\ 2 \end{bmatrix}$ $\mathbf{B} = \begin{bmatrix} 6 \\ 1 \\ 5 \end{bmatrix}$

In Exercises 4 through 6, find the transpose of the following matrices:

4. $\mathbf{A} = \begin{bmatrix} 1 & 3 & 4 \\ 2 & 1 & 2 \\ 3 & 1 & 5 \end{bmatrix}$

5. $\mathbf{B} = \begin{bmatrix} 1 & 7 & 5 \\ 3 & 4 & 6 \end{bmatrix}$

6. $\mathbf{C} = \begin{bmatrix} 1 \\ 3 \\ 2 \end{bmatrix}$

In Exercises 7 through 10, find the product of the following matrices:

7. $\mathbf{A} = \begin{bmatrix} 1 & 2 \\ 7 & 6 \end{bmatrix}$ $\mathbf{B} = \begin{bmatrix} 4 & 5 \\ 1 & 9 \end{bmatrix}$

8. $\mathbf{A} = \begin{bmatrix} 1 & 3 & 4 \\ 2 & 1 & 2 \\ 3 & 1 & 5 \end{bmatrix}$ $\mathbf{B} = \begin{bmatrix} 4 & 1 & 6 \\ 2 & 1 & 5 \\ 1 & 4 & 3 \end{bmatrix}$

9. $\mathbf{A} = \begin{bmatrix} 1 & 7 & 5 \\ 3 & 4 & 6 \end{bmatrix}$ and $\mathbf{A^T}$

10. $\mathbf{A} = \begin{bmatrix} 1 & 3 & 4 \\ 2 & 1 & 2 \\ 3 & 1 & 5 \end{bmatrix}$ $\mathbf{B} = \begin{bmatrix} 1 & 0 & 0 \\ 0 & 1 & 0 \\ 0 & 0 & 1 \end{bmatrix}$

In Exercises 11 and 12, compute the inverse of the following matrices:

11. $\mathbf{A} = \begin{bmatrix} 1 & 2 \\ 4 & 6 \end{bmatrix}$

12. $\mathbf{A} = \begin{bmatrix} 4 & 1 \\ 2 & 3 \end{bmatrix}$

In Exercises 13 through 16, use the following data:

x	y
1	5
2	6
4	9
5	10
6	14

13. Define the $\mathbf{X}$ and $\mathbf{Y}$ matrices that will be used to determine the least-squares estimates.

14. Find the least-squares estimate $\mathbf{b} = (\mathbf{X^T X})^{-1}\mathbf{X^T Y}$.

15. Find the standard error of the regression, s_e.

16. Find a 95% confidence interval estimate of β_1.

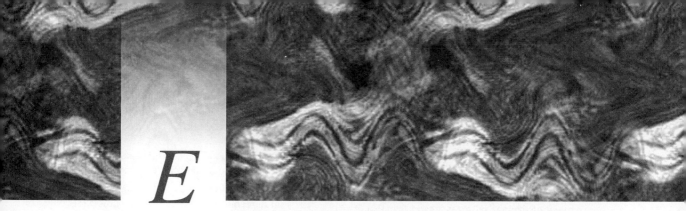

E

Solutions to Selected Odd-Numbered Exercises

2.1 Mean of HWYMPG = 28.15
Standard deviation of HWYMPG = 6.534
Median of HWYMPG = 28

3 $\mu = 3.5$ $\sigma = 1.71$

5 $\mu = 7$ $\sigma = 2.415$

7 a $\mu = 1.10$
b $110
c 0.2

9 a 0.8413
b 0.0228
c 2103.25 (approximately 2103)

11 a $z = 2.0$
b $z = 1.65$
c $z = 2.33$
d $z = 2.58$

13 $k = 84 - 2.06(7) = 69.58$. The number of months should be around 69 or 70 months.

15 a 0.0228
b $1 - 0.6826 = 0.3174$

17 a 0.0228
b 0.0
c Yes, because the population is normally distributed. This ensures that the sampling distribution of the sample mean will also be normal, even if the sample size is small.

19 0.5762

21 One way to approach this problem would be as follows. Suppose the average lifetime of the new hard drives has not changed so it is still 3250 hours. Are the data obtained consistent with this stated average lifetime? If so, then a sample mean of 3575 hours in a random sample of 50 hard drives should not be unusual. To measure how unusual this value is, find the probability of obtaining a sample mean of 3575 or more if the true average lifetime of the hard drives is 3250. If the mean lifetime of the hard drives has not changed, then the probability of finding a sample mean lifetime of 3575 hours or more in a random sample of 50 hard drives is 0.0. That is, there is virtually no chance that this should happen. But this is what we found. Therefore, we conclude that the mean lifetime of the hard drives must have changed and it must be larger than the previous mean of 3250 hours. This type of reasoning will be placed in a more structured setting in the section on hypothesis testing.

23 (5.66, 6.34)

25 (27.0846, 29.2147)

27 If a null hypothesis is rejected at the 5% level of significance, it would also be rejected at the 10% level of significance because the 10% level requires less "evidence" to reject the hypothesis than does the 5% level. In other words, a test statistic that is more extreme than the 5% critical value is also more extreme than the 10% critical value.

29   Critical value: $t(0.05, 15) = 1.753$
Test Statistic: $t = 0.5$
Decision: Do not reject H_0; standards are being met.

31   Critical value: $t(0.05, 82) \approx 1.645(z \text{ value})$
Test Statistic: $t = 2.45$
Decision: Reject H_0; there is evidence that the population average return is greater than that of the S&P 500 index.

33   (2.55, 5.45)

35    $(-4.04652, 14.17225)$

37   Critical value: $z(0.025) \approx 1.96$ is used because df is large.
Test Statistic: $t = 1.76$
Decision: Do not reject H_0; there is no difference in the population average test scores.

39   Critical value: $t(0.025, 28) = 2.048$
Test Statistic: $t = 3.97$
Decision: Reject H_0; there is a difference in mean rating scores for the two divisions.

41   Critical value: $t(0.05, 192) \approx -1.645$
Test Statistic: $t = -16.63$
Decision: Reject H_0; there is evidence that the average public school graduation rate is less than the average for private schools.

43   **a** Critical value: $t(0.025, 21) = 2.08$
Test Statistic: $t = 0.09$
Decision: Do not reject H_0; there is no difference in the average salaries.

45   **a** Critical value: $z(0.05, 51) \approx -1.645$
Test Statistic: $t = -5.83$
Decision: Reject H_0; based on the result of the test, we conclude that the average starting salary for females is less than the average for males.
   b Statistical evidence alone may not be sufficient to prove discrimination. Often, it must also be shown that there was intent to discriminate. However, when Harris Bank recognizes that a possible discriminatory situation exists, it should be concerned about correcting that situation.
   c Later in the book, certain other variables are introduced into this problem; for example, education of the employee and years of experience. Variables such as these might have some sort of moderating effect on the result of the test reported in this problem and should be considered.

47   **a** 18.2%
   b 25.0%
   c No, because 130,000 is not a limit of one of the classes. You could only approximate this percentage.
   d Between 80,000 and 89,999 (because the median represents the 50th percentile).

49   Based on the time-series plot, management may want to reconsider its decision. The time-series plot shows a pattern of increasing errors over time, which may indicate that the machine is experiencing wear and needs some type of maintenance. If the pattern continues, the subsequently drilled holes will not be of the correct diameter. Exercises 2.48 and 2.49 indicate the importance of using more than one type of graph, if appropriate, to examine data.

51    $E(X) = 220$, so they should charge $220 per policy to break even.

53   **a** $E(X) = 1.10$
   b 0.20
   c 0.95
   d 0.15
   e 0 (highest probability)

55    $x = 1000 - 1.29(25) = 967.75$ (about 968)

57   0.2514

59   0.8414

61   The interval (3.2, 4.5) is a 95% confidence interval estimate of the average sugar content. We are 95% confident that the *average* sugar content is between 3.2 and 4.5. The interval does not give us a range for 95% of the *individual* package contents however. This is not a correct interpretation of the interval.

E.2 CHAPTER 3

3.1   **b** $b_1 = 0.9107$; $b_0 = 7.1001$

3   **a** VALUE $= -50035 + 72.8$ SIZE
   b COST $= 16594 + 650$ NUMPORTS
   c STARTS $= 1726 - 22.2$ RATES

5   **a** Critical value: $t(0.025, 8) = 2.306$
Test Statistic: $t = 41.92$
Decision: Reject H_0
   b Yes. Hours of labor and number of items produced appear to be linearly related.

c Critical value: $t(0.025, 8) = 2.306$
Test Statistic: $t = -0.61$
Decision: Do not reject H_0

d The intercept of the line representing the relationship between number of items and labor is not significantly different from 0. (Note: This test does not suggest anything about whether there is or is not a relationship between number of items and labor.)

7 a Critical value: $t(0.025, 18) = 2.101$
Test Statistic: $t = 10.70$
Decision: Reject H_0

b Sales and advertising appear to be linearly related. The relationship is direct, suggesting that increases in advertising expenditures may result in an increase in sales. (Caution: See Section 3.7. There may not be a causal relationship here.)

c SALES $= -57.281 + 17.57$ ADV

d Our prediction of SALES would increase by $17.57(10) = 175.7$ or $17,570

e $(-941.19, 826.63)$

f $(14.12, 21.02)$

g Critical value: $t(0.025, 18) = 2.101$
Test Statistic: $t = -1.48$
Decision: Do not reject H_0

h The slope of the regression line is not significantly different from 20.

9 a 0.9955

b 99.55%

c Critical value: $F(0.05; 1, 8) = 5.32$
Test Statistic: $F = 1763.88$
Decision: Reject H_0

d Yes. Hours of labor and number of items produced appear to be linearly related.

11 a 86.4%

b Critical value: $F(0.05; 1, 18) = 4.41$
Test Statistic: $F = 114.54$
Decision: Reject H_0

13 a 119.4007

b $(117.36, 121.45)$

c 119.4007

d $(112.62, 126.18)$

15 a (4007, 4663) (in $100)

b

x	Point Estimate	95% Prediction Interval
20000	3457	2131, 4783
25000	4335	3043, 5627
30000	5214	3933, 6494
35000	6092	4800, 7384

Note: Both point estimates and interval limits are in $100.

17 a 10,503 ($1,050,300)
Yes. The regression equation was developed using a range of values for the x variable (ADV) of 160 to 415. The value 600 is well outside this range. Caution should be exercised if this estimate is used, because the relationship between SALES and ADV has been observed only over the range 160 to 415. It is not known whether the same relationship will serve as well outside this range.

b Disagree. The model was developed over the range of 160 to 415 for the x variable (ADV). Because ADV $= 0$ is outside this range, the resulting forecasts cannot be depended upon to make sense. Still, the least-squares method must choose a value as a y intercept. In this case, the intercept value that minimized the error sum of squares was -5700.

19 a Critical value: $t(0.025, 22) = 2.074$
Test Statistic: $t = 8.0$
Decision: Reject H_0

b Critical value: $F(0.1; 1, 22) = 2.95$
Test Statistic: $F = 64$
Decision: Reject H_0

c 0.744 or 74.4%

21

Source	DF	SS	MS	F
Regression	1	1000	1000	100
Error (Residual)	80	800	10	
Total	81	1800		

23 a Critical value: $t(0.05, 91) \approx 1.645$(z value)
Test Statistic: $t = 4.31$
Decision: Reject H_0; there is a linear relationship between salary and education.

b 17%

c 5355.8

d 5355.8

e As will be seen when this problem is continued later in this text, other factors might include experience and, unfortunately in this case, whether the employee is male or female. You can probably think of other factors that might be useful.

25 a CONS $= 2521 + 0.827$ INCOME

b 93.1%

c $(0.70, 0.96)$

d Critical value: $t(0.025, 10) = 2.228$
Test Statistic: $t = 11.63$
Decision: Reject H_0

e Critical value: $F(0.05; 1, 10) = 4.96$
Test Statistic: $F = 135.25$
Decision: Reject H_0

f No. The F test can be used only to test the two-tailed hypotheses.

g Critical value: $t(0.025, 10) = 2.228$
Test Statistic: $t = -2.43$
Decision: Reject H_0; the coefficient of INCOME is significantly different from 1.

27 a NEWCON = 368 + 43TREND

b 98.9%

c Using the R-square value, the equation fits the observed data very well. However this is no guarantee that the equation will predict new construction accurately in future years. This depends on whether the same linear increase in new construction continues into the future.

d Point prediction of y for 2002: 884.14
95% prediction interval of y for 2002: (841.68, 926.59)
Point prediction of y for 2003: 927.13
95% prediction interval of y for 2003: (882.94, 971.32)

e See part c.

29 a SHIPMENT = 1969 + 7.86 EXCHRATE

b Critical value: $t(0.025, 133) \approx 1.96$ (z value)
Test Statistic: $t = 2.09$
Decision: Reject H_0; wheat shipments and exchange rates are linearly related.

c 3.2%

d (0.49, 15.23)

E.3 CHAPTER 4

1 a COST = 51.72 + 0.95 PAPER
+ 2.47 MACHINE
+ 0.05 OVERHEAD
− 0.05 LABOR

b Critical value: $F(0.05; 4, 22) = 2.82$
Test Statistic: $F = 4629.17$
Decision: Reject H_0; at least one of the coefficients is not equal to 0. In other words, at least one of the variables is helping to explain a significant amount of the variation in y.

c 2.47; $2.47 \pm (2.074)(0.47)$

d Critical value: $t(0.025, 22) = 2.074$
Test Statistic: $t = -0.42$
Decision: Do not reject H_0; the true marginal cost of output associated with total production of paper is 1.

e 99.9%

f 99.9%

g The regression equation can be used to identify factors that are related to cost. After doing this, these factors might be useful in reducing cost. For example, machine hours are related to cost. Obviously, we cannot just start reducing machine hours, because these hours are part of the manufacturing process. But there may be a way to use the hours more efficiently, resulting in a reduction of overall hours without sacrificing production and a decrease in cost. The regression equation shows the influence of a one-unit reduction of the variables included on cost.

3 Critical value: $F(0.05; 2, 22) = 3.44$
Test Statistic: $F = 0.80$
Decision: Do not reject H_0; both coefficients are equal to 0. Neither OVERHEAD nor LABOR adds significantly to the model's ability to explain the variation in COST. Choose the REDUCED model.

5 a $RATES_i = 0.16077 + 0.97774\ RATES_{i-1}$

b Critical value: $t(0.025, 213) \approx 1.96$ (z value)
Test Statistic: $t = 97.60$
Decision: Reject H_0; there is a linear relationship.

c 97.8%

d The forecast is calculated as
$$RATES = 0.16077 + 0.97774 (6.05)$$
$$= 6.0761$$

Forecasts versus Actual Values:

Date	Forecast	Actual Value
1/03	6.0761	5.92
2/03	6.1016	5.84
3/03	6.1266	5.75
4/03	6.1510	5.81
5/03	6.1749	5.48
6/03	6.1982	5.23
7/03	6.2210	5.63
8/03	6.2433	6.26
9/03	6.2651	6.15
10/03	6.2864	5.95
11/03	6.3072	5.93
12/03	6.3276	5.91

One of the problems encountered when trying to produce forecasts for February or subsequent months is that the actual value in the previous month is not available at the end of 2002. One way around this problem is to use the forecast value for the previous month to generate the next month's forecast. To develop the 2/03 forecast, use 6.0761 as the value of the lagged variable. Then use the forecast for 2/03 as the value of the lagged variable to generate the forecast for 3/03, and so on. As forecasts are generated farther into the future, we expect them to be less accurate, in part because we are using previous month's forecasts to develop forecasts for future months. But, until we know the true values, there is not much alternative.

e Critical value: $t(0.025, 213) \approx 1.96$ (z value)
Test Statistic: $t = 1.82$
Decision: Do not reject H_0
f Critical value: $t(0.025, 213) \approx 1.96$ (z value)
Test Statistic: $t = -2.22$
Decision: Reject H_0

7 ANOVA

Source	DF	SS	MS	F
Regression	3	300	100	10
Error (Residual)	27	270	10	
Total	30	570		

9 Before drawing conclusions from the small t statistics, consider the F statistic for the overall fit of the model. Using a 5% level of significance, the test would be performed as follows:
Critical value: $F(0.05; 7, 82) \approx 2.17$
Test Statistic: $F = 4.95$
Decision: Reject H_0
The conclusion from this test is that at least one of the coefficients is not equal to 0. In other words, at least one of the variables is helping to explain a significant amount of the variation in y. This conclusion contradicts the conclusion reached from examining each individual t statistic. When the overall F test is in conflict with the t tests, this suggests that multicollinearity is likely a problem.

11 a FUELCON $= 916 - 218$ DRIVERS
$- 0.00078$ HWYMILES
$- 3.69$ GASTAX
$- 0.00549$ INCOME
b Critical value: $F(0.05; 4, 46) \approx 2.61$
Test Statistic: $F = 9.18$
Decision: Reject H_0; at least one of the coefficients is not equal to 0. In other words, at least one of the variables is helping to explain a significant amount of the variation in y.
c 44.4%
d HWYMILES
Critical value: $t(0.025, 46) \approx 1.96$ (z value)
Test Statistic: $t = -0.77$
Decision: Do not reject H_0; HWYMILES is not linearly related to FUELCON after taking account

of the effect of the other variables. At this point, the regression should be re-run with HWYMILES omitted and the remaining variables should be reassessed.

19 Pairwise correlations may not be sufficient to say whether or not multicollinearity will be a problem. There may be more complex relationships that cannot be detected by pairwise correlations. For example x_1 may be correlated with a linear combination of x_2 and x_3.

E.4 CHAPTER 5

5.1 Model summaries:

	R-square	Adjusted R-square	Standard Error
Linear Model	99.2	99.0	14.34
Second-Order Model	100.0	99.9	3.54

The second-order model appears better than the first-order model. The p value on the second-order term, RDSQR, is 0.000, so the second-order term is significant at any reasonable level of significance. The adjusted R-square is higher and the standard error is lower. The R-square on the second-order model is stated as 100%. Note that not all of the variance in the y variable is explained, but the R-square rounds off to 100%.

3 a $(0.0202)(3) = 0.0606$
b $(1.96)(1/0.94) = 2.085$
$(1.96)(1/0.95) = 2.063$
$2.085 - 2.063 = 0.022$
A rise in production volume from 0.94 to 0.95 of capacity results in a decrease in cost of 0.022, assuming INDEX remains constant.
c The point prediction is 4.6047.
d The 95% prediction interval is (4.2354, 4.9740).

5 From the scatterplots, it appears that log transformations of both the dependent and explanatory variables are necessary.

7 Using the quadratic trend model provides a better fit. The following measures of fit support the use of the quadratic trend.

	R-square	Adjusted R-square	Standard Error
Linear Model	85.2%	85.0%	0.794057
Second-Order Model	97.6%	97.5%	0.325633

Forecast for 1993:
BETS = 0.795 − 0.0490(67)
$$+ 0.00220(4489) = 7.3878$$

Forecast for 1994:
BETS = 0.795 − 0.0490(68)
$$+ 0.00220(4624) = 7.6358$$

E.5 CHAPTER 6

1 **a** Yes. It is not, however, a linear relationship.
   b $\hat{y} = 10$
   c Critical value: $t(0.005, 9) = 3.25$
   Test Statistic: $t = 0.0$
   Decision: Do not reject H_0
   Conclusion: There is no *linear* relationship between y and x.
   d There is a "strong" association. In fact, this is an exact relationship, but not a linear one. The equation expressing the relationship is: $y = x^2$

3 There is evidence that the constant variance assumption has been violated. One possible correction would be to use the log of the monthly prices as the dependent variable.

5 Critical values for DW test: 1.24 and 1.56
   Test Statistic: $d = 2.14$
   Decision: Do not reject H_0; the disturbances are not autocorrelated.

7 **a** The residuals do not appear to be randomly distributed, indicating that some assumption has been violated. It may not be clear from the residual plots which assumption has been violated, however. It is actually the linearity assumption that is violated. In this example, the violation of the linearity assumption may be mistaken as a problem with outliers. There are obviously one or two extreme points in the data set. One is the United States, which has much greater IMPORTS and GDP than the other countries in the data set. If the violation of the linearity assumption was mistaken as an outlier problem, one possible course of action would be to remove the outlier (the United States) and see if the model could be improved.
   c The removal of the United States from the data set obviously does not help improve the regression. This can be seen from the residual plots from the regression excluding the United States. The problem in this data set is the large differences between the IMPORTS figures and between the

GDP figures for the different countries, leading to a curvilinear pattern in the data. A possible correction that would narrow these large differences would be to use the natural logarithm transformation on both the dependent and explanatory variable.

9 No. You cannot compare these two models on the basis of the traditional summary statistics like R-square or standard error because the dependent variable has been transformed. To choose the preferred model, residual plots could be used. The model that appears to most closely conform to the assumptions according to residual plots would be the one chosen.

11 From the residual plots, it appears that the constant variance assumption has been violated. The residuals exhibit the cone-shaped pattern that indicates an increase in the variance as the explanatory variable (CUTTING) increases. Two possible corrections for nonconstant variance are
   (1) $\ln(\text{FATALS}) = \beta_0 + \beta_1 \text{ CUTTING} + e$
   (2) FATALS/CUTTING $= \beta_0 (1/\text{CUTTING})$
   $$+ \beta_1 + e$$
   The first of these two corrections was used.
   Four of the values of FATALS are outside the range of the natural logarithm transformation (and are treated as missing data). In this case, the values outside the range are 0s. There are various ways of handling this problem so that the natural logarithm transformation can still be used. In this problem, the 0 values were basically ignored. MINITAB will automatically assign a missing value code when a value is outside the range of a transformation. The resulting missing cases will not be used in any subsequent analyses. That is what was done in the solution. This is not the most elegant way to approach this problem, but will probably supply reasonable answers in this situation.
   One other possible solution is to increment each value of the variable FATAL by some small amount (0.01, for example), so that no 0 values are present. When this is tried in this example, some outliers are produced, which further complicates the analysis. I do not believe the regression results obtained from the incremented data are reliable.
   One alternative I did try in this situation was to change the 0s to 1s (reasoning that 1 and 0 fatalities were fairly similar numbers). The results in this case were very similar to the results with the

0s omitted, and might be a better way to approach the problem. The outlier problem is not present when this larger increment is used.

13 The scatterplot of the dependent variable (TIME) versus each of the explanatory variables (NUMBER and EXPER), the regression, and the associated residual plots represent the output necessary to determine whether any assumptions have been violated. From the residual plots it is clear that an assumption has been violated: the linearity assumption. The plot of the standardized residuals (SRES1) versus NUMBER indicates that the explanatory variable NUMBER is the one that should enter the equation in a curvilinear manner. The residual plot of SRES1 versus EXPER does not indicate any problems. Looking back at the scatterplot of TIME versus NUMBER suggests that an appropriate correction might be to use a second-order polynomial regression with both NUMBER and $NUMBER^2$ (with the second-order variable called NUMSQR). The second-order regression of TIME on NUMBER, NUMSQR, and EXPER is run.

The residual plots from the second-order regression suggest that the new model is an improvement over the original. The residuals in these plots appear to be randomly distributed. The R-square and adjusted R-square have both increased, and the standard error of the regression has decreased. These are all positive signs in favor of the new model. A t test for the coefficient of the variable EXPER, however, will show that this variable is not important in the regression. Thus, it has been removed and a final regression has been run using the explanatory variables NUMBER and NUMSQR. This is the preferred model.

15 There do not appear to be any assumptions violated.

17 From the residual plots, it is clear that an assumption has been violated. In this case, it is the linearity assumption. To correct for the violation, logs of the dependent variable, VALUE, and two of the independent variables, PHYSICAL and SIZE are used. In the revised regression, the variable DEPREC is not significant. The final model is obtained by deleting this variable.

19 VOLUME is important in explaining the number of accidents, although it explains only 16.4% of the variance. Two observations stand out as unusual in the y-direction. These are observations 30

and 31. Both of these intersections have more accidents than would be expected given their volume of traffic. The city may want to investigate whether measures could be taken to make these intersections safer (install lights, remove obstructions, etc).

E.6 CHAPTER 7

1 a Critical value: $F(0.05; 4, 88) \approx 2.53$
Test Statistic: $F = 22.98$
Decision: Reject H_0

b At least one of the coefficients is not equal to 0. At least one of the four explanatory variables is important in explaining the variation in SALARY.

c Critical value: $t(0.025, 88) \approx 1.96$ (z value)
Test Statistic: $t = 6.13$
Decision: Reject H_0

d Yes. There is a difference in salaries, on average, for male and female workers after accounting for the effects of the EDUC, EXPER, and MONTHS variables. Males' salaries are, on average, $722 higher, a statistically significant difference.

e Forecast of average salary for males with 12 years of education, 10 years of experience, and with MONTHS equal to 15: SALARY = 3526.422 + 722.461 + 90.020(12) + 1.269(10) + 23.406(15) = 5692.903

Forecast of average salary for females with 12 years of education, 10 years of experience, and with MONTHS equal to 15:
SALARY = 3526.422 + 90.020(12) + 1.269(10) + 23.406(15) = 4970.422

3 a R-square adjusted = 49.4% with interaction variable
R-square adjusted = 48.9% without interaction variable

Although the full model has a higher adjusted R-square value, the difference is very small. It is unclear from the adjusted R-square alone whether the interaction variable is necessary. The hypothesis test in part b results in the conclusion that it is not.

b Critical value: $t(0.025, 87) \approx 1.96$ (z value)
Test Statistic: $t = -1.42$
Decision: Do not reject H_0

c The interaction term is not important in this regression model. Choose the reduced model.

d From the test results, it appears that the interaction term is not useful in explaining the difference in average salaries.

5 **a** Critical value: $F(0.05; 11, 117) \approx 1.99$
Test Statistic: $F = 40.653$
Decision: Reject H_0; there is seasonal variation in furniture sales.

b Critical value: $t(0.025, 117) \approx 1.96$ (z value)
Test Statistic: $t = 6.09$ for the trend coefficient
Decision: Reject H_0; the trend variable is important.
Test Statistic: $t = 5.84$ for the coefficient of the lagged variable
Decision: Reject H_0; the lagged variable is important.

c Date	Forecast
1/2003	$1552.151 + 7.903(133)$ $+ 0.484(4678) - 643.717$ $= 4223.685$
2/2003	$1552.151 + 7.903(134)$ $+ 0.484(4223.685) - 404.481$ $= 4250.936$
3/2003	$1552.151 + 7.903(135)$ $+ 0.484(4250.936) - 80.369$ $= 4596.140$
4/2003	$1552.151 + 7.903(136)$ $+ 0.484(4596.140) - 457.714$ $= 4393.777$
5/2003	$1552.151 + 7.903(137)$ $+ 0.484(4393.777) - 178.202$ $= 4583.248$
6/2003	$1552.151 + 7.903(138)$ $+ 0.484(4583.248) - 316.796$ $= 4544.261$
7/2003	$1552.151 + 7.903(139)$ $+ 0.484(4544.261) - 278.410$ $= 4571.680$
8/2003	$1552.151 + 7.903(140)$ $+ 0.484(4571.680) - 183.319$ $= 4687.945$
9/2003	$1552.151 + 7.903(141)$ $+ 0.484(4687.945) - 357.504$ $= 4577.935$
10/2003	$1552.151 + 7.903(142)$ $+ 0.484(4577.935) - 239.408$ $= 4650.690$
11/2003	$1552.151 + 7.903(143)$ $+ 0.484(4650.690) + 6.826$ $= 4947.943$
12/2003	$1552.151 + 7.903(144)$ $+ 0.484(4947.943)$ $= 5084.987$

E.7 CHAPTER 8

1 **a** COST $= 59.43$ PAPER $+ 2.39$ MACHINE
b 99.87%
c R-square adjusted $= 0.9986$, or 99.86%
d $s_e = 11.0$
e OVERHEAD and LABOR
Overhead and labor are related to COST when examined individually. However, they add little to the ability of the equation to explain the variation in COST. The variables PAPER and MACHINE explain over 99% of the variation in COST. OVERHEAD and LABOR are probably unnecessary because most of the variation in COST is explained by the other two variables.

E.8 CHAPTER 9

1 Critical value: $F(0.05; 8, 103) \approx 2.02$
Test Statistic: $F = 14.82$
Decision: Reject H_0; there is a difference in the average number of collisions for different types of car.

3 First Test for Interaction Effects
Critical value: $F(0.05; 6, 12) = 3.00$
Test Statistic: $F = 0.20$
Decision: Do not reject H_0; there are no interaction effects. Note that this means that the F-tests for main effects can be used.
Test for Main Effects Due to Industry
Critical value: $F(0.05; 3, 12) = 3.49$
Test Statistic: $F = 49.82$
Decision: Reject H_0; the α_i are not all equal. There are differences in treatment means associated with different industries.
Test for Main Effects Due to Contact
Critical value: $F(0.05; 2, 12) = 3.89$
Test Statistic: $F = 0.03$
Decision: Do not reject H_0; the β_i are equal. There are no differences in treatment means associated with different contacts.

5 Critical value: $F(0.05; 2, 15) = 3.68$
Test Statistic: $F = 102.32$
Decision: Reject H_0; there is a difference in the average production rates.

7 **a** Critical value: $F(0.05; 2, 4) = 6.94$
Test Statistic: $F = 103.32$

Decision: Reject H_0; there is a difference in the average test scores due to training program.

b 95% confidence interval estimate of the difference between program 1 and program 2 $(-16.96, -10.38)$

9 Critical value: $F(0.05; 6, 49) \approx 2.34$
Test Statistic: $F = 29.21$
Decision: Reject H_0; there is a difference in the average sales on different days of the week.

11 a Critical value: $F(0.05; 3, 8) = 4.07$
Test Statistic: $F = 14.00$
Decision: Reject H_0; there is a difference in the average lifetimes.

b 95% confidence interval estimate of the difference between suppliers 1 and 2: $(-2.88, 0.88)$

E.9 CHAPTER 10

1 a Critical value: $z(0.025) = 1.96$
Test Statistic for coefficient of TEST 1: $z = 3.06$
Decision: Reject H_0; TEST 1 is useful.
Test Statistic for coefficient of TEST 2: $z = 1.62$
Decision: Do not reject H_0; TEST 2 is not useful.

b Use the output in Figure 10.5. The results of TEST 2 do not appear useful in helping to predict success for these employees.

1. $-43.37 + 0.4897(94) = 2.6618$
   probability 0.93
2. $-43.37 + 0.4897(80) = -4.194$
   probability 0.01
3. $-43.37 + 0.4897(82) = -3.2146$
   probability 0.04
4. $-43.37 + 0.4897(90) = 0.703$
   probability 0.67

Potential employees 1 and 4 would be classified in the "success" group, while numbers 2 and 3 would not.

E.10 CHAPTER 11

1

METHOD	MAPE	MAD	MSD
Moving Average Length 1	26	47021	3301636525
Moving Average Length 2	25	44821	2825426091
Moving Average Length 3	23	40870	2537149730
Moving Average Length 4	21	36520	2203979413
Moving Average Length 5	20	35561	2024066925
Moving Average Length 6	20	35983	2112865326
Smoothing Constant Alpha 0.0083969	20	34082	1822769972

In-sample accuracy measures are used to compare methods. No out-of-sample forecasts were generated. The best choice for WestCo to forecast envelope sales at this point appears to be simply to use the average of past sales. The naïve method (moving average length 1) is worst among the methods examined. The optimal smoothing value for exponential smoothing found by MINITAB is close to zero, suggesting that nearly equal weight be put on all past observations. This suggests that the observations have very little systematic pattern (are close to random) and that the average would be a good candidate to generate future forecasts. (The average does have the property that it is the single value that will minimize the sum of squared forecast errors).

3 Single and double exponential smoothing were considered as possible forecasting techniques. Using the standard error measures applied to the out-of-sample data, SES forecasts are more accurate than DES.

	MAD	MSD	MAPE
SES	0.142	0.029	2.38%
DES	0.288	0.129	4.88%

5 In-sample accuracy measures are used to compare methods. No out-of-sample forecasts were generated. Single and double exponential smoothing were considered. Winters' method was also considered because it was unclear if there was seasonal variation in the data. The multiplicative version of Winters' method was used with the default smoothing constants $(0.2, 0.2, 0.2)$. Using the standard error measures, SES forecasts are slightly more accurate than DES and Winters' method.

	MAD	MSD	MAPE
SES	459	389494	17%
DES	473	431732	17%
Winters'	487	420359	18%

7 SES, DES, and the naïve forecast (MA with length 1) were considered.

Period	SES Forecast	DES Forecast	Naïve	Actual
181	4.27468	4.15576	4.25	4.25
182	4.27468	4.04731	4.25	4.25
183	4.27468	3.93886	4.25	4.25
184	4.27468	3.83041	4.25	4.25
185	4.27468	3.72197	4.25	4.25
186	4.27468	3.61352	4.25	4.22
187	4.27468	3.50507	4.25	4.00
188	4.27468	3.39662	4.25	4.00
189	4.27468	3.28817	4.25	4.00
190	4.27468	3.17972	4.25	4.00
191	4.27468	3.07128	4.25	4.00
192	4.27468	2.96283	4.25	4.00

Using the standard error measures applied to out-of-sample forecasts, the naive forecasts are slightly more accurate than SES. Both SES and naïve are considerably more accurate than DES.

	MAD	MSD	MAPE
SES	0.152	0.038	3.78%
DES	0.563	0.391	13.82%
NAÏVE	0.128	0.031	3.18%

9 Because of the strong seasonal variation, Winters' method (additive and multiplicative) and decomposition (additive and multiplicative) were considered. Using the standard error measures, additive decomposition was slightly more accurate for the in-sample time period. No out-of-sample forecasts were generated. Only the default smoothing parameter values (0.2, 0.2, 0.2) were examined for the Winters' methods.

	MAD	MSD	MAPE
Winters' M	2.7456	12.6723	4.5996%
Winters' A	2.6333	12.3410	4.4866%
Decomp M	2.26314	9.29202	3.91370%
Decomp A	2.22837	9.01055	3.86912%

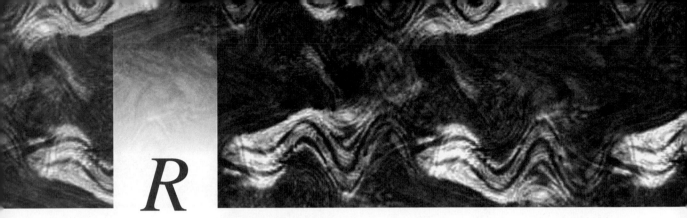

R References

Albright, S., Winston, W., and Zappe, C. *Data Analysis for Managers with Microsoft® Excel* (2nd ed.). Pacific Grove, CA: Duxbury Press, 2004.

Amemiya, T., and Powell, J. "A Comparison of the Logit Model and Normal Discriminant Analysis When Independent Variables are Binary." Technical Report No. 320, Institute for Mathematical Studies in the Social Sciences, Encina Hall, Stanford University, Stanford, CA.

Berk, K., and Carey, P. *Data Analysis with Microsoft® Excel: Updated for Office XP*. Pacific Grove, CA: Duxbury Press, 2004.

Bessler, D., and Babubla, R. "Forecasting Wheat Exports: Do Exchange Rates Really Matter?" *Journal of Business and Economic Statistics* 5(1987): 397–406.

Bowerman, B., and O'Connell, R. *Forecasting and Time Series: An Applied Approach*. (3rd ed.). Pacific Grove, CA: Duxbury Press, 1993.

Conway, D., and Roberts, H. "Regression Analyses in Employment Discrimination Cases." *Statistics and the Law*. New York, NY: Wiley, 1986.

Cravens, D., Woodruff, R., and Stamper, J. "An Analytical Approach for Evaluating Sales Territory Performance." *Journal of Marketing* 36(1972): 31–37.

Freund, R., and Littell, R. *SAS® System for Regression* (3rd ed.). Cary, NC: SAS Institute, 2000.

Freund, R., and Wilson, W. J. *Regression Analysis: Statistical Modeling of a Response Variable*. San Diego, CA: Academic Press, 1998.

Graybill, F., and Iyer, H. *Regression Analysis: Concepts and Applications*. Pacific Grove, CA: Duxbury Press, 1994.

Griffiths, W., and Surekha, K. "A Monte Carlo Evaluation of the Power of Some Tests for Heteroscedasticity." *Journal of Econometrics* 31(1986): 219–231.

Hadi, A. *Matrix Algebra as a Tool*. Pacific Grove, CA: Duxbury Press, 1996.

Jobson, J. D. *Applied Multivariate Data Analysis Volume One: Regression and Experimental Design*. New York, NY: Springer-Verlag, 1990.

Jobson, J. D. *Applied Multivariate Data Analysis Volume Two: Categorical and Multivariate Methods*. New York, NY: Springer-Verlag, 1991.

Judge, G., Griffiths, W., Hill, R., Lutkepohl, H., and Lee, T. *The Theory and Practice of Econometrics* (2nd ed.). New York, NY: Wiley, 1985.

Keller, G., and Warrack, B. *Statistics for Management and Economics* (6th ed.). Pacific Grove, CA: Duxbury Press, 2003.

Lawrence, K., and Marsh, L. "Robust Ridge Estimation Methods for Predicting U.S. Coal Mining Fatalities." *Communications in Statistics* 13 (1984): 139–149.

Lee, C., and Lynge, M. "Return, Risk, and Cost of Equity for Stock S&L Firms: Theory and Empirical Results. " *Journal of the American Real Estate and Urban Economics Association* 13(1985): 167–180.

Lunneborg, C. *Modeling Experimental and Observational Data*. Pacific Grove, CA: Duxbury Press, 1994.

McKenzie, J., and Goldman, R. The Student Edition of MINITAB™ for Windows 95/NT™, Release 12 (4th ed.). Reading, MA: Addison-Wesley, 1999.

Mendenhall, W., Beaver, R., and Beaver, B. *A Brief Course in Business Statistics*. (2nd ed.) Pacific Grove, CA: Duxbury Press, 2001.

Myers, R. *Classical and Modern Regression with Applications* (2nd ed.). Boston, MA: PWS-Kent Publishing Co., 1990.

Middleton, M. *Data Analysis Using Microsoft Excel: Updated for Office XP*. Pacific Grove, CA: Duxbury Press 2004.

Moser, B., and Stevens, G. "Homogeneity of Variance in the Two-Sample Means Test." *The American Statistician* 46(1992): 19–21.

Neter, J., Wasserman, W., and Kutner, M. *Applied Linear Statistical Models* (2nd ed.). Homewood, IL: Richard D. Irwin, Inc., 1985.

Ostle, B.; Turner, K.; Hicks, C;, and McElrath, G. *Engineering Statistics: The Industrial Experience*. Belmont, CA: Duxbury Press, 1996.

Peixoto, J. "A Property of Well Formulated Polynomial Regression Models." *The American Statistician* 44(1990): 26–30.

Ragsdale, C., and Stam, A. "Introducing Discriminant Analysis to the Business Statistics Curriculum." *Decision Sciences* 23(1992): 724–745.

Schafer, D. "Measurement-Error Diagnostics and the Sex Discrimination Problem." *Journal of Business and Economic Statistics* 5(1987): 529–537.

Shapiro, S., and Wilk, M. "An Analysis of Variance Test for Normality (complete samples)." *Biometrika* 52(1965): 591–611.

Simonoff, J. S., and Sparrow, I. R. "Predicting Movie Grosses: Winners and Losers, Blockbusters and Sleepers." *Chance*, 13(2000): 15–24.

Weiers, R. *Introduction to Business Statistics* (4th ed.). Pacific Grove, CA: Duxbury Press, 2002.

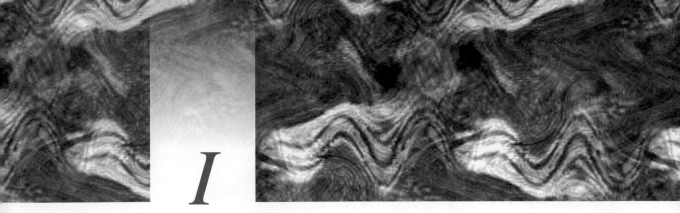

I

Index